Synthetic Coordination Chemistry

Principles and Practice

Synthetic Coordination Chemistry

Principles and Practice

J A Davies
Univ. of Toledo, USA

C M Hockensmith
Univ. of Illinois at Springfield, USA

V Yu Kukushkin
St. Petersburg State Univ., Russia

Yu N Kukushkin
St. Petersburg State Technological Institute, Russia

World Scientific
Singapore • New Jersey • London • Hong Kong

Published by

World Scientific Publishing Co Pte Ltd
P O Box 128, Farrer Road, Singapore 912805
USA office: Suite 1B, 1060 Main Street, River Edge, NJ 07661
UK office: 57 Shelton Street, Covent Garden, London WC2H 9HE

British Library Cataloguing-in-Publication Data
A catalogue record for this book is available from the British Library.

SYNTHETIC COORDINATION CHEMISTRY: PRINCIPLES AND PRACTICE

ISBN 981-02-2084-7

Printed in Singapore.

DEDICATION

To those who taught us and to those whom we teach

ACKNOWLEDGMENTS

The authors thank Professors P. Braunstein, D. Dollimore, L.-I. Elding, K. Isobe, A. Hammershøi, L. Heck, R. J. Judd, J. G. Leipoldt, J. R. Moss, Å. Oskarsson, I. Persson, and D. A. Tocher for stimulating discussions and helpful comments during the preparation of the manuscript. The technical assistance of T.-B. Bladh and D. Southard with computing problems and of R. E. Sorkina with the translation of certain materials is gratefully acknowledged.

This work was written in part while J. A. D. was on sabbatical leave, part of which was spent in St. Petersburg, Russia. Similarly, much of the text was prepared while V. Yu. K. was at the Chemical Center, Lund University, Sweden, during a period as visiting professor at the Institute for Molecular Science, Okazaki, Japan, while on sabbatical leave at the Universidad Autónoma de Madrid, Spain, and during a visiting professorship at the University of Toledo, USA. The Royal Swedish Academy of Sciences, the Academy of Sciences of Russia, the Institute for Molecular Science, the Spanish Dirección General de Investigación Científica y Técnica, and the University of Toledo Foundation are thanked for financial support. Additionally, a number of chapters were completed during release time from teaching duties granted to C. M. H. by the University of Illinois at Springfield and under the auspices of the Competitive Research Grant program of U.I.S. for which grateful acknowledgment is given. Yu. N. K. expresses his gratitude to the Russian Foundation for Basic Research for financial support and the International Science Foundation for the award of a Soros professorship.

The authors express their gratitude to their families who were patient and supportive during the writing of this book.

PREFACE

Synthesis in the western philosophical tradition is often defined as the combination of separate elements to form a coherent whole or the combination of thesis and antithesis that produces a new and higher form of being.[1] Synthesis in chemistry, however, is a concrete, tangible process in which compounds are formed from atoms, ions, radicals, and molecules. These descriptions of synthesis appear to be very different, but are they?

Coordination compounds are frequently synthesized from metal salts and ligands or through transformations of other coordination compounds. Products are formed by various processes including substitution, inner-sphere transformations of ligands, and oxidation or reduction of the central metal atoms. In a sense, a "new and higher form of being" is achieved when coordination compounds are prepared in these ways. The knowledge and preparative skills of individual chemists are applied to the synthesis of new coordination compounds. Descriptions of the compounds become part of the literature and tradition of the discipline and, based on this knowledge, new coordination compounds are synthesized. A continuing cycle of chemical thesis, antithesis, and synthesis occurs each day all over the world wherever chemists are working.

This volume is a case in point. Coordination chemistry practiced behind the old Iron Curtain had a rich, full history but much of the knowledge was not transferred easily across political barriers to the United States and other countries. The same was true in the other direction. Today, however, the barriers have fallen. We have collaborated freely on this text, often over great distances and occasionally within the same room. We have written, revised, updated, and expanded this manuscript without regard to our individual locations around the globe.

The first coordination compound synthesized in the laboratory was Berlin blue, prepared in 1704 by Disbach, so it is obvious that preparative coordination chemistry has an extensive history. At times syntheses were unpredictable, unsuccessful, or unreported. Today, coordination chemists plan their syntheses by applying fundamental chemical principles and theories. For this reason we begin with a chapter describing classical theoretical concepts that are important in the synthesis of coordination compounds before we consider other issues such as solvents, reactive intermediates, reagents, *etc.*

The manuscript was prepared using Microsoft Word 5.1a for the Macintosh®. Figures and Schemes were prepared with Chem Draw Plus® and molecular structures were generated from the original atomic coordinates with Ball&Stick®.

[1]*The American Heritage Dictionary*, 2nd edition (Houghton Mifflin, New York, 1985).

LIST OF ABBREVIATIONS

Ac	acetyl
Alk	alkyl
A_N	acceptor number
Ar	aryl
Bu	butyl
bipy	2,2'-bipyridyl
Cp	cyclopentadienyl
Cp*	pentamethylcyclopentadienyl
cod	cycloocta-1,5-diene
cot	cycloocta-1,3,5,7-tetraene
Cy	cyclohexyl
DEA	*N,N'*-diethylacetamide
DEF	*N,N'*-diethylformamide
DMA	*N,N'*-dimethylacetamide
DME	dimethoxyethane
DMF	*N,N'*-dimethylformamide
DMP	2,2-dimethoxypropane
dmpe	1,2-bis(dimethylphosphino)ethane
dmpm	bis(dimethylphosphino)methane
DMSO	dimethyl sulfoxide
D_N	donor number
dppe	1,2-bis(diphenylphosphino)ethane
dppm	bis(diphenylphosphino)methane
en	ethylenediamine
Et	ethyl
Hacac	acetylacetone
H_2acacen	the Schiff base derived from two moles acetylacetone and one mole ethylenediamine
HMP	hexamethylphosphortriamide
H_2sal	salicylic acid
H_2salen	the Schiff base derived from two moles salicaldehyde and one mole ethylenediamine
H_2salophen	the Schiff base derived from two moles salicaldehyde and one mole *o*-phenylenediamine
H_2tpp	5,10,15,20-tetraphenylporphyrin
L	ligand
M	metal or metal ion
Me	methyl
Np	naphthalene
Ph	phenyl
Pr	propyl

LIST OF ABBREVIATIONS, CONTINUED

solv	solvent molecule
phen	1,10-phenanthroline
pic	picoline (methylpyridine)
$[PPN]^+$	μ-nitrido-bis(triphenylphosphorus)(1+)
py	pyridine
TCNE	tetracyanoethylene
TCNQ	tetracyanoquinodimethane
TEOF	triethylorthoformate
THF	tetrahydrofuran
thio	thiourea
TMAO	trimethylamine *N*-oxide
TMOF	trimethylorthoformate
TMP	trimethylphosphate
TMSO	tetramethylene sulfoxide
trpy	2,2',2''-terpyridyl

CONTENTS

Chapter 1. SYNTHESIS OF COORDINATION COMPOUNDS: THEORETICAL CONSIDERATIONS

The reaction chemistry of coordination compounds is governed by a number of general principles. These principles are based largely upon experimental observations and much of the underlying theoretical basis of coordination chemistry remains essentially qualitative in nature. Nevertheless, even qualitative guidelines are valuable in designing new synthetic methods.

This Chapter deals with the basic principles used in planning syntheses of coordination compounds. Complete discussion of the underlying basis of coordination chemistry may be found in a number of texts.[1]

1.1 Labile and inert coordination compounds

Because syntheses of coordination compounds frequently involve ligand substitution reactions, an understanding of the relative lability (or its opposite, kinetic inertness) of metal complexes is important in the design of experimental procedures. Geometric isomers of labile complexes pose a special problem in synthesis because facile routes for interconversion may allow isomerization, favoring formation of the thermodynamically more stable isomer.

The reactions of thermodynamically stable, homoleptic, anionic metal cyanide complexes with $^{14}CN^-$ illustrate the fundamental difference in reactivity between labile and kinetically inert complexes,[1b] Eq. 1.1:

$$[M(CN)_6]^{3-} + 6\,^{14}CN^- \rightleftharpoons [M(^{14}CN)_6]^{3-} + 6CN^- \qquad (1.1)$$

where M = Mn, Cr

With manganese, the half life of the reaction is on the order of one hour. In contrast, with chromium, the half-life is approximately 24 days. A simple consideration of the relative ligand field stabilization energies (LFSE) of d^3 (Cr^{3+}) and d^4 (Mn^{3+}) systems and how these are affected by passage along the reaction coordinate allows rationalization of the experimental results. Regardless of whether a substitution reaction of an octahedral complex is associative (*i.e.* involving a 7-coordinate species) or dissociative (*i.e.* involving a 5-coordinate species), the changes in LFSE are significant. Basolo and Pearson have shown[2] that a low-spin, octahedral d^3 complex experiences a destabilization by 2.00 *Dq* on conversion to a 5-coordinate species and loss of 4.26 *Dq* in LFSE on conversion to a 7-coordinate species. A low-spin, octahedral d^4 complex, in contrast, suffers less destabilization. Thus, conversion to a 5-coordinate species involves loss of 1.43 Dq in LFSE and conversion to a 7-coordinate species involves destabilization by 2.98 Dq.

Consideration of changes in LFSE is of general use in predicting relative labilities. Thus, for analogous M^{2+} and M^{3+} systems, loss of LFSE during substitution will be more significant for the more highly charged ion where the field is larger.[1b] Similarly, in comparisons of the relative labilities of analogous nickel(II), palladium(II), and platinum(II) complexes, for the 3rd row ion, platinum(II), where Δ (the separation between the filled b_{2g} (d_{xy}) and empty b_{1g} ($d_{x^2-y^2}$) orbitals) is large, associative substitution reactions are slow; whereas, for the 2nd and 1st row analogs, where Δ is smaller, such substitution reactions proceed more rapidly. Indeed, it is well known that substitution reactions at palladium(II) can occur *ca.* 10^6 times faster than analogous reactions at platinum(II). For example,[3] water exchange in $[M(H_2O)_4]^{2+}$ occurs with $k(Pd)/k(Pt) = 1.4 \times 10^6$. Similarly, the anation reactions of platinum(II) and palladium(II) aqua-complexes typically occur with $k(Pd)/k(Pt) = 10^6 - 10^4$, depending on the nature of the entering ligand.[4]

Complexes for which loss of LFSE during substitution is irrelevant, *i.e.* d^0, high-spin d^5, and d^{10} systems, tend to exhibit high substitutional labilities. In the case of d^{10} complexes, such as the $[PtL_n]$ complexes formed with certain tertiary phosphines, substitution through equilibria that involve 2-, 3- and 4-coordinate species is possible. A simple valence-bond model suggests that, because the *5d* orbitals are completely filled, platinum(0) must employ the empty *6s* and *6p* orbitals to accept electrons from the ligands. Accordingly, there exists a fine balance between the advantage of maximizing the number of bonds through sp^3 hybridization and the advantage of minimizing interligand steric hindrance through reduction in coordination number *via* sp^2 or *sp* hybridization. In solution, equilibria between $[PtL_4]$, $[PtL_3]$, and $[PtL_2]$ species have been identified by NMR spectroscopy.[5] Such equilibria result in ligand scrambling between $[PtL_4]$ and $[PtL'_4]$ in solution and also prevent the isolation of optically-active [PtLL'L"L'"] complexes.

In solid-state transformations factors other than electronic and steric effects must be considered also in the evaluation of comparative rates, such as the possible restructuring of a crystal lattice (see Chapter 12). The solid state thermal transformations of the pyridinium salts of $[MCl_4]^{2-}$ (M = Pd, Pt), for example, occur at comparable rates[6] despite the fact that these reactions involve substitution at platinum(II) and palladium(II), Eq. 1.2 and 1.3:

$$[pyH]_2[PtCl_4] \rightarrow [PtCl_2(py)_2] + 2HCl \quad (1.2)$$

$$[pyH]_2[PdCl_4] \rightarrow [PdCl_2(py)_2] + 2HCl \quad (1.3)$$

1.2 Synthesis of mixed-ligand complexes by co-proportionation

In preparative coordination chemistry mixed-ligand complexes are often prepared by co-proportionation of homoleptic complexes. The reaction between $K_2[Pt(NO_2)_4]$ and $K_2[PtBr_4]$ is representative,[7] Eq. 1.4:

$$K_2[Pt(NO_2)_4] + K_2[PtBr_4] \rightarrow 2K_2[PtBr_2(NO_2)_2] \quad (1.4)$$

The relative labilities of the individual ligands within a mixed-ligand complex can differ widely. For example, in structurally similar platinum metal complexes, anionic ligands that are the conjugate bases of strong protic acids (*e.g.* Cl^-, Br^-) are typically more labile than neutral Lewis bases, such as amines. Thus, it is possible to prepare platinum(II) complexes containing three different ligands through co-proportionation of two different $[PtX_2(amine)_2]$ complexes, *e.g.* Eq. 1.5:

$$trans\text{-}[Pt(NO_2)_2(NH_3)_2] + trans\text{-}[PtBr_2(NH_3)_2] \rightarrow 2trans\text{-}[PtBr(NO_2)(NH_3)_2] \quad (1.5)$$

If the amine ligands present in the two reactants are different, then the preparation of platinum(II) complexes with four different ligands becomes possible. Octahedral platinum(IV) complexes with six different ligands are also well known and syntheses of such compounds can be designed only when the *trans*-effects of the individual ligands are taken into account (see Section 1.4).

In 1920, based on relatively few experimental data (only about 10 reactions were known), Chugaev[8] concluded that Magnus salts, $[PtL_4][PtX_4]$ (X = Cl, Br; L = amines, thioethers), were subject to solid-state rearrangement upon heating, Eq. 1.6:

$$[PtL_4][PtX_4] \rightarrow 2[PtX_2L_2] \quad (1.6)$$

At the present time more than 100 compounds of the type $[PtX_2L_2]$ have been prepared by this general route.[9,10] If thermal decomposition of the Magnus salt does not take place upon heating, then, in all cases, $[ML_4][MX_4]$ salts (M = Pt, Pd; X = Cl, Br, I) rearrange with liberation of energy to produce $[MX_2L_2]$. The higher the *trans*-effect of L, the lower the temperature required for the transformation to take place. The palladium(II) complexes, $[PdL_4][PdX_4]$, are subject to thermal rearrangement at much lower temperatures than the analogous Magnus salts of platinum(II).

The formation of mixed-ligand complexes by co-proportionation of homoleptic complexes has been considered theoretically. The crucial issue is the relationship between the equilibrium constants for formation of the homoleptic and mixed-ligand complexes. This relationship largely controls the outcome of ligand redistribution. In cases where ligands have no influence on one another and behave essentially independently, the co-proportionation constant (*vide infra*) is determined solely on a statistical basis. The random redistribution of ligands favors the formation of mixed-ligand complexes. Thus, in the case of a single mixed-ligand complex, the system is more "homogeneous" than when different homoleptic complexes are present. Marcus and Elizer have shown[11] that for the equilibrium reaction, Eq. 1.7:

$$jMX_m + iMY_m \rightleftharpoons m\,MX_jY_i \tag{1.7}$$

the co-proportionation constant is given by Eq. 1.8:

$$K = [m!/j!\,i!] \tag{1.8}$$

Equation 1.8 suggests that the co-proportionation constant, calculated on a statistical basis, will always be above unity. Thus, the statistical redistribution of ligands will favor the formation of mixed-ligand complexes.

Deviations in ligand redistribution patterns from the statistical prediction may be due not only to the *trans*- and *cis*-effects (see Sections 1.4 and 1.5) but also to so-called *stacking-effects*. This term refers to the higher stabilities of mixed-ligand complexes, as compared to those of homoleptic systems, and results from unusual ligand-ligand interactions found in certain mixed-ligand complexes.[12,13]

1.3 The chelate effect

If a ligand has more than one donor group it can occupy more than one coordination site and so function as a bridge between metal centers or participate in ring closure at a single metal center. Ligands that participate in ring-closure reactions are known as chelating ligands and the complexes that form are called chelate complexes.

Many reports have shown that coordination compounds with chelating ligands are thermodynamically more stable than those with closely related ligands than do not chelate.[1b] For example, the tris(ligand) complex of nickel(II) with ethylenediamine is thermodynamically more stable than the hexa(ligand) complex formed with methylamine, Scheme 1.1.

Chugaev concluded that five- or six-membered chelate rings are the most favored in coordination compounds[14] and this conclusion is supported by quantitative data. Compounds with seven-membered rings are close in thermodynamic stability to analogous coordination compounds without

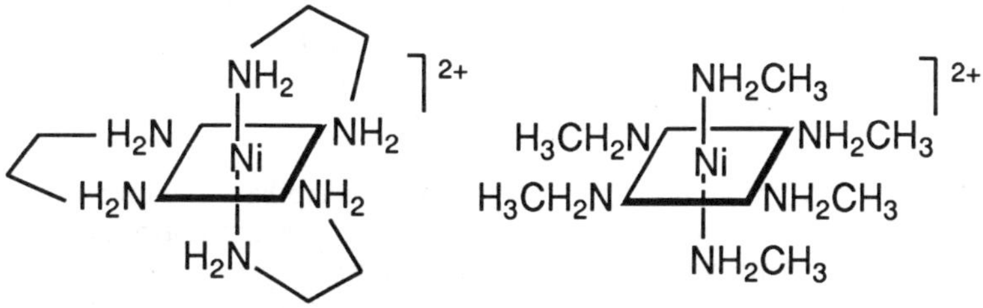

Scheme 1.1

chelating ligands and those with eight-membered rings are typically less stable than similar complexes without chelating ligands. Swarzenbach proposed the term *chelate effect* to describe the difference in thermodynamic stability between a chelate complex and a similar complex without chelating ligands. When complexes are formed, for example in aqueous solution, one or more coordinated water molecules is substituted by the incoming ligand. The number of water molecules liberated is typically equal to the number of incoming ligands when monodentate ligands like NH_3 are employed, Eq. 1.9:

$$[Ni(H_2O)_6]^{2+} + 6NH_3 \rightarrow [Ni(NH_3)_6]^{2+} + 6H_2O \qquad (1.9)$$

The number of liberated water molecules is typically twice the number of incoming ligands when bidentate ligands like ethylenediamine are employed, Eq. 1.10:

$$[Ni(H_2O)_6]^{2+} + 3en \rightarrow [Ni(en)_3]^{2+} + 6H_2O \qquad (1.10)$$

As a result of this difference, the entropy component of the standard free energy change is affected, Eq. 1.11:

$$\Delta G° = -RT \ln K = \Delta H° - T\Delta S° \qquad (1.11)$$

Thus, the standard free energy change associated with Eq. 1.10 is more negative than that associated with Eq. 1.9 and so the tris(ethylenediamine) complex is thermodynamically more stable than the hexa(ammino) complex. The enthalpy component of the free energy change may also be a contributing factor in the chelate effect.[15, 16]

Observations on the chelate effect and, in particular, the importance of ring size, enable predictions to be made concerning the formation of stable coordination compounds with polydentate ligands that contain a variety of individual functional groups. Work on the synthesis of organic ligands which are analytical reagents for the detection and/or separation of certain metal ions, for example, has made use of such observations.

1.4 The *trans*-effect

The *trans*-effect refers to the ability of a ligand to labilize (*i.e.* increase the rate of substitution) of a ligand in the *trans*-position in a square planar or octahedral complex. This kinetic effect is important in planning multi-step syntheses of mixed-ligand complexes and is distinctly different from the *trans*-influence, which is a ground-state property that refers to the ability of a ligand to affect the bond distance between the metal center and a ligand in the *trans*-position. The *trans*-effect was formulated by Chernyaev in 1926[17] and, in general, plays a more significant role in governing reaction kinetics than the related *cis*-effect (see Section 1.5).[18] A *trans*-effect series of common ligands is:

$$CN^- > C_2H_4 > CO, NO_2^- > S{=}C(NH_2)_2 >$$

$$R_2S, R_3P, I^- > Br^- > Cl^- > F^-, NH_3 > H_2O$$

Long before the *trans*-effect was recognized by Chernyaev, empirical rules (Peyrone's rule, Jörgensen's rule, and Kurnakov's rule) were known in coordination chemistry that were based upon a large body of experimental data and that enabled predictions to be made concerning substitution reactions at platinum(II). Chernyaev was able to formulate a chemical basis for these rules and incorporate them into a single, comprehensive framework. Each of the three empirical rules is discussed in the paragraphs that follow.

Peyrone's rule: Amines, A, react with $[PtX_4]^{2-}$ (where X = Cl, Br, I, SCN, CN, NO_2) to form *cis*-$[PtX_2A_2]$ complexes, Scheme 1.2:

Scheme 1.2

Experimentally, this generalization was demonstrated by Jörgensen and given its stereochemical interpretation by Werner. In memory of Peyrone, who was the first to prepare *cis*-$[PtCl_2(NH_3)_2]$, this rule was later called Peyrone's rule.

Jörgensen's rule: Based upon a large body of experimental data, Jörgensen found that protic acids, HX, react with $[PtA_4]^{2+}$ (where A = amine) to produce *trans*-$[PtX_2A_2]$, Scheme 1.3:

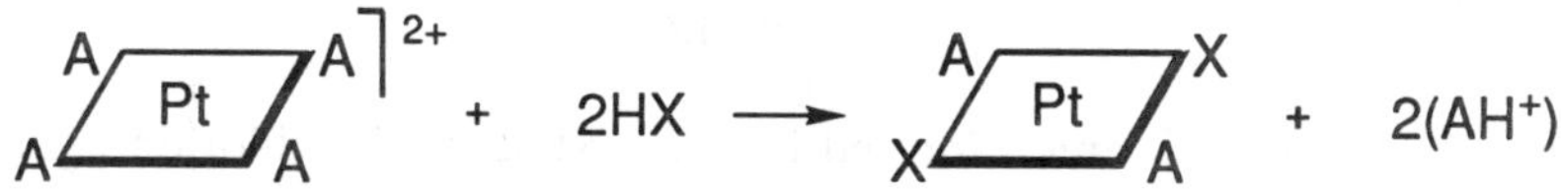

Scheme 1.3

The solid-state thermolysis reactions of the tetra(ammino) complexes, $[Pt(NH_3)_4]X_2$, proceed similarly, for example, Scheme 1.4:

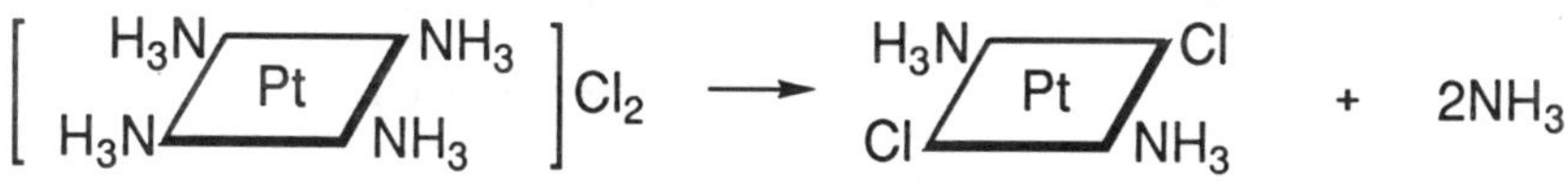

Scheme 1.4

Kurnakov's rule: Kurnakov found that isomeric *cis*- and *trans*-$[PtCl_2(NH_3)_2]$ complexes react with thiourea (thio) to form different products. Later, Grinberg found that the reactions between thiourea and other aminohalide complexes proceed similarly. The general reactions are as follows, Scheme 1.5:

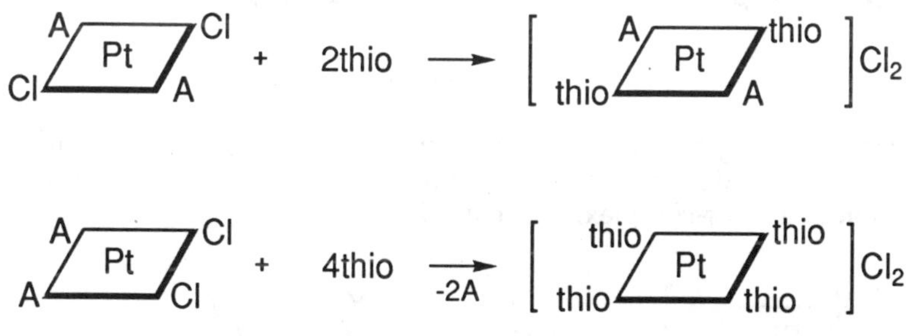

Scheme 1.5

Based upon the concept of the *trans*-effect, Chernyaev was able to explain each of these rules. Thus, referring to Scheme 1.2, when an amine reacts with $K_2[PtCl_4]$, any one of the four, equivalent, halide ligands can undergo substitution, Scheme 1.6:

Scheme 1.6

In the $[PtCl_3(amine)]^-$ anion formed in this first step, one chloride ligand is *trans* to an amine and is, therefore, less labile than either of the other, mutually *trans*, chloride ligands. Accordingly, substitution of one of the mutually *trans* chloride ligands then occurs and a *cis*-product results, Scheme 1.7:

Scheme 1.7

In Scheme 1.3, any of the four, equivalent, amine ligands in the tetramine complex, $[PtA_4]^{2+}$, is subject to substitution upon treatment with a hydrohalic acid, Scheme 1.8:

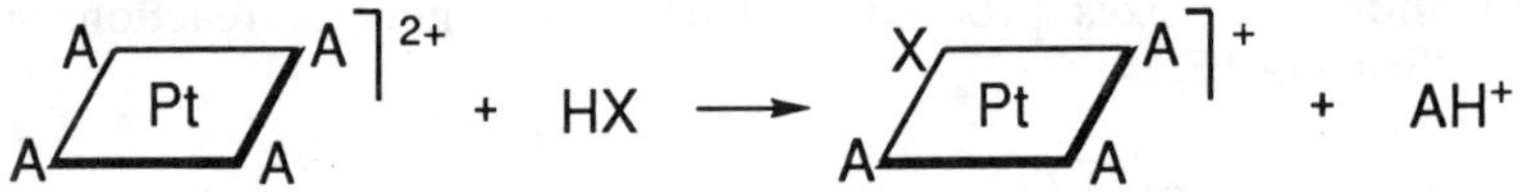

Scheme 1.8

Because halide ligands exhibit higher *trans*-effects than amines, the amine *trans* to the halide ligand is then subject to substitution, leading to the formation of a *trans*-complex, Scheme 1.9:

Scheme 1.9

The first step in the formation of $[Pt(thio)_4]X_2$ by treatment of *cis*-$[PtX_2A_2]$ with thiourea (Kurnakov's rule, Scheme 1.5) is substitution of an amine ligand, which is *trans* to a halide. The high *trans*-effect of thiourea next leads to substitution of the *trans* halide. Repetition of these steps leads to formation of $[Pt(thio)_4]^{2+}$. Treatment of *trans*-$[PtX_2A_2]$ with thiourea (Scheme 1.5) leads to an initial substitution of one of the two mutually *trans* halides, followed by substitution of the halide *trans* to thiourea. Thus, *trans*-$[PtA_2(thio)_2]X_2$ is produced. Under different reaction conditions, *i.e.* at elevated temperatures and over extended periods of time, substitution of the amine ligands of *trans*-$[PtA_2(thio)_2]X_2$ can be accomplished also.

It has been demonstrated that the *trans*-effect is widely applicable in the chemistry of square-planar and/or octahedral complexes of platinum(II), platinum(IV), palladium(II), rhodium(III), iridium(III), cobalt(III), and chromium(III).[19] An example of how the *trans*-effect can be usefully applied in synthesis is the preparation of platinum(IV) complexes containing six different ligands. Essen[20] has reported the multi-step synthesis of such a complex, shown in Figure 1.1:

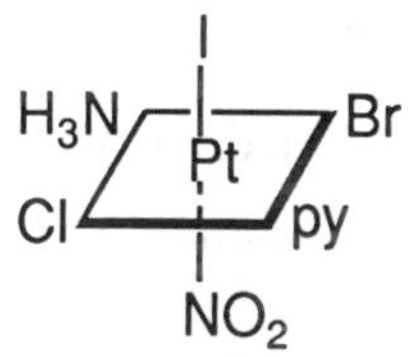

Figure 1.1

Starting with $K_2[PtCl_4]$ the stepwise synthesis begins as follows, Scheme 1.10:

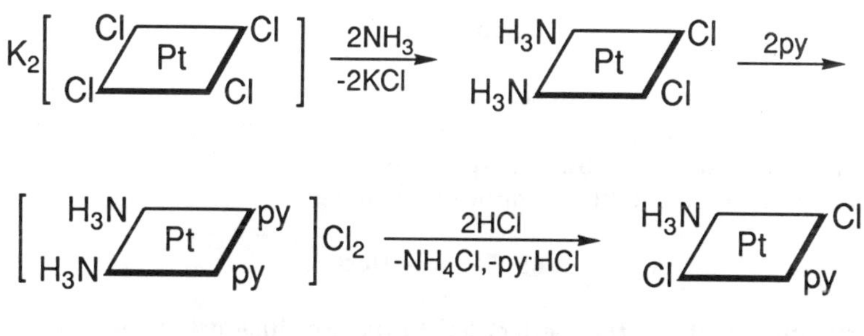

Scheme 1.10

The complex *trans*-$[PtCl_2(py)(NH_3)]$ is then treated with an equimolar amount of KBr, to generate a single isomer of a platinum(II) complex with four different ligands, Scheme 1.11:

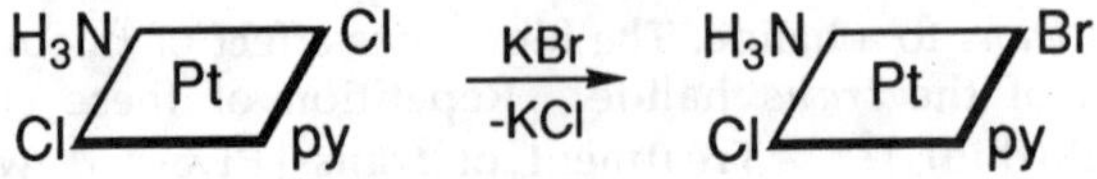

Scheme 1.11

Treatment with nitric acid leads to oxidation of platinum(II) to platinum(IV), Scheme 1.12:

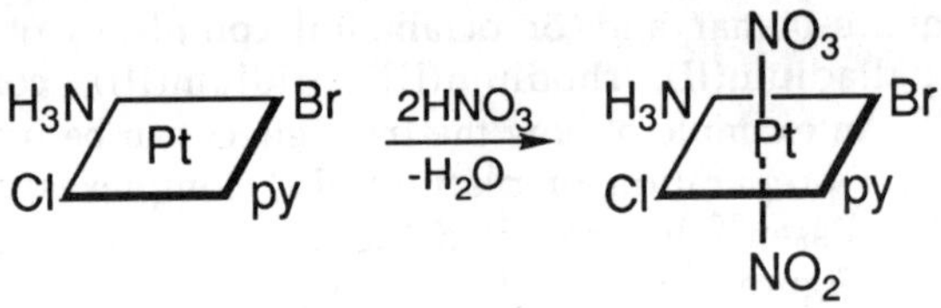

Scheme 1.12

Finally, the nitrate ligand is substituted, *e.g.* by iodide, Scheme 1.13:

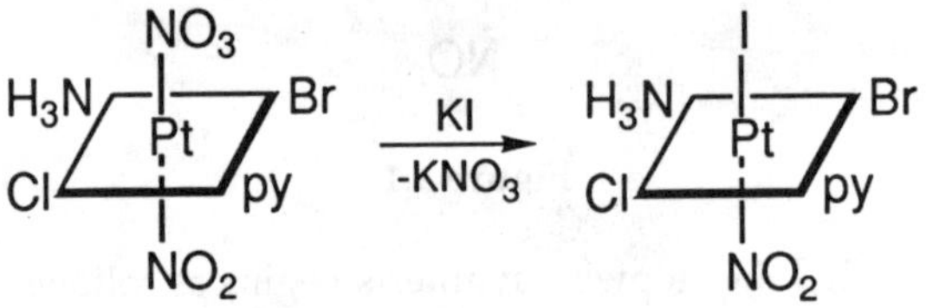

Scheme 1.13

Although the steps represented in Schemes 1.10 – 1.13 appear quite straightforward, finding the appropriate conditions under which to run each reaction (*i.e.* concentration, temperature, time, solvent, p*H*, *etc.*) in a sequence of this type can be a significant challenge.

1.5 The *cis*-effect

Early studies of the *trans*-effect led to the conclusion that the lability of a ligand is affected not only by the ligand in the *trans*-position, but also by the ligands in the *cis*-positions. Indeed, Grinberg noted the possibility of the existence of a *cis-effect* in his 1945 monograph.[21]

Grinberg and Kukushkin established experimentally the *cis*-effects of chloride and ammonia ligands on the substitution kinetics of chloride ligands in platinum(II) complexes.[22] They showed that the substitution of chloride by ammonia in the $[PtCl_4]^{2-}$ ion, Scheme 1.14:

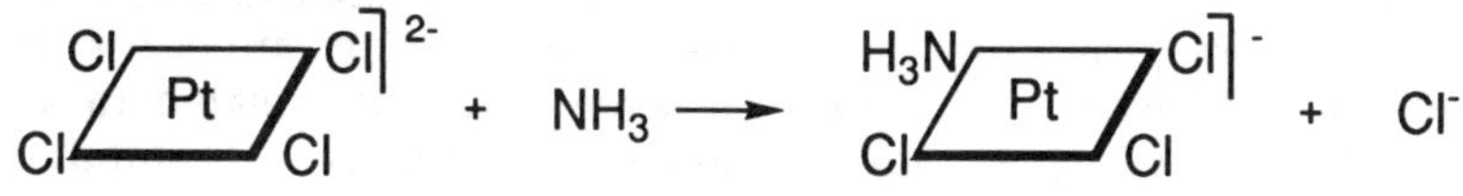

Scheme 1.14

was significantly slower than in $[PtCl_3(NH_3)]^-$, Scheme 1.15:

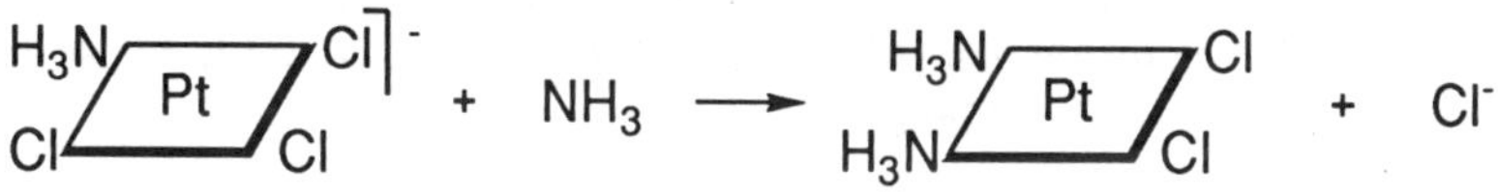

Scheme 1.15

In the absence of any *cis*-effect the rates of ligand substitution in these complexes would be approximately equal because substitution occurs *trans* to chloride in both cases. Further, as the ions have different charges, identical rates would not be anticipated but, once this difference is accounted for, the *cis*-effect becomes evident.

In general, the higher the *trans*-effect of a ligand, the lower is its *cis*-effect. However, the series are not completely reversible. For example, dimethyl sulfoxide, coordinated to platinum(II) through sulfur, exhibits both a very high *trans*-effect (comparable with that of ethylene) and also a high *cis*-effect. A *cis*-effect series based upon comparable data is:

$$C_2H_4 < Cl^- < NH_3, \text{py} < S\text{-DMSO}$$

The *cis*-effects of ligands are an aid to understanding and predicting some properties of coordination compounds, but they are not as important in planning syntheses as the *trans*-effects.

1.6 Geometric isomerization of square-planar platinum(II) and palladium(II) complexes

Geometric isomerism is an important feature of the chemistry of square-planar compounds and has been studied in detail for palladium(II) and, especially, platinum(II) complexes. A pair of isomeric compounds will differ in thermodynamic stability and, under appropriate conditions, a thermodynamically less-stable isomer can spontaneously convert into the thermodynamically more-stable form. Many examples of the isomerization of platinum(II) and palladium(II) complexes, both in solution and in the solid state, have been described. Because geometric isomers often differ in chemical and biological properties, isomerization reactions are of significance in synthetic chemistry. This practical issue is directly related to a more fundamental question — how to assess the relative thermodynamic stabilities of a pair of isomers and so predict the direction of an isomerization reaction.

A simple rule has been formulated to predict the outcome of an isomerization reaction of a platinum(II) or palladium(II) complex.[23] Thus, *isomerization reactions typically favor formation of the isomer in which a neutral ligand with a weak* trans-*effect is* trans *to the ligand with the weakest* trans-*effect*. For example, *cis*-[MX_2(amine)$_2$] complexes tend to isomerize into the *trans*-isomers,[24] Scheme 1.16:

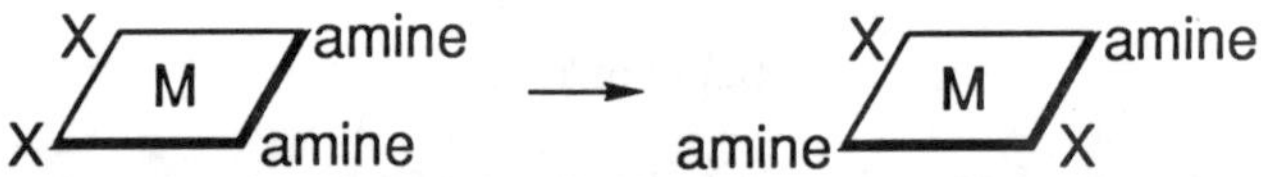

Scheme 1.16

Amines possess lower *trans*-effects than halide ligands, X^-, and, therefore, formation of *trans*-[MX_2(amine)$_2$] is favored. In [$PtCl_2$(amine)(R_2SO)] complexes the ligands of lowest *trans*-effect are the amine and chloride ligands. Thus, isomerization of the *trans*-isomer into the *cis*-form is favored, Scheme 1.17:

Scheme 1.17

This simple rule is broadly applicable to the chemistry of palladium(II) and platinum(II) and also to isomerization reactions of square-planar iridium(I), rhodium(I) and gold(III) complexes. However, there are exceptions. In complexes of the type [MX_2L_2] (M = Pd, Pt; X = halide; L = phosphine, thioether) the favored direction of isomerization is highly dependent upon the nature of the phosphine or thioether. The favored direction of isomerization reactions in solution is, of course, dependent upon solvent polarity in cases where a pair of isomers exhibit different dipole moments.

1.7 The effective atomic number concept

The concept of an effective atomic number, developed by Sidgwick[25] in the 1920s, can be useful in the prediction of composition of coordination compounds. Sidgwick found that in certain complexes the metal center attains an electron count equal to that of the noble gas of the same period by accepting electrons from the coordinated ligands. For example, the cobalt(III) ion (24 electrons) accepts six electron pairs from the ammine ligands in $[Co(NH_3)_6]^{3+}$ to attain the krypton configuration (36 electrons). Sidgwick called this number of electrons the effective atomic number.

The concept of the effective atomic number applies particularly well to carbonyl and nitrosyl compounds of the *d*-block elements. For example, the composition of mononuclear nickel(0) and iron(0) carbonyl complexes may be rationalized in terms of effective atomic number. To attain the krypton configuration, nickel (28 electrons) and iron (26 electrons) need to accept four and five electron pairs, respectively. Thus, $[Ni(CO)_4]$ and $[Fe(CO)_5]$ are the predicted compositions. Linear nitrosyls are three-electron donors and so binding of a $[Co(CO)_3]$ fragment (33 electrons) to a single NO ligand to form $[Co(NO)(CO)_3]$ would result in the krypton electron configuration.

The noble gases in each of the three long periods have eighteen valence electrons (*i.e.* $(n\text{-}1)s^2$, $(n\text{-}1)p^6$, nd^{10}) and extension of the effective atomic number concept, therefore, suggests that many stable coordination compounds will similarly possess 18 valence electrons.

1.8 The 16-/18-electron rule

In 1972, Tolman[26] formulated the 16-/18-electron rule which has proved to be valuable in understanding structure and reactivity in organometallic chemistry and, especially, in homogeneous catalysis. In general, organometallic compounds of the mid-transition metals adopt structures that allow the metal center to attain a closed-shell, 18-valence-electron configuration. The late transition metals, *e.g.* palladium(II) and platinum(II), form complexes where ligand-field effects favor four-coordinate, square-planar structures and typically attain 16-valence-electron configurations. Early transition metals often support electron-deficient structures because steric effects limit the coordination number.

1.9 The hard-soft acid-base (HSAB) principle

Aquated metal ions tend to group into two categories in their reactions with halide ions.[27a] One class favors reactions with small, unpolarizable bases (*e.g.* fluoride ion) and the other favors reactions with large, polarizable bases (*e.g.* iodide ion). Swarzenbach referred to these classes of metal ions as class **A** acceptors and class **B** acceptors, respectively.[27b] In 1958, Ahrland, Chatt and Davies[27c] extended the categories of metal ions through consideration of reactions with ligands other than halides. Metal ions forming the most stable complexes with ligands whose donor atoms belong to the first row, *e.g.* N, O, and F, were assigned to class **a**. Metal ions forming the most stable complexes with ligands whose donor atoms belong to the second or subsequent rows, *e.g.* P, S, Se, Cl, Br, and I, were assigned to class **b**. The classification system recognized also the existence of borderline Lewis acids, which do not show typical class **a** or class **b** behavior.

In 1963, Pearson[28] devised new classifications for Lewis acids and bases, including examples from both organic and inorganic chemistry based upon thermodynamic and kinetic considerations. The new classes were named hard (approximately parallel to class **a**) and soft (approximately parallel to class **b**). Hard species tend to possess closely-held orbitals and are small in size (indicating low polarizability), while soft species tend to be large, with less closely-held orbitals (indicating high polarizability). A borderline category was recognized also in this classification system.

The hard-soft acid-base (HSAB) principle states that *hard acids prefer to associate with, and react readily with, hard bases while soft acids prefer to associate with, and react readily with, soft bases.* The HSAB principle embodies both kinetic and thermodynamic meaning. Thus, interaction between a Lewis acid and a Lewis base of comparable hardness or softness is predicted to proceed readily and result in the formation of a thermodynamically stable product. Applications of the HSAB principle to coordination chemistry abound.[29] For example, DMSO is an ambidentate ligand with both hard (oxygen) and soft (sulfur) donor sites. When complexes are formed with platinum(II), a soft acid, DMSO will typically coordinate *via* sulfur, while, with the harder acid nickel(II), coordination *via* oxygen is favored.[30]

Pearson has described[28] how the concepts of hardness and softness can be understood through application of density functional theory (DFT) to chemical systems and, further, how the results correlate with molecular orbital theory. A firm theoretical basis for the HSAB principle is evolving from these studies that links chemical hardness to absolute electronegativity which, in this sense, refers to the electronic chemical potential of a system rather than the electronegativity of a single atom within a molecule.

1.10 Factors affecting the acid/base properties of coordination compounds

The studies of Werner's coworker, Pfeiffer, on the chromium(III) complex, $[Cr(py)_2(H_2O)_4]Cl_3$, were central to the development of an understanding of the acid-base properties of metal complexes.[31] An aqueous solution of the complex, when treated with ammonia, precipitates the relatively insoluble compound, $[Cr(OH)_2(py)_2(H_2O)_2]Cl$. The reaction was found to be accompanied by a dramatic change in the color of the solution, from red-violet to gray-green. Ammonia does not function as a ligand in the reaction but, rather, plays the role of a base, Eq. 1.12:

$$[Cr(py)_2(H_2O)_4]Cl_3 + 2NH_3 \rightleftharpoons [Cr(OH)_2(py)_2(H_2O)_2]Cl + 2NH_4Cl \quad (1.12)$$

The complex $[Cr(OH)_2(py)_2(H_2O)_2]Cl$ equilibrates with the protonated form, $[Cr(py)_2(H_2O)_4]Cl_3$, in aqueous solution, Eq. 1.13 and 1.14:

$$[Cr(OH)_2(py)_2(H_2O)_2]^+ + H_2O \rightleftharpoons [Cr(OH)(py)_2(H_2O)_3]^{2+} + OH^- \quad (1.13)$$

$$[Cr(OH)(py)_2(H_2O)_3]^{2+} + H_2O \rightleftharpoons [Cr(py)_2(H_2O)_4]^{3+} + OH^- \quad (1.14)$$

The addition of acid to the solution leads to shifts in the equilibria that favor the aqua-form of the complex. Thus, Pfeiffer, in 1906, established the reversible transformation of aqua-complexes into hydroxo-complexes. Werner later used this observation in the development of acid-base concepts in coordination chemistry.[32] Werner's work in this field is now largely of historical interest but two points, emphasized by Werner, remain especially noteworthy, *i.e.* the importance of solvent in acid-base equilibria and that bases may be considered as proton acceptors. These statements were incorporated into the acid-base theory developed by Brønsted and Lowry in 1923.

Bjerrum and Brønsted noted that the acidic properties of aqua-complexes depend on the formal charge on the metal center and on the charge on the complex ion. Thus, the acidic properties of water coordinated to platinum(IV) are more pronounced than those of water coordinated to platinum(II) in complexes with equal overall charges. For example, of the equilibria represented by Eq. 1.15 and 1.16:

$$[Pt(NH_3)_3(H_2O)]^{2+} \rightleftharpoons [Pt(OH)(NH_3)_3]^+ + H^+ \quad (1.15)$$

$$[PtCl_2(NH_3)_3(H_2O)]^{2+} \rightleftharpoons [PtCl_2(OH)(NH_3)_3]^+ + H^+ \quad (1.16)$$

the latter lies further to the right. Between the platinum(IV) complexes, $[PtCl_2(NH_3)_3(H_2O)]^{2+}$ and $[Pt(NH_3)_5(H_2O)]^{4+}$, the latter, with the higher overall charge, is the stronger acid because the equilibrium, Eq. 1.17:

$$[Pt(NH_3)_5(H_2O)]^{4+} \rightleftharpoons [Pt(OH)(NH_3)_5]^{3+} + H^+ \qquad (1.17)$$

lies further to the right than that represented by Eq. 1.18:

$$[PtCl_2(NH_3)_3(H_2O)]^{2+} \rightleftharpoons [PtCl_2(OH)(NH_3)_3]^{+} + H^+ \qquad (1.18)$$

In 1915, Chugaev found that $[PtCl(NH_3)_5]Cl_3$ is deprotonated by sodium hydroxide with formation[33] of the relatively insoluble amido complex, $[PtCl(NH_2)(NH_3)_4]Cl_2$, Eq. 1.19:

$$[PtCl(NH_3)_5]Cl_3 + NaOH \rightleftharpoons [PtCl(NH_2)(NH_3)_4]Cl_2 + NaCl + H_2O \qquad (1.19)$$

In terms of the observed acid-base chemistry, the amido-complex, $[PtCl(NH_2)(NH_3)_4]Cl_2$, bears a close resemblance to the hydroxo-complex, $[Cr(OH)_2(py)_2(H_2O)_2]Cl$, described above. Thus, a basic solution is formed by dissolution of the former in water, Eq. 1.20:

$$[PtCl(NH_2)(NH_3)_4]^{2+} + H_2O \rightleftharpoons [PtCl(NH_3)_5]^{3+} + OH^- \qquad (1.20)$$

Kirmreuthyer[34] reported a similar reaction some years before Chugaev's report. He synthesized several platinum(II) sulfamate complexes through reactions of platinum salts with sulfamic acid, NH_2SO_3H. The acid dissociation constant of sulfamic acid, K_a, is 0.1. Kirmreiter found that aqueous solutions of the isomeric compounds, *cis*- and *trans*-$K_2[PtCl_2(NH_2SO_3)_2]$, react with bases to form $K_4[PtCl_2(NHSO_3)_2]$, containing deprotonated sulfamate ions.

The work of Chugaev allowed Grinberg to extend the acid-base concepts of Bjerrum and Brønsted to metal complexes with other potential proton-donor ligands. In particular, based upon the chemistry of complexes of ammonia, methylamine, ethylenediamine, dimethylglyoxime, hydroxylamine and related ligands, Grinberg established a functional relationship between the acidic properties of metal complexes with proton-donor ligands (X-H) and factors related to structure and bonding,[35] Eq. 1.21:

$$\mathbf{A} = f(\mathbf{H}, \mathbf{E}, \mathbf{a}, \mathbf{b}, \mathbf{n}, \mathbf{G}) \qquad (1.21)$$

where **A** characterizes the acidic properties of a complex containing X-H as a ligand.

H is the force field of the metal ion, determined by its charge (here, synonymous with oxidation state), radius, electronic structure and polarizability.

E is the overall charge on the complex.

a is the degree of dissociation of free X-H.

b is the degree to which dissociation of X-H is affected by coordination.

n is the number of coordinated X-H ligands.

G is the geometry of the complex.

Thus, all other consideration being equal, proton dissociation from coordinated X-H is promoted by small, highly charged metal ions. Table 1.1 illustrates acid dissociation constants for similarly sized ions with different overall charges.

Table 1.1

Acid dissociation constants of platinum(IV) ammino-complexes in water[14,35]

$[Pt(NH_3)_6]^{4+}$	1.5×10^{-7}
$[PtCl(NH_3)_5]^{3+}$	7.9×10^{-9}
cis-$[PtCl_2(NH_3)_4]^{2+}$	3.5×10^{-10}
trans-$[PtCl_2(NH_3)_4]^{2+}$	6×10^{-12}

The importance of electronic structure is evident in a comparison[35] of the acidic properties of $[Pt(en)_3]^{4+}$ and $[Os(en)_3]^{4+}$. For the platinum complex, $K_1 = 3.5 \times 10^{-6}$ and $K_2 = 1.8 \times 10^{-10}$. In the osmium case, in the first dissociation step, the complex acts as a strong acid and in the second step $K_2 = 1.6 \times 10^{-6}$. The ionic radii of platinum(IV) and osmium(IV) are quite similar and the formal oxidation states and overall charges on the complexes are identical yet the d^6 and d^4 complexes exhibit significantly different behavior as protic acids.

The relative ordering of the acidities of free molecules, X-H, is generally maintained upon coordination to a metal center. Thus, coordinated water is deprotonated before a coordinated ammine. The acid dissociation constants of platinum(IV) aqua-complexes have been found to be 10^3-10^5 times larger than those of the analogous ammine complexes. A comparison of $[Pt(NH_3)_6]^{4+}$ and $[Pt(NH_3)_5(H_2O)]^{4+}$ provides acid dissociation constants of 1.5 $\times 10^{-7}$ and *ca.* 10^{-4}, respectively.

Brønsted has shown that acidic properties depend on the number of coordinated proton-donor ligands, Table 1.2:

Table 1.2

Acid dissociation constants of cobalt(III) complexes in water[14,35]

Complex	K
$[Co(NH_3)_5(H_2O)]^{3+}$	2.04×10^{-6}
$[Co(NH_3)_4(H_2O)_2]^{3+}$	6.03×10^{-6}
$[Co(NH_3)_3(H_2O)_3]^{3+}$	1.88×10^{-5}
$[Co(NH_3)_2(H_2O)_4]^{3+}$	4.00×10^{-4}

An increase in the number of water molecules coordinated to cobalt(III) is accompanied by an increase in the first acid dissociation constant and this is attributed to proton repulsion by clustering of the proton-donor groups. In such cases, the complex geometry may be important since a *cis*-arrangement of X-H ligands in an octahedral complex may give rise to more effective proton repulsion than a *trans*-arrangement (*vide infra*).

Grinberg and Ryabtchikov[36] were the first to study the influence of complex geometry on acidic properties of aqua-complexes. Based on a study of isomeric $[Pt(NH_3)_2(H_2O)_2]^{2+}$ complexes, they concluded that K_1 of the *trans*-isomer is larger than K_1 of the *cis*-isomer. Further, K_1/K_2 of the *cis*-isomer is much larger than the same ratio for the *trans*-isomer. Later, such relationships were found for other diaquadiamine-complexes, Table 1.3. These data demonstrate a less favorable first proton-dissociation step in *cis*-isomers as compared to *trans*-isomers of diaquadiamine-complexes. According to Grinberg, the high *trans*-influence of hydroxide, as compared to H_2O, results in a significant difference between K_1 and K_2 in *trans*-$[Pt(NH_3)_2(H_2O)_2]^{2+}$.

Table 1.3

Acid dissociation constants of isomeric platinum(II) diaquadiamine-complexes in water[35]

Complex	*Isomer*	K_1	K_2	K_1/K_2
$[Pt(NH_3)_2(H_2O)_2]^{2+}$	*cis-*	2.8×10^{-6}	4.8×10^{-8}	58
	trans-	4.8×10^{-5}	4.2×10^{-8}	1147
$[Pt(NH_2Me)_2(H_2O)_2]^{2+}$	*cis-*	2.7×10^{-6}	7.3×10^{-8}	37
	trans-	3.4×10^{-5}	8.0×10^{-8}	425
$[Pt(NH_2Et)_2(H_2O)_2]^{2+}$	*cis-*	2.9×10^{-6}	8.2×10^{-8}	35
	trans-	3.4×10^{-5}	10.0×10^{-8}	340
$[Pt(py)_2(H_2O)_2]^{2+}$	*cis-*	3.2×10^{-5}	5.8×10^{-7}	55
	trans-	1.5×10^{-4}	3.7×10^{-7}	406

Thus, the first proton dissociation occurs from a water molecule coordinated *trans* to water but the second dissociation occurs from a water molecule coordinated *trans* to the newly formed hydroxide. Similar effects are found in comparisons of first and second acid dissociation constants of the isomers of $[Co(en)_2(H_2O)_2]^{3+}$ and $[Cr(en)_2(H_2O)_2]$.[35]

Grinberg[35] has shown that octahedral platinum(IV) amine complexes of *cis*-geometry are stronger protic acids than the corresponding *trans*-isomers. For example, the *cis*-isomer of $[PtCl_2(pn)_2]^{2+}$ (where pn = propylenediamine) has $K_1 = 5.4 \times 10^{-9}$ and $K_2 = 3.6 \times 10^{-11}$ whereas, for the *trans*-isomer (only the first acid dissociation constant has been measured) $K_1 = 1.8 \times 10^{-11}$. The same relationship exists for isomeric $[PtCl_2(NH_3)_4]^{2+}$ complexes where for the *cis*-isomer $K_1 = 3.5 \times 10^{-10}$ and $K_2 = 5.6 \times 10^{-11}$ while the *trans*-isomer has a first acid dissociation constant that could barely be measured potentiometrically, *i.e.* $K_1 = 6 \times 10^{-12}$. Grinberg has attributed these effects to enhanced proton repulsion in octahedral platinum(IV) complexes of *cis*-geometry in comparison to that found for *trans*-isomers.

The effect of electron configuration on the acidity of metal aqua- and ammino-complexes is illustrated by a comparison of similar compounds from the same period.[14] For example, the d^5 complex, $[Ru(NH_3)_5(H_2O)]^{3+}$, has pK_a *ca.* 4 while the d^6 complex, $[Rh(NH_3)_5(H_2O)]^{3+}$, has pK_a 6.9.

Similarly, for the d^4 complex, *trans*-$[OsCl_2(NH_3)_4]^{2+}$, the pK_a is *ca*. 4, but for the d^6 complex, *trans*-$[PtCl_2(NH_3)_4]^{2+}$, the pK_a is 11.2. For $[Os(NH_3)_6]^{4+}$ and $[Pt(NH_3)_6]^{4+}$ the pK_a values are < 0 and 6.9, respectively. Thus, complexes with t_{2g} orbitals that are not fully occupied are stronger protic acids than complexes where these orbitals are fully occupied. The deprotonated form of the X-H ligand may be stabilized by π-donation in cases where the t_{2g} orbitals are not fully occupied.

In 1955, Leden and Chatt[37] determined the acid dissociation constant of $[PtCl_2(C_2H_4)(H_2O)]$. The value was some two orders of magnitude higher than expected, *ca*. 10^{-5}. Such high acid dissociation constants (10^{-4} to 10^{-6}) are more typical of dicationic aqua-complexes, such as $[Pt(amine)_2(H_2O)_2]^{2+}$ (Table 1.3). The neutral aqua-complex $[PtCl_2(C_2H_4)(H_2O)]$ differs from the $[Pt(amine)_2(H_2O)_2]^{2+}$ complexes studied because it contains coordinated ethylene, a ligand with significant π-acceptor character. Subsequently, it was established that the presence of a π-acceptor ligand gives rise to a dramatic increase in the acidic character of coordinated X-H ligands. For example, Grinberg and coworkers[35,38] showed that substitution of ammonia by pyridine in platinum(II) aqua-compounds enhances proton dissociation from coordinated X-H ligands. Similarly, it has been found that coordination of pyridine promotes proton dissociation from ammonia coordinated to platinum(IV).[39-41] Results of Gelfman and Smolenskaya[42] on the acidic properties of the dicationic platinum(II) aqua-complex, *trans*-$[Pt(C_2H_4)(NH_3)_2(H_2O)]^{2+}$, confirmed the earlier observation that strong π-acceptor ligands, such as ethylene, promote proton dissociation from coordinated water.

The complex $K[PtCl_3(DMSO)]$[43,44] has been found to have many properties similar to those of Zeise's salt, $K[PtCl_3(C_2H_4)]$. The similarity of the two complexes may be attributed, in part, to the fact that ethylene and DMSO both possess significant π-acceptor capabilities. The acid dissociation constant of $[PtCl_2(H_2O)(DMSO)]$ is 1.4×10^{-5}, very similar to the value found for $[PtCl_2(C_2H_4)(H_2O)]$.

1.11 Ligand effects on redox potentials of coordination compounds

In 1898, Peters[45] observed the effects of complex formation on the Fe(II)/Fe(III) redox potential. He found that, in aqueous solution, the redox potential of a system consisting of iron(III) and iron(II) ions depended on the concentration of hydrochloric acid. Carter and Clews,[46] in 1924, explained this phenomenon in terms of metal ion complexation. Michaelis and coworkers[47,48] carried out a systematic study of the influence of anions on the

Fe(II)/Fe(III) redox potential. They showed that replacement of one anion by another resulted in potential changes over the range +0.7 to -0.2 *V*.

Complexation affects the position of the equilibrium between the oxidized and reduced forms of a metal ion. If the ligands are neutral and the coordination numbers of both forms of the complex are the same, the equilibria can be expressed as follows, Scheme 1.18:

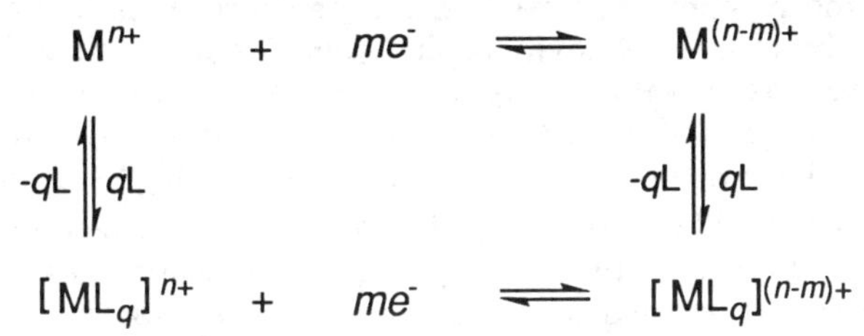

Scheme 1.18

If the reduced form of the metal ion is more stabilized by complexation than the oxidized form, then, according to the Nernst equation, the redox potential will increase. If the oxidized form is stabilized more than the reduced form, then the opposite is found.

Stabilization by complexation depends upon a large number of inter-related factors which might depend upon the nature of the metal ion, the properties of the available ligands, or both. For example, one important factor is coordination number which, although largely determined by the electronic configuration of the metal ion, also may be controlled by the properties of the available ligands.

Pairs of geometric isomers differ from each other in their standard free energies of formation. However, the differences are typically not large. In the case of labile complexes, one isomer will be transformed into the thermodynamically favored form as equilibrium is reached. However, in the case of inert complexes, the conversion of one isomer into the other can be inhibited. In such cases, it may prove possible to study redox equilibria without the problem of an accompanying isomerization. Platinum and ruthenium chemistry has provided many examples of such systems.

Following the work of Gelman and Ryabtchikov,[49] who observed that Zeise's salt, $K[PtCl_3(C_2H_4)]$, is not oxidized by potassium permanganate while $K[PtCl_3(NH_3)]$ is readily quantitatively oxidized, Ahrland and Chatt[50] suggested that the redox properties of metal complexes depend upon the π-acceptor abilities of the ligands.

The relationship between the ability of a coordination compound to function as an oxidant and the nature of the coordinated ligands was investigated by Gazo[51] who observed the reduction of copper(II) chloride to copper(I) chloride in aqueous acetone, with concomitant chlorination of the acetone, Eq. 1.22:

$$2CuCl_2 + CH_3COCH_3 \rightarrow 2CuCl + CH_3COCH_2Cl + HCl \qquad (1.22)$$

The rate of the reaction increased in the presence of potential π-acceptor ligands, *e.g.*, thiourea, triphenylphosphine, and organonitriles. It was shown that these ligands complex copper(II). Complexation was followed by homolytic cleavage of the Cu–Cl bond and formation of a chlorine radical. Free radical chlorination of acetone results. Gazo suggested that homolytic cleavage of the Cu–Cl bond is favored by complexation of π-acceptor ligands.

Systematic studies of the redox properties of platinum complexes with anionic ligands were carried out by Grinberg and coworkers.[29] They showed that the Pt(IV)/Pt(II) redox potential, measured in aqueous solution, exhibits a significant anion dependence, as shown in the series:

$$NO_2^- > Cl^- > Br^- > SCN^- > CN^- > I^-$$

Thus, nitrite complexes are characterized by the highest redox potentials. The nature of the supporting electrolyte is of importance here. For example, potentials measured in aqueous HCl are much higher than those measured in aqueous NaCl. Apparently protonation of coordinated nitrite may be significant.

Pyridine differs from ammonia and aliphatic amines in its influence on the redox potentials of metal complexes, Table 1.4:

Table 1.4

Redox potentials of platinum amine complexes (*vs.* Ag/AgCl, 25 °C)[52]

Redox couple	*E°, V (±0.003 V)*	
	1M NaCl	*1M HCl*
$[PtCl_6]^{2-} + 2e^- \rightleftharpoons [PtCl_4]^{2-} + 2Cl^-$	—	0.734
$[PtCl_5(NH_3)]^- + 2e^- \rightleftharpoons [PtCl_3(NH_3)]^- + 2Cl^-$	0.731	0.694
$[PtCl_5(py)]^- + 2e^- \rightleftharpoons [PtCl_3(py)]^- + 2Cl^-$	0.763	0.751
$[PtCl_2(NH_3)_4]^{2+} + 2e^- \rightleftharpoons [Pt(NH_3)_4]^{2+} + 2Cl^-$	0.619	0.600
$[PtCl_2(py)_4]^{2+} + 2e^- \rightleftharpoons [Pt(py)_4]^{2+} + 2Cl^-$	0.932	0.847

In general, if substitution of chloride by ammonia leads to lower redox potentials, then substitution by pyridine results in higher potentials.

Through ligand substitution in complexes of the same general type it is sometimes possible to observe additivity patterns. For example, the redox potentials of analogous systems increase in an additive fashion in sequential substitution of ammonia by pyridine in *trans*-$[PtCl_2(NH_3)_4]^{2+}$. Unfortunately, such correlations are not always observed. Nevertheless, in many cases, when one ligand is replaced by another, it is possible to estimate the redox potential of the new system. Independent of the type of complex, the influence of amines of different types on redox potentials remains essentially the same, Table 1.5:

Table 1.5

Redox potentials for:

$$[PtCl_2(NH_3)_2(amine)_2]^{2+} + 2e^- \rightleftharpoons [Pt(NH_3)_2(amine)_2]^{2+} + 2Cl^-$$

(*vs.* Ag/AgCl, 1*M* aq. NaCl)[52]

Amine	*E, V*	*Amine*	*E, V*
$PhNH_2$	0.569	Me_2NH	0.683
$(HOC_2H_4)_2NH$	0.593	1/2(bipy)	0.685
1/2(en)	0.597	Et_2NH	0.701
$MeNH_2$	0.615	4Me-py	0.713
NH_3	0.619	1/2(phen)	0.718
$EtNH_2$	0.625	3Me-py	0.726
$HOC_3H_6NH_2$	0.636	Quinoline	0.746
$HOC_2H_4NH_2$	0.641	py	0.776
$PhCH_2NH_2$	0.648	1/2($Me_2NC_2H_4NMe_2$)	0.786
n-$PrNH_2$	0.654	2Me-py	0.801
n-$BuNH_2$	0.661		

Because amine complexes of the type shown in Table 1.5 are readily available and freely soluble in water, these redox systems have been extensively investigated. The amines in Table 1.5 are arranged in order of increasing redox potential for the Pt(IV)/Pt(II) couple. Table 1.5 shows that these potentials exhibit a strong dependence on the nature of the amine ligand. The difference in the effects of pyridine and ammonia on the potentials of Pt(IV)/Pt(II) couples has been attributed to the π-acceptor ability of pyridine.

A good π-acceptor ligand will stabilize platinum(II) more than platinum(IV) and, thus, lead to an increase in the redox potentials.

The π-acceptor abilities of dialkyl sulfoxide ligands similarly result in high redox potentials in Pt(IV)/Pt(II) systems. For example, studies[52] of the redox couple shown in Eq. 1.23:

$$[PtCl_5L]^- + 2e^- \rightleftharpoons [PtCl_3L]^- + 2Cl^- \quad (1.23)$$

gave rise to the data presented in Table 1.6. For comparison, Table 1.6 includes potentials of related systems with amine ligands measured under conditions comparable to those employed for the sulfoxide complexes.

Table 1.6

Redox potentials for:
$[PtCl_5L]^- + 2e^- \rightleftharpoons [PtCl_3L]^- + 2Cl^-$ (*vs.* Ag/AgCl, 1*M* aq. NaCl)[52]

L	*E, V*	*L*	*E, V*
$MeNH_2$	0.728	$C_5H_{10}NH$	0.755
NH_3	0.730	py	0.763
$PhCH_2NH_2$	0.720	Me_2SO	0.825
$EtNH_2$	0.737	Et_2SO	0.830

The sulfoxide complexes exhibit higher potentials than even the pyridine analog.

The influence of thioethers on the redox properties of platinum complexes, as compared to the effect of amines, has been investigated.[52] The substitution of ammonia by methylpyridine, pyridine, or thioethers leads to a considerable increase in the redox potential of the Pt(IV)/Pt(II) couple, Table 1.7. The π-acceptor abilities of thioethers are reflected in increased redox potentials, which exceed even those of the pyridine analog. Thus, the experimental data suggest that complexes containing π-acceptor ligands usually exhibit high redox potentials. Undoubtedly, differences in solvation of the oxidized and reduced forms will contribute to differences observed in the potential values and so ligand effects on redox potentials are best studied when both forms of a complex have the same charge.

Table 1.7

Redox potentials for:

$[PtCl_3(en)L]^+ + 2e^- \rightleftharpoons [PtCl(en)L]^+ + 2Cl^-$ (*vs.* Ag/AgCl, 1*M* aq. HCl)[52]

L	*E, V*	*L*	*E, V*
NH_3	0.609	$O(C_2H_4)_2S$	0.690
3Me-py	0.654	Et_2S	0.700
py	0.657	$EtSC_2H_4OH$	0.721
Me_2S	0.669	$(HOC_2H_4)_2S$	0.731

Complexes of the Werner-Miolati series are examples where the oxidized and reduced forms bear identical charges, Eq. 1.24 and 1.25:

$$[PtCl_6]^{2-} + 2e^- \rightleftharpoons [PtCl_4]^{2-} + 2Cl^- \tag{1.24}$$

$$[PtCl_5(NH_3)]^- + 2e^- \rightleftharpoons [PtCl_3(NH_3)]^- + 2Cl^- \tag{1.25}$$

However, in such systems, where the coordination numbers of the reduced and oxidized forms differ, special consideration must be paid to the stepwise nature of the overall process. For example, electron transfer to platinum(IV), as shown in Eq. 1.21, might be followed by a rate-limiting bond cleavage step that leads to extrusion of chloride ion. The processes observed, for example by cyclic voltammetry, may be under kinetic control and thermodynamic parameters may not be readily obtained from experimental data.

Lever has described[53] the electrochemical parameterization of metal complex redox couples based upon a Ru(III)/Ru(II) couple to generate an electrochemical series of ligands. Values of the parameter, $E_L(L)$, defined as one sixth of the value of the Ru(III)/Ru(II) potential for $[RuL_6]$ in acetonitrile, were presented for over 200 ligands, and additional parameters S_M and I_M, which depend upon the metal and redox couple, as well as spin state and stereochemistry, were also tabulated. The calculated redox potential of a M(n)/M(n-1) couple was defined as $S_M[\Sigma E_L(L)] + I_M$ and the model tested over a range of complexes, largely restricted to examples where an electron was added to, or removed from, the t_{2g} subset in octahedral systems. The approach has proved to be useful both in prediction of electrochemical properties and as an aid in the interpretation of experimental data.[53]

1.12 Reactivity of coordinated ligands

The change in reactivity of ligands as a result of their coordination to a metal center forms the basis for the use of metal complexes as stoichiometric reagents and homogeneous catalysts in organic chemistry.

Among the factors governing the reactivity of coordinated ligands, the charge and electron-acceptor properties of the metal ion are the most important. The increase in acidity of X-H ligands upon coordination has been discussed in Section 1.10. Electrophilicity also may be significantly affected by coordination. Coordination to a positively charged metal center generally leads to an increase in the electrophilicity of a ligand due to transfer of electron density to the metal center.

Olefins typically undergo an increase in electrophilicity upon coordination. Free olefins undergo electrophilic addition reactions, for example, with hydrohalic acids. When olefins are coordinated to a positively charged metal center, attack by nucleophilic reagents becomes possible. Because nucleophilic attack on a coordinated ligand is favored by a positive charge on the complex, the use of neutral supporting ligands, rather than anionic ligands, may promote the reaction.

The increase in electrophilicity of coordinated ligands commonly leads to more favorable hydrolysis reactions. There are many reports on hydrolysis reactions of coordinated carboxamides, halogenated alkylamines, Schiff bases, thioamides, nitriles, *etc.*[54] For example, thiourea and thioacetamide complexes of platinum metals decompose on heating in basic solution with formation of the corresponding metal sulfides, Eq. 1.26 and 1.27:

$$[Pt]–S{=}C(NH_2)_2 + OH^- \rightarrow [Pt]–S–C(OH)(NH_2)_2 \qquad (1.26)$$

$$[Pt]–S–C(OH)(NH_2)_2 + OH^- \rightarrow PtS + O{=}C(NH_2)_2 + H_2O \qquad (1.27)$$

This process has been used for the isolation and concentration of platinum metals as well as their gravimetric determination.

Lewis acid catalysis of the hydrolysis of peptide bonds by metal ions and complexes has been investigated[55] with regard to the hydrolytic stability of proteins. An amide linkage in a protein has a half-life of about seven years at neutral p*H* and 25 °C. In model systems, the hydrolysis rate can be increased by *ca.* 10^8 through coordination of the amide carbonyl group to a cationic metal complex. The coordination to cobalt(III) and subsequent Lewis acid assisted hydrolysis of the amide group in the simple amide, 4-formylmorpholine is illustrated in Scheme 1.19:[55]

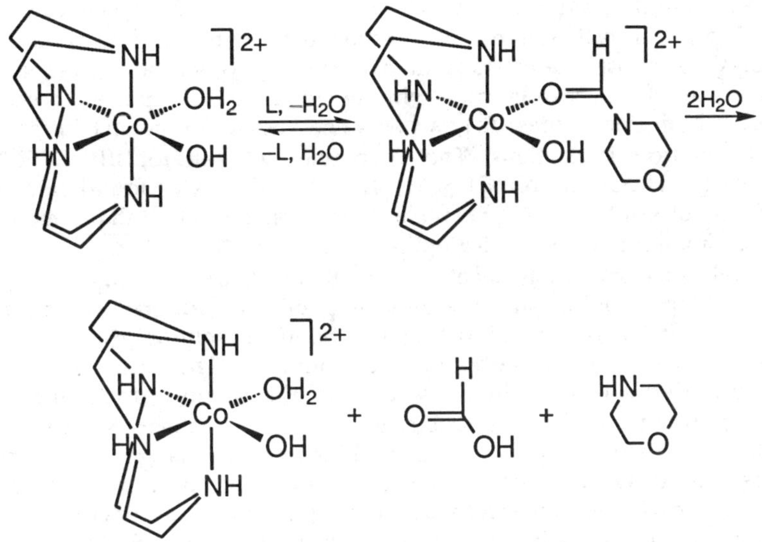

Scheme 1.19

The relative rates of hydrolysis of free 4-formylmorpholine, 4-formylmorpholine coordinated to a substitutionally inert $[Co(NH_3)_5]^{3+}$ moiety, and 4-formylmorpholine coordinated to the metal-hydroxide group, $[CoL_4(OH)]^{2+}$, are shown in Scheme 1.20:[55]

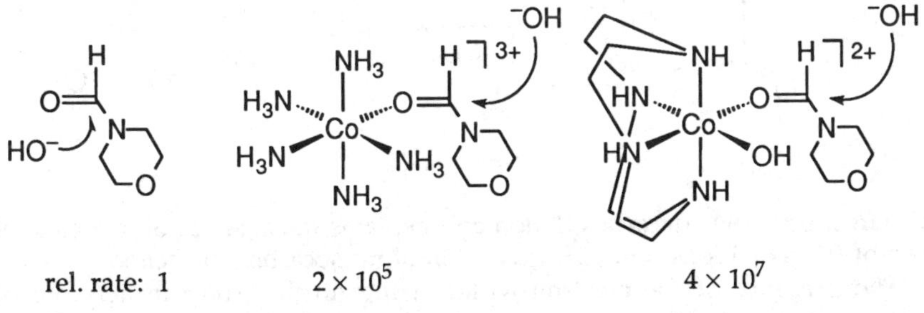

Scheme 1.20

The presence of the metal-hydroxide group provides additional rate enhancement over coordination to a substitutionally inert Lewis acidic center.[55] Cobalt(III) complexes of the type $[CoL_4(H_2O)_2]^{3+}$ have been found to be effective reagents for the hydrolysis of a number of important functional groups, including organic esters,[56] nitriles,[57] and phosphate esters,[58] at neutral p*H*. Iron(III)-ligand systems have been found[59] to promote the hydrolysis of phosphate esters in the weakly acid to neutral p*H* range. Such studies are of interest because simple aquairon(III) species precipitate insoluble hydrated oxides in this p*H* range and so are not useful as Lewis acids for hydrolysis reactions. The 1:1 complex between iron(III) and $EDDA^{2-}$ (H_2EDDA = $HOOCCH_2NHCH_2CH_2NHCH_2COOH$) was found to promote hydrolysis of 4-nitrophenylphosphate with a *ca*. 160-fold enhancement over the unpromoted reaction under optimal conditions.[59]

In related work, it has been shown[60] that palladium(II) aqua-complexes react with methionine sulfur atoms in peptides and that this promotes regioselective hydrolysis of the amide bonds involving the carboxylate groups of such palladium-bound methionines. The rate constants for hydrolysis depend upon the steric bulk of the leaving group and so the palladium-promoted cleavage process exhibits some sequence-selectivity.[60]

Decarboxylation reactions similarly illustrate the change in reactivity of ligands caused by the redistribution of electron density upon coordination. Thus, the catalytic decarboxylation of polycarboxylic acids involves coordination of a carboxyl group to a metal ion. Steinberger and Westheimer[61] found that coordination of one carboxyl group leads to a change in electron density at the carboxyl group not involved in coordination, Scheme 1.21:

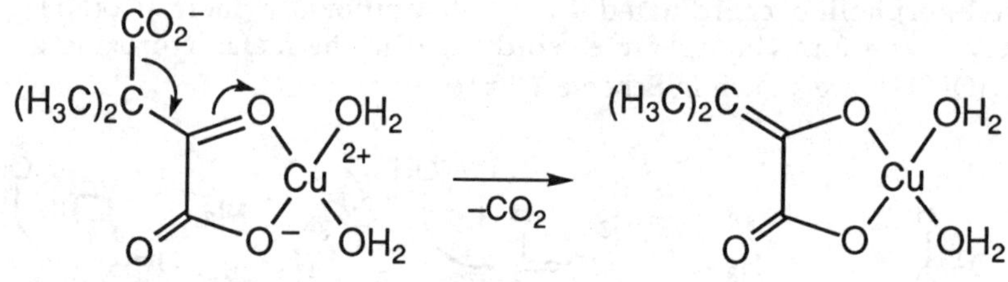

Scheme 1.21

This effect promotes decarboxylation and explains the absence of any catalytic effect of the metal ions on decarboxylation of monocarboxylic acids.

The presence of the electron-withdrawing nitrile group in acrylonitrile leads to polarization ($^{\delta+}CH_2{=}CH{-}C{\equiv}N^{\delta-}$) that promotes nucleophilic attack at the methylene carbon atom. The coordination of acrylonitrile to a metal center through the nitrile group results in enhanced polarization and further promotes reactions with nucleophilic reagents. For example, acetylacetonate ion reacts with the acrylonitrile complex, $[Co(NH_3)_5N{\equiv}CCH{=}CH_2]^{3+}$, in aqueous solution at p$H \approx 9$, Eq. 1.28:[62]

$$[Co(NH_3)_5N{\equiv}CCH{=}CH_2]^{3+} + {}^{-}CH(COMe)_2 \rightarrow [Co(NH_3)_5N{\equiv}C(CH_2)_2CH(COMe)_2]^{3+} + OH^- \quad (1.28)$$

Within five seconds the reaction occurs with a yield of 50%. The reaction of free acrylonitrile with acetylacetonate ion under similar conditions requires seven hours in order to obtain a 50% yield. Thus, the reaction of coordinated acrylonitrile proceeds *ca.* 10^4 times faster than the reaction of free acrylonitrile.

The charge on a complex is an important factor that influences the reactivity of coordinated ligands. The *N*-alkylation of the quinoline group of the chelated heterocyclic ligand in the copper(II) complex shown below is an example, Scheme 1.22:[62]

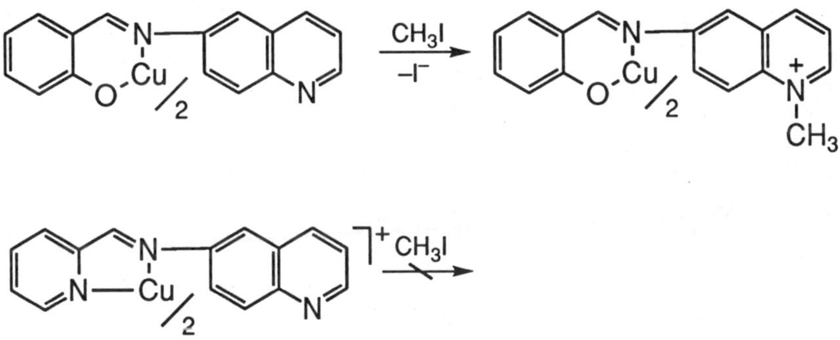

Scheme 1.22

Alkylation of the tertiary nitrogen with methyl iodide in a neutral complex proceeds rapidly. However, a similar, but positively charged, complex does not react under the same conditions.

The reactivity of coordinated ligands depends upon the nature of the metal center, even in series that are electronically analogous. For example, nickel(II) complexes of tertiary phosphines undergo solid state oxidation of the coordinated phosphine to the phosphine oxide, Eq. 1.29:[63]

$$[NiX_2(PR_3)_2] + O_2 \rightarrow [NiX_2(OPR_3)_2] \quad (1.29)$$

The reaction is exoenergetic and is accompanied by the expected mass increase. Unlike these nickel(II) compounds, phosphine complexes of palladium(II) and platinum(II), $[MX_2(PR_3)_2]$, do not undergo such a reaction. While phosphine oxidation might take place at high temperatures, it is obscured by decomposition of the phosphine ligand. Kukushkin and

Antonov[64] showed that substitution of one chloride ligand in $[PdCl_2(PR_3)_2]$ by $SnCl_3^-$ promotes oxidation of the coordinated phosphine. However, in the platinum(II) complex, $[PtCl(SnCl_3)(PR_3)_2]$, phosphine oxidation has not been observed under the same reaction conditions. In $[Pd(SnCl_3)_2(PR_3)_2]$ phosphine oxidation occurs at a lower temperature than in $[PdCl(SnCl_3)(PR_3)_2]$. In the presence of two $SnCl_3^-$ ligands the oxidation of phosphines will take place even in the platinum(II) complexes, $[Pt(SnCl_3)_2(PR_3)_2]$. Differences in the reactivity of the phosphine complexes of nickel(II), palladium(II) and platinum(II) with oxygen can be understood in terms of the HSAB principle. Thus, nickel(II), a harder acid than palladium(II) or platinum(II), will favor the formation of M–O bonds and so facilitate coordination of the phosphine oxide.

References

1. (a) F. A. Cotton and G. Wilkinson, *Advanced Inorganic Chemistry* (John Wiley & Sons, New York, 1988, 5th Ed.); (b) J. E. Huheey, E. A. Keiter and R. L. Keiter *Inorganic Chemistry, Principles of Structure and Reactivity* (Harper Collins, New York, 1993, 4th Ed.); (c) W. L. Jolly, *Modern Inorganic Chemistry* (McGraw-Hill, New York, 1991, 2nd Ed.); (d) G. L. Miessler and D. A. Tarr, *Inorganic Chemistry* (Prentice Hall, Englewood Cliffs, NJ, 1991); (e) D. F. Shriver, P. W. Atkins, and C. H. Langford, *Inorganic Chemistry* (W. H. Freeman, New York, 1990); (f) Yu. N. Kukushkin, *The Chemistry of Coordination Compounds* [in Russian] (Vishaya Shkola, Moscow, 1985); (g) R. B. King, *Encyclopedia of Inorganic Chemistry* (Wiley, New York, 1994) [8 volumes]; (h) *Comprehensive Coordination Chemistry* eds. G. Wilkinson, R. D. Gillard, and J. A. McCleverty (Pergamon, Oxford, 1987) [8 volumes].

2. F. Basolo and R. G. Pearson, *Mechanisms of Inorganic Reactions* (John Wiley & Sons, New York, 1967, 2nd Ed.).

3. (a) L. Helm, L. I. Elding, and A. E. Merbach, *Helv. Chim. Acta* **67** (1984) 1453; (b) L. Helm, L. I. Elding, and A. E. Merbach, *Inorg. Chem.* **24** (1985) 1719.

4. L. I. Elding, *Inorg. Chim. Acta* **28** (1978) 255.

5. B. E. Mann and A. Musco, *J. Chem. Soc., Dalton Trans.* (1980) 776.

6. (a) Yu. N. Kukushkin, G. N. Sedova, and R. A. Vlasova, *Zh. Neorg. Khim.* **23** (1978) 737; (b) Yu. N. Kukushkin, G. N. Sedova, and A. N. Beliaev, *Zh. Neorg. Khim.* **23** (1978) 2121; (c) Yu. N. Kukushkin, G. N. Sedova, R. A. Vlasova, and L. Ya. Pizova, *Zh. Neorg. Khim.* **24** (1979) 733; (d) Yu. N. Kukushkin, G. N. Sedova, and A. N. Beliaev, *Zh. Neorg. Khim.* **26** (1981) 429.

7. *Synthesis of Platinum Group Metal Complexes.* ed. I. I. Chernyaev [in Russian] (Nauka, Moscow, 1964) p. 13.

8. L. A. Chugaev and N. K. Pshenitsin, *Zh. Russ. Phys.-Khim. Obstch.* **52** (1920) 47.

9. Yu. N. Kukushkin, G. N. Sedova, and V. F. Budanova, *Zh. Neorg. Khim.* **25** (1980) 200.

10. Yu. N. Kukushkin, V. F. Budanova, and G. N. Sedova, *Solid-Phase Thermal Conversions of Coordination Compounds* [in Russian] (LGU [Leningradskii Gosudarstvennii Universitet — Leningrad State University], Leningrad, 1981) p. 92.

11. Y. Marcus and I. Elizer, *Coord. Chem. Rev.* **4** (1969) 273.

12. H. Sigel and C. F. Naumann, *J. Am. Chem. Soc.* **98** (1976) 730.

13. V. V. Lukachina, *Koord. Khim.* **8** (1982) 419.

14. V. M. Zamyatkina, Yu. N. Kukushkin, and A. A. Makarenya, *Lev Aleksandrovich Chugaev* [in Russian] (Nauka, Moscow, 1973) p. 102.

15. J. M. Lehn, *Pure Appl. Chem.* **49** (1977) 857.

16. K. B. Yatsimirskii and Ya. D. Lampeka, *Physicochemistry of Metal Complexes Containing Macrocyclic Ligands* [in Russian] (Naukova Dumka, Kiev, 1985) p. 256.

17. I. I. Chernyaev, *Izv. Inst. Platiny* **4** (1926) 243.

18. Meeting on *Trans*-Influence Regularity, *Izv. Inst. Platiny* **28** (1954) 12.

19. Yu. N. Kukushkin and R. I. Bobohodzhaev, *Chernyaev's Trans-Influence Regularity* [in Russian] (Nauka, Moscow, 1977) p. 183.

20. L. N. Essen, *Geometrical Isomers of Platinum and the Trans-Influence* [in Russian] (Nauka, Moscow, 1969) p. 75.

21. A. A. Grinberg, *Introduction to the Chemistry of Complex Compounds* [in Russian] (Khimia, Moscow, 1966).

22. A. A. Grinberg and Yu. N. Kukushkin, *Zh. Neorg. Khim.* **2** (1957) 106.

23. Yu. N. Kukushkin, *Platinum Metals Review* **35** (1991) 28 and references therein.

24. See, for example: (a) S. P. Roe, J. O. Hill, and R. J. Magel, *Inorg. Chim. Acta* **115** (1986) L15; (b) G. Faraglia, L. Sindellari, and S. Sitran, *Thermochim. Acta* **115** (1987) 229; (c) G. Faraglia, L. Sindellari, and V. Cherch, *Transit. Met. Chem.* **11** (1986) 98 and references therein.

25. N. V. Sidgwick, *The Electronic Theory of Valency* (Oxford University, Oxford, 1929) p. 163.

26. C. A. Tolman, *Chem. Soc. Rev.* **1** (1972) 337.

27. (a) J. Bjerrum, *Chem. Rev.* **46** (1950) 381; (b) G. Swarzenbach, *Experientia, Suppl.* **5** (1956) 162; (c) S. Ahrland, J. Chatt, and N. R. Davies, *Quart. Rev.* **12** (1958) 265.

28. R. G. Pearson, *Coord. Chem. Rev.* **100** (1990) 403 and references therein.

29. J. A. Davies and F. R. Hartley, *Chem. Rev.* **81** (1981) 79.

30. (a) H. B. Kagan and B. Ronan, *Rev. Heteroatom Chem.* **7** (1992) 92; (b) J. A. Davies, *Adv. Inorg. Chem. Radiochem.* **24** (1981) 115; (c) G. Mestroni, E. Alessio, M. Calligaris, W. M. Attia, F. Quadrifoglio, S. Cauci, G. Sava, S. Zorzet, S. Pacor, C. Monti-Bragadin, M. Tamaro, and L. Dolzani, *Prog. Clinical Biochem. and Medicine* **10** (1989) 71; (d) N. Farrell in *Platinum, Gold, and Other Metal Chemotherapeutic Agents,* ed. S. Lippard (ACS Symposium Series 209, American Chemical Society, Washington, DC, 1983) p. 279.

31. P. Pfeiffer, *Ber.* **39** (1906) 1864.

32. *The Chemistry of the Coordination Compounds*, ed. J. C. Bailar (Reinhold, New York, 1956).

33. L. A. Chugaev, *Izv. Inst. Izuch. Platiny* **4** (1926) 3.

34. H. Kirmreuthyer, *Ber.* **44** (1911) 3115.

35. A. A. Grinberg, *Introduction to the Chemistry of Complex Compounds* [in Russian] (Khimia, Moscow, 1966).

36. A. A. Grünberg and D. I. Rjabtcikov, *Acta phys.-chim.* **3** (1935) 555.

37. I. Leden and J. Chatt, *J. Chem. Soc.* (1955) 2936.

38. A. A. Grinberg, Kh. I. Gildengershel, and E. P. Panteleeva, *Zh. Neorg. Khim.* **8** (1963) 2226.

39. A. A. Grinberg, Kh. I. Gildengershel, and V. F. Budanova, *Zh. Neorg. Khim.* **11** (1966) 2519.

40. A. A. Grinberg, Kh. I. Gildengershel, and V. F. Budanova, *Zh. Neorg. Khim.* **13** (1968) 1365.

41. A. A. Grinberg, Kh. I. Gildengershel, and V. F. Budanova, *Zh. Neorg. Khim.* **13** (1968) 3040.

42. M. I. Gelfman and D. B. Smolenskaya, *Dokl. Akad. Nauk SSSR* **174** (1967) 1118.

43. Yu. N. Kukushkin, Yu. E. Vyazmensky, L. I. Zorina, and Yu. L. Pazukhina, *Zh. Neorg. Khim.* **13** (1968) 1595.

44. Yu. N. Kukushkin, G. S. Krilova, and M. A. Kirilova, *Koord. Khim.* **2** (1976) 350.

45. R. Peters, *Z. Phys. Chem.* **26** (1898) 193.

46. S. R. Carter and F. H. Clews, *J. Chem. Soc.* (1924) 1880.

47. L. Michaelis and F. Friedheim, *J. Biol. Chem.* **91** (1931) 343.

48. L. Michaelis, *Oxydations-Reductions-Potentiale mit Besonderer Berücksichtigung ihrer Physiologischen Bedlutung. Zweite Auflage* (Springer, Berlin, 1933).

49. A. D. Gelman and D. I. Ryabtchikov, *Dokl. AN SSSR* **33** (1941) 462.

50. S. Ahrland and J. Chatt, *J. Chem. Soc.* (1957) 1379.

51. J. Gazo, *Abstr. XIII Internat. Conf. on Coord. Chem.*, 1970.

52. Yu. N. Kukushkin, *Koord. Khim.* **7** (1981) 335.

53. A. B. P. Lever, *Inorg. Chem.* **29** (1990) 1271.

54. (a) E. C. Constable, *Metals and Ligand Reactivity* (First Edition: Ellis Horwood, Chichester, 1990; Second Edition: Verlag Chemie, 1995); (b) Yu. N. Kukushkin, *Reactivity of Coordination Compounds* [in Russian] (Khimia, Leningrad, 1987) p. 288.

55. B. K. Takasaki, J. H. Kim, E. Rubin, and J. Chin, *J. Am. Chem. Soc.* **115** (1993) 1157.

56. J. Chin and M. Banaszczyk, *J. Am. Chem. Soc.* **111** (1989) 2724.

57. J. H. Kim and J. Britten, *J. Am. Chem. Soc.* **115** (1993) 3618.

58. D. H. Vance and A. W. Czarnik, *J. Am. Chem. Soc.* **115** (1993) 12165.

59. N. P. Sadler, C.-C. Chuang, and R. M. Milburn, *Inorg. Chem.* **34** (1995) 402.

60. L. Zhu and N. M. Kostic, *J. Am. Chem. Soc.* **115** (1993) 4566.

61. R. Steinberger and F. H. Westheimer, *J. Am. Chem. Soc.* **73** (1951) 429.

62. F. P. Dwyer and D. P. Mellor, *Chelating Agents and Metal Chelates* (Academic, New York, 1963).

63. Yu. N. Kukushkin and G. N. Sedova, *Koord. Khim.* **6** (1980) 27.

64. Yu. N. Kukushkin and P. G. Antonov, *Sov. Sci. Rev. B. Chem.* **10** (1987) 275.

Chapter 2. THE SOLUBILITY OF COORDINATION COMPOUNDS: RELATIONSHIP TO COMPOSITION AND STRUCTURE

The study of preparative coordination chemistry necessitates an understanding of the factors governing the solubilities of metal complexes in water and non-aqueous solvents. However, a rigorous theory of the solubility of chemical compounds has not been devised as yet. Even when the solubilities of analogous materials are known, it remains extremely difficult to predict the solubility of a given substance. Semiempirical approaches to the prediction of solubilities appear in the literature.[1-5] In particular, the theory of Hildebrand and Scott,[5a] which relates solubility to the energy of vaporization and the molar volume of a solute, is effective for non-electrolytes. There are also approaches in which solubility is related to the melting point[1] of the solute or the polarity[2,3] of the solvent. Nonetheless, most chemists still use the rule of thumb that *like dissolves like.*

In general, the solubility of a substance is determined by the relationship between its crystal-lattice energy and its energy of solvation in a particular solvent. Table 2.1 illustrates the influence of crystal lattice energies, hydration enthalpies, and hydration entropies on solubilities with values obtained by quantitative measurements of saturated solutions at known temperatures.

Table 2.1[4]

Crystal lattice energies, hydration enthalpies, hydration entropies, and solubilities of some compounds

Compound	U (kJ/mol)	H_{hyd} (kJ/mol)	S_{hyd} (J/mol·K)	Sol (291 K) (g/100g)
LiF	-1031	-1006	-276	0.27
NaCl	-780	-760	-185	35.7
KF	-814	-819	-207	92.3
RbF	-776	-793	-200	130.6
$SrCl_2$	-2107	-2161	-400	43.5[a]
AlF_3	-6140	-6160	-929	0.559
$AlCl_3$	-5460	-5770	-753	69.9

a. At 273 K.

In preparative chemistry there are a number of principles by which predictions of solubility may be linked to the composition and structure of a compound. In the case of metal complexes a large number of interacting factors influence solubility. While the dissolution of a complex appears to be a simple chemical phenomenon, it is, in fact, complicated. The purpose of this Chapter is to establish a relationship between composition, structure, and the solubility of coordination compounds.

2.1 Hydrophilic and lipophilic coordinated ligands

Feigl has suggested that systematic modification of the solubilities of coordination compounds may be achieved by alteration of the hydrophilic-lipophilic balance (HLB) of the molecules.[5b] Hydrophilic-lipophilic balance is a specific relationship between polar and nonpolar groups in a compound. By increasing the lipophilicity of a molecule, decreased solubility in water and increased solubility in organic solvents usually results. Similarly, increasing the hydrophilicity of a molecule usually results in decreased solubility in organic solvents and increased solubility in water

For example, it is known that the complex, $[Ru(bipy)_3][ClO_4]_2$, dissolves readily in water but is largely insoluble in nonpolar, aprotic solvents. An analog with the structure shown in Figure 2.1 has been synthesized:[6]

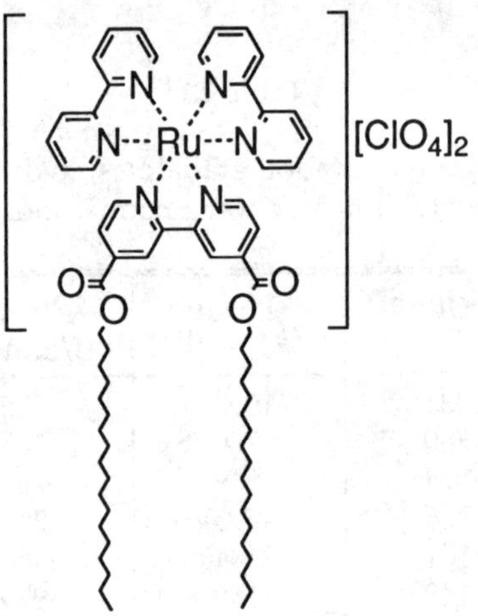

Figure 2.1

Functionalization of the bipyridyl ligand with lipophilic hydrocarbon groups causes the complex to become virtually insoluble in water and highly soluble in chloroform.

The salts, $[PPN]_n[M(CN)_6]$ (M = Fe, n = 3 or 4; M = Ru, n = 4; M = Co, n = 3) and $[PPN][Fe(CN)_5(NO)]$, have been prepared by metathetical exchange reactions in aqueous solution between [PPN]Cl and the corresponding alkali metal salts of the cyanometallates.[7] Unlike alkali metal salts, the $[PPN]^+$ salts are very soluble in chlorinated solvents and in most polar, aprotic solvents including acetonitrile, acetone, and dimethyl sulfoxide.

One important factor which influences the solubilities of coordination compounds is their ability to form hydrogen bonds with solvent molecules. Thus, if ligands contain functional groups capable of forming hydrogen bonds, the complexes containing these ligands are, as a rule, soluble in polar, protic solvents.[8] Solubility occurs because such compounds have high energies of solvation. For example, Il'inskii's complex, shown in Figure 2.2 (R = H), is sparingly soluble in water.

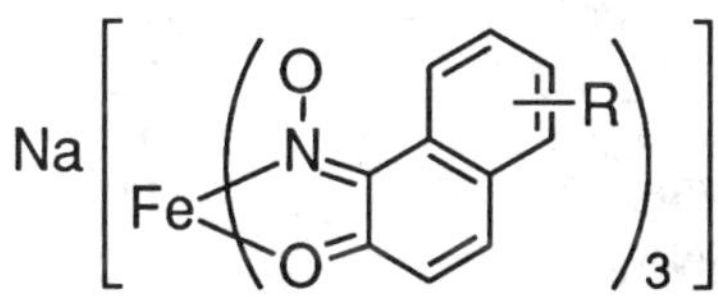

Figure 2.2

The introduction of substituents R (R = NH_2, OH, CO_2H) which are capable of forming hydrogen bonds results in increased solubilities for these nitrosonaphtholato ($\rightleftharpoons$ quinonemonoximato) complexes in water or ethanol.

The two examples discussed above (Figures 2.1 and 2.2) illustrate that conversion of ligands into more lipophilic or more hydrophilic groups results in an alteration in the energies of solvation of the resultant complexes.

Several approaches to the evaluation of the hydrophilic-lipophilic properties of chemical compounds are known from the literature. For example, in order to evaluate the HLB of surfactants, Davies developed a system based on the analysis of group numbers.[9] The "group number" characterizes the contribution of each specific functional group to the energy that would be required if a solvent molecule were changed from water to an organic solvent.

Group numbers are employed in the empirical formula shown, Eq. 2.1:

$$\mathrm{HLB} = \Sigma(\text{hydrophilic group numbers}) + \Sigma(\text{hydrophobic group numbers}) + 7 \qquad (2.1)$$

Table 2.2 shows the Davies group numbers for a range of common substituents.

Table 2.2

Davies group numbers

Functional group	Group number
Hydrophilic	
$-SO_3Na$	38.7
$-O-$	1.3
$-CO_2K$	21.1
$-CO_2H$	2.1
$-CO_2Na$	19.1
$-OH$	1.9
$-NR_3^+$	9.4
Lipophilic	
HC (*tert*)	-0.475
$-CH_2-$	-0.475
$-CH_3$	-0.475
$=CH-$	-0.475
Amphiphilic	
$-C_2H_4O-$	0.33
$-C_3H_6O-$	-0.15

The Davies system provides a general picture of the solubility properties of compounds based upon substituent contributions, but rigorous quantitative calculations of the HLB are not possible with this system. The system is useful in the evaluation of the HLB for individual ligands in a complex but the sum of the HLB's of all ligands cannot be used to evaluate the HLB of a coordination compound. Deviations from additivity are caused by the nature of the metal ion, the redistribution of electron density in the ligands upon complexation, the symmetry of the molecule, the conformations of coordinated ligands, and many other factors. However, for a series of similar complexes, the higher the lipophilicity or hydrophilicity of the ligands that make up the complex, the higher the corresponding lipophilicity or hydrophilicity of that complex will be.

2.2 Solubility of ionic complexes: counterion effects

Chugaev pioneered the use of a wide range of organic reagents in the synthesis of metal complexes.[10] Many of the ligands he studied were sparingly soluble in water while the metal ion sources were largely insoluble in common organic solvents. For example, in order to react various ligands with the $[PtCl_4]^{2-}$ anion in non-aqueous solutions, Chugaev synthesized the tri(*n*-propyl)ammonium salt, $[(n\text{-}Pr)_3NH]_2[PtCl_4]$, which is sufficiently soluble in ethanol, acetone, and chloroform to be a useful reagent.[11] The synthesis of this compound was apparently the first example of counterion replacement in $K_2[PtCl_4]$ for the purpose of increasing solubility in non-aqueous solvents.

Replacement of counterions often takes place as a result of an ion exchange reaction in aqueous, aqueous/alcoholic, or alcoholic solutions (see Chapter 5). The complexes, $Q_2[PtCl_4]$ (Q = $[Ph_3PCH_2Ph]$ or $[(Ph_3P)_2N]$), precipitate from an aqueous solution of $K_2[PtCl_4]$ when an ethanolic solution of $[Ph_3PCH_2Ph]Cl$ or $[(Ph_3P)_2N]Cl$ is added, Scheme 2.1:

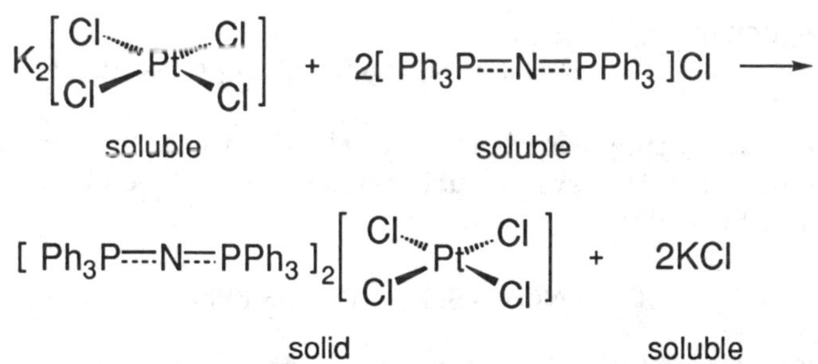

Scheme 2.1

Unlike potassium tetrachloroplatinate(II), which is readily soluble only in water, compounds of the type $Q_2[PtCl_4]$ are freely soluble in a range of organic solvents, including dichloromethane and nitromethane.[12,13] Further, counterions with different hydrophilic-lipophilic properties have been used in the preparation of salts of $[M(CN)_8]^{3-}$ (M = Mo, W) that are soluble in either water or organic solvents.[14] $K_3[M(CN)_8]$ has been prepared in aqueous solution *via* nitric acid oxidation of $K_4[M(CN)_8]$. The addition of $(n\text{-Bu})_3N$ to this solution leads to precipitation of $[(n\text{-Bu})_3NH]_3[M(CN)_8]$, Eq. 2.2:

$$K_3[M(CN)_8](aq) + 3(n\text{-Bu})_3N(aq) + 3H^+ \rightarrow [(n\text{-Bu})_3NH]_3[M(CN)_8]\ (s) + 3K^+ \quad (2.2)$$

Unlike the potassium salts, the $[(n\text{-Bu})_3NH]_3[M(CN)_8]$ complexes are soluble in methanol, ethanol, acetonitrile and benzene and are quantitatively converted, in methanol, to $Cs_3[M(CN)_8]$ by treatment with CsCl, Eq. 2.3:

$$[(n\text{-Bu})_3NH]_3[M(CN)_8]_{soluble} + 3CsCl_{soluble} \rightarrow Cs_3[M(CN)_8]_{solid} + 3[(n\text{-Bu})_3NH]Cl_{soluble} \quad (2.3)$$

In contrast to $[(n\text{-Bu})_3NH]_3[M(CN)_8]$, the cesium salts are insoluble in many organic solvents, but, like their precursors, the potassium salts, they are soluble in water.

Preparation of the platinum(II) organonitrile complexes, *trans*-$[PtMe(RCN)L_2][X]$ ($L = PMe_2Ph, AsMe_3$), has been described, Eqs. 2.4 and 2.5:

$$[PtMeClL_2] + AgX + ROH \rightarrow trans\text{-}[PtMe(ROH)L_2][X] + AgCl \quad (2.4)$$

$$[PtMe(ROH)L_2]X + RCN \rightarrow trans\text{-}[PtMe(RCN)L_2][X] + ROH \quad (2.5)$$

The platinum(II) nitrile complexes crystallize only with difficulty from alcoholic solutions.[15] However, solubilities may be decreased by changing the counterions in the order:

$$BF_4^- > NO_3^- > SbF_6^- > PF_6^- >> BPh_4^-$$

Thus, $[PtMe(RCN)L_2]^+$, on addition of $NaBPh_4$, forms a precipitate of $[PtMe(RCN)L_2][BPh_4]$ in methanol. These $[PtMe(RCN)L_2][BPh_4]$ salts are freely soluble in dichloromethane. Clearly, the precipitation of the complexes from aqueous solution by lipophilic counterions and the simultaneous increase in solubility of the complexes in organic solvents are closely related to one another. Accordingly, it is helpful to consider existing ideas about the solubilities of coordination compounds ionized in water.

Yatsimirskii has determined[16] the conditions under which the water solubility of simple salts and metal complexes is minimized. Those conditions[17] are summarized in the following statements:

1. The ratio between the radii of the ions comprising the salt must approximate a specific value, characteristic of the particular type of salt. For MX salts, the minimum solubility is achieved when the ratio of the radius of the cation to the radius of the anion is equal to 0.7. For MX_2 salts, this ratio is equal to 1.1. A change in the ratio between the radii of the ions has a greater effect on solubility if the charges on the ions are high.

2. All other conditions being equal, the higher the charges on the ions, the lower the solubilities of the salts.

3. The ligands in complex ions should be lipophilic and not contain functional groups capable of forming hydrogen bonds with water molecules. Polar groups should form only *intramolecular* hydrogen bonds.

Basolo[18a] has stated the general principle: "*Solid salts precipitate from aqueous solutions most readily when the salt is either a small cation — small anion or large cation — large anion pair preferably in a system with the same but opposite charges on the counterions*". The formation of the salt, MX, Eq. 2.6:

$$M^{+}_{soluble} + X^{-}_{soluble} \rightarrow MX_{solid} \tag{2.6}$$

is determined by the relationship between the crystal-lattice energy of the salt, MX (or its solvate), and the energies of hydration of the cation and anion. Accordingly, the driving force for the formation of the solid compound, MX, in a *small cation — small anion* system is the large crystal-lattice energy.[18a] In a *large cation — large anion* combination, the driving force for the formation of MX is the small energies of hydration of the large ions. Therefore, an examination of the energies of hydration of *small* and *large* ions and the entropies of hydration of these ions (Table 2.3) is instructive.

Because the radii of complex ions are typically large, such ions are characterized by low values of the ionic potential, which is determined by the ratio of the total charge to the radius of a given ion. Therefore, the crystal-lattice energies fail to reach the high values found for the *small anion — small cation* combinations. Thus, large complex ions are most easily precipitated from aqueous solutions by large counterions.

Basolo[18a] has shown that complexes present in solution at low equilibrium concentrations can be isolated as solids when they are precipitated from solution as a result of ion exchange. For example, the dissolution of copper(II) chloride in an aqueous solution rich in chloride ions results mainly in the formation of the complex $[CuCl_3]^-$.

Table 2.3

Crystallographic ionic radii (r_c, Å), heats of hydration (ΔH°_h, kJ/mol), and heats of transfer between water and organic solvents (ΔH°_{tr}, kJ/mol) for some cations and anions[8]

Ion	r_c (Å)	$-\Delta H^\circ_h$, kJ/mol	ΔH°_{tr}, kJ/mol		
			MeOH	Propylene carbonate	DMSO
H^+		1103		43.9	-25.5
Li^+	0.93	533	-18.9	3.1	-26.4
Na^+	1.17	417	-20.4	-10.2	-27.7
K^+	1.49	333	-20.0	-21.9	-34.9
Rb^+	1.64	307	-19.7	-24.6	-33.5
Cs^+	1.83	289	-15.5	-26.8	-30.0
Et_4N^+	2.7	127	5.9	0.7	- 2.8
Ph_4P^+	4.15	46		-13.4	- 9.3
Ph_4As^+	4.3	42	3.3	-14.6	-11.9
F^-	1.16	502			
Cl^-	1.64	366	7.9	26.4	18.8
Br^-	1.80	335	3.8	13.6	3.5
I^-	2.04	294	-1.6	-3.3	-12.8
$CF_3CO_2^-$			10	32.6	
ClO_4^-	2.3		-4.6	-16.4	-19.2
BPh_4^-	4.05	47	3.3	-14.6	-11.9

When the chromium(III) complex, $[Cr(NH_3)_6]^{3+}$, is added to such a solution, $[Cr(NH_3)_6][CuCl_5]$ precipitates. Because copper(II) complexes are substitutionally labile, equilibria that involve chloride exchange can be established easily, Scheme 2.2. The driving force for the formation of the stable solid, $[Cr(NH_3)_6][CuCl_5]$, is the low solubility of this compound.[18a]

The sparingly soluble, bridged compound, $[(n\text{-}Pr)_4N]_2[Pt_2(\mu\text{-}Cl)_2Cl_4]$, is precipitated when aqueous solutions of potassium tetrachloroplatinate(II) and tetra(*n*-propyl)ammonium chloride are mixed and subsequently heated.[18b] The "expected" compound, $[(n\text{-}Pr)_4N]_2[PtCl_4]$, is not produced. Accordingly, caution is necessary when forming conclusions about the composition of a complex in solution when those conclusions are based on knowledge of the composition of a solid.

In an aqueous medium the di-μ-amido-bis[tetrammineplatinum(IV)] complex, **A**, is readily deprotonated, Scheme 2.3.

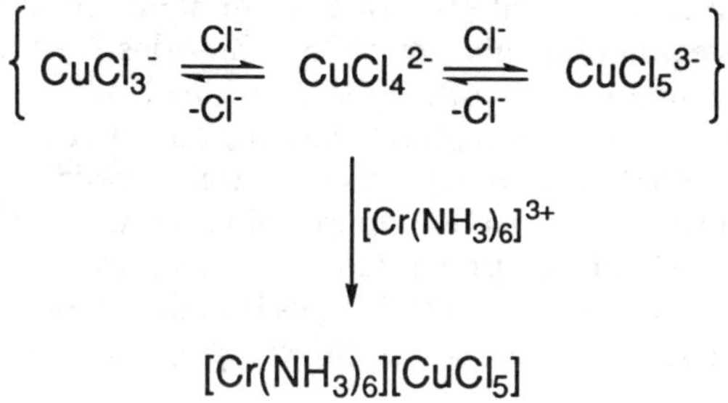

Scheme 2.2

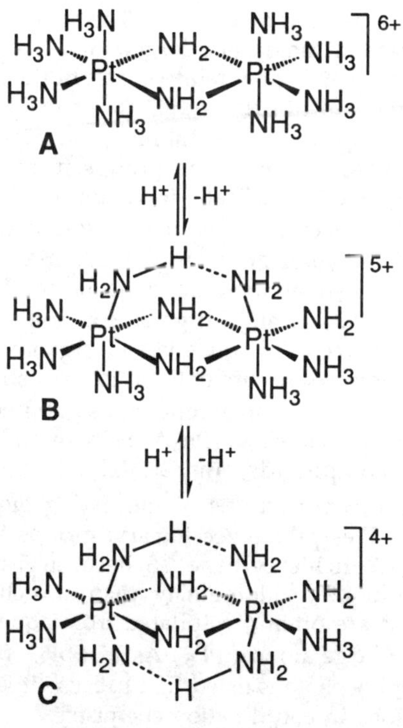

Scheme 2.3

The perchlorate salt of the di-μ-amido-μ-amidoammine-bis[triammine-platinum(IV)] cation, **B**, has been isolated and characterized by X-ray crystallography.[18c] In contrast to the ease of isolation of the perchlorate of **B**, attempts to precipitate the chloride of **B** from solution failed. The product obtained was an equimolar mixture of the chlorides of **A** and **C**, which are assumed to be much less soluble than the chloride of **B**.[18d]

Quaternary ammonium, phosphonium, and arsonium salts serve as sources of large lipophilic cations which may be employed in precipitation reactions. In addition, bis(triphenylphosphine)iminium salts, $[(Ph_3P)_2N]X$, have been employed for the precipitation of complex anions from aqueous solution.[19] One advantage of the $[(Ph_3P)_2N]^+$ cation is that it often combines with anionic complexes to form crystalline compounds suitable for analysis by X-ray diffraction.

Systematic investigations of the solubilities of onium complexes that contain a range of different cations have not been carried out. However, Bekkevoll, Svorstøl, Høiland, and Songstad have investigated the relative solubilities of the onium thiocyanates, $[Me_4N]SCN$, $[Bu_4N]SCN$, $[Ph_4As]SCN$, and $[(Ph_3P)_2N]SCN$, in dichloromethane.[20] The solubility of $[Bu_4N]SCN$ was found to be four orders of magnitude larger than that of the corresponding tetramethylammonium salt. The relative solubilities in dichloromethane parallel the reported solubility data for the same compounds in dichloroethane.[21] Changes in the solubilities of tetraalkylammonium salts that occur as the number of methylene groups in the alkyl substituents is increased are the result of significant increases in the lipophilicity of the cation as a whole.[20] However, a simple relationship between the length of the hydrocarbon chain of the alkyl group in the cation and the solubility of the corresponding salts in hydrophobic solvents is not apparent. In fact, in processes involving the extraction of organic compounds,[22] an increase in the partition coefficients between the aqueous and organic phases as the number of methylene groups increases occurs only up to a certain point.

It has been shown[20] that the exceptional solubilities of phenyl- or other aryl-containing onium salts, such as $[Ph_4As]SCN$ or $[(Ph_3P)_2N]SCN$, and also non-ionic aromatic compounds, in CH_2Cl_2, is attributable to specific interactions between the π-electrons of the aryl groups and the hydrogen atoms of the solvent. This preference for aryl groups has been observed for CH_2Cl_2 but not for 1,2-dichloroethane. Salts of aryl-containing cations are generally less soluble in 1,2-dichloroethane than in dichloromethane.

Complex cations are often precipitated from aqueous solution with the aid of salts or acids of large anions: PF_6^-, AsF_6^-, SbF_6^-, BF_4^-, BPh_4^-, ClO_4^-, and $CF_3SO_3^-$. Hexafluorophosphate is listed first because it is commonly employed in precipitation reactions in coordination chemistry.

An important condition for the replacement of counterions is that one reaction product possess high solubility so that the other product is driven to precipitate. Reactions that involve the replacement of counterions *in non-aqueous solvents* are usually carried out in such a manner that the desired complex remains in solution, while the secondary reaction product precipitates. For example, the metathesis of the complex [PdI(trenMe$_6$)]I (trenMe$_6$ = tris(2-(dimethylamino)ethyl)amine) with [Et$_4$N][PF$_6$] in acetone proceeds as shown in Eq. 2.7:

$$[PdI(trenMe_6)]I_{soluble} + [Et_4N][PF_6]_{soluble} \rightarrow [PdI(trenMe_6)][PF_6]_{soluble} + [Et_4N]I_{solid} \quad (2.7)$$

This reaction[23] results in precipitation of [Et$_4$N]I and formation of [PdI(trenMe$_6$)][PF$_6$] in solution. The product is obtained by filtration of [Et$_4$N]I, concentration of the filtrate, and addition of hexane to cause precipitation.

The addition of stoichiometric amounts of AgPF$_6$ to [PdX(trenMe$_6$)]X (X = Cl or SCN) in acetone results in the precipitation of AgX and the formation of [PdX(trenMe$_6$)][PF$_6$] in solution.[23] Halide and pseudohalide *counterions* are removed from the complexes instead of coordinated halide and pseudohalide *ligands*.

Large counterions provide stabilization for complex species both in the solid state and in solution. Increased stabilization is a function of the counterion[24] as demonstrated by the increase in resistance to oxidative decomposition of the carbonyl complexes, M[FeH(CO)$_4$], M[V(CO)$_6$], and M[μ-H{Cr(CO)$_5$}$_2$] (M^+ = the cation of an alkali metal), when M^+ is replaced by [(Ph$_3$P)$_2$N]$^+$. The cation M^+ is associated with a carbonyl ligand in the complexes listed above. For example, interaction of Na^+ with one of the "equatorial" carbonyl oxygens in the complex, Na[FeH(CO)$_4$], occurs as shown in Figure 2.3:

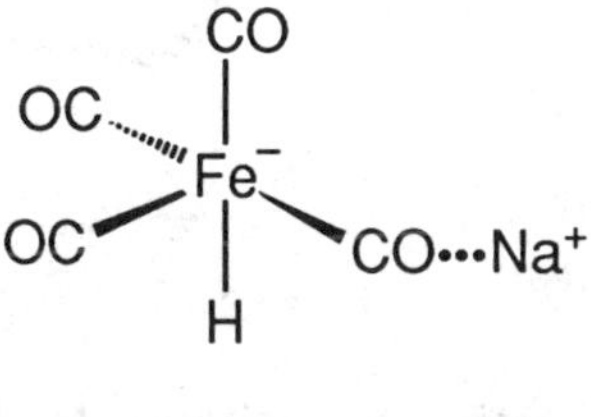

Figure 2.3

This interaction has a dramatic effect on the sensitivity of [FeH(CO)$_4$]$^-$ towards oxygen. When Na^+ is replaced by [(Ph$_3$P)$_2$N]$^+$, there is no interaction of the cation with the CO ligands and the compound $[(Ph_3P)_2N][FeH(CO)_4]$ becomes more resistant to oxidation.

Counterions are not always chemically inert. They can participate in various chemical reactions and the products of these reactions may themselves act as ligands. For example, the $[(Ph_3P)_2N]^+$ ion reacts with hydroxide or methoxide ions in methanol to form the compound $Ph_2P(O)NPPh_3$, which can act as a ligand in complexation reactions.[25,26] In particular, the reaction of the complex $[W(CO)_5(THF)]$ with $[(Ph_3P)_2N]Cl$ and sodium methoxide results in the formation of $[W(CO)_5OPPh_2NPPh_3]$ which has been structurally characterized, Figure 2.4:

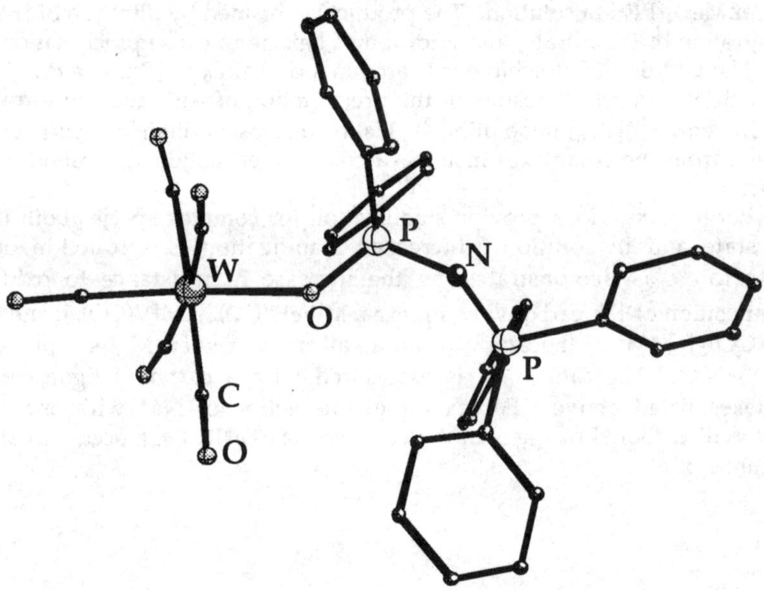

Figure 2.4

Phosphonium salts are easily hydrolyzed to form the corresponding phosphine oxides and when they are reacted with strong reducing agents,[27] the salts are converted into phosphines which can act as ligands, Eqs. 2.8 and 2.9:

$$R_4P^+X^- + NaOH \rightarrow R_3PO + RH + NaX \quad (2.8)$$

$$R_4P^+X^- + 2Na \rightarrow R_3P + RNa + NaX \quad (2.9)$$

In an acidic medium the BPh_4^- anion is converted into BPh_3 and benzene.[28] Further, the complexes, $[M(dien)H_2O]^{2+}$ (M = Pd, Pt; dien = diethylenetriamine), are known[29] to react in aqueous solutions with BPh_4^- to form the compounds $[M(dien)Ph][BPh_4]$. It has been shown[29] that the reactions producing these compounds are similar to those of BPh_4^- in aqueous acidic solutions. Other examples of phenylation of metal ions by BPh_4^- are compiled in Table 2.4. Phenyl migration from the tetraphenylborate anion to platinum(II) is illustrated[33,34] by the conversion of $[PtCl(Me_2S)_3][BPh_4]$ into $[PtPhCl(Me_2S)_2]$, a reaction that proceeds both in the solid state and in solution, Scheme 2.4:

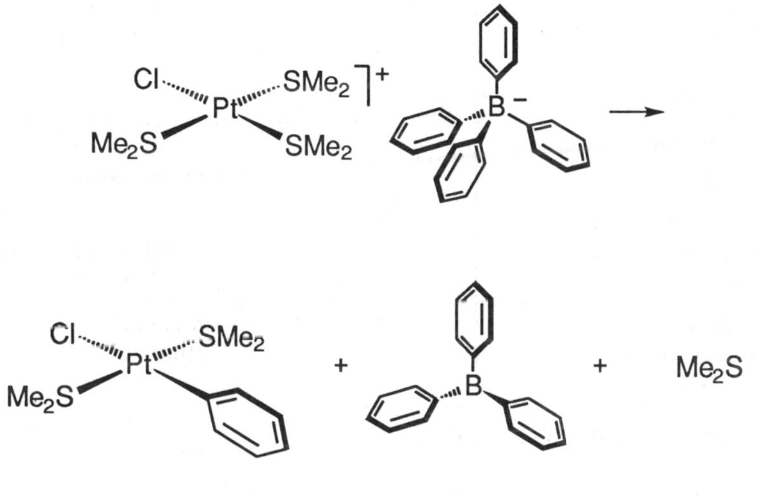

Scheme 2.4

It has been shown[41] that the reaction of $[ReMe(CO)_5]$ with Ph_3CPF_6 or $Et_2O{\cdot}HPF_6$ in dichloromethane produces the complex, $[(CO)_5Re(\mu\text{-}F)Re(CO)_5][PF_6]$. The source of the bridging fluoride ligand is the PF_6^- ion. Other examples of fluoride abstraction from the normally inert PF_6^- and BF_4^- ions,[42-45] as well as examples of facile PF_6^- hydrolysis,[46] are found in the literature.

Table 2.4

Phenyl migration from BPh_4^- to a metal center[15,30-40]

Starting materials	Products	Notes and references
$NaBPh_4$ and $[PtCl_2(PEt_3)_2]$	$[PtPh_2(PEt_3)_2]$	THF, refluxed, 40 hours [15].
$NaBPh_4$ and $[PtMe(solv)(PMe_2Ph)_2]$ solv = MeOH, RCN	$[PtPh_2(PMe_2Ph)_2]$	Starting complexes prepared *in situ* [15, 30].
$[PtH(C_2H_4)(PEt_3)_2][BPh_4]$	$[PtPh_2(PEt_3)_2]$	Slow reaction in MeOH [31].
$NaBPh_4$ and $[PtH(NO_3)(PEt_3)_2]$	$[PtPh_2(PEt_3)_2]$, $[(PEt_3)_2HPt(\mu\text{-}H)Pt\text{-}Ph(PEt_3)_2]^+$	Slow reaction in MeOH [31, 32].
$NaBPh_4$ and $[M(dien)(H_2O)]^+$ M = Pt, Pd	$[MPh(dien)][BPh_4]$	Starting complex prepared *in situ* [29(b), 29(c)].
$[PtCl(R_2S)_3][BPh_4]$ R_2S = dimethyl sulfide, thioxane	$[PtPhCl(R_2S)_2]$	Reactions proceed in solid state and in solution [33, 34].
$[Pt(R_3P)_2(solv)_2]^{2+}$ R = Me, Et, Ph; solv = MeOH, H_2O	$[PtPh(R_3P)_2(solv)]^+$	Mechanism was studied [35].
$[Rh(\eta^6\text{-}PhBPh_3)(dppe)]$	$[RhPh(dppe)]$	Heated in acetone [36].
$NaBPh_4$ and $[RuCl(CO)_2Cp]$	$[RuPh(CO)_2Cp]$	Refluxed in EtOH [37].
$NaBPh_4$ and $[NiX(L)][BPh_4]$ X = Cl, Br; L = $N(C_2H_4AsPh)_3$	$[NiPh(L)][BPh_4]$	Refluxed in butanol [38].
$NaBPh_4$ and $[M(NO)_2(BF_4)Cp]$ M = Cr, Mo, W	$[MPh(NO)_2Cp]$	Stirred in CH_2Cl_2 for 1 h, 20-25 °C [39, 40].

2.3 Changes in solubility through modification of cationic counterions

The solubilities of ionic compounds of the type MX (where M is an alkali metal cation) in organic solvents are known to increase significantly in the presence of certain macrocyclic ligands, particularly in the presence of macrocyclic ethers. The compound MX passes into solution in the form [M(macrocyclic ligand)]X. The cation and the donor atoms of the macrocyclic ligand that interact with the cation are shielded from the organic solvent by peripheral hydrocarbon groups of the ligand. The lipophilicity of the molecule is thereby enhanced and its solubility in organic solvents increases as a result.

Ionic diameters range[47] from 1.56 to 3.30 Å for the series Li^+ through Cs^+ but complexation with macrocyclic ligands results in an increase in cation diameter of 6–7 Å in comparison to the cesium salt. Therefore, ions of the type $[M(\text{macrocyclic ligand})]^+$ have smaller surface charge densities than the M^+ ions themselves. The ion-ion and ion-dipole interactions of these cations with an X^- anion or solvent molecule are, accordingly, less effective than those associated with the alkali metal ions alone.

A large amount of experimental and theoretical work on this subject appeared following the reported complexation of macrocyclic ethers with ions of alkali and alkaline earth metals. The results of these investigations have been presented in monographs and reviews.[47-61] Of the various aspects of crown ether chemistry described by Luk'yanenko, Bogatskii and Kirichenko[62] their applications to synthesis are among the most important. Here, our emphasis is on preparative coordination chemistry in aqueous solutions with macrocyclic ethers. The first synthetic macrocyclic ethers were prepared by Pedersen in 1967.[63] He named the compounds crown ethers because of their shape. In particular, he synthesized the compounds shown in Figure 2.5. Pedersen found that considerable increases in the solubilities of inorganic and complex salts in nonpolar aprotic solvents occurred in the presence of crown ethers. For example, salts such as $KMnO_4$ and $K_2[PdCl_4]$ were found to dissolve in aromatic hydrocarbons in the presence of dicyclohexyl-18-crown-6.

Soon after the publication of Pedersen's work,[63] Lehn[47,64,65] described work on polycyclic aminopolyethers which he called cryptands. The compounds shown in Figure 2.6 were synthesized.

Ovchinnikov, Ivanov and Shkrab have drawn an analogy[49] between the complexation of an alkali-metal cation with a macrocyclic ligand and the alteration of the solvation shell of a cation upon replacement of one solvent by a second solvent of greater solvating ability. The analogy was based on data[49] from a large number of X-ray structural investigations of complexes of

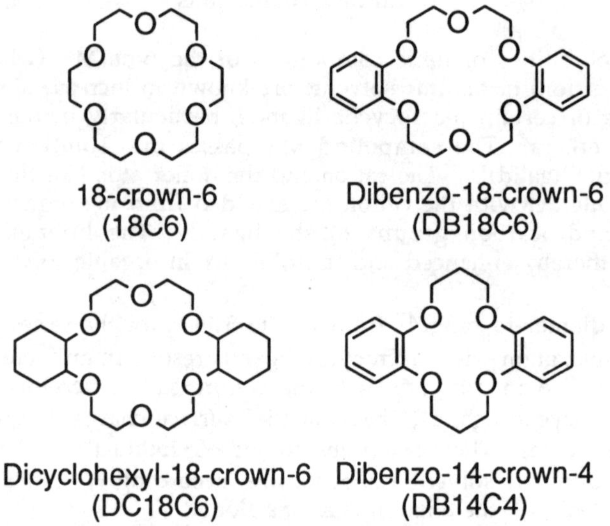

Figure 2.5

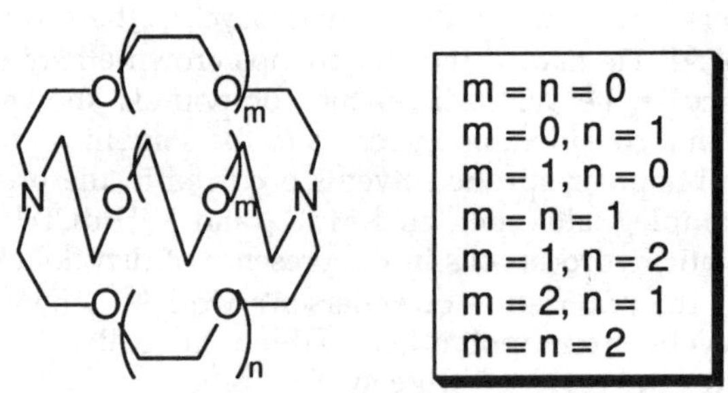

Figure 2.6

macrocyclic compounds and crystalline solvates. In both types of compounds cations were surrounded by the negatively charged ends of the dipoles of the solvating groups and the distance between each cation and donor atom was close to the sum of the crystallographic radii. Ovchinikov, Ivanov and Shkrab also noted[49] that with "intramolecular solvation" of a cation by a macrocyclic ligand, the number of donor atoms surrounding the cation depends primarily on the structure and conformation of the macrocyclic ligand rather than on the character of the metal ion itself.

Yatsimirskii and Lampeka have shown[51] that complexation with macrocyclic ligands is governed by the nature of the metal ion that enters the ligand cavity. Indeed, the cavity may undergo adjustment so that especially stable complexes are obtained when the *maximum size of the cavity in the macrocycle corresponds to the size of the metal ion.* Effective complexation[51] requires the following conditions:

1. The mean diameter of the cavity must be equal to the diameter of the metal ion.

2. The donor atoms of the macrocycle must have specific electron-donor and electrostatic characteristics.

3. Correspondence between the number of donor atoms in the cavity and the coordination number characteristic of the particular metal (for example, 4 or 5 in the case of Na^+ and 6, 7, or 8 in the cases of K^+, Rb^+, and Cs^+) is necessary.

4. A definite geometric arrangement of the donor atoms in the cavity and conformational flexibility of the macrocycle is needed.

Few systematic investigations have been carried out to evaluate the effects of crown ethers on the solubilities of anionic complexes in organic solvents. The use of $K[CoL]\cdot 2H_2O$ (H_4L = EDTA) as a model compound for evaluation of the effects of crown ethers on the solubilities of metal complexes has been proposed.[66] This compound is solubilized by crown ethers in a variety of organic solvents such as acetone, acetonitrile, ethanol, and methanol. The salt is quite stable in these organic solvents and the solubility can be monitored easily spectrophotometrically. The solubilities of $K[CoL]\cdot 2H_2O$ in acetonitrile in the presence of the macrocyclic ligands 12C4, 15C5, DC18C6, and 18C6 have been measured.[66] The compound is sparingly soluble in pure acetonitrile, but solubility is increased strikingly by the addition of crown ethers. The solubility of $K[CoL]\cdot 2H_2O$ in MeCN increases by two orders of magnitude in the presence of 18C6. The crown ethers investigated formed the series 12C4 < DC18C6 ~15C5 < 18C6 according to their abilities to increase the solubility of $K[CoL]\cdot 2H_2O$. Further, a parallel exists between the stability constants for the complexation of potassium ion with crown ethers and the values of the solubilities of $K[CoL]\cdot 2H_2O$ in the presence of the corresponding crown ethers.[66]

The synthesis of complexes of the type [M(crown ether)]X is usually carried out by one of two methods.[48] The first method involves the interaction of the salt MX with the crown ether in a particular organic solvent. For example, the complex [K(18C6)]OH is obtained in dichloromethane[67] by stirring a suspension of 18C6 and potassium hydroxide in a 1:10 ratio in anhydrous CH_2Cl_2. This synthetic approach is not always effective because solutions with a sufficiently high concentration of the complex cannot always be obtained in a given organic solvent.

In the second synthetic method the salt, MX, is dissolved in the crown ether in MeOH and then the solvent is removed under vacuum. Subsequently the complex is dissolved in a second solvent. It is believed that a sharp increase in the solubilities of the complexes in methanol, in contrast to aprotic, nonpolar solvents, is the result not only of the solvation of the anions by methanol, but also the solvation of the complex cation.[49] Methanol is often found tightly bound to the cyclic ether of the complex formed and is not removed when the solvent is evaporated *in vacuo*.

When methanol does not interfere with subsequent reactions, the synthesis of complexes of macrocyclic ligands can be performed without isolation but with a small amount of methanol added to the solvent system. For example, potassium hydroxide dissolves in a mixture of benzene with methanol (99:1 in volume) in the presence of DC18C6 forming solutions with concentrations as high as 1 *M*.[48]

2.4 Solubility of complexes in mixed solvents

Studies of the solvation of ions or polar molecules in binary solvents have established that the ratio between the components of a solvent mixture in the solvation shell of a solute usually differs from that in the bulk of the solution, *i.e.*, the composition of a microscopic solvation shell differs from the macroscopic composition of the solvent. This phenomenon has been termed *selective solvation*. For this reason mixed solvents sometimes display specific properties with respect to the solubilities of complex compounds. This situation is illustrated by the thiocyanate analog of Magnus salt, $[Pt(NH_3)_4][Pt(SCN)_4]$. Grinberg has reported that the compound is only slightly soluble in water or in acetone, but is highly soluble in a mixture of the two.[68] Demidov found that the 2-aminopyridine complex of platinum(II), *trans*-$[PtCl_2(2\text{-}NH_2C_5H_4N)_2]$, is highly soluble in aqueous acetone, but is insoluble in either water or acetone alone.[69] The rhodium(III) complex, $[RhCl(C_2O_4)(py)_3]$, is freely soluble in a mixture of pyridine and water, but is quite insoluble in water or pyridine alone. Reichardt[70] believes that only the simultaneous selective solvation of the pyridine ligand by the pyridine in the solvent and of the oxalate ligand by water in the solvent provides a sufficiently large heat of solvation to overcome the crystal-lattice energy. In his opinion, the selective solvation of individual regions of a complex by the components of a mixed solvent is probably the most common reason for the increased solubility of coordination compounds in such mixed solvent systems.

In contrast, mixed solvent systems are used widely in preparative coordination chemistry for the precipitation of compounds[71] from solution. This operation is based on the compound's decreased solubility in the mixed solvent.

An increase in the solubility of a complex is anticipated when the complex has acid properties and the solvent, or one of its components, has basic properties. For example, the increased solubility of inert amine complexes of transition metals in liquid ammonia or amines is known.[17] Complexes with basic properties are possible in principle and, although they are encountered only rarely in practice, such complexes should have increased solubility in acidic solvents.

The formation of outer-sphere complexes also alters the solubilities of compounds. The solubility method is one of the few techniques which can be used to study outer-sphere complexes and their stepwise formation.[72-74] For example, addition of perchlorate ion to aqueous solutions of hexamminecobalt(III) perchlorate first decreases the solubility, according to the law of mass action, but then, at higher concentrations of perchlorate, the solubility increases much more than changes in ionic strength would require.[75] These experiments were used to demonstrate the ability of perchlorate ions to form ion pairs with the cobalt(III) cations and to determine the association constant.

In conclusion, complexes undergo interactions with solvent molecules forming new species in solution. Such interactions can be detected by observation of increased solubility. The potential for such interactions must be carefully considered when a solvent is selected for a given metal complex.

References

1. *Chemical Encyclopedic Dictionary* [in Russian] (Sov. Encyclop., Moscow, 1983) p. 494.
2. Yu. V. Gurikov, *Zh. Phys. Khim.* **54** (1980) 1223.
3. H. F. Herbrandson and F. R. Neufeld, *J. Org. Chem.* **31** (1966) 1140.
4. B. R. Eggins, *Chemical Structure and Reactivity* (MacMillan, London, 1972) p. 140.
5. (a) J. H. Hildebrand and R. L. Scott, *The Solubility of Nonelectrolytes* (Reinhold, New York, 1950); (b) F. Feigl, *Chemistry of Specific, Selective and Sensitive Reactions* (Academic, New York, 1949) p. 411.
6. (a) G. Sprintschnik, H. W. Sprintschnik, P. P. Kirsch, and D. G. Whitten, *J. Am. Chem. Soc.* **98** (1976) 2337; (b) G. Sprintschnik, H. W. Sprintschnik, P. P. Kirsch, and D. G. Whitten, *J. Am. Chem. Soc.* **99** (1977) 4947.
7. D. A. Cauzzi, G. Mori, G. Predieri, A. Tiripicchio, and F. Cavarota, *Inorg. Chim. Acta* **204** (1993) 181.
8. S. Ahrland, *Pure Appl. Chem.* **51** (1979) 2019.
9. A. A. Abramzon and G. Gaeva, eds., *Surfactants* [in Russian] (Khimia, Moscow-Leningrad, 1979) p. 14.
10. V. M. Zamiatkina, Yu. N. Kukushkin, and A. A. Makarenia, *Lev Aleksandrovich Chugaev* [in Russian] (Nauka, Moscow, 1973).
11. L. Tchugaeff [Chugaev], *C. R. Acad. Sci.* **159** (1914) 188.
12. V. Yu. Kukushkin and N. P. Kiseleva, *Koord. Khim.* **14** (1988) 693.
13. L. I. Elding, Å. Oskarsson, and V. Yu. Kukushkin, *Inorg. Synth.* **31** (1995) in press.
14. L. D. C. Bok, J. G. Leipoldt, and S. S. Basson, *Z. Anorg. Allg. Chem.* **415** (1975) 81.
15. H. C. Clark and L. E. Manzer, *Inorg. Chem.* **10** (1971) 2699.
16. K. B. Yatsimirskii, *The Thermochemistry of Complex Compounds* [in Russian] (Izd. Akad. Nauk SSSR, Moscow, 1951).
17. Yu. N. Kukushkin, *Chemistry of Coordination Compounds* [in Russian] (Vyshaya Shkola, Moscow, 1985) p. 379.
18. (a) F. Basolo, *Coord. Chem. Rev.* **3** (1968) 213; (b) P. L. Goggin, R. J. Goodfellow, and F. J. S. Reed, *J. Chem. Soc., Dalton Trans.* (1972) 1298; (c) L. Heck, M. Ardon, A. Bino, and J. Zapp, *J. Am. Chem. Soc.* **101** (1988) 2691; (d) Th. Raber and L. Heck, *personal communication.*

19. (a) J. K. Ruff and W. J. Schlientz, *Inorg. Synth.* **15** (1974) 84; (b) V. Yu. Kukushkin and A. I. Moiseev, *Inorg. Chim. Acta* **176** (1990) 79; (c) F. J. Lalor and S. Chaona, *J. Organomet. Chem.* **344** (1988) 163.

20. S. Bekkevoll, I. Svorstøl, H. Høiland, and J. Songstad, *Acta Chem. Scand., Ser. B.* **37** (1983) 935.

21. M. H. Abraham, A. F. Danil de Namur, and R. A. Schultz, *J. Solut. Chem.* **5** (1976) 529.

22. I. M. Korenman, *Extraction in the Analysis of Organic Compounds* [in Russian] (Khimia, Moscow, 1977) p. 83.

23. S. N. Bhattacharya and C. V. Senoff, *Inorg. Chem.* **22** (1983) 1607.

24. M. Darensbourg, H. Barros, and C. Borman, *J. Am. Chem. Soc.* **99** (1977) 1647.

25. D. J. Darensbourg and M. Pala, *ACS Meeting, Division of Inorg. Chem.* September 8-13, 1985 Abstr. 357.

26. D. J. Darensbourg, M. Pala, and A. L. Rheingold, *Inorg. Chem.* **25** (1986) 125.

27. D. E. C. Corbridge, *Phosphorus: An Outline of its Chemistry, Biochemistry and Technology,* (Elsevier, Amsterdam, 1990, 4th Ed.).

28. J. N. Cooper and R. E. Powell, *J. Am. Chem. Soc.* **85** (1963) 1590.

29. (a) F. R. Hartley and J. A. Davies, *Rev. Inorg. Chem.* **4** (1982) 27; (b) R. M. Alcock, F. R. Hartley, and D. E. Rogers, *J. Chem. Soc., Dalton Trans.* (1973) 1070; (c) P. K. F. Chin and F. R. Hartley, *Inorg. Chem.* **15** (1976) 982.

30. H. C. Clark and J. D. Ruddick, *Inorg. Chem.* **9** (1970) 1226.

31. A. J. Deeming, B. F. G. Johnson, and J. Lewis, *J. Chem. Soc., Dalton Trans.* (1973) 1848.

32. G. Bracker, D. M. Grove, L. M. Venanzi, F. Bachechi, P. Mura, and L. Zambonelli, *Angew. Chem.* **90** (1978) 826.

33. V. Yu. Kukushkin, A. Olsson, Å. Oskarsson, and L. I. Elding, *unpublished results.*

34. V. Yu. Kukushkin, K. Lövqvist, B. Nóren, Å Oskarsson, and L. I. Elding, *Phosphorus, Sulfur, and Silicon* **73** (1992) 253.

35. K. Siegmann, P. S. Pregosin, and L. M. Venanzi, *Organometallics* **8** (1989) 2659.

36. P. Albano, M. Aresta, and M. Manassero, *Inorg. Chem.* **19** (1980) 1069.

37. R. J. Haines and A. L. du Preez, *J. Am. Chem. Soc.* **93** (1971) 2820.

38. L. Sacconi, P. Dapporto, and P. Stoppioni, *Inorg. Chem.* **15** (1976) 325.

39. P. Legzdins and D. T. Martin, *Organometallics* **2** (1983) 1785.

40. P. Legzdins, G. B. Richter-Addo, F. W. B. Einstein, and R. H. Jones, *Organometallics* **9** (1990) 431.

41. K. Raab and W. Beck, *Chem. Ber.* **118** (1985) 3830.

42. M. L. H. Green, A. K. Hughes, P. Lincoln, J. J. Martin-Polo, P. Mountford, A. Sella, L.-L. Wong, J. A. Bandy, T. W. Banks, K. Prout, and D. J. Watkin, *J. Chem. Soc., Dalton Trans.* (1992) 2063.

43. V. Desobry and E. P. Kundig, *Helv. Chim. Acta* **61** (1961) 1288.

44. (a) H. Nakazawa, K. Morimasa, Y. Yushi, and H. Yoneda, *Organometallics* **7** (1988) 458; (b) B. Demersman, G. Bouquet, and M. Bigorge, *J. Organomet. Chem.* **35** (1972) 125.

45. R. H. Morris, J. F. Sawyer, C. T. Schweitzer, and A. Sella, *Organometallics* **8** (1989) 2099.

46. R. Fernández-Galán, B. R. Manzano, A. Otero, M. Lanfranchi, and M. A. Pellinghelli, *Inorg. Chem.* **33** (1994) 2309.

47. J. M. Lehn, *Pure Appl. Chem.* **49** (1977) 857.

48. C. J. Pedersen and H. F. Frensdorf, *Angew. Chem. Int. Ed. Engl.* **11** (1972) 16.

49. Yu. A. Ovchinnikov, V. T. Ivanov, and A. M. Shkrab, *Membrane-Active Complexing Agents* [in Russian] (Nauka, Moscow, 1974).

50. K. B. Yatsimirskii and Ya. D. Lampeka, *Physical Chemistry of Complexes of Metals with Macrocyclic Ligands* [in Russian] (Naukova Dumka, Kiev, 1985).

51. M. R. Truter and C. J. Pedersen, *Endeavour* **30** (1971) 142.

52. M. R. Truter, *Chem. Brit.* **7** (1971) 203.

53. (a) J. M. Lehn, *Struct. Bonding (Berlin)* **16** (1973); (b) M. R. Truter, *Struct. Bonding (Berlin)* **16** (1973); (c) W. Simon, W. E. Morf, and P. Ch. Meier, *Struct. Bonding (Berlin)* **16** (1973); (d) R. M. Izatt, D. J. Eatough, and J. J. Christensen, *Struct. Bonding (Berlin)* **16** (1973).

54. J. J. Christensen, D. J. Eatough, and R. M. Izatt, *Chem. Rev.* **74** (1974) 351.

55. D. J. Cram and J. M. Cram, *Science* **183** (1974) 803.

56. N. S. Poonia, *J. Sci. Ind. Res.* **36** (1977) 268.

57. D. J. Cram and J. M. Cram, *Acc. Chem. Res.* **11** (1978) 8.

58. I. M. Kolthoff, *Anal. Chem.* **51** (1979) R1.

59. R. M. Izatt and J. J. Christensen, eds., *Progress in Macrocyclic Chemistry*, vol. 1, (Wiley, New York, 1979).

60. R. M. Izatt, J. S. Bradshaw, S. A. Nielsen, J. D. Lamb, J. J. Christensen, and D. Sen, *Chem. Rev.* **85** (1985) 271.

61. A. Yu. Tsivadze, A. A. Varnek, and V. E. Khutorskii, *Coordination Compounds of Metals with Crown Ligands* [in Russian] (Nauka, Moscow, 1991).

62. N. G. Luk'yanenko, A. V. Bogatskii, and T. I. Kirichenko, *Zh. Vses. Khim. Obstch.* **30** (1985) 571.

63. C. J. Pedersen, *J. Am. Chem. Soc.* **89** (1967) 7017.

64. B. Dietrich, J. M. Lehn, and J. P. Sauvage, *Tetrahedron Lett.* (1969) 2885.

65. B. Dietrich, J. M. Lehn, and J. P. Sauvage, *Tetrahedron Lett.* (1969) 2889.

66. T. Taura, *Inorg. Chim. Acta* **98** (1985) L15.

67. T. Endo, T. Miyazava, S. Shiihasi, and M. Olawara, *J. Am. Chem. Soc.* **106** (1984) 3877.

68. A. A. Grinberg and V. M. Shulman, *Izv. Inst. Plat.* **11** (1933) 111.

69. V. N. Demidov, *personal communication.*

70. C. Reichardt, *Solvent Effects in Organic Chemistry*, vol. 3, Monographs in Modern Chemistry (Verlag Chemie, Deerfield Beach, Florida, 1979).

71. *Salting Out Substances from Solutions.* Proceedings of the Second All-Union Symposium [in Russian], Kaunas, 1970.

72. V. E. Mironov, *Uspekhi Khimii* **39** (1970) 702.

73. M. T. Beck, *Chemistry of Complex Equilibria* (Van Nostrand Reinhold, New York, 1970).

74. V. V. Kuznetsov, *Uspekhi Khimii* **55** (1986) 1409.

75. (a) L. Heck, *Inorg. Nucl. Chem. Lett.* **7** (1971) 709; (b) L. Heck, *personal communication.*

Chapter 3. GENERATION OF SOLVENTO-COMPLEXES

The second half of the twentieth century has brought a resurgence of interest in coordination compounds and coordination chemistry generated by the results of research initiated during World War II and by new, intriguing applications for coordination compounds in industry, manufacturing, medicine, and consumer goods. The use of coordination complexes as precatalysts (*i.e.* catalyst precursors) in organic syntheses is one example of such a contemporary application.

The synthesis of coordination compounds often involves reactions in which ligand replacement and exchange occur in solution. The solvent chosen for such reactions is important; clearly, solvent/substrate and solvent/product interactions cannot be too strong or the solvent will interfere and function as a reagent. Typically, solvents are chosen for their relatively poor donor abilities so that they interact only weakly with reagents and products over the course of a chemical transformation. In the synthesis of coordination compounds, the choice of solvent is particularly important because intermediate solvento-complexes may be involved. The term solvento-complex describes a compound containing coordinated solvent molecules as well as other coordinated ligands. If the coordination sphere of the complex contains only solvento-ligands, the complex is referred to as a homoleptic solvento-complex. Solvents are purposely chosen for their relatively weak donor abilities, so they are easily replaced by other ligands.

Metal-solvent interactions can be conveniently considered in terms of the hard-soft acid-base (HSAB) principle (Chapter 1). For example, palladium(II) is a soft metal center and so the hard oxygen donor solvent, diethyl ether, interacts only poorly with it. Simple valence bond models have been presented that adequately explain such soft metal-hard base interactions.[1] In this chapter complexes containing coordinated halocarbons are treated in a separate section (3.7) from those that contain other "hard bases" since such species have only been recognized as well-defined complexes in recent years and their potential for exploitation in coordination chemistry merits special attention.

When solvent molecules are introduced as ligands in the inner sphere of a complex, the chemical properties of the complex may be altered. For example, if an anionic ligand is replaced by a neutral solvent molecule, the ionic charge on the complex is changed. With the formation of such a solvento-complex, the rate of subsequent ligand replacement becomes rapid, particularly when the substituting ligand carries a charge opposite to that of the complex ion.

A process referred to as "aging" may occur when a complex in solution has been allowed to stand for some time. Ligand exchange reactions often occur more rapidly within an aged solution than when that solution is freshly prepared. This effect is attributed to the formation of a solvento-complex intermediate in solution. The solvento-complex is spontaneously formed as time passes and equilibrium is achieved. Induction periods, often seen not only in ligand exchange reactions but also in homogeneous catalytic processes, are frequently the result of this "aging effect".

The equilibrium generation of solvento-complexes is based on reaction between the solvent and the complex, Scheme 3.1:

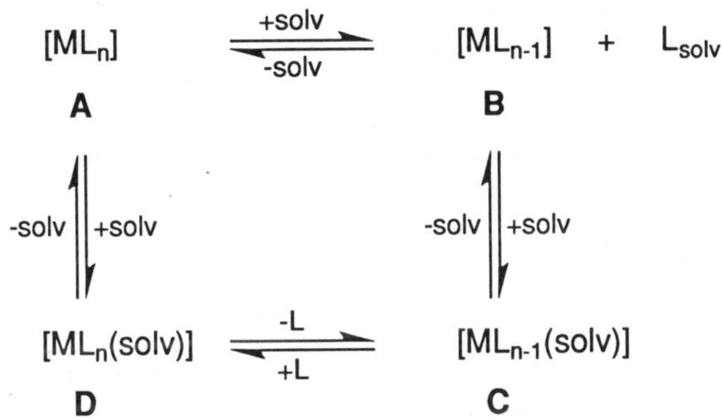

Scheme 3.1

The pathway followed will depend upon the nature of the complex and the solvent. If the complex is coordinatively saturated, a dissociative pathway (*i.e.* **A** $\rightleftharpoons$ **B**) might be followed with the right hand side of the equilibrium favored by solvation of the displaced ligand, **L**. If the solvent is a sufficiently powerful donor, then the coordinatively unsaturated intermediate, **B**, will react to produce the solvento-complex, **C**. If the solvent is not a sufficiently good donor to promote formation of **C**, then complex **B** will typically enter into other reactions (*e.g.* dimerization) that result in formation of coordinatively saturated products. Alternatively, if **A** is coordinatively unsaturated, then an associative pathway (**A** $\rightleftharpoons$ **D** $\rightleftharpoons$ **C**) might be followed that leads to the generation of equilibrium concentrations of the solvento-complex, **C**.

As a rule, the coordinated solvent molecules in complex **C** are easily replaced by other ligands, **L***, forming complexes of type **E**. The equilibrium concentration of complex **E** may be low when it is generated from complex **A** *via* complex **C**, Scheme 3.2:

$$\underset{\mathbf{A}}{[ML_n]} + xL^* \xrightleftharpoons{solv} \underset{\mathbf{E}}{[ML_{n-x}L^*_x]} + xL$$

Scheme 3.2

Given this equilibrium situation, substitution of **L** by **L*** might be favored by high concentrations of **L***, low solubility of complex **E**, removal of displaced **L**, *etc*. For example, the interaction of $Na_3[RhCl_6]\cdot 12H_2O$ and dimethyl sulfoxide at a molar ratio of 1:2 is known to result in the formation of $Na[RhCl_4(DMSO)_2]$.[2] With a tenfold excess of DMSO, another chloride ion can be displaced and $[RhCl_3(DMSO)_3]$ is formed.[3] Further substitution is possible only under forcing conditions, *e.g.* where $AgClO_4$ is used as a reagent for halide abstraction.[4] In the presence of $AgClO_4$, the reaction proceeds in dimethyl sulfoxide by Eq. 3.1:

$$[RhCl_3(DMSO)_3] + 2AgClO_4 + 2DMSO \rightarrow [RhCl(DMSO)_5][ClO_4]_2 + 2AgCl \qquad (3.1)$$

Abstraction of the last chloride ligand from $[RhCl(DMSO)_5][ClO_4]_2$ has been unsuccessful even with an excess of silver perchlorate.[4]

The nature of L dramatically affects the equilibrium concentration of the solvento-complex, **C**. This is clearly illustrated in the substitution of chloride[5] by CD_3CN in a series of ruthenium(II) complexes, $[RuCl(PR_3)_2Cp]$ (PR_3 = various phosphines and phosphites). On dissolution of the complexes in acetonitrile-d_3 an equilibrium is established, Scheme 3.3. The solvolysis rate and the degree of conversion of the starting material depend strongly on the basicity of the coordinated PR_3 ligand. Treichel and Vincenti[5] have stated that the introduction of more basic PR_3 ligands into the complex leads to an increase in the electron density on the metal center and, hence, to a decrease in the strength of the metal-halide bond. When the bond strength is reduced, solvolysis is favored. In fact, the maximum extent of solvolysis, 83%, is reached when the very basic ligand trimethylphosphine is employed.

Various approaches can be adopted to perturb the equilibrium shown in Scheme 3.3 and to favor formation of $[Ru(CD_3CN)(PR_3)_2Cp]^+$. For example, the interaction of $[RuCl(PR_3)_2Cp]$ and chloride ion acceptors, such as silver, sodium, or ammonium hexafluorophosphates, results in precipitation of AgCl, NaCl, or NH_4Cl and 100% conversion of the starting compound into the solvento-complex. The reaction with $NaPF_6$ is illustrated in Scheme 3.4.

Widely used in preparative coordination chemistry, labile solvento-complexes are generated by a number of different approaches. Aspects of the

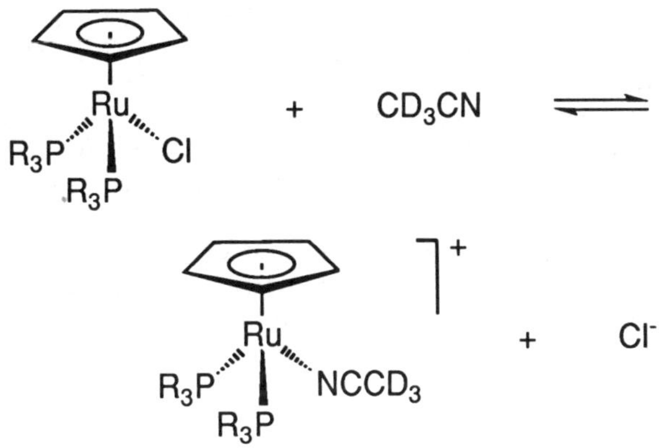

Scheme 3.3

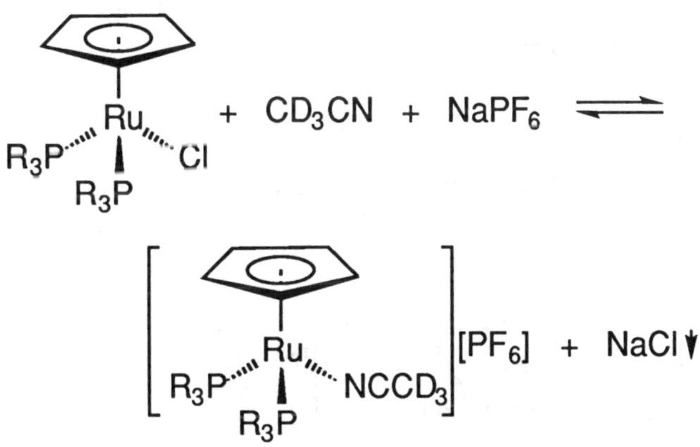

Scheme 3.4

chemistry of solvento-complexes and related compounds have been discussed in review articles.[1,6-8]

3.1 Halide abstraction reactions

The most common halide abstraction method utilizes silver salts of poorly coordinating anions. A large number of such reactions have been described in the literature. For example, *cis*-$[PtCl_2(NH_2Me)_2]$ cannot be prepared through the direct reaction of methylamine with potassium tetrachloroplatinate(II) because the major product is $[Pt(NH_2Me)_4][PtCl_4]$. Additionally, while *cis*-$[PtI_2(NH_2Me)_2]$ is obtained in good yield by the reaction of methylamine with $K_2[PtI_4]$, the transformation of *cis*-$[PtI_2(NH_2Me)_2]$ into the chloride analog by treatment with KCl cannot be accomplished directly. However, this transformation becomes possible *via* intermediate formation of a solvento-complex.[9] Two mole equivalents of silver nitrate are added to *cis*-$[PtI_2(NH_2Me)_2]$ in water and the reaction proceeds as shown in Scheme 3.5:

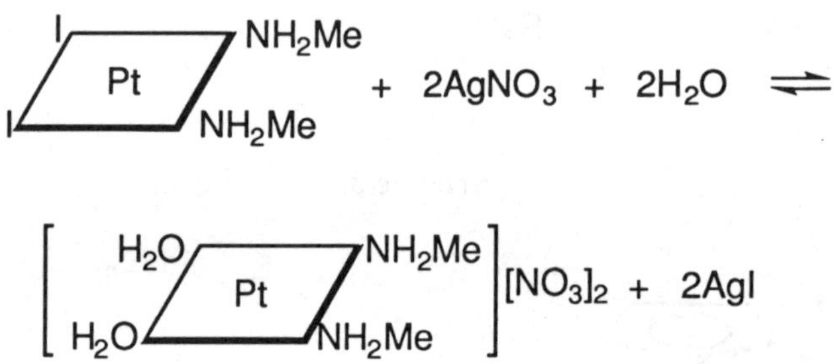

Scheme 3.5

After removal of the AgI precipitate by filtration, potassium chloride is added to the solution resulting in precipitation of the desired product as a solid, Scheme 3.6:

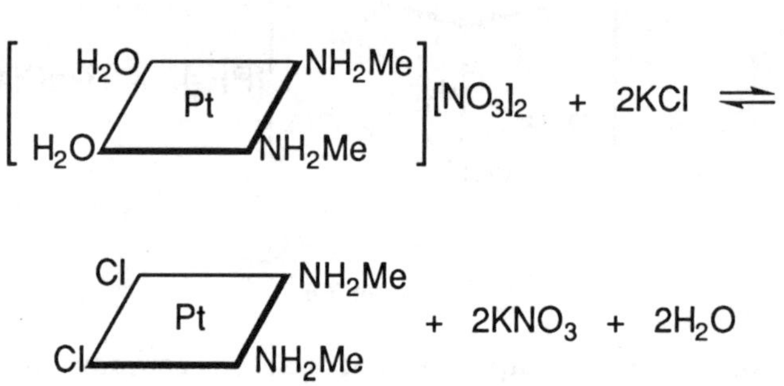

Scheme 3.6

The synthesis of *cis*-$[PtCl_2(amine)_2]$ complexes described above was developed by Grinberg and Gildengerschel[9] in 1951. In 1969 the antitumor activity of *cis*-$[PtCl_2(NH_3)_2]$ and other platinum-amine complexes was discovered.[10-12] Since that time, such complexes have been studied extensively by both chemists and biologists. At present, many of these compounds[13-15] are prepared using Grinberg and Gildengerschel's method *via* intermediate formation of platinum(II) aqua-complexes of the type $[Pt(amine)_2(H_2O)_2]^{2+}$. During the past 10-15 years this method of halide abstraction/replacement has been extended to compounds of other metals both in aqueous and non-aqueous media.[16]

The synthesis of the homoleptic platinum(II) aqua-dication *via* halide abstraction illustrates the utility of silver salts in such processes. For many years workers were unable to prepare $[Pt(H_2O)_4]^{2+}$ and Livingstone[17] contended that the complex does not form. However, in 1976 Elding prepared $[Pt(H_2O)_4]^{2+}$ *via* chloride abstraction with a silver salt,[18] demonstrating that when $K_2[PtCl_4]$ is reacted with an excess of silver perchlorate in water, the reaction proceeds as shown in Scheme 3.7:

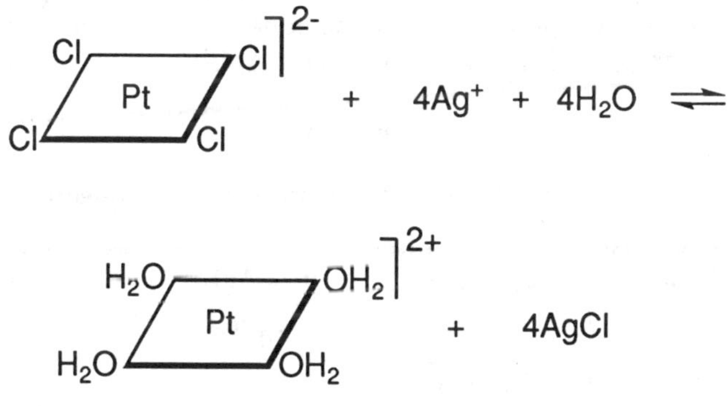

Scheme 3.7

The AgCl precipitate is removed by filtration and any excess silver ions are electrochemically removed. Other workers[19] have utilized the $[Pt(H_2O)_4]^{2+}$ ion in a number of substitution reactions.

Seigmann, Pregosin, and Venanzi[20] have carried out related reactions in which removal of coordinated chloride ions leads to the formation of complexes containing solvento-ligands (solv = H_2O or MeOH), Eq. 3.2:

$$[PtCl_2(PMe_3)_2] + 2AgSO_3CF_3 + 2solv \rightarrow [Pt(PMe_3)_2(solv)_2][SO_3CF_3]_2 + 2AgCl \quad (3.2)$$

Silver salts have been widely used for halide abstraction in non-aqueous systems. Thus, $AgPF_6$, $AgAsF_6$, $AgSbF_6$, $AgSO_3CF_3$, $AgBF_4$, $AgBPh_4$, $AgClO_4$, and other silver salts, with at least some solubility in organic solvents, are used in such reactions. For example, the reaction between [RuCl(PMe$_3$)(PPh$_3$)(η^5-pentadienyl)] and $AgBF_4$ in methanol yields [Ru(MeOH)(PMe$_3$)(PPh$_3$)(η^5-pentadienyl)][BF$_4$]. In this compound, the coordinated solvent molecule is replaced easily by CO, PMe_3 or BuNC, all better ligands for ruthenium(II) than methanol.[21]

Similarly, [NbCl$_2$ICp$_2$] reacts with two equivalents of silver perchlorate in acetone[22] to form [NbCl(Me$_2$CO)$_2$Cp$_2$][ClO$_4$]$_2$. The compound has been isolated in the solid state and characterized by elemental analysis, as well as IR and ^{1}H NMR spectroscopies. Both 2,2'-bipyridyl and 1,10-phenanthroline readily displace the coordinated acetone molecules.

When halide abstraction reactions are performed with silver salts in water or in non-aqueous solutions, halide ions are abstracted according to the solubility product of the silver halide (Table 3.1) with the least soluble salts being the first to be precipitated. Abstraction follows the sequence $I^- > Br^- > Cl^- > F^-$. Accordingly, then, the interaction of [NbCl$_2$ICp$_2$] with one mole equivalent of silver perchlorate in acetone leads to the formation of [NbCl$_2$(Me$_2$CO)Cp$_2$][ClO$_4$] and precipitation of AgI.[22]

Table 3.1

Solubility products[23] of silver halides in some non-aqueous solvents

Solvent	$K_{sp}(AgCl)^a$	$K_{sp}(AgBr)$	$K_{sp}(AgI)$
DMSO	$10^{-10.4}$	$10^{-10.6}$	$10^{-12.0}$
MeCN	$10^{-12.4}$	$10^{-13.2}$	$10^{-14.2}$
MeOH	$10^{-13.0}$	$10^{-15.2}$	$10^{-18.2}$
Me_2CO	$10^{-16.4}$	$10^{-18.7}$	$10^{-20.9}$
$EtNO_2$	$10^{-21.1}$	$10^{-21.8}$	$10^{-22.6}$

a. $K_{sp}(AgX) = [Ag^+][X^-]$ at 23 °C.

The preferential abstraction of halide ions from chromium complexes has been studied,[24] Scheme 3.8:

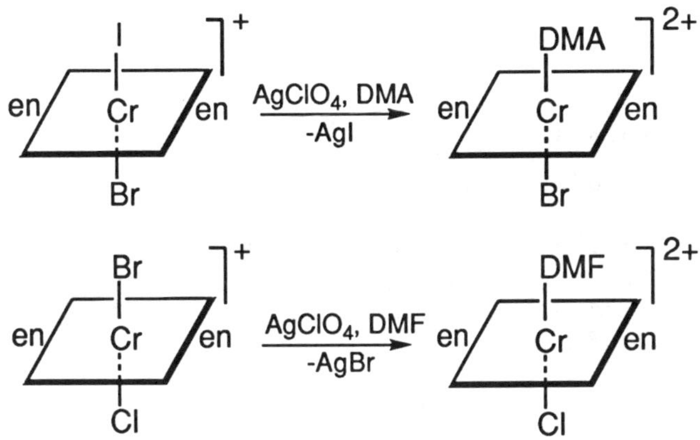

Scheme 3.8

In these reactions, the less soluble silver halide is precipitated and a solvent molecule occupies the free coordination site. Halide abstraction from a related, mixed, chloride-iodide complex with silver perchlorate results in precipitation of the less-soluble AgI and formation of the corresponding solvento-complex, Scheme 3.9:

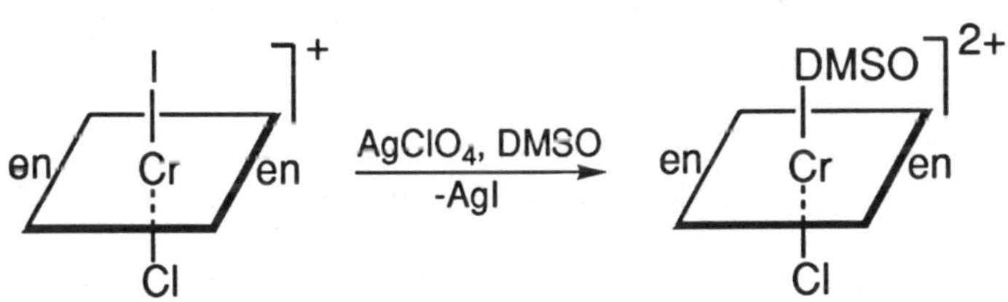

Scheme 3.9

Reagents for halide abstraction are not limited to silver salts of poorly coordinating anions; similar abstractions are performed with salts of other metals and other Lewis acids. Thus, reactions with thallium(I) salts are fairly common. $[WCl(MeCN)(PMe_3)_2(C_2H_4)_2][BF_4]$, for instance, is formed by abstraction of chloride ion from $[WCl_2(PMe_3)_2(C_2H_4)_2]$ with $TlBF_4$ in acetonitrile,[25] Eq. 3.3:

$$[WCl_2(PMe_3)_2(C_2H_4)_2] + TlBF_4 + MeCN \rightarrow [WCl(MeCN)(PMe_3)_2(C_2H_4)_2][BF_4] + TlCl \quad (3.3)$$

Mercury(II) salts have been employed as reagents for the abstraction of halides and pseudohalide ligands from complexes of transition metals,[26] Eq. 3.4:

$$2[CoCl(NH_3)_5]^{2+} + 2Hg^{2+} + H_2O + NO_3^- \rightarrow [Co(NO_3)(NH_3)_5]^{2+} + [Co(NH_3)_5(H_2O)]^{3+} + 2HgCl^+ \quad (3.4)$$

When chloride ion is removed from the ruthenium(II) complex [RuCl(dppe)Cp] in a methanol-acetonitrile mixture with NH_4PF_6, the driving force for the process[27] is the formation of the sparingly soluble salt, ammonium chloride, Eq. 3.5:

$$[RuCl(dppe)Cp] + NH_4PF_6 + MeCN \rightarrow [Ru(MeCN)(dppe)Cp][PF_6] + NH_4Cl \quad (3.5)$$

Lewis acids are convenient reagents for halide abstraction. For example, the reaction of $TiCl_4$ with $SbCl_5$ in MeCN solution[28] leads to the formation of $[TiCl_3(MeCN)_3][SbCl_6]$. Halide abstraction[29] from $[TiCl_2Cp_2]$ with $FeCl_3$ in acetonitrile results in the formation of $[TiCl(MeCN)Cp_2][FeCl_4]$. In addition, when metal chlorides, MCl_3, (M = Ti, V) are heated with $ZnCl_2$ in THF, the product is a solvento-complex,[30,31] $[MCl_2(THF)_4][ZnCl_3(THF)]$. The cationic solvento-complexes formed in these reactions are stabilized by the counterions generated during the halide abstraction reaction. The copper(I) complex $[Cu(MeCN)_4][BF_4]$ has been employed as a reagent for halide abstraction,[32] Eq. 3.6:

$$[Rh_2Cl_2(C_2H_4)_4] + 2[Cu(MeCN)_4][BF_4] \rightarrow 2[Rh(MeCN)_2(C_2H_4)_2][BF_4] + 2CuCl + 4MeCN \quad (3.6)$$

Nitrosonium salts, more typically employed as oxidants (see below and Chapter 8), also function as reagents for halide abstraction in specific cases,[33] *e.g.* Scheme 3.10.

Silver(I) salts are good oxidizing agents (Chapter 8) especially in non-aqueous, weak donor solvents. Oxidation and halide abstraction can be competing processes during reactions with transition metal halide complexes. When it is necessary to avoid oxidation, the advantage of halide abstraction with Tl(I) salts or other Lewis acids becomes quite evident. For example, in the reaction between $[W(H)_2Cl_2(PMe_3)_4]$ and $AgBF_4$ the product is the result of a one-electron oxidation[34] not of halide abstraction, Eq. 3.7:

$$[W(H)_2Cl_2(PMe_3)_4] + AgBF_4 \rightarrow [W(H)_2Cl_2(PMe_3)_4][BF_4] + Ag \quad (3.7)$$

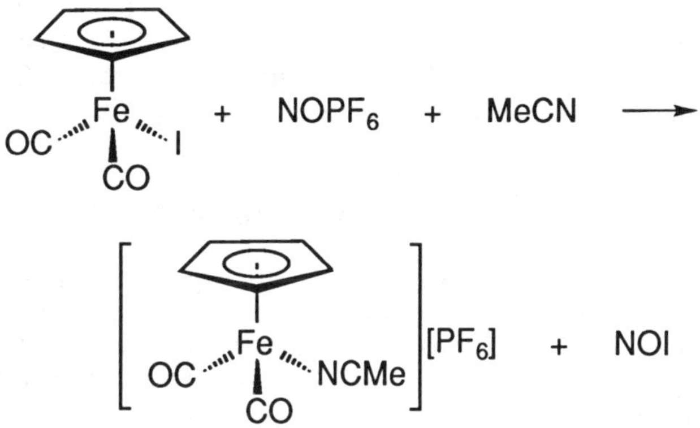

Scheme 3.10

In contrast, the reaction of the same tungsten(IV) complex with $TlBF_4$ or $AlCl_3$ results in chloride abstraction without oxidation. The reaction with aluminum chloride is described by Eq. 3.8:

$$[W(H)_2Cl_2(PMe_3)_4] + AlCl_3 + MeCN \rightarrow [W(H)_2Cl(MeCN)(PMe_3)_4][AlCl_4] \quad (3.8)$$

McNeese and Mueller have shown[35] that [VCl(Schiff Base)] complexes lose halide in reactions with thallium tetrafluoroborate in acetone. The corresponding solvento-complexes are generated along with thallium chloride. The use of silver salts in such reactions could lead to oxidation of the starting material.

The use of silver salts (AgY) as reagents for halide abstraction can lead to coordination of the weakly-coordinating anion, Y. Despite the fact that the perchlorate ion is a poor ligand for soft metal ions, the interaction between $[PdCl_2(dppe)]$ and two equivalents of silver perchlorate in methanol or tetrahydrofuran[36] does not result in the isolation of the corresponding solvento-complexes (although these are likely to be present in solution) but, rather, isolation of $[Pd(OClO_3)_2(dppe)]$. The synthesis of complexes containing anions of very strong acids is discussed further in Chapter 5. Here the emphasis is on avoiding the coordination of the anion, Y, and so the reaction must be conducted in a solvent that is able to compete with Y for a vacant coordination site. Thus, the reaction of $[PdCl_2(dppe)]$ with two equivalents of $AgClO_4$ in DMSO leads to formation[37] of the oxygen-bonded sulfoxide complex $[Pd(DMSO)_2(dppe)][ClO_4]_2$ because DMSO is a better ligand for palladium(II) than the perchlorate anion. In related reactions, abstraction of the halide from $[CoX(NH_3)_5][ClO_4]_2$ (X = Cl, I) with silver perchlorate was performed in dimethyl sulfoxide[38] and in acetonitrile.[39] Perchlorate ions cannot compete with these solvents as ligands for cobalt(III) and so the corresponding solvento-complexes were generated, Eq. 3.9:

$$[CoX(NH_3)_5][ClO_4]_2 + AgClO_4 + solv \rightarrow [Co(NH_3)_5(solv)][ClO_4]_3 + AgX \quad (3.9)$$

Introduction of a solvent molecule into a complex by halide abstraction occurs initially by interaction between the coordinated halide and the halide acceptor (Ac). It is likely that the halide acceptor forms an intermediate complex with a M–halide–Ac bond where the halide functions as a bridging ligand. This leads to weakening of the M–halide bond[40,41] and, hence, to the replacement of the halide by the incoming solvent molecule. This general scheme for halide abstraction/replacement has been confirmed by several detailed studies of the interaction of halide acceptors with metal halide complexes. For example, the reaction of [RuH(Cl)(PP^3)] (where PP^3 = $P(CH_2CH_2PPh_2)_3$) in CH_2Cl_2 with $TlPF_6$ in water generated a complex[42] containing a Ru–Cl–Tl bridge. The structure of the complex has been determined by X-ray crystallography, Figure 3.1:

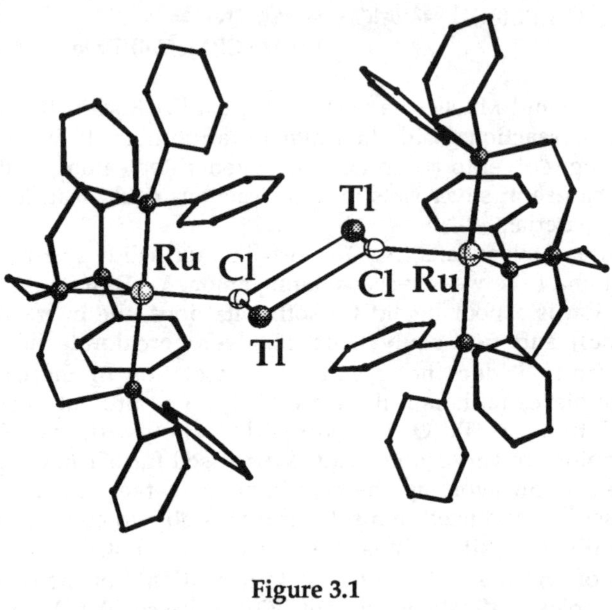

Figure 3.1

In the case of halide abstraction by silver salts, the reaction is anion dependent.[43] Thus, $AgClO_4$ reacts with $[IrCl(CO)(PPh_3)_2]$ rapidly to produce AgCl whereas $Ag(B_{11}CH_{12})$ forms a 1:1 adduct with Ir–Ag bonding. In contrast, $Ag(B_{11}CH_{12})$ reacts with $[FeI(CO)_2Cp]$ to produce a 1:1 adduct that appears to involve Ag–I–Fe bonding. This adduct ultimately deposits AgI and liberates $[Fe(B_{11}CH_{12})(CO)_2Cp]$, which has been characterized by X-ray crystallography.

3.2 Solvento-complexes from redox transformations of coordination compounds

3.2.1 Redox transformations of N-donor ligands

The most common examples of redox reactions of N-donor ligands are transformations of coordinated nitrosyl (NO) groups. The electrophilic character of the linear NO ligand is manifested in its reactivity. Electron donors react *via* nucleophilic attack on the nitrogen atom and/or by electron transfer (reduction) reactions.[44,45] Only reactions in which a reagent attacking the nitrosyl-containing complex acts as both nucleophile and reductant will be discussed here. The interaction of NO-containing substrates with such reagents often leads to the transformation of the NO ligand into a leaving group that is subsequently expelled from the complex. The vacant coordination site is filled by a solvent molecule.

In 1895, Hofmann[46] studied the reduction of coordinated nitrosyl groups. The interaction of sodium nitroprusside with hydrazine in an aqueous solution of NaOH formed the corresponding aqua-complex with the explosive liberation of dinitrogen. The overall process is described by Eq. 3.10:

$$4Na_2[Fe(CN)_5NO] + 4NaOH + 3N_2H_4 \rightarrow 4Na_3[Fe(CN)_5(H_2O)] + 5N_2 + 4H_2O \quad (3.10)$$

Additionally, Hofmann's work showed that the reaction of sodium nitroprusside with hydroxylamine leads to formation of $Na_3[Fe(CN)_5(H_2O)]$ and to generation of nitrous oxide. The kinetics of this reaction have been discussed by several workers.[47,48] Based on the experimental data reported, the process appears to begin with nucleophilic attack on the coordinated nitrosyl ligand by the hydroxylamine N-atom. Then, the decomposition of the resultant intermediate leads to formation of nitrous oxide and the aqua-complex.

Interaction between nitrosyl ligands and hydroxylamine is not limited to reactions with sodium nitroprusside. For example, reaction[49] with $K[IrX_5(NO)]$ (X = Cl, Br) in aqueous solution results in formation of $[IrX_5(H_2O)]^{2-}$ aqua-complexes and liberation of N_2O.

Azide ion also enters into redox interactions with linear nitrosyls. The reaction between $[RuCl(NO)(bipy)_2]^{2+}$ and KN_3 was first discussed by Miller and Meyer.[50] The reaction, Eq. 3.11, proceeds fairly rapidly in water, acetonitrile, or acetone. The corresponding solvento-complex is formed and the gaseous reaction products, dinitrogen and nitrous oxide, are liberated. Because the coordinated acetone molecule is extremely labile, it is displaced easily by water, pyridine, or chloride ion in $[RuCl(bipy)_2(Me_2CO)]^+$.

$$[RuCl(NO)(bipy)_2]^{2+} + N_3^- + solv \rightarrow [RuCl(bipy)_2(solv)]^+ + N_2 + N_2O \quad (3.11)$$

More recently it has been established[51] that the affinity of acetone for ruthenium(II) in $[RuCl(bipy)_2(Me_2CO)]^+$ is so weak that, over time, the compound, in an acetone-methanol solvent mixture, spontaneously eliminates the solvento-ligand and dimerization of the coordinatively unsaturated intermediate occurs, Eq. 3.12:

$$2[RuCl(bipy)_2(Me_2CO)]^+ \rightarrow [Ru(\mu\text{-}Cl)(bipy)_2]_2^{2+} + 2Me_2CO \quad (3.12)$$

The reactions of coordinated linear nitrosyls with azide ions provide a method for the synthesis of a variety of solvento-complexes.[52] For example, the ruthenium(II) compounds, $[RuXY(bipy)_2]$, where X and Y are various anionic ligands, can be synthesized through redox and acid-base transformations of the starting material, $[Ru(NO_2)(bipy)_2(NO)]^{2+}$, Scheme 3.11:

$$[L_2Ru(NO)(NO_2)]^{2+} \xrightarrow[-N_2O,\ -N_2]{N_3^-,\ solv} [L_2Ru(NO_2)(solv)]^+ \xrightarrow[-\ solv]{X^-}$$

$$[L_2Ru(X)(NO_2)] \xrightarrow[-2H_2O]{2H^+} [L_2Ru(X)(NO)]^{2+} \xrightarrow[-N_2O,\ -N_2]{N_3^-,\ solv}$$

$$[L_2Ru(X)(solv)]^+ \xrightarrow[-\ solv]{Y^-} [L_2RuXY]$$

Scheme 3.11

$[RuCl(diars)_2(^{15}NO)]^{2+}$ (diars = *o*-phenylenebis(dimethylarsine)), which reacts with excess sodium azide in water,[53] has been studied by Douglas and

Feltham. The product, [RuCl(N_3)(diars)$_2$], was shown to form under the experimental conditions employed and the liberated nitrous oxide consisted of equal amounts of $^{14}N^{14}NO$ and $^{14}N^{15}NO$. These results led Douglas and Feltham to conclude that the reaction proceeds *via* the formation of a cyclic intermediate which then decomposes into N_2 and N_2O, Figure 3.2:

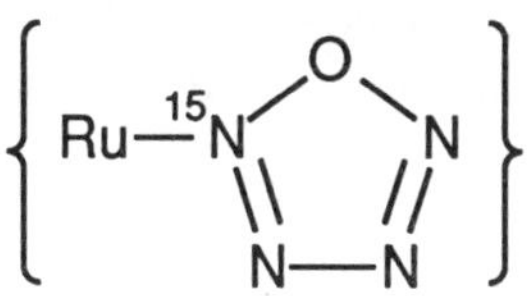

Figure 3.2

A similar conclusion regarding the formation of a cyclic intermediate was reached by Bottomley and Mukaida[54] who studied the interaction between sodium azide and [RuCl(NO)(py)$_4$][ClO_4]$_2$.

The reaction of nitrosyl complexes with N_3^- ions is not limited to coordination complexes of particular metal ions. Instead, the reaction seems to be specific for compounds containing nitrosyl ligands exhibiting electrophilic reactivity, *i.e.* the linear nitrosyls. As discussed above, sodium nitroprusside[37] and iridium nitrosyls[48] react with azide ion in the same manner as ruthenium(II) nitrosyls.

Another avenue for the preparation of solvento-complexes is the redox transformation of azide ligands in complexes by treatment with nitrosylating agents. At least formally this class of reactions is similar to the interaction between nitrosyl-containing complexes and azide ion. The reaction of $[CoN_3(NH_3)_5]^{2+}$, an azide-containing complex, and NO^+ was reported for the first time by Haim and Taube[55] in 1963, almost ten years before the work of Miller and Meyer[50] mentioned above. Although both reactions are often discussed together, the treatment of an azide-containing complex with NO^+ is conceptually different from the treatment of an NO-containing complex with azide ion.

Haim and Taube reacted $[CoN_3(NH_3)_5]^{2+}$ with nitrosyl chloride in water[55] which led to the formation of $[Co(NH_3)_5(H_2O)]^{3+}$, dinitrogen, and nitrous oxide. $[Co(N_3)Cl(en)_2]^+$ underwent a similar reaction in water[56,57] with formation of $[CoCl(en)_2(H_2O)]^{2+}$. Following this early work, several methods were developed that similarly employed azide ligand elimination as the result of redox transformations in non-aqueous solvents. The complex $[CoN_3(NH_3)_5]^{2+}$ and nitrosonium salts were reacted in trimethylphosphate,[58] tetramethylene sulfone,[59] or organonitriles[60,61] with the formation of the corresponding $[Co(NH_3)_5(solv)]^{3+}$ solvento-complexes.

Solvento-complexes also may be generated through redox transformations of other types of N-donor ligands.[62,63] For example, the reduction at an electrode of the arenediazenate ligand in the ruthenium(II) complex, $[RuCl(bipy)_2(N_2C_6H_4Me)]^{2+}$, in MeCN solution leads to formation of $[RuCl(bipy)_2(MeCN)]^+$ in addition to toluene and dinitrogen.

Kukushkin and Stefanova[64] demonstrated the elimination of nitro groups in their study of the reactions of $K_2[Pd(NO_2)_4]$ with sulfamic acid in aqueous solution. Both the mono- and bis(aqua)complexes were formed, Scheme 3.12:

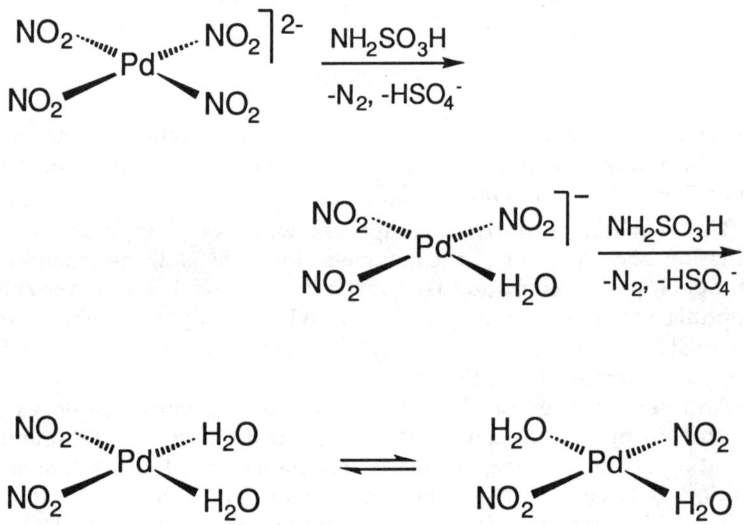

Scheme 3.12

Wood and Balch[65] have duplicated the reactions shown in Scheme 3.12. The palladium product, *cis*-$[Pd(NO_2)_2(H_2O)_2]$, has been identified by NMR spectroscopy and was found to be subject to *cis-trans* isomerization, which is consistent with general observations on the geometric isomerization of related platinum(II) and palladium(II) complexes.[66,67]

Titanium(III) sulfate has been employed successfully to generate a solvento-complex of platinum(II) by reduction of the nitro ligand[68] in $[Ph_3PCH_2Ph]_2[PtCl_3(NO_2)]$. The reaction, performed in a mixture of MeCN and H_2O, produced the acetonitrile complex, $[Ph_3PCH_2Ph][PtCl_3(MeCN)]$, in addition to nitric oxide and $[TiO(SO_4)]$. The platinum(II) complex $[Ph_3PCH_2Ph][PtCl_3(MeCN)]$ also has been prepared from $[Ph_3PCH_2Ph]_2[PtCl_4]$ and $BF_3{\cdot}OEt_2$ in acetonitrile.[68]

3.2.2 *Redox transformations with other classes of ligands*

The formation of solvento-complexes as a result of redox transformations of coordinated ligands is not restricted to N-donor ligands; reactions with other ligands are well known. For example, oxidation of the dimethyl sulfoxide ligand in $[Co(NH_3)_5(DMSO)]^{3+}$ can be effected by treatment with $KMnO_4$[69,70] or Cl_2 in aqueous solution.[71] The sulfoxide oxidizes to the sulfone which exhibits a much lower donor ability than Me_2SO and, as a result of hydrolysis, is expelled from the complex generating the aqua-complex.

Sargeson and Taube[72] have provided an interesting example of the formation of a solvento-complex. Nitrosylation of $[Co(OC(O)NH_2)(NH_3)_5]^{2+}$ is known to lead to a redox-induced elimination and coordination of a water molecule at the vacant site, Eq. 3.13:

$$[Co(OC(O)NH_2)(NH_3)_5]^{2+} + NO^+ \rightarrow [Co(H_2O)(NH_3)_5]^{3+} + N_2 + CO_2 \quad (3.13)$$

The synthesis of zirconium solvento-complexes through oxidation of one of the alkyl groups in $[ZrCp_2R_2]$ with silver tetraphenylborate has been demonstrated by Jordan and coworkers[73-75] (R = Me, CH_2Ph and solv = MeCN, THF), Eq. 3.14:

$$[ZrCp_2R_2] + AgBPh_4 + solv \rightarrow [Zr(solv)Cp_2R][BPh_4] + Ag + 1/2R\text{–}R \quad (3.14)$$

NMR spectroscopy was used to monitor ethane liberated in the synthesis of $[Zr(MeCN)Cp_2Me][BPh_4]$ by this method.[74] Oxidation of the methyl ligand apparently proceeds *via* intermediate formation of $AgCH_3$. The X-ray crystal structure of the acetonitrile solvento-complex, $[Zr(NCMe)Cp_2(CH_2Ph)]^+$, has been determined, Figure 3.3.

The same approach has been used to generate acetonitrile complexes of titanium utilizing $AgBPh_4$ as the oxidant and $[Ti(CH_2Ph)_2Cp_2]$ as the starting material.[76]

Hydrogenation[77] of coordinated dienes in rhodium(I) diene complexes has been investigated. For example, when the mixed-ligand rhodium(I) complex, $[Rh(Ph_2PC_3H_6PPh_2)(diene)]^+$, is reacted with hydrogen, the diene ligand is lost as the corresponding alkane and the vacant sites are occupied by solvent molecules, *e.g.* acetone or methanol, Eq. 3.15.

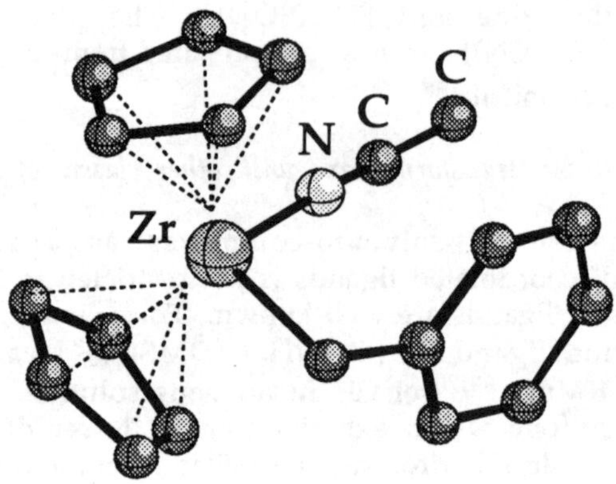

Figure 3.3

$$[Rh(Ph_2PC_3H_6PPh_2)(diene)]^+ + 2H_2 + 2solv \rightarrow [Rh(Ph_2PC_3H_6PPh_2)(solv)_2]^+ + alkane \quad (3.15)$$

Similarly, a cyclic diene is eliminated as a cycloalkane by hydrogenation[78,79] of $[Ir(PPh_3)_2(cod)][BF_4]$ where oxidative addition of H_2 leads to the generation of $[Ir(H)_2(PPh_3)_2(solv)_2][BF_4]$ (solv = tetrahydrofuran, acetonitrile, acetone, or alcohols).

3.2.3 Oxidation or reduction of central atoms

Ligands that are good π-acceptors have an affinity for metals in low oxidation states. Such ligands bind less strongly when the oxidation state of the metal increases and, then, solvolysis may be possible. For example, the reaction of $[Mo(CO)_6]$ with nitrosonium salts, NOX (X = BF_4, PF_6), in acetonitrile or nitromethane results in oxidation of the central atom, the loss of the carbonyl ligands,[80,81] and formation of mixed-ligand, $[Mo(NO)_2(solv)_4][X]_2$, solvento-complexes. In another example, the platinum(II) complex, $[Pt(MeCN)_2(PPh_3)_2][BF_4]_2$, was prepared *via* oxidation of the platinum(0) complex $[Pt(PPh_3)_2(C_2H_4)]$ with nitrosonium tetrafluoroborate in acetonitrile.[82] Oxidation of platinum(0) to platinum(II) results in poorer binding of ethylene and, hence, its elimination and replacement by solvent molecules. $[Rh(MeCN)_2(cod)][X]$ and $[Rh(cod)_2][X]$ similarly react with NOX (X= BF_4, PF_6) in MeCN to yield the rhodium(III)[83] complexes, $[Rh(NO)(MeCN)_4][X]_2$.

The nitrosyl cation can be used to oxidize metallic palladium[84] in the presence of acetonitrile and generate $[Pd(MeCN)_4]^{2+}$, a reaction described in detail in Chapter 4, Eq. 3.16:

$$Pd + 2NOBF_4 + 4MeCN \rightarrow [Pd(MeCN)_4][BF_4]_2 + 2NO \quad (3.16)$$

Similar oxidations of other metals have been reported (see Chapter 4) including, for example, the reaction of metallic copper[85] with nitrosyl perchlorate to generate $[Cu(MeCN)_4]^{2+}$ and $[Cu(MeNO_2)_4]^{2+}$. One disadvantage of the NO^+ oxidizing system is illustrated[86] by the reaction of $NOBF_4$ with $[PtO_2(PPh_3)_2]$ in poorly coordinating solvents (*e.g.* benzene, dichloromethane, nitromethane). The extreme moisture-sensitivity of NO^+ and the presence of traces of moisture lead to reactions in which H^+ is produced. Protonation of the dioxygen ligand is followed by hydrolysis, extrusion of H_2O_2, and dimerization, Scheme 3.13:

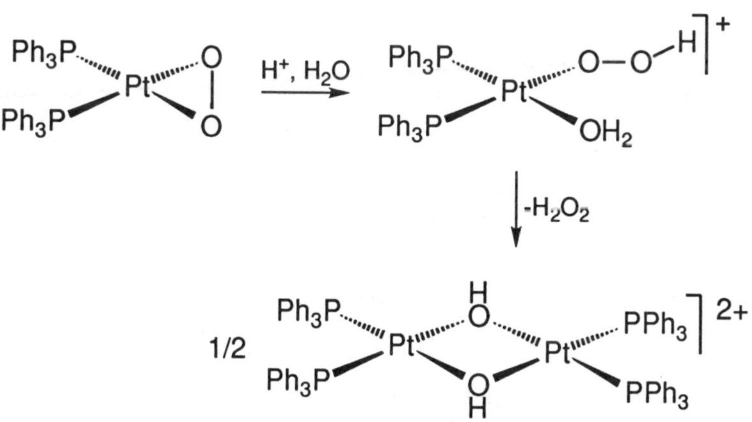

Scheme 3.13

In some cases, a metal ion exhibits different coordination numbers in different oxidation states. It may be possible, then, to take advantage of the change in coordination number to form solvento-complexes. For example, platinum(II) complexes typically exhibit a coordination number of four, whereas, the coordination number is usually six for platinum(IV) complexes. Electrochemical oxidation of $[PtCl_2(PR_3)_2]$ (PR_3 = various phosphines) in acetonitrile was once thought[87] to lead to the formation of

$[PtCl_2(MeCN)_2(PR_3)_2]^{2+}$ but it is now known[88] that halide transfer from the platinum(II) precursor to the electrochemically generated platinum(IV) center prevents this methodology from being usefully exploited. The electrochemical oxidation of *cis*-$[PtPh_2(PEt_3)_2]$ in MeCN does produce *cis, cis, trans*-$[PtPh_2(MeCN)_2(PEt_3)_2]^{2+}$ (an X-ray crystal structure determination of the methyl analog has confirmed the geometry) but the reaction is not general.[89] Thus, electrochemical oxidation of *trans*-$[Pt(CH_2Ph)Cl(PEt_3)_2]$ and *cis*-$[Pt(CH_2Ph)_2(PEt_3)_2]$ results in cleavage of the Pt–C bonds and not formation of platinum(IV) solvento-complexes.[89] Similarly, electrochemical oxidation of *trans*-$[PtHCl(PEt_3)_2]$ does not lead to the generation of a platinum(IV) solvento-complex. The two-electron oxidation entails the formal conversion of H^- into H^+ and not platinum(II) into platinum(IV).[90]

In certain cases solvento-complexes may be prepared by reduction of the central metal atom. Thus, [VO(acacen)], a pentacoordinate compound, is transformed into the hexacoordinate solvento-complex [VCl(THF)(acacen)] by reduction of the vanadium center with $[TiCl_3(THF)_3]$ in tetrahydrofuran.[91,92]

3.3 Formation of solvento-complexes through reactions of coordinated ligands (non-redox)

It has been shown[93,94] that an acetate ligand may be abstracted from a metal center on reaction with a powerful methylating or acylating reagent in a process that involves no redox transformations. The reactions between $[Co(OC(O)Me)(NH_3)_5]^{2+}$ and methyltrifluoromethanesulfonate/TMP or acetyl perchlorate/DMF are examples of this synthetic method. These reactions are described by Eqs. 3.17 and 3.18:

$$[Co(OC(O)Me)(NH_3)_5]^{2+} + MeOSO_2CF_3 + OP(OMe)_3 \rightarrow [Co(NH_3)_5(OP(OMe)_3)]^{3+} + MeOC(O)Me + CF_3SO_3^- \quad (3.17)$$

$$[Co(OC(O)Me)(NH_3)_5]^{2+} + MeCOClO_4 + DMF \rightarrow [Co(DMF)(NH_3)_5]^{3+} + (MeCO)_2O + ClO_4^- \quad (3.18)$$

Electrophilic attack on the carbonyl oxygen of the acetate ligand appears to be the route by which these reactions proceed. The attack is carried out by acetyl cations, methyl cations, or cation equivalents. As a result the acetate group is transformed into a ligand easily replaced by the solvent molecule.

Methylation of dimethyl sulfoxide[68] in *cis*-$[PtCl_2(DMSO)_2]$ with one equivalent of methyl-3-nitrobenzenesulfonate in acetonitrile leads to the formation of an acetonitrile-containing complex, Eq. 3.19:

$$cis\text{-}[PtCl_2(DMSO)_2] + MeOSO_2C_6H_4NO_2 + MeCN \rightarrow cis\text{-}[PtCl_2(DMSO)(MeCN)] + [Me_2SOMe][SO_3C_6H_4NO_2] \quad (3.19)$$

Jackson, Jurisson, and O'Leary have reported preparation of the $[Co(MeSO_2)(NH_3)_5]^{2+}$ cation[95] containing *O*-coordinated methanesulfinate, $MeSO_2^-$. This complex is formed by reaction of $[Co(NH_3)_5(H_2O)]^{3+}$ with MeSOCl in the presence of 2,6-lutidine in trimethylphosphate solution. The $[Co(MeSO_2)(NH_3)_5]^{2+}$ is alkylated by MeI in dimethyl sulfoxide to generate the solvento-complex, $[Co(NH_3)_5(DMSO)]^{3+}$, and free dimethyl sulfone. No intermediates were detected for this reaction.

An additional pathway for the formation of solvento-complexes is the decomposition by protic acids of coordinated carbonate or oxalate ligands. There are several examples of such reactions reported in the literature. Thus, the ruthenium(II) bis(aqua)-complex has been prepared[96] by the reaction shown in Eq. 3.20:

$$[RuCO_3(bipy)_2] + 2H^+ + H_2O \rightarrow [Ru(bipy)_2(H_2O)_2]^{2+} + CO_2 \quad (3.20)$$

The complex $[Co(CO_3)(H_2O)_2(py)_2][ClO_4]$, containing coordinated water molecules, has been prepared[97] *via* decomposition of one carbonate ligand in $K[Co(CO_3)_2(py)_2]$ by 1.0 *M* $HClO_4$. The complex $[Co(CO_3)(NH_3)_4][ClO_4]$ rapidly reacts with trifluoromethanesulfonic acid, HSO_3CF_3, forming $[Co(OSO_2CF_3)_2(NH_3)_4][SO_3CF_3]$. The bis(aqua)-product is then formed when the anionic ligands are replaced by water molecules.[98] Acidolysis of $[Ru(C_2O_4)(NH_3)_4]^+$, containing a bidentate oxalate ligand, occurs in aqueous trifluoromethanesulfonic acid. The reaction[99] produces $[Ru(H_2O)_2(NH_3)_4]^{3+}$. The kinetics of decarboxylation and the mechanism of the reaction of $[Co(CO_3)(H_2O)(py)_3][ClO_4]\cdot H_2O$ with aqueous $HClO_4$ have been studied.[100] The cobalt(III) complex, $[Co(H_2O)_3(py)_3][ClO_4]_3\cdot 2H_2O$, is the reaction product and the mechanism incorporates both the protonation of the C=O group on the bidentate carbonate ligand and subsequent cleavage of the chelating ligand.[100]

Abstraction of hydride ligands is an additional method for the synthesis of solvento-complexes of good donor solvents. This process is discussed in detail in Chapter 8 and only three illustrative examples are presented here. In most cases the abstraction of H^- ligands is performed using Lewis acids; *e.g.* Alk^+, Ph_3C^+, H^+, Ag^+, *etc.* For example, when the hydride-containing complex, $[RuH(PMe_3)(\eta^6\text{-}C_6H_6)]^+$, is reacted with triethyloxonium tetrafluoroborate in acetone, the solvento-complex $[Ru(Me_2CO)(PMe_3)(\eta^6\text{-}C_6H_6)]^{2+}$ is generated and ethane is liberated.[82] The vanadium complex, $[V(CO)_3(THF)Cp]$, is synthesized by hydride abstraction from the anion $[VH(CO)_3Cp]^-$ with the Ph_3C^+ cation in THF.[101] Finally, two Mo–H bonds in $[Mo(H)_4(PMe_2Ph)_4]$ are cleaved[102] by silver tetrafluoroborate (or $HBF_4{\cdot}OEt_2$) in acetonitrile, Eq. 3.21:

$$[Mo(H)_4(PMe_2Ph)_4] + 2AgBF_4 + 2MeCN \rightarrow [Mo(H)_2(MeCN)_2(PMe_2Ph)_4][BF_4]_2 + 2Ag(s) + H_2 \quad (3.21)$$

Cleavage is followed by extrusion of molecular hydrogen and formation of a solvento-complex.

3.4 Dehydration reactions of aqua-complexes and formation of solvento-complexes

So and Boudjouk[103] have reported that trimethylchlorosilane, Me_3SiCl, is an efficient dehydrating agent. It interacts with hydrated metal chlorides, $MX_n{\cdot}xH_2O$, to produce MX_n, $Me_3SiOSiMe_3$ and HCl. When tetrahydrofuran was used as the solvent, solvento-complexes were obtained. Thus, for example, $CrCl_3{\cdot}6H_2O$ in THF solution rapidly reacted at room temperature with Me_3SiCl to give $[CrCl_3(THF)_3]$ which, after work-up, was isolated in 89% yield. It was pointed out that the formation of both HCl and the very strong silicon-oxygen bond in hexamethyldisiloxane drives the reaction.

Dehydration of aqua-complexes is commonly performed with zeolites or molecular sieves. $[Co(NH_3)_5(H_2O)][ClO_4]_3$, for example, dehydrates in the presence of zeolites and the vacant coordination site is occupied by a solvent molecule (*e.g.* RCN).[104] When the same cobalt(III) aqua-complex is dehydrated[105] in dimethylacetamide, $MeCONMe_2$, the solvento-complex, $[Co(NH_3)_5(MeCONMe_2)][ClO_4]_3$, is formed. If this product is further reacted in *p*-toluenesulfonamide,[105] then $[Co(NHSO_2C_6H_4Me)(NH_3)_5][ClO_4]_2$ is formed by substitution. Dehydration of $[Rh(NH_3)_5(H_2O)][ClO_4]_3$ in a DMA/acetamide mixture generates $[Rh(NHCOMe)(NH_3)_5][ClO_4]_2$ in the presence of zeolites.[106] Molecular sieves are especially useful in the preparation of homoleptic solvento-complexes. For example, removal of H_2O from the complexes $[M(H_2O)_6][ClO_4]_2$ (M = Mn, Cu)[107,108] in acetonitrile leads to the formation of the hexa(acetonitrile) derivatives, $[M(MeCN)_6][ClO_4]_2$. These compounds cannot be synthesized by replacement of coordinated water in the absence of a dehydrating agent even when a large excess of MeCN is present.

A common method for the preparation of homoleptic solvento-complexes is dehydration with 2,2-dialkoxypropanes or trialkyl-orthoformates. A full discussion of this synthetic method is provided in Chapter 4 but one example is presented here. The uranium solvento-complexes, $[M][UO_2(MeCO_2H)_4]$, (M = Mg, Ba, Ca, and Zn), have been isolated in good yields[109] from the reactions of $[M(O_2CMe)_2]\cdot nH_2O$ and $[UO_2(O_2CMe)_2]\cdot 2H_2O$ with acetic anhydride in acetic acid. The trapping of water by acetic anhydride leads to the generation of vacant coordination sites which, subsequently, are occupied by solvent molecules.

3.5 Decarbonylation reactions and removal of other supporting ligands (olefins, acetylenes, arenes, dinitrogen)

Chemical (Chapter 8), electrochemical (Chapter 7), thermal, and photochemical elimination of CO ligands from carbonyl-containing complexes are common methods for the generation of solvento-complexes and such reactions are discussed extensively in the literature. In 1968 a comprehensive account of the topic was published.[110] Since then many examples have been reported along with studies of synthetic pathways and mechanisms. This section is confined to a description of several examples of decarbonylation based on thermal and photochemical means. More detailed discussions of these subjects are found elsewhere.[111-114]

Chromium, molybdenum or tungsten complexes of the type $[M(CO)_3(MeCN)_3]$ are generated by thermolysis of the corresponding $[M(CO)_6]$ compounds[115] in acetonitrile, Eq. 3.22:

$$[M(CO)_6] + 3MeCN \rightarrow [M(CO)_3(MeCN)_3] + 3CO \qquad (3.22)$$

In $[M(CO)_3(MeCN)_3]$ the acetonitrile ligands are weak donors and are easily replaced by more powerful ligands.[116] Such reactions have been widely employed and illustrate the synthetic importance of carbonyl-containing solvento-complexes as synthetic intermediates. It has been established[112] that the rate of thermolysis of $[M(CO)_6]$ increases with an increase in the boiling point of the RCN solvent in the order Me < Et < Pr.

Elimination of CO by photolysis is frequently employed in the preparation of solvento-complexes. For example, UV-irradiation of $[Re(CO)_3Cp]$ in ether[117] or THF[118] produces $[Re(CO)_2(solv)Cp]$. Interaction between $[Re(CO)_2(solv)Cp]$ and aryldiazonium tetrafluoroborate leads to the replacement of the solvento-ligand and formation of the rhenium(I) complex, $[Re(CO)_2(N_2Ar)Cp][BF_4]$.[118] Irradiation of $[Mn(CO)_3Cp]$ in THF is used to prepare $[Mn(CO)_2(THF)Cp]$ which can be reacted with KCN[119] to form $K[Mn(CN)(CO)_2Cp]$.

While thermolysis and photolysis of carbonyl-containing compounds result in the elimination of gaseous CO, with the vacant coordination site then occupied by a solvent molecule, more generally such methodology is employed for the elimination of olefins,[120-126] acetylenes,[127] arenes,[128-130] and dinitrogen[131,132] from coordination complexes. In these cases, elimination of the ligands in the presence of donor solvents leads to the generation of solvento-complexes.

3.6 Application of solvento-complexes in the synthesis of bridged hetero- and homopolynuclear complexes

Solvent molecules are typically weak donors and can be replaced readily by other ligands. In most cases, these substitution reactions proceed so easily that solvento-complexes can often be considered as coordinatively unsaturated intermediates. However, the importance in synthesis of solvento-complexes is dramatically illustrated by substitution reactions employed in the synthesis of bridged hetero- and homopolynuclear coordination compounds.

Reaction of [Mo(CO)$_3$(diglyme)] with [Co(cot)Cp] in THF[133] results in facile replacement of the diglyme and formation of [CpCo(μ-cot)Mo(CO)$_3$], which has been isolated in the solid state. X-ray crystal structure analysis of the complex shows that the cyclooctatetraene ligand bridges the [CoCp] and [Mo(CO)$_3$] fragments.

In another example,[134] the coordinated tetrahydrofuran ligand in [Cr(CO)$_5$(THF)] is easily displaced by [W(CO)$_2$(N$_2$Ph)Cp] in THF generating the heterobinuclear μ-N$_2$Ph complex, Eq. 3.23:

$$[Cr(CO)_5(THF)] + [W(CO)_2(N_2Ph)Cp] \rightarrow [(CO)_5Cr(\mu\text{-}N_2Ph)W(CO)_2Cp] + THF \quad (3.23)$$

The interaction between [Au(PPh$_3$)(THF)]$^+$ and the polyhydride [Ir(H)$_3$(PPh$_3$)$_3$] results in the generation of a bridged Au–H–Ir compound,[135] Figure 3.4:

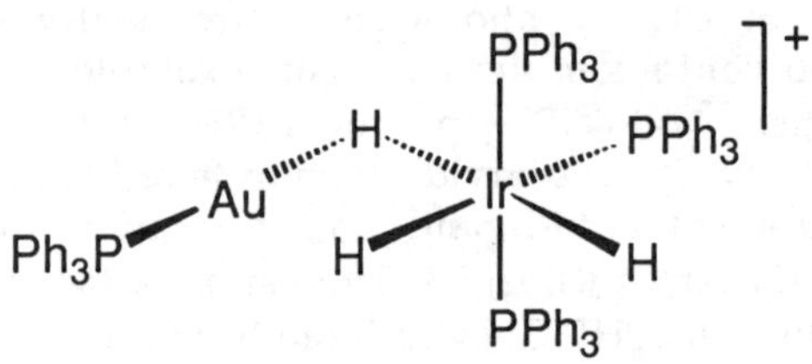

Figure 3.4

Similarly, an unusual octahedral copper(I) complex, involving coordination to six rhenium-bound hydride ligands,[136] is generated in the following reaction, Eq. 3.24:

$$[Cu(MeCN)_4][PF_6] + 2[Re(H)_5(PMePh_2)_3] \rightarrow [\{Re(H)_5(PMePh_2)_3\}_2Cu][PF_6] + 4MeCN \quad (3.24)$$

Here, both $[Re(H)_5(PMePh_2)_3]$ fragments function as tridentate ligands with respect to the copper(I).

Synthesis of binuclear complexes with metal-metal bonds[137] is also possible with a solvento-complex as intermediate, Eq. 3.25:

$$[Co(CS)(PMe_3)Cp] + [Mn(CO)_2(THF)Cp] \rightarrow [Cp(PMe_3)Co(\mu\text{-}CS)(\mu\text{-}CO)Mn(CO)Cp] + THF \quad (3.25)$$

Homobinuclear complexes also can be formed if weakly-bound ligands are substituted by a bifunctional ligand. Thus, the interaction between $[Mn(CO)_2(THF)Cp^*]$ and 4,4'-bipyridyl in a 2:1 molar ratio leads to the formation of the bridged compound,[138] Figure 3.5:

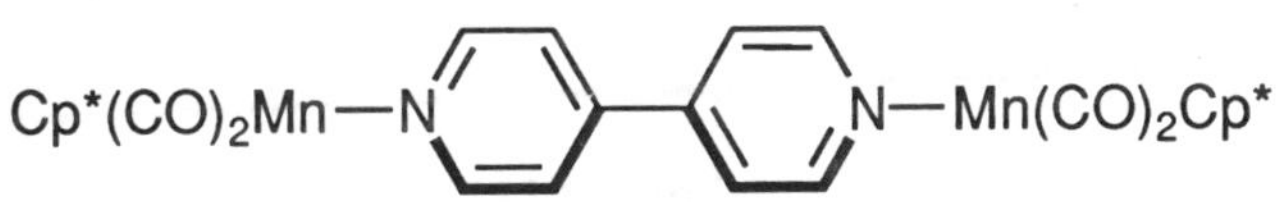

Figure 3.5

Similarly, the dimers, $[\{M(CO)_5\}_2(\mu\text{-}H_2NC_2H_4SEt)]$ (M = Cr, W), have been synthesized by reaction of $[M(CO)_5(THF)]$ and the bifunctional ligand $H_2NC_2H_4SEt$ in THF solution.[139]

3.7 Complexes containing coordinated halocarbons

It is well known that certain metal halide complexes react with silver salts of classically noncoordinating anions in the presence of methylene chloride or chloroform to deposit silver halide and generate a soluble complex. Seldom has the exact nature of the soluble complex been defined; coordination of the classically noncoordinating anion was possible, as was formation of aqua (or hydroxy) complexes, by adventitious water. Coordination of the halocarbon solvent as a ligand was considered a real possibility when, in 1972, Cook, Dahl and Dickerhoof[140] reported the reaction of $[PtCl_4]_4$ with Ph_3CCl in CH_2Cl_2. X-ray crystallographic results supported assignment of the product as $[CPh_3][PtCl_5(CH_2Cl_2)]$ but definitive conclusions as to the structure were frustrated by disorder problems. It has been observed[141] that if the monodentate coordination of CH_2Cl_2 to platinum(IV) in $[PtCl_5(CH_2Cl_2)]^-$ were confirmed, the complex would be the first isolated, characterized halocarbon complex.

In 1978 Beck, Schloter, Sünkel, and Urban reported the synthesis of compounds formulated on the basis of elemental analysis and IR spectroscopy as $[M(CO)_3(CH_2Cl_2)Cp][PF_6]$ (where M = Mo, W).[142,143] Bands assigned to ν(CCl) were consistent with dichloromethane acting as a monodentate σ-donor ligand in both complexes. It has also been shown that above 15 °C the molybdenum complex, $[Mo(CO)_3(CH_2Cl_2)Cp][PF_6]$, dissociates dichloromethane to yield a compound containing a coordinated FPF_5^- anion, Scheme 3.14:

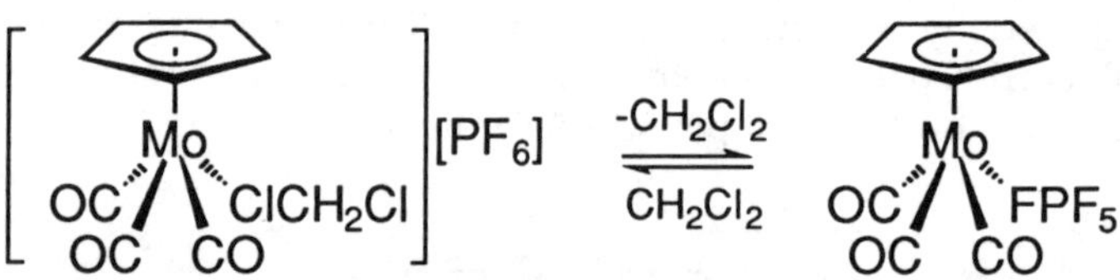

Scheme 3.14

Further indirect evidence for halocarbon coordination has been presented in several papers[144-146] but the first unambiguous report of an isolable halocarbon complex was not published until 1982, when the synthesis and structure of $[IrH_2(PPh_3)_2(o\text{-}C_6H_4I_2)][BF_4]$ were described.[147] The acetone ligands in the solvento-complex *cis,cis,trans*-$[IrH_2(Me_2CO)_2(PPh_3)_2][BF_4]$ were substituted by the halocarbons *o*-diiodobenzene or *o*-dibromobenzene to give *cis,trans*-$[IrH_2(PPh_3)_2(o\text{-}C_6H_4X_2)][BF_4]$.

The structure of the $C_6H_4I_2$ complex was determined by X-ray crystallography which showed that $C_6H_4I_2$ was chelated to iridium by Ir–I bonds. The core of the $[IrH_2(PPh_3)_2(o\text{-}C_6H_4X_2)]^+$ cation, where carbon atoms of the phenyl groups, other than the ones directly bound to phosphorus, have been omitted for clarity, is shown in Figure 3.6. Since that time, the X-ray crystal structures of a number of iodocarbon complexes, including those of $[IrH_2(PPh_3)_2(MeI)_2][SbF_6]$,[148] $[Re(NO)(PPh_3)(Me_3SiCH_2I)Cp][BF_4]$,[149,150] and the ruthenium(II) complex, $[Ru(t\text{-}BuCN)(PPh_3)(MeI)Cp][PF_6]$,[151] have been reported.

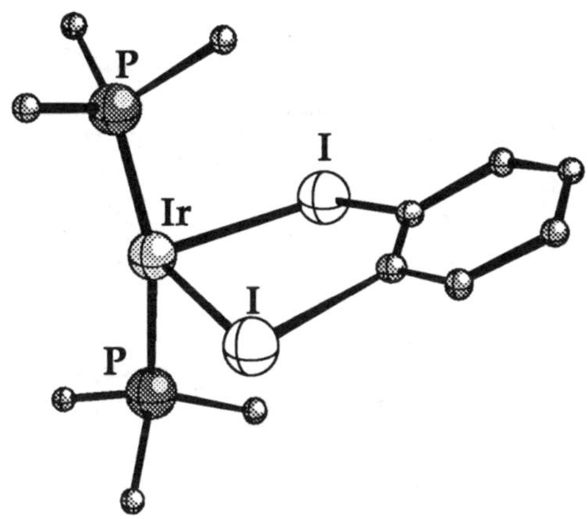

Figure 3.6

A range of coordinated chlorocarbon-metal complexes have been studied[152,153] and it has been shown that the reaction of the methyl complex, [Re(NO)(PPh_3)(Me)Cp], with $HBF_4{\cdot}OEt_2$ or $HPF_6{\cdot}OEt_2$ in CH_2Cl_2 at -78 °C produced [Re(NO)(PPh_3)($ClCH_2Cl$)Cp]$^+$, Scheme 3.15:

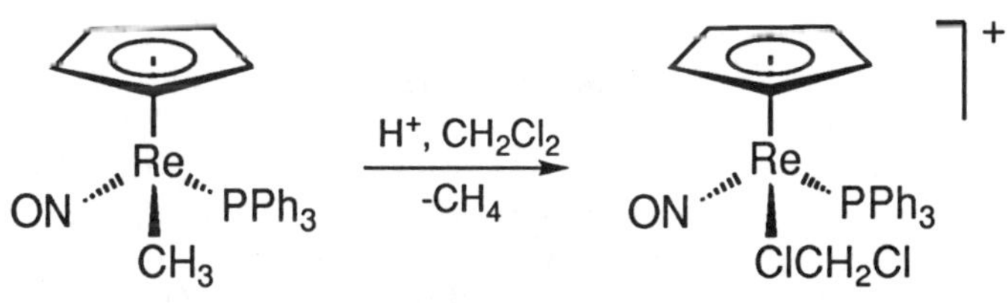

Scheme 3.15

Low temperature ^{13}C NMR data suggested that the reaction occurs by initial protonation of the metal, followed by reductive elimination of CH_4 to give [Re(NO)(PPh_3)($ClCH_2Cl$)Cp]$^+$. The coordinated CH_2Cl_2 ligand can be substituted by other weak donor molecules such as tetrahydrofuran, dimethyl ether, and diethyl ether[151] which have higher thermodynamic binding affinities than dichloromethane for the metal fragment [Re(NO)(PPh_3)Cp]$^+$, Scheme 3.16:

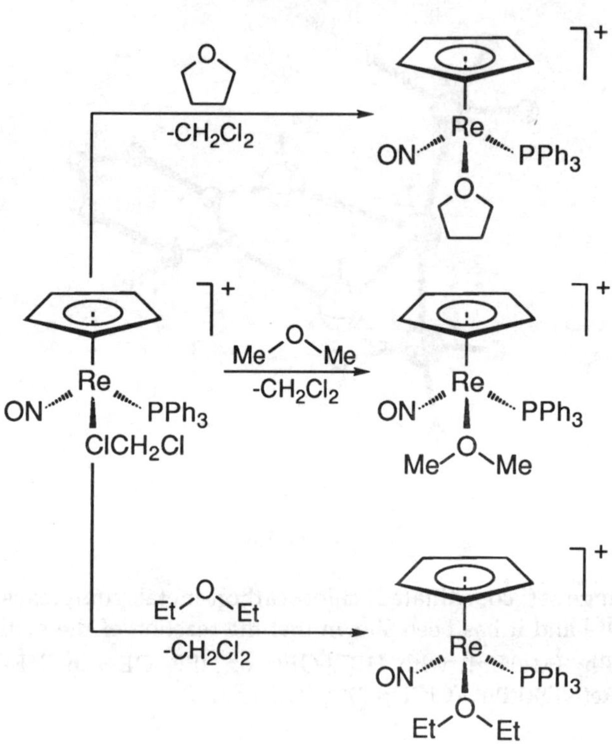

Scheme 3.16

The interaction of $AgOTeF_5$ with dichloromethane and also with 1,2-dichloroethane has been investigated.[141,154] After dissolution of this material in halocarbons, followed by removal of solvents, the very moisture-sensitive compounds [$AgOTeF_5(CH_2Cl_2)$] and [$AgOTeF_5(1,2\text{-}C_2H_4Cl_2)$] were isolated. The structure of [$AgOTeF_5(1,2\text{-}C_2H_4Cl_2)$] was determined by X-ray crystallography, Figure 3.7. The complex [$AgOTeF_5(1,2\text{-}C_2H_4Cl_2)$] is the first unambiguous example of a structurally characterized compound containing a chlorocarbon coordinated to a metal center. Other structures have been reported more recently, including those of $[Ag(CH_2Cl_2)(OTeF_5)]_2$, $[Ag_2(CH_2Cl_2)_4Pd(OTeF_5)]_4$, $[Ag_2(1,2\text{-}C_2H_4Cl_2)_4Pd(OTeF_5)_4]$, $[Ag(CH_2Cl_2)_3]_2$-$[Ti(OTeF_5)_6]$, $[Ag(CH_2Br_2)_3][Nb(OTeF_5)_6]$,[141] and $[(\eta^6\text{-}C_6Me_6)_2Ru_2H_2(CH_2Cl_2)$-$RuB_{10}H_8(OEt_2)_2]$. The latter complex contains dichloromethane as a bridging ligand. The binding of the CH_2Cl_2 ligand to the Ru_3 triangle in $[(\eta^6\text{-}C_6Me_6)_2$-$Ru_2H_2(CH_2Cl_2)RuB_{10}H_8(OEt_2)_2]$ is illustrated in Figure 3.8.[155]

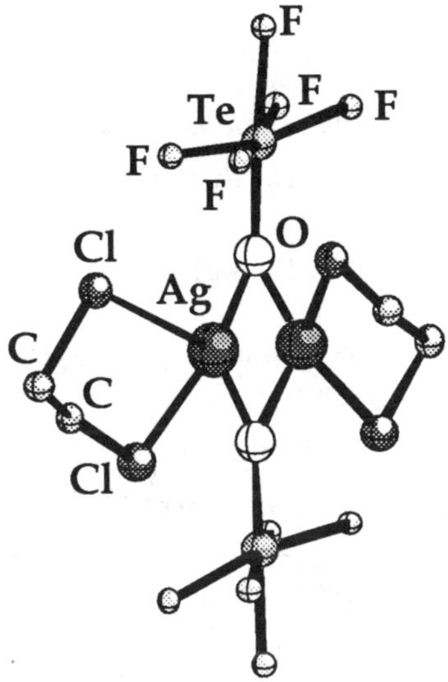

Figure 3.7

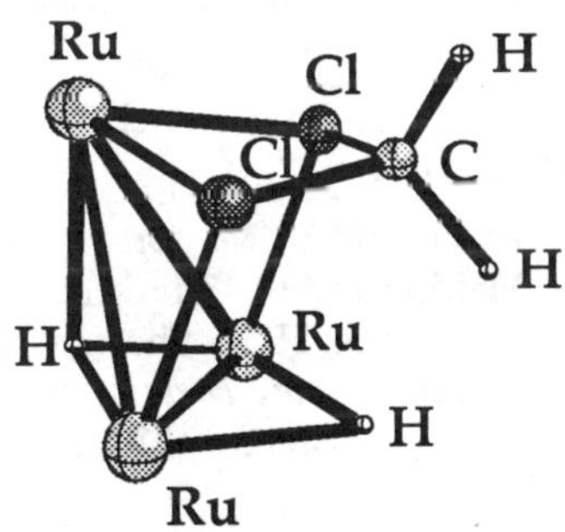

Figure 3.8

In conclusion, the generation of solvento-complexes as reactive intermediates is extensively used in synthesis. Three distinct approaches have been developed to provide solvent-binding sites: *(i)* conversion of ligands into products with low solubility or their transformation into volatile compounds, *(ii)* chemical transformations of coordinated ligands where the products of the reaction are poorer donors and are expelled from the coordination sphere of the complex, and *(iii)* manipulation of the metal oxidation state to affect ligand-binding properties.

References

1. (a) J. A. Davies and F. R. Hartley, *Chem. Rev.* **81** (1981) 79; (b) F. R. Hartley and J. A. Davies, *Rev. Inorg. Chem.* **4** (1982) 27; (c) J. A. Davies, *Adv. Inorg. Chem. Radiochem.* **24** (1981) 115.
2. Yu. N. Kukushkin and N. D. Rubtsova, *Zh. Neorg. Khim.* **14** (1969) 1867.
3. Yu. N. Kukushkin and N. D. Rubtsova, *Zh. Neorg. Khim.* **14** (1969) 2124.
4. Yu. N. Kukushkin, N. D. Rubtsova, and N. V. Ivannikova, *Zh. Neorg. Khim.* **15** (1970) 2001.
5. P. M. Treichel and P. J. Vincenti, *Inorg. Chem.* **24** (1985) 228.
6. K. R. Dunbar, *Comments Inorg. Chem.* **13** (1992) 313.
7. (a) W. Beck and K. Sünkel, *Chem. Rev.* **88** (1988) 1405; (b) S. H. Strauss, *Chem. Rev.* **93** (1993) 927 and *Chemtracts* **6** (1994) 1; (c) G. A. Lawrance, *Adv. Inorg. Chem.* **34** (1989) 145.
8. (a) B. D. Berezin, *Sov. J. Coord. Chem. (Engl. Transl.)* **17** (1991) 597; (b) B. D. Berezin and O. A. Golubtchikov, *Coordination Chemistry of Solvento-Complexes of Transition Metal Salts* [in Russian] (Nauka, Moscow, 1992) p. 236; (c) B. D. Berezin, *Russ. Chem. Rev. (Engl. Transl.)* **60** (1990) 996.
9. A. A. Grinberg and H. I. Gildengerschel, *Izv. Sektora Platiny IONKh AN SSSR* **26** (1951) 115.
10. (a) M. Green, M. Garner, and D. M. Orton, *Transit. Met. Chem.* **17** (1992) 164; (b) I. Haiduc and C. Silvestru, *Coord. Chem. Rev.* **99** (1990) 253; (c) C. A. McAuliffe, H. L. Sharma, and N. D. Tinker in *Chemistry of the Platinum Group Metals; Recent Developments* , ed. F. R. Hartley (Elsevier, Amsterdam, 1991) p. 546; (d) B. Lippert, *BioMetals* **5** (1992) 195.
11. For general reviews see (a) N. Farrell, *Transition Metal Complexes as Drugs and Chemotherapeutic Agents* in *Catalysis by Metal Complexes*, eds. B. R. James and R. Ugo, Vol. 11, (Reidel-Kluwer, Dordrecht, 1989); (b) N. Farrell, *Cancer Investigation* **11** (1993) 578.
12. (a) J. Reedijk, A. H. J. Fichtinger-Schepman, A. T. Van Oosteron, and P. Van de Putte, *Struct. Bond.*, **67** (1987) 53; (b) J. Reedijk, *Inorg. Chim. Acta* **198-200** (1992) 873; (c) E. L. M. Lempers and J. Reedijk, *Adv. Inorg. Chem.* **37** (1991) 175.
13. T. Solin, K. Matsumoto, and K. Fuwa, *Inorg. Chim. Acta* **65** (1982) L171.

14. R. C. Harrison and C. A. McAuliffe, *Inorg. Chim. Acta* **46** (1980) L15.

15. J. Kuduk-Jaworska and B. Jezowska-Trzebiatowska, *Polish J. Chem.* **55** (1981) 1143.

16. J. M. Mayer, D. L. Thorn, and T. H. Tulip, *J. Am. Chem. Soc.* **107** (1985) 7454.

17. S. E. Livingstone, *The Chemistry of Ruthenium, Rhodium, Palladium, Osmium, Iridium and Platinum*, Chapter 43 of *Comprehensive Inorganic Chemistry* (Pergamon, London, 1973).

18. L. I. Elding, *Inorg. Chim. Acta* **20** (1976) 65.

19. (a) U. Frey, S. Elmroth, B. Moullet, L. I. Elding, and A. E. Merbach, *Inorg. Chem.* **30** (1991) 5033; (b) L. I. Elding and Å. Oskarsson, *Inorg. Chim. Acta* **103** (1985) 127; (c) Z. Bugarcic, B. Norén, Å. Oskarsson, C. Stålhandske, and L. I. Elding, *Acta Chem. Scand.* **45** (1991) 361; (d) H. Toftlund, *J. Chem. Soc., Chem. Commun.* (1979) 837; (e) O. Simonsen and H. Toftlund, *Inorg. Chem.* **20** (1981) 4044; (f) B. Hellquist, L. A. Bengtsson, B. Holmberg, B. Hedman, I. Persson, and L. I. Elding, *Acta Chem. Scand.* **45** (1991) 449.

20. K. Seigmann, P. S. Pregosin, and L. Venanzi, *Organometallics* **8** (1989) 2659.

21. J. R. Bleeke and D. J. Rauscher, *J. Am. Chem. Soc.* **111** (1989) 8972.

22. P. Royo, A. Antinolo, M. Fajardo, J. A. Gomez, and A. Otero, *IV FECHEM Conf. Organomet. Chem.*, Abstr., S.I., s.a., A-46.

23. D. C. Luehrs, R. T. Iwamoto, and J. Kleinberg, *Inorg. Chem.* **5** (1966) 201.

24. W. W. Fee, J. N. Mac B. Harrowfield, and W. G. Jackson, *J. Chem. Soc., Ser. A* (1970) 2612.

25. P. R. Sharp, *Organometallics* **3** (1984) 1217.

26. (a) R. Banerjee, *Coord. Chem. Rev.* **68** (1985) 152; (b) S. E. Miller, H. Wen, D. A. House, and W. T. Robinson, *Inorg. Chim. Acta* **184** (1991) 111.

27. F. Morandini, G. Consiglio, G. Ciani, and A. Sironi, *Inorg. Chim. Acta* **82** (1984) L27.

28. P. P. K. Claire, G. R. Willey, and M. G. B. Drew, *J. Chem. Soc., Chem. Commun.* (1987) 1100.

29. M. G. Meirim and E. W. Neuse, *Transit. Metal Chem.* **9** (1984) 337.

30. K. Folting, J. C. Huffman, R. L. Bansemer, and K. G. Caulton, *Inorg. Chem.* **23** (1984) 3289.

31. K. Folting, J. C. Huffman, R. L. Bansemer, K. G. Caulton, J. L. Martin, and P. D. Smith, *Inorg. Chem.* **23** (1984) 4589.

32. F. Maspero, E. Perrotti, and F. Simonetti, *J. Organomet. Chem.* **38** (1972) 43.

33. N. G. Connelly and J. D. Davies, *J. Organomet. Chem.* **38** (1972) 385.

34. P. R. Sharp and K. G. Frank, *Inorg. Chem.* **24** (1985) 1808.

35. T. J. McNeese and T. E. Mueller, *Inorg. Chem.* **24** (1985) 2981.

36. J. A. Davies, F. R. Hartley, and S. G. Murray, *J. Chem. Soc., Dalton Trans.* (1980) 2246.

37. J. A. Davies, F. R. Hartley, and S. G. Murray, *J. Chem. Soc., Dalton Trans.* (1979) 1705.

38. N. E. Dixon, W. G. Jackson, W. Marty, and A. M. Sargeson, *Inorg. Chem.* **21** (1982) 688.

39. D. A. Buckingham, F. R. Keene, and A. M. Sargeson, *J. Am. Chem. Soc.* **95** (1973) 5649.

40. V. Yu. Kukushkin, V. K. Belsky, V. E. Konovalov, R. R. Shifrina, A. I. Moiseev, and R. A. Vlasova, *Inorg. Chim. Acta.* **183** (1991) 57.

41. V. Yu. Kukushkin, V. M. Tkachuk, I. A. Krol, Z. A. Starikova, B. V. Zhadanov, and N. P. Kiseleva, *Zh. Obsch. Khim.* **61** (1991) 51.

42. C. Bianchini, D. Masi, K. Linn, C. Mealli, M. Peruzzini, and F. Zanobini, *Inorg. Chem.* **31** (1992) 4036.

43. D. J. Liston, Y. J. Lee, W. R. Scheidt, and C. A. Reed, *J. Am. Chem. Soc.* **111** (1989) 6643.

44. J. A. McCleverty, *Chem. Rev.* **79** (1979) 53.

45. Yu. N. Kukushkin, *Reactivity of Coordination Compounds* [in Russian] (Khimia, Leningrad, 1987) p.161.

46. K. A. Hofmann, Z. *Anorg. Chem.* **10** (1895) 262.

47. S. Lunak and J. Veprek-Siska, *Coll. Czech. Chem. Commun.* **39** (1974) 2719.

48. S. K. Wolfe, C. Andrade, and J. H. Swinehart, *Inorg. Chem.* **13** (1974) 2567.

49. F. Bottomley, S. G. Clarkson, and S.-B. Tong, *J. Chem. Soc., Dalton Trans.* (1974) 2344.

50. F. J. Miller and T. J. Meyer, *J. Am. Chem. Soc.* **93** (1971) 1294.

51. E. C. Johnson, B. P. Sullivan, D. J. Salmon, S. A. Adeyemi, and T. J. Meyer, *Inorg. Chem.* **17** (1978) 2211.

52. S. A. Adeyemi, F. J. Miller, and T. J. Meyer, *Inorg. Chem.* **11** (1972) 994.

53. P. G. Douglas and R. D. Feltham, *J. Am. Chem. Soc.* **94** (1972) 5254.

54. F. Bottomley and M. J. Mukaida, *J. Chem. Soc., Dalton Trans.* (1982) 1933.

55. A. Haim and H. Taube, *Inorg. Chem.* **2** (1963) 1199.

56. W. L. Reynolds, N. Morey, S. C. McFarlan, M. Collins, and M. Rigney, *J. Am. Chem. Soc.* **97** (1975) 7165.

57. D. A. Loeliger and H. Taube, *Inorg. Chem.* **5** (1966) 1376.

58. R. B. Jordan, A. M. Sargeson, and H. Taube, *Inorg. Chem.* **5** (1966) 1091.

59. J. K. Hurst and H. Taube, *J. Am. Chem. Soc.* **90** (1968) 1174.

60. W. R. Ellis and W. L. Pursell, *Inorg. Chem.* **21** (1982) 834.

61. R. J. Balahura, *Can. J. Chem.* **52** (1974) 1762.

62. W. L. Bowden, W. F. Little, and T. J. Meyer, *J. Am. Chem. Soc.* **99** (1977) 4340.

63. W. L. Bowden, W. F. Little, and T. J. Meyer, *J. Am. Chem. Soc.* **95** (1973) 5084.

64. Yu. N. Kukushkin and O. V. Stefanova, *Zh. Neorg. Khim.* **22** (1977) 3375.

65. F. E. Wood and A. L. Balch, *Inorg. Chim. Acta* **76** (1983) L63.

66. Yu. N. Kukushkin, *Platinum Metals Review* **35** (1991) 28.

67. (a) Yu. N. Kukushkin, *J. Serb. Chem. Soc.* **55** (1990) 681; (b) Yu. N. Kukushkin, *Koord. Khim.* **5** (1979) 1856; (c) Yu. N. Kukushkin, *Koord. Khim.* **8** (1982) 201.

68. (a) I. A. Krol, V. Yu. Kukushkin, Z. A. Starikova, B. V. Zhadanov, and V. M. Tkachuk, *Zh. Obsch. Khim.* **58** (1988) 2625; (b) V. Yu. Kukushkin, E. Yu. Pankova, T. N. Fomina, and N. P. Kiseleva, *Koord. Khim.* **14** (1988) 1110; (c) V. Yu. Kukushkin, I. A. Krol, Z. A. Starikova, V. M. Tkachuk, *Koord. Khim.* **16** (1990) 1406.

69. L. A. De Oliveira, H. E. Toma, and E. Giesbrecht, *Inorg. Nucl. Chem. Lett.* **12** (1976) 195.

70. W. L. Reynolds, *Inorg. Chem.* **14** (1975) 680.

71. (a) J. MacB. Harrowfield, A. M. Sargeson, B. Singh, and J. C. Sullivan, *Inorg. Chem.* **14** (1975) 2864; (b) W. G. Jackson and A. M. Sargeson, *Inorg. Chem.* **17** (1978) 1348.

72. A. M. Sargeson and H. Taube, *Inorg. Chem.* **5** (1966) 1094.

73. R. F. Jordan, *Adv. Organomet. Chem.* **32** (1992) 325.

74. (a) R. F. Jordan, C. S. Bajgur, R. Willet, and B. Scott, *J. Am. Chem. Soc.* **108** (1986) 7410; (b) R. F. Jordan, C. S. Bajgur, and W. E. Dasher, *Organometallics* **6** (1987) 1041; (c) R. F. Jordan, W. E. Dasher, and S. F. Echols, *J. Am. Chem. Soc.* **108** (1986) 1718; (d) R. F. Jordan and A. S. Guram, *Organometallics* **9** (1990) 2116; (e) R. F. Jordan, D. F. Taylor, and N. C. Baenziger, *Organometallics* **9** (1990) 1546.

75. R. F. Jordan, R. E. LaPointe, C. S. Bajgur, S. F. Echols, and R. Willett, *J. Am. Chem. Soc.* **109** (1987) 4111.

76. S. L. Borkowski, N. C. Baenziger, and R. F. Jordan, *Organometallics* **12** (1993) 486.

77. (a) R. R. Schrock and J. A. Osborn, *J. Am. Chem. Soc.* **98** (1976) 2134; (b) R. R. Schrock and J. A. Osborn, *J. Am. Chem. Soc.* **98** (1976) 2143.

78. R. H. Crabtree, P. C. Demou, D. Eden, J. M. Michelcic, C. A. Parnell, J. M. Quirk, and G. E. Morris, *J. Am. Chem. Soc.* **104** (1982) 6994.

79. J. R. Shapley, R. R. Schrock, and J. A. Osborn, *J. Am. Chem. Soc.* **91** (1969) 2816.

80. S. Clamp, N. G. Connelly, G. E. Taylor, and T. S. Louttit, *J. Chem. Soc., Dalton Trans.* (1980) 2162.

81. P. Legzdins and J. C. Oxley, *Inorg. Chem.* **23** (1984) 1053.

82. H. Werner and R. Werner, *Chem. Ber.* **118** (1985) 4543.

83. N. G. Connelly, P. T. Draggett, M. Green, and T. A. Kuc, *J. Chem. Soc., Dalton Trans.* (1977) 70.

84. (a) R. F. Schramm and B. B. Wayland, *J. Chem. Soc., Chem. Commun.* (1968) 898; (b) R. R. Thomas and A. Sen, *Inorg. Synth.* **28** (1990) 63.

85. B. J. Hathaway and A. E. Underhill, *J. Chem. Soc.* (1960) 3705.

86. D. A. Phillips, M. Kubota, and J. Thomas, *Inorg. Chem.* **15** (1976) 118.

87. G. Mazzocchin, G. Bontempelli, M. Nicolini, and B. Crociani, *Inorg. Chim. Acta* **18** (1976) 159.

88. L. Chen, J. A. Davies, and R. J. Staples, *Inorg. Chim. Acta* **163** (1989) 11.

89. J. A. Davies, L. Chen, C. T. Eagle, and R. J. Staples in *Molecular Electrochemistry of Inorganic, Bioinorganic, and Organometallic Compounds*, eds. A. J. L. Pombeiro and J. A. McCleverty, NATO ASI Series, Series C, Vol. 385 (Kluwer, Dordrecht, 1993) p. 351.

90. L. Chen and J. A. Davies, *Inorg. Chim. Acta* **175** (1990) 41.

91. M. Mazzanti, S. Gambarotta, C. Floriani, A. Chiesi-Villa, and C. Guastini, *Inorg. Chem.* **25** (1986) 2308.

92. M. Mazzanti, C. Floriani, A. Chiesi-Villa, and C. Guastini, *Inorg. Chem.* **25** (1986) 4158.

93. L. M. Jackman, R. M. Scott, R. H. Portman, and J. F. Dormish, *Inorg. Chem.* **18** (1979) 1497.

94. L. M. Jackman, J. F. Dormish, R. M. Scott, R. H. Portman, and R. D. Minard, *Inorg. Chem.* **18** (1979) 1503.

95. W. G. Jackson, S. S. Jurisson, and M. A. O'Leary, *Inorg. Chem.* **32** (1993) 445.

96. J. M. Calvert and T. J. Meyer, *Inorg. Chem.* **21** (1982) 3978.

97. J. F. Glenister, K. E. Hyde, and G. Davies, *Inorg. Chem.* **21** (1982) 2331.

98. S. Balt and C. R. Piriz MacColl, *Z. Anorg. Allg. Chem.* **513** (1984) 208.

99. D. M. Stanbury, D. Gaswick, G. M. Brown, and H. Taube, *Inorg. Chem.* **22** (1983) 1975.

100. T. Mallen and K. E. Hyde, *Inorg. Chim. Acta* **130** (1987) 177.

101. M. Hoch and D. J. Rehder, *J. Organomet. Chem.* **228** (1985) C25.

102. L. F. Rhodes, J. D. Zubkowski, K. Folting, J. C. Huffman, and K. G. Caulton, *Inorg. Chem.* **21** (1982) 4185.

103. (a) J.-H. So and P. Boudjouk, *Inorg. Chem.* **29** (1990) 1592; (b) P. Boudjouk and J.-H. So, *Inorg. Synth.* **29** (1992) 108.

104. (a) D. Pinnell, G. B. Wright, and R. B. Jordan, *J. Am. Chem. Soc.* **94** (1972) 6104; (b) Yu. P. Manuylov, *Koord. Khim.* **5** (1979) 1827.

105. J. L. Laird and R. B. Jordan, *Inorg. Chem.* **21** (1982) 855.

106. A. W. Zanella and P. C. Ford, *Inorg. Chem.* **14** (1975) 42.

107. M. J. Sisley, Y. Yano, and T. W. Swaddle, *Inorg. Chem.* **21** (1982) 1141.

108. T. N. Sorrell and D. L. Jameson, *Inorg. Chem.* **21** (1982) 1014.

109. K. K. Bhasin, S. Jindal, S. Singh, and R. D. Verma, *Synth. and React. Inorg. and Metal-org. Chem.* **15** (1985) 733.

110. J. P. Candlin, K. A. Taylor, and D. T. Thompson, *Reactions of Transition Metal-Complexes* (Elsevier, Amsterdam, 1968).

111. M. O. Albers and N. J. Coville, *Coord. Chem. Rev.* **53** (1984) 227.

112. G. J. Kubas, *Inorg. Chem.* **22** (1983) 692.

113. (a) R. D. Adams in *The Chemistry of Metal Cluster Complexes*, eds. D. F. Shriver, H. D. Kaesz, and R. D. Adams (VCH, New York, 1990) p. 123. (b) G. Lavigne in *The Chemistry of Metal Cluster Complexes*, eds. D. F. Shriver, H. D. Kaesz, and R. D. Adams (VCH, New York, 1990) p. 204.

114. (a) S. K. Nayak, G. J. Farrell, and T. J. Burkey, *Inorg. Chem.* **33** (1994) 2236; (b) G. L. Geoffroy and M. S. Wrighton, *Organometallic Photochemistry* (Academic, New York, 1979).

115. A. J. Deeming, M. Karim, and N. I. Powell, *J. Chem. Soc., Dalton Trans.* (1990) 2321.

116. B. N. Storhoff and H. C. Lewis, *Coord. Chem. Rev.* **23** (1977) 1.

117. W. A. Shenk, U. Karl, M. R. Horn, and S. Müssig, *Z. Naturforsch.* **45** (1990) 239.

118. C. F. Barrientos-Penna, F. W. B. Einstein, T. Jones, and D. Sutton, *Inorg. Chem.* **21** (1982) 2578.

119. P. Le Maux, G. Simonneaux, and G. Jaouen, *J. Organomet. Chem.* **217** (1981) 61.

120. (a) D. M. Haddleton and R. N. Perutz, *J. Chem. Soc., Chem. Commun.* (1985) 1372; (b) S. T. Belt, D. M. Haddleton, R. N. Perutz, B. P. H. Smith, and A. J. Dixon, *J. Chem. Soc., Chem. Commun.* (1987) 1347.

121. A. J. Carmichael and A. McCamley, *J. Chem. Soc., Dalton Trans.* (1995) 3125.

122. K. R. Pope and M. S. Wrighton, *Inorg. Chem.* **24** (1985) 2792.

123. J. Auffret, P. Courtot, R. Pichon, R. Rumin, and J. Y. Salaun, *J. Organomet. Chem.* **262** (1984) C19.

124. P. Courtot, R. Rumin, and A. J. Peron, *J. Organomet. Chem.* **144** (1978) 357.

125. Yu. N. Kukushkin and I. V. Pakhomova, *Zh. Neorg. Khim.* **15** (1970) 1882.

126. I. V. Pakhomova, Yu. N. Kukushkin, L. V. Konovalov, and V. V. Strukov, *Zh. Neorg. Khim.* **29** (1984) 1000.

127. A. D. Allen and T. Theophanides, *Can. J. Chem.* **42** (1964) 1551.

128. (a) T. P. Gill and K. R. Mann, *Inorg. Chem.* **22** (1983) 1986; (b) T. P. Gill and K. R. Mann, *Organometallics* **1** (1982) 485.

129. T. Karlen, A. Hauser, and A. Ludi, *Inorg. Chem.* **33** (1994) 2213.

130. A. M. McNair, J. L. Schenk, and K. R. Mann, *Inorg. Chem.* **23** (1984) 2633.

131. A. J. L. Pombeiro, J. C. Jeffery, C. J. Pickett, and R. L. Richards, *J. Organomet. Chem.* **227** (1984) C7.

132. T. Ishida, Y. Mizobe, and M. Hidai, *Chem. Lett.* (1989) 2077.

133. A. Salzer, T. Egolf, L. Linowsky, and W. Petter, *J. Organomet. Chem.* **221** (1981) 339.

134. G. L. Hillhouse, B. L. Haymore, S. A. Bistram, and W. A. Herrmann, *Inorg. Chem.* **22** (1983) 314.

135. H. Lehner, D. Matt, P. S. Pregosin, L. M. Venanzi, and A. Albinati, *J. Am. Chem. Soc.* **104** (1982) 6825.

136. L. F. Rhodes, J. C. Huffman, and K. G. Caulton, *J. Am. Chem. Soc.* **105** (1983) 5137.

137. O. Kolb and H. Werner, *Angew. Chem.* **94** (1982) 207.

138. P. L. Gaus, N. Marchant, and M. A. Marsinek, *ACS Division of Inorg. Chem., Winter Newsletter* (1984) Abstr. 187.

139. E. W. Ainscough, A. M. Brodie, N. G. Larsen, and M. R. Stevens, *Inorg. Chim. Acta* **50** (1981) 215.

140. P. M. Cook, L. F. Dahl, and D. W. Dickerhoof, *J. Am. Chem. Soc.* **94** (1972) 5511.

141. (a) M. R. Colsman, T. D. Newbound, L. J. Marshall, M. D. Noirot, M. M. Miller, G. P. Wulfsberg, J. S. Frye, O. P. Anderson, and S. H. Strauss, *J. Am. Chem. Soc.* **112** (1990) 2349; (b) D. M. Van Seggen, P. K. Hurlburt, O. P. Anderson, and S. H. Strauss, *Inorg. Chem.* **34** (1995) 3453.

142. W. Beck and K. Schloter, *Z. Naturforsch.* **33B** (1978) 1214.

143. K. Sünkel, G. Urban, and W. Beck, *J. Organomet. Chem.* **252** (1983) 187.

144. E. Horn and M. R. Snow, *Austral. J. Chem.* **33** (1980) 2369.

145. E. K. G. Schmidt and C. H. Thiel, *J. Organomet. Chem.* **220** (1981) 87.

146. B. M. Mattson and W. A. G. Graham, *Inorg. Chem.* **20** (1981) 3186.

147. R. H. Crabtree, J. W. Faller, M. F. Mellea, and J. M. Quirk, *Organometallics* **1** (1982) 1361.

148. M. J. Burk, B. Segmuller, and R. H. Crabtree, *Organometallics* **6** (1987) 2241.

149. C. H. Winter, A. M. Arif, and J. A. Gladysz, *J. Am. Chem. Soc.* **109** (1987) 7560.

150. C. H. Winter, W. R. Veal, C. M. Garner, A. M. Arif, and J. A. Gladysz, *J. Am. Chem. Soc.* **111** (1989) 4766.

151. F. M. Conroy-Lewis, A. D. Redhouse, and S. J. Simpson, *J. Organomet. Chem.* **366** (1989) 357.

152. (a) C. H. Winter and J. A. Gladysz, *J. Organomet. Chem.* **354** (1988) C33; (b) J. M. Fernández and J. A. Gladysz, *Organometallics* **8** (1989) 207; (c) J. J. Kowalczyk, S. K. Agbossou, and J. A. Gladysz, *J. Organomet. Chem.* **397** (1990) 333.

153. T.-S. Peng, C. H. Winter, and J. A. Gladysz, *Inorg. Chem.* **33** (1994) 2534 and references therein.

154. M. R. Colsman, M. D. Noirot, M. M. Miller, O. P. Anderson, and S. H. Strauss, *J. Am. Chem. Soc.* **110** (1988) 6886.

155. M. Bown and J. M. Waters, *J. Am. Chem. Soc.* **112** (1990) 2442.

Chapter 4. HOMOLEPTIC SOLVENTO-COMPLEXES: STARTING MATERIALS FOR PREPARATIVE COORDINATION CHEMISTRY

4.1 Introduction

The increasing use of non-aqueous solvents in preparative coordination chemistry makes the choice of starting materials for synthesis by ligand substitution particularly important. Ideally, starting materials must be readily available, freely soluble in non-aqueous solvents, and contain ligands that are relatively poor electron-donors and, thus, are replaced easily. Solvento-complexes, *i.e.* complexes whose ligands are more commonly encountered as solvents, satisfy these requirements. In this chapter we consider homoleptic solvento-complexes. For simplicity, metal complexes with ambidentate ligands coordinated in more than one way to the metal center are not routinely differentiated. The platinum(II) and palladium(II) tetrakis(dimethyl sulfoxide) complexes, *cis*-$[M(O\text{-}DMSO)_2(S\text{-}DMSO)_2]^{2+}$ are typical examples.[1,2] Figure 4.1 illustrates the structure of the *cis*-$[Pt(O\text{-}DMSO)_2(S\text{-}DMSO)_2]^{2+}$ cation.[1]

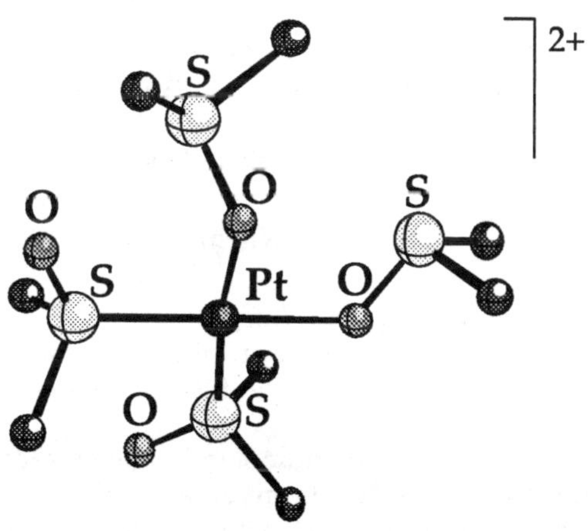

Figure 4.1

Although a homoleptic solvento-complex will often exhibit good solubility in a solvent that also functions as the ligand (a reflection of the well known empirical rule, *similia similibus*), other considerations may also be important. Thus, solvento-complexes in which the solvento-ligand is a molecule with a small dielectric constant may exhibit high solubility in a weakly solvating solvent where the dielectric constant is large. Some solvent molecules exhibit poor electron-donor properties (Chapters 1 and 3) and, thus, can be replaced easily by stronger donors. For example, the benzonitrile ligands in $[PdCl_2(PhCN)_2]$ are replaced[3] so easily that this complex may be considered as a soluble form of palladium dichloride. Replacement of the benzonitrile ligands in $[PdCl_2(PhCN)_2]$ has been employed in many syntheses.[4,5] The reaction between this complex and the thioether, $S(CH_2CO_2H)_2$, in benzene results in rapid replacement of PhCN and formation of $[PdCl_2\{S(CH_2CO_2H)_2\}_2]$. In the reaction of $[PdCl_2(PhCN)_2]$ with $NR_4^+Br^-$ salts (where NR_4^+ represents alkylammonium ions), the unusual complexes $[NR_4]_2[PdBr_2Cl_2]$, where the coordination sphere contains both bromide and chloride ions, are formed. The complex, $[N_2H_6]_2[PdCl_4]$,[4,5] which contains a cation derived from the strong reducing agent hydrazine, was prepared in a similar manner. In summary, work on the replacement of benzonitrile ligands in $[PdCl_2(PhCN)_2]$ has demonstrated that the complex may be considered as "coordinatively unsaturated" for most preparative purposes.

Solvento-complexes are effective reagents for the preparation of coordination compounds. For example, the first syntheses of nickel(II) diaminates and triaminates were performed with water as the reaction medium. The reactions were complete within several hours and the resulting compounds contained water of crystallization in various amounts depending on the techniques used for drying the samples.[6,7] A similar reaction[6,7] has been performed with ethanol as the solvent, Eq. 4.1:

$$NiBr_2{\cdot}4EtOH \quad + \quad 3en \quad \rightarrow \quad [Ni(en)_3]Br_2 \quad + \quad 4EtOH \tag{4.1}$$

The use of the solvento-complex, $NiBr_2{\cdot}4EtOH$, as the starting material and ethanol as the solvent permits a rapid and convenient synthesis of $[Ni(en)_3]Br_2$ with a yield of 98%. The resulting compound has no water of crystallization.

In some cases the use of homoleptic solvento-complexes as starting materials in non-aqueous solvents involves a diversion from the common reaction pathway. The interaction between platinum(II) and palladium(II) complexes with potentially tetradentate thioether ligands,[8-11] Figure 4.2, has been studied. The reactions of the tetrahalide complexes $[MX_4]^{2-}$, where M = Pt, Pd; X = Cl, Br, I, with the thioether ligands in an ethanol-dichloromethane mixture lead to polymeric products of the type $[M_2X_4(\mu\text{-}L)]_n$. The use of the homoleptic solvento-complexes, $[M(MeCN)_4][ClO_4]_2$, however, makes possible isolation of the monomeric $[ML][ClO_4]_2$.

Water-soluble copper(I) salts are known to rapidly disproportionate to the hydrated Cu^{2+} cation and metallic copper. The instability of the copper(I) ion[12] has been attributed to the high lattice energy of metallic copper and the high formation constants for copper(II) solvates in water compared with those of copper(I).

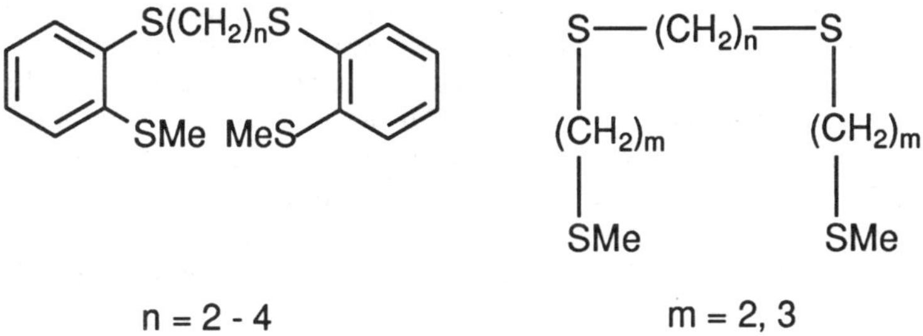

Figure 4.2

The use of $[Cu(MeCN)_4]^+$ as a starting material simplifies the preparation of copper(I) complexes of substituted imidazoles (imid = 1-, 2-, or 4-methylimidazole),[13] Eq. 4.2:

$$[Cu(MeCN)_4][BF_4] + 2imid \rightarrow [Cu(imid)_2][BF_4] + 4MeCN \quad (4.2)$$

In this synthesis, stoichiometric quantities of the solvento-complex and the heterocyclic amine were combined followed by solvent removal *in vacuo*. After crystallization, yields of the final products were 60-70%.

The examples described above illustrate the utility of solvento-complexes in preparative coordination chemistry. The balance of this chapter reviews methods for the synthesis of homoleptic solvento-complexes.

4.2 Synthesis of solvento-complexes by metal oxidation in non-aqueous solvents

In order to prepare a homoleptic solvento-complex by metal oxidation in a non-aqueous medium, the reduction products must not compete with the solvent molecules as ligands. Nitrosonium perchlorate, $NOClO_4$, is a suitable oxidant.[14,15] Reactions have been performed in acetonitrile, nitromethane and ethylacetate. In acetonitrile this oxidant converts copper powder into the copper(I) and copper(II) homoleptic solvento-complexes, $[Cu(MeCN)_4][ClO_4]$ and $[Cu(MeCN)_4][ClO_4]_2$. On boiling an acetonitrile solution of these complexes with metallic copper, reduction of the copper(II) complex occurs with the formation of $[Cu(MeCN)_4][ClO_4]$ as the sole product. With nitromethane and ethyl acetate as the solvents, the oxidation of copper leads to complexes with the following proposed formulae based on characterization data: $[Cu(ClO_4)_2(MeNO_2)_2]$ and $[Cu(ClO_4)_2(EtOAc)_{2.5}]$. It is reasonable to assume that perchlorate ion coordination occurs.

Metallic palladium and nitrosonium tetrafluoroborate have been used[16,17] in the synthesis of $[Pd(MeCN)_4][BF_4]_2$. One equivalent of palladium sponge is mixed with 2.2 equivalents of $NOBF_4$ in acetonitrile and a vacuum applied to remove oxygen and NO liberated during the reaction. After nine hours, the reaction mass is filtered and $[Pd(MeCN)_4][BF_4]_2$ precipitated by the slow addition of diethyl ether.[16] The acetonitrile complex of europium, $[Eu(MeCN)_x]Y_3$ ($Y = PF_6$ or BF_4), is prepared in a similar manner.[18,19] It is important to note that NO, formed by reduction of the NO^+ cation, can participate in the formation of nitrosyl complexes (see Chapter 8), thus complicating potential synthetic routes.

Silver salts containing anions of strong acids are suitable reagents for synthesis of solvento-complexes by metal oxidation in non-aqueous solvents. Metallic gold in acetonitrile reacts with silver perchlorate[20] to produce $[Au(MeCN)_4][ClO_4]$, Eq. 4.3:

$$Au + AgClO_4 + 4MeCN \rightarrow [Au(MeCN)_4][ClO_4] + Ag \qquad (4.3)$$

The synthesis of $[Cu(MeCN)_4][NO_3]$ was accomplished in a similar manner[21] with $AgNO_3$. The liberated silver was separated easily from the reaction mixture.

The complexes, $[Fe(py)_4]X_2$ (X = Br, I), have been prepared by oxidation of metallic iron with bromine or iodine in methanol. Pyridine is added to the resultant solution and the pyridine complexes precipitated.[22] Interaction between metallic zinc or cadmium and an excess of arsenic pentafluoride in liquid sulfur dioxide leads to the formation of solvento-complexes containing two or four SO_2 ligands depending on the experimental conditions.[23,24] The reactions studied are summarized as follows, Eq. 4.4:

$$M + 3AsF_5 + xSO_2 \rightarrow [M(SO_2)_x][AsF_6]_2 + AsF_3 \qquad (4.4)$$

where M = Zn, Cd; $x = 2$ or 4

Homoleptic solvento-complexes can be prepared by electrochemical oxidation of metals in non-aqueous solvents. For example, $[V(DMSO)_6][BF_4]_2$ and $[In(MeCN)_6][BF_4]_3$ have been synthesized[25,26] by vanadium or indium oxidation in dimethyl sulfoxide or acetonitrile in the presence of HBF_4. Gram quantities of these solvento-complexes may be obtained with yields of 80-90% based on the dissolved metal.

Additionally, metal oxidation can be performed with protic acids in non-aqueous solvents. For instance, the interaction between metallic titanium and HBF_4 in acetonitrile leads to the formation of the titanium(III) complex,[25,26] $[Ti(MeCN)_6][BF_4]_3$, Eq. 4.5:

$$Ti + 3HBF_4 + 6MeCN \rightarrow [Ti(MeCN)_6][BF_4]_3 + 1.5H_2 \qquad (4.5)$$

The preparation of homoleptic solvento-complexes by metal oxidation with ArN_2^+ or Ph_3C^+ salts in acetonitrile has been proposed.[27] However, such procedures are not recommended[27] as routine preparative methods.

Certain low-valent metal carbonyls are oxidized easily with protic acids in non-aqueous media. Thus, $[Mo(CO)_6]$ in diglyme is oxidized by 100% formic acid with the formation of $[Mo_2(O_2CH)_4]$. In a related reaction, $[Mo(SO_3CF_3)_3]$ is prepared by boiling a $[Mo(CO)_6]$ suspension in an excess of HSO_3CF_3.[28] It is reasonable to assume that if a similar reaction was performed in the presence of a strong donor solvent, the reaction would lead to the formation of a homoleptic solvento-complex.

4.3 Synthesis of solvento-complexes by dissolution of salts in non-aqueous solvents

Gutmann has discussed the process of salt dissolution[29] in terms of a donor-acceptor reaction that results in coordination of solvent molecules to the metal ion. Further, the interaction between salts, MX, and solvent, solv, has been described by Burger[30] as shown in Scheme 4.1:

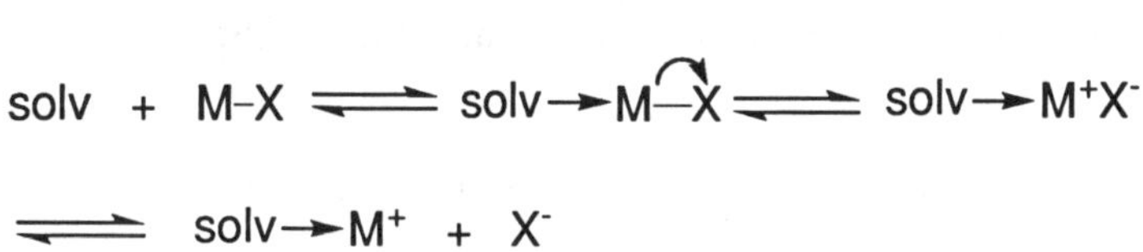

Scheme 4.1

Nucleophilic attack occurs on the electrophilic center of MX. The formation of the solv→M coordinate bond facilitates polarization of electron density in the M→X direction. The first equilibrium is shifted to the right when the solvent, solv, has a large donor number, D_N, or is characterized by a large value of D_S.[31] Thus, the solubility of $NiCl_2$ has been shown to increase with increasing values of D_N or D_S for each solvent[29] in the series shown in Table 4.1:

Table 4.1

D_N and D_S parameters of selected solvents

Solvent	D_N	D_S
$MeNO_2$	2.7	9
MeCN	14.1	12
H_2O	18.0	17
TMP	23.0	23
DMSO	29.8	27.5

It has been shown[30] that solvents such as dimethylformamide and dimethyl sulfoxide, that have nonbonding electrons localized to a considerable degree on the oxygen atoms, freely solvate cations of soluble salts. In contrast, nitromethane, in which nonbonding electrons are more delocalized, solvates cations poorly.

An analysis of solvent nucleophilicity[29] suggests that steric factors should be taken into account in addition to electronic properties. For example, the D_N and D_S values for acetonitrile (D_N 14.1 and D_S 12) and propanediol-1,2-carbonate (D_N 15.1 and D_S 12) alone do not explain why acetonitrile dissolves salts more effectively than propanediol-1,2-carbonate. However, a consideration of steric effects suggests, at least qualitatively, that the small, rod-shaped acetonitrile molecule will be more effective in solvation than the larger, bulky propanediol-1,2-carbonate molecule.

In Burger's description the equilibrium shown in Scheme 4.2:

$$\text{solv}\rightarrow M^+X^- \rightleftharpoons \text{solv}\rightarrow M^+ + X^-$$

Scheme 4.2

depends on the dielectric constant of the solvent and the nature of the M–X bond. The latter point is illustrated by the following example: the dissolution of $AlCl_3$ in methanol[32] leads to the generation of two compounds: $[AlCl_2(MeOH)_4]^+$ and $[AlCl(MeOH)_5]^{2+}$, whereas, the dissolution of $Al(ClO_4)_3$ in methanol produces[32] only the hexasolvate $[Al(MeOH)_6]^{3+}$. This example suggests that dissolution of salts can be used to generate homoleptic solvento-complexes with ligands having relatively high donor numbers and low steric hindrances. For synthetic purposes the use of salts containing anions of strong acids, which are weak electron donors, is most effective.

For example, to prepare $[In(DMAA)_6][ClO_4]_3$, $In(ClO_4)_3$ is dissolved in dimethylacetamide. The resulting compound is isolated as a solid by removal of excess solvent.[33] The salt, $[Ag(DMF)_2][NO_3]$, is prepared in a similar manner by dissolution of silver nitrate in DMF followed by isolation of the product as a solid by the addition of benzene.[34] On dissolution of $Cu(BF_4)_2$ in acetonitrile in the presence of metallic copper as a reductant, $[Cu(MeCN)_4][BF_4]$ is formed.[35]

In certain cases, homoleptic solvento-complexes containing halides as counterions have been prepared (see Table 4.2). In general, preparation of such homoleptic solvento-complexes is of limited use due to the difficulties inherent in the preparation of the necessary anhydrous salts. In recent years it has been found that trifluoromethanesulfonate salts are easily prepared in anhydrous form. The $CF_3SO_3^-$ ion only rarely forms isolable complexes in which it is coordinated to a metal ion and is normally a weaker donor than the perchlorate anion. Unlike the common perchlorate salts, metal trifluoromethanesulfonates are not typically explosive under normal laboratory conditions. Thus, anhydrous metal trifluoromethanesulfonates are very useful precursors in the preparation of homoleptic solvento-complexes.

4.4 Synthesis of solvento-complexes by dehydration of aqua-complexes and hydrated salts

Many metal salts contain water of crystallization or coordinated water molecules. The interaction of such salts with non-aqueous solvents is a complex process involving competition for the metal ion between the solvent molecules, water, and the anion. To prepare homoleptic solvento-complexes, salts of strong acids, whose anions are poor electron donors, *e.g.* $HClO_4$, HBF_4, HPF_6, HSO_3CF_3, HSO_3Ar *etc.*, are employed. This essentially eliminates competition from the X^- anion for coordination sites at the metal center.

The removal of water can be difficult because water easily forms not only strong homoleptic and mixed ligand aqua-complexes, but oxo- and hydroxo-complexes as well. Acid anhydrides have been usefully employed to remove water. For example, the complex $[Ni(MeCN)_6][ClO_4]_2$ has been prepared[36,37] by dehydration of $[Ni(H_2O)_6][ClO_4]_2$ with P_4O_{10} in acetonitrile. Fifteen grams of $[Ni(H_2O)_6][ClO_4]_2$ are stirred with three one-gram amounts of P_4O_{10} in 200 mL of acetonitrile at room temperature. Complete dehydration is achieved by boiling the solution for 1.5 hours over 15 g of fresh P_4O_{10}. Copper(II) and cobalt(II) acetonitrile complexes have been prepared similarly.[38]

The synthesis of $[M(AcOH)_6]X_2$ (M = Mg, Mn, Co, Ni, Cu, Zn; X = ClO_4, NO_3, BF_4) by reaction of $[M(H_2O)_6]X_2$ with stoichiometric quantities of acetic anhydride in the absence of a solvent[39] has been reported, Scheme 4.3:

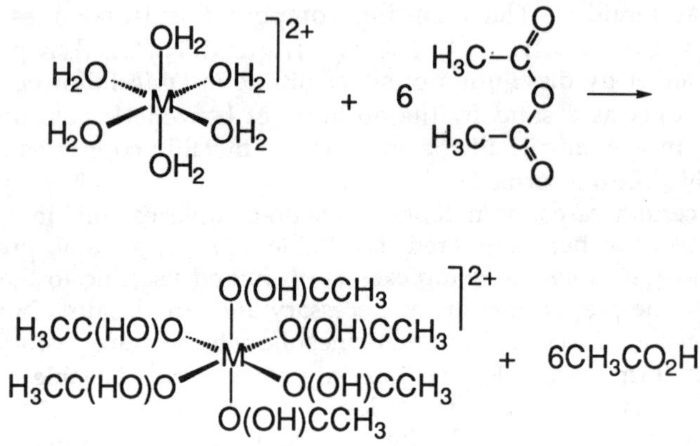

Scheme 4.3

It has been shown that the acetic acid ligands in $[M(AcOH)_6]X_2$ complexes[39] are not deprotonated but are coordinated through the carbonyl oxygen atoms.

Acid anhydrides have commonly been neglected as dehydrating agents in the synthesis of homoleptic solvento-complexes. Applications of acid anhydrides in the synthesis of anhydrous salts[40,41] or complexes with different ligands[42-44] are more typical.

2,2-dialkoxypropanes are useful dehydrating reagents[45,46] and their interaction with water is described by Eq. 4.6:

$$(RO)_2C(Me)_2 + H_2O \rightarrow Me_2CO + 2ROH \qquad (4.6)$$

These reagents have been utilized[47] in the synthesis of homoleptic solvento-complexes, *e.g.* $[Ni(MeOH)_6][ClO_4]_2$ is prepared from $[Ni(H_2O)_6][ClO_4]_2$ and a ten-fold excess of $(MeO)_2C(Me)_2$ in methanol. The reaction is performed at 4 °C over 10 hours.

It has been established[48-50] that trialkylorthoformates are more effective dehydrating reagents than 2,2-dialkoxypropanes. The orthoesters of formic acid react with water according to Eq. 4.7:

$$(RO)_3CH + H_2O \rightarrow 2ROH + ROOCH \qquad (4.7)$$

The reaction of $[Ni(H_2O)_6][ClO_4]_2$ with a slight excess of $(MeO)_3CH$ starts immediately upon addition of the orthoester and is complete within a few minutes.[50] An extensive series of homoleptic solvento-complexes of various metals has been prepared using this technique.

Molecular sieves are used to remove water from hydrated salts or aqua-complexes. For example, $[Mn(MeCN)_6][ClO_4]_2$ and $[Fe(MeCN)_6][ClO_4]_2$ are prepared from the corresponding aqua-complexes using molecular sieves.[51,52] When zeolites are used in such dehydration processes, the reaction may proceed only slowly. Thus, hydrated aluminum perchlorate is transformed into $[Al(MeOH)_6][ClO_4]_3$ using a zeolite in methanol[53] but the reaction takes six days.

The removal of coordinated water or water of crystallization from aqua-complexes and hydrated salts can be accomplished by azeotropic distillation in favorable situations. This technique is used to prepare $[Fe(MeCN)_6][ClO_4]_2$ from $[Fe(H_2O)_6][ClO_4]_2$ in acetonitrile and in the presence of molecular sieves.[54]

Hydrated metal halides have been used to prepare solvento-complexes. Toluene complexes of copper(I) have been synthesized from hydrated copper(II) halides by reduction with copper metal in the presence of a suitable Lewis acid,[55] Eqs. 4.8 and 4.9:

$$CuF_2{\cdot}2H_2O + Cu + 2BF_3 + 6C_6H_5Me \rightarrow 2[CuBF_4(C_6H_5Me)_3] + 2H_2O \quad (4.8)$$

$$CuF_2{\cdot}2H_2O + Cu + 2PF_5 + 6C_6H_5Me \rightarrow 2[CuPF_6(C_6H_5Me)_3] + 2H_2O \quad (4.9)$$

The reactions are performed in boiling toluene with continuous azeotropic distillation.

4.5 Synthesis of solvento-complexes by substitution of ligated solvent molecules

The exchange of one type of coordinated solvent molecule for another is conveniently represented by Scheme 4.4:

$$[M(solv)_x]A_y + x\,solv^* \longrightarrow [M(solv^*)_x]A_y + x\,solv$$

Scheme 4.4

A strong affinity of solv* for M as compared to that of solv is one of the main driving forces for this process. Such ligand substitution is common when the donor number of solv* is higher than that of solv. Thus, synthetic methods based on substitution of solvento-ligands are used routinely for the preparation of homoleptic solvento-complexes containing TMP, TEP, DMF, DMAA, DMSO, or HMP, each of which has a high donor number. The substitution of solv by solv* is favored by a large excess of solv* and by formation of a low-solubility product. For example,[56,57] dissolution of $[M(H_2O)_6][ClO_4]_n$ in acetone followed by addition of excess DMSO results in the precipitation of $[M(DMSO)_6][ClO_4]_n$ (M = Co, Ni, Mn, Fe, Zn, n = 2; M = Ga, Fe, n = 3). Cotton and Francis[58] have prepared $[M(DMSO)_6][ClO_4]_n$ by dissolution of hydrated perchlorates in dimethyl sulfoxide. The resultant complexes were isolated by removal of excess solvent *in vacuo*. However, it is often impossible to remove water completely from products prepared in this manner.[58]

The complexes, $[PdL_4][BF_4]_2$ (L = DMSO, N-methylacetamide, N-methylformamide), have been prepared by replacement of the acetonitrile ligands in $[Pd(MeCN)_4][BF_4]_2$ through reaction with excess L.[16] The compound $[CuBF_4(C_6H_5Me)_3]$ reacts readily with acetonitrile to produce[55] $[Cu(MeCN)_4][BF_4]$. A modified method of dehydration of aqua-complexes with triethylorthoformate (see Table 4.2) in the synthesis of homoleptic solvento-complexes containing solvento-ligands with high donor numbers has been developed. For example, in the preparation of $[Y(TMP)_6][ClO_4]_2$ one equivalent of hydrated yttrium perchlorate and 20 equivalents of triethylorthoformate are heated to 60 °C over one hour resulting in the formation of the corresponding ethanol complex *in situ*. Addition of seven equivalents of trimethyl phosphate to the reaction mixture leads to ligand substitution and formation of $[Y(TMP)_6][ClO_4]_2$ with an isolated yield[59] of about 90%.

4.6 Synthesis of solvento-complexes by halide abstraction from salts and complexes

In Chapter 3 we considered methods of formation of solvento-complexes and illustrated the various ways these compounds are prepared. Halide abstraction techniques that utilize silver salts are commonly employed in the synthesis of mixed-ligand solvento-complexes and such techniques are used also for the transformation of halide complexes into homoleptic solvento-complexes.

The conversion[60-62] of $K_2[PtCl_4]$ into $[Pt(H_2O)_4][ClO_4]_2$ provides an interesting example. Although the preparation of aqua-complexes is not the major subject of this chapter, the technique may be used with non-aqueous solvents. The method is reported[60] as follows: 3.5 g (8.5 mmol) of $K_2[PtCl_4]$ is added to 400 mL of 1.00 *M* $HClO_4$. The mixture is purged with argon to prevent oxidation, warmed to 70 °C, and protected from light. In 1.5 hours the potassium tetrachloroplatinate(II) dissolves completely. In small batches, 400 mL of 0.2 *M* $AgClO_4$ is added to the solution and the resultant mixture stirred for six days. Silver chloride is removed by filtration and the large excess of Ag^+ ions is removed by electrolysis, Scheme 4.5:

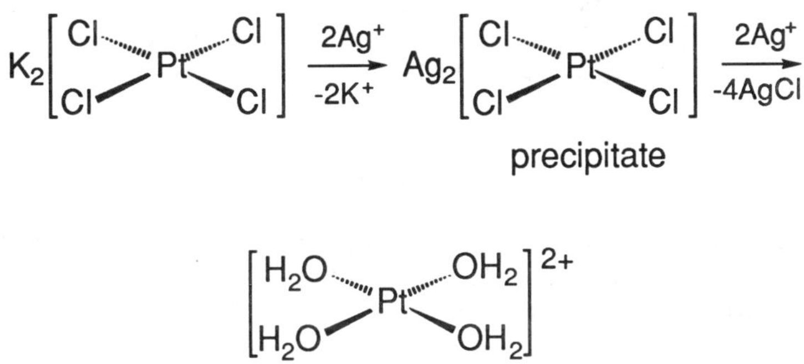

Scheme 4.5

The use of mercury perchlorate as a reagent for halide abstraction[60] from $K_2[PtCl_4]$ has been suggested. A later study,[63] also dealing with the preparation of $[Pt(H_2O)_4][ClO_4]_2$, has shown that the reaction of $K_2[PtCl_4]$ with $Hg(ClO_4)_2$ initially proceeds rapidly but the Hg^{2+} ion does not remove all of the halide ligands from $[PtCl_4]^{2-}$. In contrast, halide abstraction with an excess of silver perchlorate leads to the complete removal of the chloride ligands. However, the reaction is complicated by precipitation of the slightly soluble $Ag_2[PtCl_4]$ and, therefore, proceeds only slowly. The use of mercury and silver perchlorates in combination for the synthesis of $[Pt(H_2O)_4][ClO_4]_2$ has been described[63] but, although the reaction proceeds fairly rapidly, the resultant solutions are contaminated with Hg^{2+}.

There are many examples of the synthesis of homoleptic solvento-complexes by halide abstraction from compounds containing both coordinated halides and solvento-ligands. Palladium chloride is known to dissolve in nitriles, dimethylformamide, and dimethyl sulfoxide with the formation of complexes of the type $[PdCl_2(solv)_2]$ which can be isolated as solids.[3,64] The reaction of $[PdCl_2(RCN)_2]$ with an excess of silver ions[65] is a convenient method for the synthesis of $[Pd(RCN)_4]^{2+}$. The tetra(acetonitrile)-platinum(II) and tetra(propionitrile)platinum(II) salts have been prepared in this way[66,67] (Scheme 4.6). Wayland and Schramm[16] have synthesized homoleptic palladium(II) complexes with DMSO, DMF, and DMAA in a similar manner.

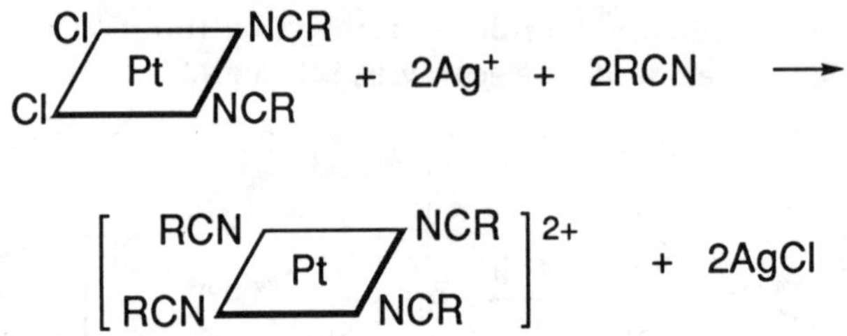

Scheme 4.6

In addition, anhydrous metal halides can be used for the preparation of solvento-complexes *via* halide abstraction. For example, halide abstraction from nickel(II) bromide[68] is used for the synthesis of $[Ni(EtOH)_6]X_2$ (X = ClO_4, NO_3), Eq. 4.10:

$$NiBr_2 + 2AgX + 6EtOH \rightarrow [Ni(EtOH)_6]X_2 + 2AgBr \qquad (4.10)$$

Studies[69,70] dealing with halide abstraction from $CoCl_2$ and $CuCl_2$ in methanol provide additional examples of the preparation of homoleptic solvento-complexes by this method.

A technique has been developed (see Table 4.2) for the synthesis of homoleptic solvento-complexes by halide abstraction from metal halides by Lewis acids, *i.e.* boron(III), aluminum(III), gallium(III), indium(III), thallium(III), tin(IV), antimony(V), and iron(III) halides.[71-77]

An additional technique is based on the interaction of metal oxides with acids or acid anhydrides in non-aqueous solvents.[78,79] This method is convenient but has not been widely used for the synthesis of homoleptic solvento-complexes in preparative coordination chemistry.

Table 4.2 describes routes to homoleptic solvento-complexes. The compounds are arranged by the metal's group number in the Periodic Table. Within each group, complexes are listed in descending order of the metal. For each metal, ligands are arranged in descending order of donor number.

Some classes of metal complexes are described by general formulae in Table 4.2, so it is best to examine all entries in a given group to ensure appropriate coverage. Table 4.2 is not intended to be exhaustive and is limited to complexes containing commonly-encountered solvents as ligands. It does, however, contain many compounds of importance in synthesis. Some selected examples of compounds containing metal-metal bonds and M=O functionalities are also included. To facilitate the use of Table 4.2 references are numbered differently from those in the main text of this Chapter and are collected separately (T1–T229).

Table 4.2

Homoleptic solvento-complexes and preparation methods

COMPOUNDS [REFS]	METHODS AND REAGENTS/NOTES
Cu	
$[Cu(HMP)_4][ClO_4]_2$ [T1, T2]	Dehydration of hydrated Cu(II) perchlorate by TEOF or DMP followed by addition of HMP. Yield of 91% [T1].
$[CuA_4][ClO_4]$	
A = py, γ-pic [T3]	Substitution of H_2O in $[Cu(H_2O)_6][ClO_4]_2$ in a solution of A followed by reduction with metallic Cu.
A = py [T4]	Substitution of MeCN in $[Cu(MeCN)_4][ClO_4]$ by py.
A = py [T5]	Reaction of $[Cu(MeCN)_4][ClO_4]$ with py. X-ray structure has been reported. X-ray absorption edge and EXAFS studies of $[Cu(py)_4]^+$ have been reported [T6].

$[CuA_4]B_2$ [T7] A = py, γ-pic B = CF_3CO_2, CCl_3CO_2	Dissolution of CuB_2 in Et_2O followed by addition of stoichiometric amounts of A. Crystallization from $CHCl_3$.
$[Cu(DMSO)_4][CuX_4]$ [T8, T9] X = Cl, Br	Dissolution of anhydrous CuX_2 in DMSO followed by precipitation of the complex by addition of C_6H_6.
$[Cu(DMSO)_4][ClO_4]_2$ [T10 - T12]	Dissolution of hydrated Cu(II) perchlorate in Me_2CO followed by addition of DMSO; refluxing of hydrated Cu(II) perchlorate in DMSO followed by solvent removal *in vacuo* [T11, T12]. Crystallization from DMSO [T12].
$[Cu(DMSO)_6][BF_4]_2$ [T13]	Dissolution of hydrated Cu(II) tetrafluoroborate in DMSO followed by solvent removal *in vacuo*. X-ray absorption edge and EXAFS studies of both $[Cu(DMSO)_4]^+$ and $[Cu(DMSO)_6]^{2+}$ have been reported [T6].
$[Cu\{OPMe(OPr)_2\}_4][ClO_4]_2$ [T14]	Dehydration of hydrated Cu(II) perchlorate by TEOF followed by addition of the ligand. Complex soluble in $MeNO_2$ and Me_2CO; sparingly soluble in $CHCl_3$ and CCl_4.
$[Cu(MeCN)_n]X$ n = 4, X = BF_4, ClO_4 n = 6, X = NO_3 [T15 - T18]	Oxidation of metallic Cu by $NOBF_4$ [T15, T16], Ph_3CBF_4, ArN_2BF_4 [T18], or $AgNO_3$ [T17] in MeCN. Oxidation by salts of Ph_3C^+ and ArN_2^+ is not recommended [T18].

[Cu(MeCN)$_4$]X X = ClO_4, BF_4, PF_6 [T18 - T22]	Reaction of Cu_2O with HX or $BF_3{\cdot}OEt_2$ [T18] in MeCN. Preparation of [Cu(MeCN)$_4$][PF_6] is described in *Inorganic Syntheses* [T20]. [Cu(MeCN)$_4$]$^+$ is soluble in Me_2CO [T23] and in MeOH/MeCN [T24]; X-ray crystal structure [T25], X-ray absorption edge, and EXAFS studies [T6] have been reported.
[Cu(MeCN)$_4$]X X = ClO_4, BF_4 [T26 - T28]	Reaction of anhydrous salts with MeCN in the presence of metallic Cu.
[Cu(MeCN)$_4$][BF_4] [T29]	Substitution of toluene in [Cu(BF_4)(C_6H_5Me)$_3$] by MeCN.
[Cu(MeCN)$_4$]X X = BCl_4, $AlCl_4$, $GaCl_4$, $InCl_4$, $TlCl_4$, $InBr_4$, $TlBr_4$, 1/2[$SnCl_6$], 1/2[$Sn(NCS)_6$] [T30 - T35]	Halide abstraction from anhydrous salts by the appropriate Lewis acids in MeCN.
[Cu(MeCN)$_4$]X [T36] X = PF_6, TaF_6	Reaction of anhydrous CuF_2 with PF_5 or TaF_5 followed by reduction of the Cu(II) species by metallic Cu.
[Cu(RCN)$_4$][BF_4] [T18] R = Et, CH_2Ph, Ph	Reaction of Cu_2O with $BF_3{\cdot}OEt_2$ in RCN.

$[Cu(MeCN)_n][ClO_4]_2$	
n = 4 [T37]	Dehydration of hydrated Cu(II) perchlorate by TEOF followed by addition of MeCN.
n = 6 [T38]	Dehydration of $[Cu(H_2O)_6][ClO_4]_2$ by P_4O_{10} in MeCN.
$[Cu(RCN)_6][SbCl_6]_2$ [T39, T40]	
R = Pr, Bu, Ph	Halide abstraction from $CuCl_2$ by $SbCl_5$ in RCN.
$[Cu(MeOH)_6][ClO_4]_2$ [T41]	Halide abstraction from $CuCl_2$ by $AgClO_4$ in MeOH. Compound formed in solution without isolation as the solid; composition established by NMR spectroscopy.
$[Cu(MeOH)_6][SO_3CF_3]_2$ [T42]	Dehydration of hydrated Cu(II) triflate in TMOF.
$[Cu(EtOH)_6][ClO_4]_2$ [T43]	Dehydration of hydrated Cu(II) perchlorate by TEOF.
$[Cu(AcOH)_6]X_2$ [T44]	
X = ClO_4, NO_3, BF_4	Dehydration of hydrated salts by acetic anhydride. Acetic acid ligands coordinate to copper through the carbonyl oxygens.
$[CuL_6][SbCl_6]$ [T45]	
L = EtOAc, Me_2CO	Substitution of $MeNO_2$ in $[Cu(MeNO_2)_6][SbCl_6]$ by L.
Ag, Au	
$[Ag(HMP)_2][ClO_4]$ [T2]	Dissolution of $AgClO_4$ in excess HMP.

[Ag(HMP)][ClO$_4$] [T2]	Interaction of $AgClO_4$ with a stoichiometric amount of HMP in Me_2CO followed by precipitation of the complex with Et_2O.
[Ag(py)$_2$]X [T46, T47] X = ClO_4, NO_3	Reaction of AgX salts with excess py [T46] or with py/MeCN [T47].
[Ag(py)$_4$][ClO$_4$] [T48]	Dissolution of $AgClO_4$ in py. Compound generated *in situ*. LAXS study has been reported [T48]. X-ray structure of [Ag(py)$_4$][ClO$_4$] has been reported [T5].
[Ag(DMSO)$_2$][ClO$_4$] [T49]	Addition of excess DMSO to a solution of $AgClO_4$ in Me_2CO followed by solvent removal *in vacuo* [T49]. X-ray structure and a LAXS study in DMSO solution have been reported [T49]. **Warning**: the compound is explosive [T50].
[AgL$_2$][NO$_3$] [T51] L = DMF, DMAA	Dissolution of $AgNO_3$ in excess ligand followed by precipitation of product on addition of C_6H_6.
[Ag(MeCN)$_4$]X [T52] X = NO_3, ClO_4	Dissolution of AgX in MeCN. Nitrate complex present in solution at concentrations of less than 4*M* [T52]. X-ray structure of [Ag(MeCN)$_4$][ClO$_4$] [T53] and LAXS studies [T54] have been reported.
[Ag(MeCN)$_4$]X [T55] X = BF_4, ClO_4	Oxidation of metallic Ag by NOX in MeCN.
[Ag(MeCN)$_4$][PF$_6$] [T56]	Reaction of Ag_2O with HPF_6 in MeCN.

[Ag(RCN)$_n$]X R = Me, n = 2, X = $TlCl_4$ [T30] R = Me, n = 4, X = $SbCl_6$ [T31 - T33] R = Me, Et, CH_2Ph, n = 4, X = BF_4 [T18] R = Me, n = 4, X = BF_4 [T18]	 Halide abstraction from AgCl by the appropriate Lewis acid in acetonitrile. Reaction of Ag_2O with $BF_3{\cdot}OEt_2$ in RCN. Oxidation of metallic Ag by Ph_3CBF_4 or ArN_2BF_4 in MeCN; not recommended as a preparative method.
[Ag($C_4H_8O_2$)$_3$][ClO_4] [T57]	Reaction of $AgClO_4$ with dioxane in suspension. Yield is almost quantitative.
[AgL$_n$][BF_4] [T18] n = 3, L = $C_4H_8O_2$, Et_2O, $MeOC_2H_4OMe$ n = 4, L = PhCN	 Dissolution of $AgBF_4$ in the appropriate ligand. Synthesis of $AgBF_4$ described in [T58].
[Au(MeCN)$_n$][$SbCl_6$] [T31 - T33, T59] n = 2, 3	 Halide abstraction from AgCl by $SbCl_5$ in MeCN.
[Au(MeCN)$_2$][SbF_6] [T60]	Recrystallization of [Au(CO)$_2$][Sb_2F_{11}] from acetonitrile; X-ray structure has been reported.
[Au(MeCN)$_4$][ClO_4] [T61]	Oxidation of metallic Au by $AgClO_4$ or $NOClO_4$ in MeCN. The solution structure has been determined by EXAFS [T61b].

Be	
$[BeL_4]X_2$ L = DMSO, DMF, DMAA, X = ClO_4 [T62 - T65] L = DEAA, X = ClO_4 [T66] L = DMF, X = ClO_4, NO_3 [T67]	Dehydration of $[Be(H_2O)_4]X_2$ by TEOF or DMP followed by addition of ligand. Yield *ca*. 90%; soluble in $MeNO_2$ [T66].
$[Be(MeCN)_n]X_2$ n = 4, X = $TlCl_4$ [T30] n = 4, X = $1/2[SnCl_6]$ [T31 - T33]	Halide abstraction from $BeCl_2$ by the appropriate Lewis acid in MeCN.
Mg, Ca, Sr, Ba	
$[M(HMP)_4]X_2$ M = Mg, Ca, Sr, Ba, X = ClO_4 [T1, T2]	Dehydration of hydrated salts by TEOF or DMP in Me_2CO followed by addition of HMP.
$[M(py)_6][CF_3SO_3]_2$ M = Sr, Ba [T68]	Dissolution of $M(CF_3SO_3)_2$ in py. LAXS and EXAFS studies have been reported.
$[Mg(DMSO)_6][MoFeS_4Cl_2]$ [T69]	X-ray structure has been reported. No information on synthesis has been found.
$[Mg(DMF)_6][(FeCl_3)O]$ [T69]	X-ray structure has been reported. No information on synthesis has been found.

$[MgL_6][ClO_4]_2$ [T70 - T72] L = DMF, TMP	Dehydration of hydrated $Mg(ClO_4)_2$ by TEOF followed by addition of ligand.
$[Mg(DMF)_6][ClO_4]_2$ [T73]	Reaction of anhydrous $Mg(ClO_4)_2$ with DMF. Compound has been structurally characterized.
$[M\{OPMe(OPr)_2\}_4][ClO_4]_2$ [T14] M = Mg, Ca, Sr	Dehydration of hydrated perchlorates by TEOF followed by addition of $OPMe(OPr_2)_2$. Sparingly soluble in $CHCl_3$ and CCl_4; soluble in $MeNO_2$, Me_2CO, EtOH, and dioxane.
$[M(MeCN)_n]X_2$ M = Mg, n = 6, X = BCl_4, $AlCl_4$, $GaCl_4$, $InCl_4$, $TlCl_4$, $FeCl_4$, $InBr_4$, $TlBr_4$, $FeBr_4$ [T30, T35] M = Ca, n = 6, X = $InCl_4$, $TlCl_4$, $InBr_4$; n = 7, X = $AlCl_4$, $GaCl_4$, $FeCl_4$; n = 8, X = BCl_4 [T30, T35] M = Sr, n = 7, X = $AlCl_4$, $InCl_4$; n = 8, X = $TlCl_4$, $FeCl_4$ [T30] M = Ba, n = 8, X = $GaCl_4$, $InCl_4$, $TlCl_4$, $FeCl_4$ [T30] M = Mg, Ca, n = 6, 7, X = 1/2$[SnCl_6]$ [T31, T33] M = Sr, n = 6, 8, X = 1/2$[SnCl_6]$ [T31, T33] M = Mg, n = 6, X = 1/2$[Sn(NCS)_6]$ [T34]	Halide abstraction by the appropriate Lewis acid in MeCN [T30, T35]. X-ray structure of $[Mg(MeCN)_6][SbCl_4]$ has been reported [T74].

$[Mg(RCN)_6][SbCl_6]_2$ [T39] R = Pr, Bu, Ph	Halide abstraction from $MgCl_2$ by $SbCl_5$ in RCN.
$[Mg(MeOH)_6][ClO_4]_2$ [T75]	Dehydration of $Mg(ClO_4)_2{\cdot}2H_2O$ by DMP. Compound recrystallized from a $MeOH/CH_2Cl_2$ mixture.
$[Mg(ROH)_6][ClO_4]_2$ [T76 - T78] R = Me, Et	Dissolution of $Mg(ClO_4)_2$ in ROH. Complexes generated *in situ;* composition determined by NMR spectroscopy.
$[M(ROH)_6]X_2$ M = Mg, R = Me, X = ClO_4, NO_3, BF_4 [T79] M = Mg, R = Et, X = ClO_4, NO_3 [T43] M = Ca, R = Et, X = NO_3 [T43]	Dehydration of hydrated salts by TEOF (R = Et) or TMOF (R = Me).
$[Mg(THF)_6][FeCl_5]$ [T80]	Halide abstraction from $[MgCl_2(THF)_2]$ by $FeCl_3$ in THF.
$M(ClO_4)_2{\cdot}2C_4H_8O_2$ [T81] M = Sr, Ba	Reaction of $M(ClO_4)_2$ with dioxane. Sparingly soluble in Et_2O; soluble in ROH, Me_2CO, and THF. Structures not reported.
$[Mg(AcOH)_6]X_2$ [T44] X = ClO_4, NO_3, BF_4	Dehydration of hydrated salts by acetic anhydride. Acetic acid ligands coordinate to magnesium through the carbonyl oxygens.
$[M(EtOAc)_6][SbCl_6]_2$ [T82] M = Mg, Ca	Halide abstraction by $SbCl_5$ in $EtOAc/MeNO_2$.

$[M(Me_2CO)_6][M'Cl_4]_2$ [T83] M = Mg, Ca, Sr M′ = In, Fe	Halide abstraction from anhydrous metal chlorides by the appropriate Lewis acid in $MeNO_2/Me_2CO$.
$[M(RNO_2)_6]X_2$ M = Mg, Ca, R = Ph, X = $SbCl_6$ [T84] M = Mg, Ca, Sr, R = Me, X = $SbCl_6$ [T85] M = Mg, R = Me, X = $FeCl_4$ [T86]	Halide abstraction from anhydrous metal chlorides by the appropriate Lewis acid in $PhNO_2/MeNO_2$ or in $MeNO_2$.
Zn, Cd, Hg	
$[M(HMP)_4]X_2$ M = Zn, X = BF_4, BPh_4, ClO_4 [T1, T87 - T89] M = Cd, X = BF_4, ClO_4 [T87]	Dehydration of hydrated salts by TEOF or DMP followed by addition of HMP.
$[ZnA_n]B_2$ n = 4, A = py, γ-pic, B = CF_3CO_2, CCl_3CO_2 [T7] n = 6, A = py, B = ClO_4 [T90]	Reaction of ZnB_2 with four equivalents of A in Et_2O. Compounds recrystallized from $CHCl_3$. Dissolution of hydrated $Zn(ClO_4)_2$ in excess py.
$[Hg(py)_6][CF_3SO_3]_2$ [T91]	Dissolution of $Hg(CF_3SO_3)_2$ in py. X-ray structures (298 and 183 K), TG and DSC analyses have been reported.

$[Zn(DMSO)_n][ClO_4]_2$ n = 4 [T10], 6 [T10 - T12, T49]	Reaction of $Zn(ClO_4)_2$ with an excess of DMSO [T11] or a mixture of DMSO with Me_2CO [T49] followed by removal of solvent *in vacuo* or crystallization of the complex at -20 °C [T49]. Method [T11] not recommended, see [T49]. Compound recrystallized from DMSO [T12]. X-ray structure and LAXS studies of $[Zn(DMSO)_6][ClO_4]_2$ have been reported; all DMSO ligands are *O*-bonded [T92].
$[M(DMSO)_6][BF_4]_2$ [T13, T93, T94] M = Zn, Cd	Reaction of hydrated $Zn(BF_4)_2$ with excess DMSO followed by removal of solvent *in vacuo* [T13]. Electrosynthesis [T93, T94]. Yields of 80-90% based on dissolved metal.
$[Cd(DMSO)_6][ClO_4]_2$ [T49]	Dehydration of hydrated $Cd(ClO_4)_2$ by DMP in Me_2CO solution followed by addition of DMSO. X-ray structure and a LAXS study in DMSO solution have been reported [T95]. X-ray structures of $[Cd(DMSO)_6][CdI_3(DMSO)]_2$ and $[Cd(DMSO)_6][CdI_4]$ have been reported also [T96].
$[Cd(DMSO)_6][CF_3SO_3]_2$ [T97]	Reaction of one equivalent of $Cd(CO_3)_2$ and two equivalents of HSO_3CF_3 in DMSO (> six equivalents) followed by evaporation of excess DMSO and crystallization on cooling. Recrystallized from DMSO.

$[Hg(DMSO)_6][ClO_4]_2$ [T49]	Dissolution of $Hg(ClO_4)_2{\cdot}4DMSO$ in MeOH followed by addition of DMSO.
$[Hg(DMSO)_6][ClO_4]_2$ [T98]	Dissolution of $Hg(ClO_4)_2{\cdot}4DMSO$ in DMSO followed by cooling of the solution. X-ray structure has been reported. All DMSO ligands are *O*-bonded.
$[ML_6][BF_4]_2$ [T99] M = Zn, L = DMF M = Cd, L = DMAA	Dehydration of hydrated salts by TEOF followed by addition of ligand.
$[ZnL_n][ClO_4]_2$ n = 6, L = DMF, DMAA [T100] n = 5, L = TMP [T89]	Dehydration of hydrated $Zn(ClO_4)_2$ by TEOF followed by addition of ligand. Yield of 75-90%.
$[M(MeCN)_6]X_2$ M = Zn , X = $GaCl_4$, $InCl_4$, $TlCl_4$, $FeCl_4$, $InBr_4$, $TlBr_4$, $1/2[Sn(NCS)_6]$ [T30, T34, T35] M = Cd, X = $TlCl_4$, $FeCl_4$, $TlBr_4$ [T30, T34]	Halide abstraction from MX_2 by the appropriate Lewis acid in MeCN.
$[Zn(RCN)_6][SbCl_6]_2$ [T39] R = Pr, Bu, Ph	Halide abstraction from anhydrous $ZnCl_2$ by $SbCl_5$ in RCN.
$[Hg(MeCN)_4][TlCl_4]_2$ [T30]	Halide abstraction from $HgCl_2$ by $TlCl_3$ in MeCN.
$[M(ROH)_6]X_2$ [T43, T79] M = Zn, Cd, R = Me, Et, X = ClO_4, NO_3, BF_4	Dehydration of hydrated salts by TEOF or TMOF.

$[Zn(MeOH)_6][ClO_4]_2$ [T101]	Dissolution of anhydrous $Zn(ClO_4)_2$ in MeOH. Complex generated *in situ*. Composition determined by NMR spectroscopy.
$[Zn(C_4H_8O_2)_6][ClO_4]_2$ [T90, T102]	Dissolution of hydrated $Zn(ClO_4)_2$ in excess dioxane.
$[Zn(AcOH)_6]X_2$ [T44] $X = ClO_4, NO_3, BF_4$	Dehydration of hydrated salts by acetic anhydride. Acetic acid ligands coordinate to zinc through the carbonyl oxygens.
$[ZnL_6]X_2$ L = EtOAc, X = $SbCl_6$ [T82] L = Me_2CO, X = $FeCl_4$, $InCl_4$ [T83] L = $MeNO_2$, X = $SbCl_6$ [T85] L = $PhNO_2$, X = $SbCl_6$ [T84]	Halide abstraction from $ZnCl_2$ by the appropriate Lewis acid in a mixture of ligand and $MeNO_2$.
Group IIIB and the Lanthanides	
$[M(HMP)_6][ClO_4]_3$ [T103, T104] M = Sc, Y, La , Ce - Lu	Halide abstraction from the appropriate chlorides by $AgClO_4$ in EtOH followed by addition of HMP [T103]. Dissolution of hydrated perchlorates in EtOH followed by addition of HMP [T104].
$[M(DMSO)_{7.5}][CF_3SO_3]_3$ [T105] M = La - Lu, Y	Dehydration of $M(CF_3SO_3)_3{\cdot}9H_2O$ by TEOF in DMSO. IR spectroscopic evidence indicates that the triflate ions are not coordinated.

$[M(DMSO)_n][ClO_4]_3$ [T106] n = 7, M = Sm, Gd, Y n = 8, M = La, Ce, Pr, Nd	Dissolution of hydrated perchlorates in MeOH followed by addition of DMSO.
$[M_n(TMSO)_m(NO_3)_p]$ [T107] M = La, n = 1, m = 4, p = 3 M = Nd, Sm, n = 2, m = 7, p = 6 M = Gd, Dy, Eu, Yb, n = 2, m = 6, p = 6	Dehydration of hydrated nitrates by DMP in EtOH followed by addition of TMSO. There is the possibility that NO_3^- coordinates to the metal ions.
$[Sc(DMSO)_6][ClO_4]_3$ [T108]	Addition of DMSO to a solution of $Sc(ClO_4)_3{\cdot}7H_2O$ in Me_2CO.
$M(PF_6)_3{\cdot}8DMF$ [T109] M = La - Lu, Y	Reaction of concentrated aqueous solutions of metal hexafluorophosphates with excess DMF followed by dehydration with TEOF. Structures not reported. Compounds are soluble in $MeNO_2$, MeCN, Me_2CO, MeOH, and H_2O, but insoluble in C_6H_6.
$[YL_6][ClO_4]_3$ [T110] L = TMP, DMAA, DEAA	Dehydration of hydrated $Y(ClO_4)_3$ by TEOF followed by addition of the ligand.
$[LuL_6][ClO_4]_3$ [T111] L = DMAA, DEAA	Dehydration of hydrated $Lu(ClO_4)_3$ followed by addition of the ligand.

$[ScL_6][ClO_4]_3$ L = DMSO, DMF, DEF, DMAA, DEAA [T112 - T114] L = TMP [T115, T116] L = $OPMe(OMe)_2$ [T117]	Dissolution of hydrated $Sc(ClO_4)_3$ in an excess of ligand followed by removal of the solvent [T112]. Dehydration of hydrated $Sc(ClO_4)_3$ by TEOF followed by addition of the ligand [T115 - T117].
$La(MeCN)_9[SbCl_6]_3$ [T118]	Halide abstraction from anhydrous $LaCl_3$ by $SbCl_5$ in MeCN. Structure not reported.
$[Nd(MeCN)_4][ClO_4]_3$ [T119, T120]	Dissolution of $Nd(ClO_4)_3$ in acetonitrile.
$[Eu(MeCN)_x]Y_3$ [T120, T121] Y = BF_4, PF_6 , value of x not specified	Oxidation of Eu by NOY in acetonitrile.
$M(ClO_4)_3 \cdot x MeCN$ [T122] M = Er, Pr; value of x not specified	Dehydration of hydrated salts by molecular sieves in MeCN. Composition has not been established.
U	
$[U(DMSO)_8][ClO_4]_4$ [T123]	Electrolytic reduction of an aqueous solution of hydrated perchlorate followed by addition of DMSO and filtration of the precipitate formed.
$[UO_2(HMP)_4][ClO_4]_2$ [T124]	Dehydration of hydrated perchlorate by TEOF followed by addition of HMP.

$[UO_2L_5][ClO_4]_2$ [T125 - T130] L = DMSO, DMF, DMAA, DEAA, TMP, TEP	Dehydration of hydrated perchlorate by TEOF followed by addition of the ligand. X-ray structure of the DMSO complex has been reported [T130].
Al, In, Ga	
$[M(HMP)_4][ClO_4]_3$ M = Al [T2] M = Al, In [T1]	Dehydration of hydrated perchlorates by DMP or TEOF followed by addition of HMP; yield of 89%.
$[M(DMSO)_6][ClO_4]_3$ M = Al [T131] M = Ga, In [T10, T131, T132]	Dehydration of hydrated perchlorates by TEOF followed by addition of DMSO [T131]. Addition of DMSO to solutions of the hydrated perchlorates in Me_2CO [T10, T131, T132]. X-ray structure of $[In(DMSO)_6][ClO_4]_3$ has been reported. The ligands are all *O*-bonded [T133].
$[In(DMSO)_6][BF_4]_3$ [T93, T94]	Electrosynthesis. Yields of 80-90% based on dissolved metal.
$[InL_6][ClO_4]_3$ [T134] L = DMSO, DMAA	Reaction of $In(ClO_4)_3$ with L.

$[M(DMF)_6][ClO_4]_3$	
M = Al, Ga [T135, T136]	Dehydration of hydrated perchlorates by DMP followed by addition of DMF.
M = Al, Ga, In [T131, T132]	Reaction of DMF with a solution of the hydrated perchlorate in Me_2CO. Recrystallized from dimethylformamide.
$[Al(DMF)_6][SbCl_6]_3$ [T137]	Halide abstraction from $AlCl_3$ by $SbCl_5$ in DMF.
$[M(TMP)_6][ClO_4]_3$ [T138]	
M = Al, Ga, In	Dehydration of hydrated perchlorates by TEOF followed by addition of TMP.
$M(MeCN)_9[SbCl_6]_3$ [T118]	
M = Al, Ga, In	Halide abstraction from the appropriate chlorides by $SbCl_5$ in MeCN. No structural information available.
$[In(MeCN)_6][BF_4]_3$ [T93, T94]	Electrosynthesis. Yields of 80-90% based on the dissolved metal.
$[Al(MeOH)_6]X_3$	
X = ClO_4, NO_3 [T139]	Dissolution of AlX_3 in MeOH. Complexes prepared *in situ*. Composition of complexes in solution determined by NMR spectroscopy.
X = ClO_4 [T101]	Dehydration of hydrated $Al(ClO_4)_3$ by molecular sieves for 150 hours.

Tl	
$[Tl(H_2O)_6][ClO_4]_3$ [T140 - T142]	Generated *in situ* by anodic oxidation of $TlClO_4$ in water. X-ray diffraction measurements of aqueous solutions have been reported [T140 - T141]. Generated *in situ* by dissolution of Tl_2O_3 in concentrated $HClO_4$ at *ca.* 50 °C [T142]
$[Tl(MeCN)_2][SbCl_6]$ [T31 - T33]	Halide abstraction from TlCl by $SbCl_5$ in MeCN.
Ti	
$[Ti(DMF)_6][SO_3CF_3]_3$ [T143]	Dissolution of anhydrous $Ti(SO_3CF_3)_3$ in TEOF followed by addition of dimethylformamide and precipitation with anhydrous Et_2O.
$[Ti(MeCN)_6][BF_4]_3$ [T93, T94]	Oxidation of metallic Ti by HBF_4 in MeCN.
$[TiO(DMSO)_5][SO_3CF_3]_2$ [T144]	Dissolution of anhydrous $Ti(SO_3CF_3)_3$ in TEOF followed by addition of excess DMSO and filtration of the precipitate formed. The oxidation of Ti(III) to form oxotitanium(IV) is due to oxygen abstraction from DMSO by Ti(III).

Sn, Pb	
$[SnL_n][ClO_4]_2$ [T145] L = DMSO, n = 4 L = DMF, n = 3	Dehydration of hydrated $Sn(ClO_4)_2$ by TEOF in EtOH followed by addition of ligand, L. Complexes soluble in DMF, MeCN, and $MeNO_2$ but insoluble in $CHCl_3$, EtOH, and THF.
$Sn(ClO_4)_2 \cdot 2MeCN$ [T146]	Electrosynthesis. No structural information available.
$[Pb(HMP)_3][ClO_4]_2$ [T2]	Dehydration of hydrated $Pb(ClO_4)_2$ by DMP in Me_2CO followed by addition of HMP. Yield of 90%.
$Pb(ClO_4)_2 \cdot 2C_4H_8O_2$ [T81]	Interaction of $Pb(ClO_4)_2$ with dioxane. No structural information available. Complex sparingly soluble in Et_2O and soluble in ROH, Me_2CO, and THF.
V	
$[V(DMSO)_6][BF_4]_2$ [T93, T94]	Electrosynthesis. Yield of 80-90% based on dissolved metal.
$[V(DMSO)_6][ClO_4]_3$ [T147]	Addition of DMSO to aqueous $V(ClO_4)_3$, partial evaporation of solvent followed by filtration of precipitate formed. Compound recrystallized from DMSO.
$V(SbCl_6)_3 \cdot 9MeCN$ [T118]	Halide abstraction from VCl_3 by $SbCl_5$ in MeCN. No structural information given.

$VX_2 \cdot 6ROH$ [T148, T149] X = Br, R = Et X = I, R = Me	Dehydration of hydrated vanadium halides by TEOF or TMOF. ROH molecules may be coordinated.
$[V(MeOH)_6]Cl_2$ [T150]	Reaction of $[V_2Cl_3(THF)_6][AlCl_2Et_2]$ with MeOH.
$[V(MeOH)_6]Cl_3$ [T151]	Dissolution of VCl_3 in MeOH followed by removal of solvent *in vacuo*.
Cr, Mo	
$[Cr(HMP)_6][ClO_4]_3$ [T2]	Dehydration of hydrated $Cr(ClO_4)_3$ by DMP followed by addition of HMP. Yield of 81%.
$[Cr(DMSO)_6][BF_4]_3$ [T93, T94]	Electrosynthesis. Yield of 80-90% based on dissolved metal.
$[Cr(DMSO)_6]X_3$ X = ClO_4, NO_3, Br [T11, T152 - T155]	Dissolution of CrX_3 in DMSO followed by removal of solvent by distillation.
X = Cl [T156] X = Br, I [T157]	Gaseous HCl bubbled through a solution of $[Cr(DMSO)_6]X_3$ [T156] or addition of aqueous solutions of KBr or KI [T157] to a solution of $[Cr(DMSO)_6]X_3$, where X = ClO_4, NO_3, in DMSO.
$[CrL_6][ClO_4]_3$ [T158] L = DMSO, DMF	Hydrated $Cr(ClO_4)_3$ heated in a mixture of ligand/isopropanol.
$[Cr(DMF)_6][ClO_4]_3$ [T159]	Dissolution of anhydrous $Cr(ClO_4)_3$ in DMF at 100 °C. Complex recrystallized twice from DMF.

$[Cr(DMAA)_6][ClO_4]_3$ [T160]	Dehydration of hydrated $Cr(ClO_4)_3$ by DMP followed by addition of DMAA.
$[Cr(MeCN)_n][BF_4]_3$ [T93, T94] n = 4 or 6	 Electrosynthesis.
$Cr(SbCl_6)_3\cdot 9MeCN$ [T118]	Halide abstraction from $CrCl_3$ by $SbCl_5$ in MeCN. No structural information available.
$[Cr(CD_3OD)_6][O_3SCF_3]_3$ [T161]	Reaction of CD_3OD and CF_3SO_3H with $Cr(OAc)_3$; removal of solvent *in vacuo*; process repeated twice.
$[Cr(THF)_6][BF_4]_3$ [T162]	Halide abstraction from $[CrCl_3(THF)_3]$ by $AgBF_4$ in tetrahydrofuran. Complex obtained *in situ*; decomposes when isolated as solid.
$[Mo_2(MeCN)_8][SO_3CF_3]_4$ [T163]	Dissolution of $[Mo_2(O_3SCF_3)_2(H_2O)_4][O_3SCF_3]_2$ in MeCN.
$[Mo_2(MeCN)_8($*ax*$\text{-}MeCN)_{0.5}][BF_4]_4$ [T164, T165]	Reaction of $[Mo_2(O_2CMe)_4]$ with $HBF_4\cdot Et_2O$ in $MeCN/CH_2Cl_2$ followed by drying the compound formed at 35 °C. X-ray structure has been reported.
Mn, Tc, Re	
$[Mn(HMP)_4]X_2$ X = ClO_4 [T1, T2] X = BF_4 [T87]	Dehydration of hydrated salts by TEOF or DMP followed by addition of HMP.

$[MnA_4]B_2$ [T7] A = py, γ-pic, B = CF_3CO_2, CCl_3CO_2	Reaction of MnB_2 with four equivalents of A in Et_2O. Complexes recrystallized from $CHCl_3$.
$[Mn(DMSO)_6][BF_4]_2$ [T93, T94]	Oxidation of metallic Mn by HBF_4 in DMSO.
$[Mn(DMSO)_6]X_2$ X = ClO_4 [T10, T12, T71, T152] X = Br, I [T155] X = BF_4 [T13]	Dissolution of hydrated salts in Me_2CO followed by addition of or dissolution in DMSO and removal of solvent *in vacuo.*
$[Mn(DMSO)_4][MnCl_4]$ [T8, T9]	Dissolution of $MnCl_2$ in dimethyl sulfoxide and precipitation of complex by EtOH.
$[MnL_6][ClO_4]_3$ [T166, T167] L = DMSO, DMF	Dissolution of $Mn(OAc)_3{\cdot}2H_2O$ in the ligand followed by precipitation of the complex by aqueous $HClO_4$.
$[Mn(MeCN)_6][ClO_4]_2$ [T168]	Dehydration of $[Mn(H_2O)_6][ClO_4]_2$ by molecular sieves in MeCN.
$[Mn(MeCN)_6]X_2$ X = $AlCl_4$, $GaCl_4$, $InCl_4$, $TlCl_4$, $FeCl_4$ [T30] X = $InBr_4$, $TlBr_4$, $FeBr_4$ [T35] X = 1/2$[Sn(NCS)_6]$ [T34] X = 1/2$[SnCl_6]$ [T31 - T33]	Halide abstraction from anhydrous chlorides by the appropriate Lewis acid in MeCN.
$[Mn(MeCN)_6][MnI_4]$ [T169]	Dissolution of MnI_2 in MeCN.

$[Mn(ROH)_6]X_2$ [T43, T79] R = Me, Et X = ClO_4, NO_3, BF_4	Dehydration of $[Mn(H_2O)_6]X_2$ by TEOF or TMOF.
$[Mn(AcOH)_6]X_2$ [T44] X = ClO_4, NO_3, BF_4	Dehydration of $[Mn(H_2O)_6]X_2$ by acetic anhydride. Acetic acid ligands are coordinated to Mn through the carbonyl oxygens.
$[MnL_6]X_2$ L = EtOAc, X = $SbCl_6$ [T82] L = Me_2CO, X = $InCl_4$, $FeCl_4$ [T83] L = $MeNO_2$, X = $SbCl_6$, $FeCl_4$ [T85, T86] L = $PhNO_2$, X = $SbCl_6$ [T84]	Halide abstraction from $MnCl_2$ by Lewis acid in a mixture of ligand/$MeNO_2$.
$[Tc_2(MeCN)_{10}][BF_4]_4$ [T170a]	Reaction of $[Tc_2Cl_4(PR_3)_4]$ with $HBF_4 \cdot Et_2O$ in a mixture of MeCN/CH_2Cl_2; reduction of $[Bu_4N][TcCl_6]$ with either Bu_3SnH or Zn, followed by treatment with $HBF_4 \cdot Et_2O$ in MeCN/CH_2Cl_2.
$[Re_2(MeCN)_{10}][X]_n$ [T170b] X = BF_4, n = 4 X = Mo_6O_{19}, n = 2	Reaction of $[(n\text{-}Bu)_4N]_2[Re_2Cl_8]$ or $[Re_2Cl_4\{P(n\text{-}Pr)_3\}_4]$ with $HBF_4 \cdot OEt_2$ in a mixture of MeCN/CH_2Cl_2. Yields are 15% and 80%, respectively. X-ray structure of $[Re_2(MeCN)_{10}]$-$[Mo_6O_{19}]_2 \cdot 4MeCN \cdot 2H_2O$ has been reported.

Fe, Ru, Os	
$[Fe(HMP)_n]X_m$ n = 4, m = 2, 3, X = ClO_4 [T1, T2] n = 4, m = 2, X = BPh_4, BF_4 [T87, T171] n = 6, m = 3, X = ClO_4 [T2]	Dehydration of hydrated salts by TEOF or DMP followed by addition of HMP. IR spectroscopic, X-ray powder diffraction [T1] and Mössbauer spectroscopic studies have been reported [T171a].
$[FeA_4]Cl_2$ [T171, T172] A = py, γ-pic, β-pic, isoquinoline	Addition of A, or a solution of A in methanol, to a concentrated aqueous solution of $FeCl_2$. An electronic absorption spectroscopic study has been reported [T172].
$[FeA_4]X_2$ [T173] A = py, X = Br, I A = γ-pic, X = I	Oxidation of metallic Fe by X_2 in MeOH followed by addition of four equivalents of A.
$[Fe(\gamma\text{-pic})_4]Cl_2$ [T174]	Dehydration of $FeCl_2{\cdot}4H_2O$ by DMP in deoxygenated ethanol over iron filings followed by addition of γ-pic. A Mössbauer spectroscopic study has been reported.

[Fe(DMSO)$_6$]X$_m$·nDMSO X = ClO_4, n = 0, m = 2 [T10] X = BF_4, n = 0, m = 2 [T13] X = ClO_4, n = 0, m = 3 [T10, T152, T155] X = NO_3, n = 0, m = 3 [T155] X = ClO_4, n = 1, m = 3 [T11] X = [$Cl_2Fe(MoOS_3)$], [$Cl_2Fe(MoOS_4)$], m = 1, n = 0 [T175]	Dissolution of hydrated salts in DMSO followed by removal of excess solvent *in vacuo.* X-ray structures of [Fe(DMSO)$_6$][$Cl_2Fe(MoOS_3)$] and [Fe(DMSO)$_6$][$Cl_2Fe(MoOS_4)$] have been reported [T175]. All ligands are *O*-bonded.
[Fe(DMSO)$_6$][ClO$_4$]$_2$ [T176]	Dehydration of hydrated $Fe(ClO_4)_2$ by TEOF followed by addition of anhydrous DMSO, filtration, and recrystallization from DMSO/$MeNO_2$.
[Fe(DMSO)$_6$][BF$_4$]$_3$ [T93, T94]	Electrosynthesis. Yields of 80-90% based on dissolved metal.
[Fe(DMF)$_6$]X$_2$ X = ClO_4 [T177] X = $SbCl_6$ [T137] X_2 = [$Cl_2Fe(WS_4)$] [T175]	Dissolution of hydrated $Fe(ClO_4)_2$ in DMF, removal of H_2O by distillation followed by precipitation of compound with Et_2O or hexane. X-ray structure has been reported [T178]. Halide abstraction from $FeCl_2$ by $SbCl_5$ in DMF/CCl_4 [T137]. X-ray structure of [Fe(*O*-DMF)$_6$][$Cl_2Fe(WS_4)$] has been reported [T175].

[Fe(MeCN)$_n$]X$_2$ [T30 - T35] n = 6, X = BCl_4, $AlCl_4$, $GaCl_4$, $InCl_4$, $TlCl_4$, $FeCl_4$, $InBr_4$, $TlBr_4$, $FeBr_4$, 1/2[Sn(NCS)$_6$], 1/2[$SnCl_6$] n = 7, X = 1/2[$SnCl_6$]	Halide abstraction from anhydrous salts by the appropriate Lewis acid in MeCN.
[Fe(MeCN)$_6$][ClO$_4$]$_2$ [T179]	Dehydration of hydrated $Fe(ClO_4)_2$ by molecular sieves in MeCN.
[Fe(MeCN)$_6$][I$_3$]$_2$ [T180]	Interaction of FeI_2 with I_2 in acetonitrile.
[Fe(EtOH)$_6$][ClO$_4$]$_2$ [T43]	Dehydration of hydrated $Fe(ClO_4)_2$ by TEOF.
[FeL$_6$]X$_2$ [T82 - T86] L = EtOAc, X = $SbCl_6$ L = Me_2CO, X = $FeCl_4$, $InCl_4$ L = $MeNO_2$, X = $FeCl_4$, $SbCl_6$ L = $PhNO_2$, X = $SbCl_6$	Halide abstraction from $FeCl_2$ by the appropriate Lewis acid in a mixture of ligand/$MeNO_2$.
[FeL$_6$][FeX$_4$] [T180, T181] L = MeCN, X = Cl, Br, I L = EtOAc, $MeNO_2$, X = Br	Oxidation of metallic Fe by X_2 in a solution of ligand.
[Ru(py)$_6$][PF$_6$]$_2$ [T182]	Zn reduction of [$RuCl_2Cp^*$]$_n$ in EtOH, Me_2CO or THF followed by addition of py and KPF_6, evaporation to dryness, and extraction with CH_2Cl_2. Yield is *ca.* 10% based on Ru.

[Ru(DMSO)$_6$]X$_n$	
n = 2, X = ClO_4 [T183a]	Halide abstraction from [$RuCl_2(DMSO)_4$] by $AgClO_4$ in DMSO/ethanol. X-ray structure of [Ru(*S*-DMSO)$_3$(*O*-DMSO)$_3$][BF_4]$_2$ has been reported [T183b].
n = 3, X = Cl [T184, T185]	Hydrated $RuCl_3$ heated in DMSO/ethanol.
n = 3, X = ClO_4 [T185]	Removal of halide ions from [Ru(DMSO)$_6$]Cl_3 by $AgClO_4$ in DMSO.
[Ru(DMF)$_6$][O_3SCF_3]$_n$ [T186]	
n = 2, 3	Dehydration of [Ru(H_2O)$_6$][O_3SCF_3]$_2$ by TEOF in DMF (n = 2). Oxidation of [Ru(DMF)$_6$][O_3SCF_3]$_2$ by $AgSO_3CF_3$ in DMF (n = 3).
[Ru(DMF)$_6$][O_3SCF_3]$_n$ [T187]	
n = 2	Reduction of [Ru(DMF)$_6$][O_3SCF_3]$_3$ by H_2 in DMF in the presence of Pt black.
n = 3	Reduction of $RuCl_3{\cdot}xH_2O$ by Sn in DMF followed by addition of $Pb(SO_3CF_3)_2$ and stirring in air.
	X-ray structures of both [Ru(DMF)$_6$][O_3SCF_3]$_n$ (n = 2, 3) have been reported. Coordination of DMF to Ru(II) or Ru(III) is through oxygen. The same conclusion was reached for both complexes in DMF solution based upon NMR spectroscopy.

$[Ru(DMF)_6][O_3SC_6H_4Me]_2$ [T187, T188]	Dehydration of $[Ru(H_2O)_6][O_3SC_6H_4Me]_2$ [T188] by TEOF in DMF at room temperature under Ar [T187]. Reduction of $RuCl_3 \cdot xH_2O$ by Sn in dimethylformamide followed by addition of $TlO_3SC_6H_4Me$ [T187].
$[Ru(H_2O)_6][O_3SC_6H_4Me]_2$ [T189]	Oxidation of $RuO_2 \cdot xH_2O$ by $NaIO_4$ to produce RuO_4 followed by reaction with metallic Pb to form $[Ru(H_2O)_6]^{2+}$. Ion exchange and crystallization gives the tosylate salt in 70 - 80 % yield. Compound recrystallized from an aqueous solution of *p*-toluenesulfonic acid [T190].
$[Ru(MeCN)_6][O_3SCF_3]_2$ [T187, T191]	Dissolution of $[Ru(H_2O)_6][O_3SCF_3]_2$ in MeCN followed by stirring at 80 °C and filtration of the precipitate formed [T191]. The compound is one of the most stable and inert complexes of Ru(II) known. Attempts to oxidize it chemically to form $[Ru(MeCN)_6]^{3+}$ were unsuccessful. When $[Ru(H_2O)_6][O_3SCF_3]_3$ was used as the starting material, $[Ru(MeCN)_6]^{2+}$ was again obtained [T187].
$[Ru(MeCN)_6][PF_6]_2$ [T192a]	Reaction of $K_3[Ru_2(\mu\text{-}SO_4)_4(H_2O)_2]$ with acetonitrile followed by addition of HO_3SCF_3, reflux, and addition of $[Bu_4N][PF_6]$. Yield is 60%. X-ray structure has been reported.
$[Os(MeCN)_6][PF_6]_2$ [T192b]	Photolysis of $[Os(\eta^6\text{-}C_6H_6)_2][PF_6]_2$ in MeCN.

Co	
$[Co(HMP)_4]X_2$ [T87, T88] $X = ClO_4, BF_4$	Dehydration of hydrated salts by TEOF or DMP followed by addition of HMP. Soluble in $CHCl_3$.
$[CoA_4]B_2$ [T7] A = py, γ-pic, $B = CF_3CO_2, CCl_3CO_2$	Reaction of CoB_2 with four equivalents of A in Et_2O. Complex recrystallized from $CHCl_3$.
$[Co(DMSO)_6]X_2$ [T10 - T13, T152, T155] $X = ClO_4, NO_3, BF_4, I$	Dissolution of hydrated salts in DMSO followed by removal of excess solvent *in vacuo*.
$[Co(DMSO)_4][CoCl_4]$ [T8, T9]	Dissolution of $CoCl_2$ in dimethyl sulfoxide and precipitation of the complex by addition of EtOAc. Yield of 96%. Recrystallized from $MeNO_2$/EtOAc.
$[Co(DMSO)_6][ClO_4]_2$ [T176]	Dehydration of hydrated $Co(ClO_4)_2$ by TEOF followed by addition of anhydrous DMSO, filtration, and recrystallization from DMSO/$MeNO_2$.
$[Co(DMSO)_6][BF_4]_3$ [T93, T94]	Electrosynthesis. Yield 80-90% based on dissolved metal.

$[CoL_6]X_2$ L = DMF, DEF, DMAA, X = ClO_4, BF_4 [T99, T193] L = DMF, X = $SbCl_6$ [T137] L = DMAA, X = ClO_4 [T194]	 Dehydration of hydrated salts by TEOF or DMP followed by addition of ligand. Halide abstraction from $CoCl_2$ by $SbCl_5$ in DMF. Reaction of $[Co(H_2O)_6][ClO_4]_2$ with excess DMAA followed by solvent removal *in vacuo*. Pure compound prepared with three repetitions of procedure.
$[Co(MeCN)_6][ClO_4]_2$ [T38, T195, T196]	Dehydration of $[Co(H_2O)_6][ClO_4]_2$ by P_4O_{10} [T38] or DMP in MeCN.
$[Co(MeCN)_6]X_2$ [T30 - T35] X = BCl_4, $AlCl_4$, $GaCl_4$, $InCl_4$, $TlCl_4$, $FeCl_4$, 1/2$[Sn(NCS)_6]$, 1/2$[SnCl_6]$	 Halide abstraction from anhydrous Co(II) halides by the appropriate Lewis acid in MeCN.
$[Co(MeCN)_6][CoI_4]$ [T169]	Dissolution of CoI_2 in MeCN.
$[Co(RCN)_6][SbCl_6]_2$ [T39] R = Pr, Bu, Ph	 Halide abstraction from anhydrous $CoCl_2$ by $SbCl_5$ in RCN.
$[Co(ROH)_6]X_2$ [T43, T79] R = Me, Et X = ClO_4, NO_3, BF_4	 Dehydration of $[Co(H_2O)_6]X_2$ by TMOF or TEOF without solvent.
$[Co(MeOH)_6][ClO_4]_2$ [T197]	Halide abstraction from $CoCl_2$ by $AgClO_4$ in MeOH.

$[Co(AcOH)_6]X_2$ [T44] X = ClO_4, NO_3, BF_4	Dehydration of $[Co(H_2O)_6]X_2$ by acetic anhydride. Acetic acid ligands coordinate to cobalt through the carbonyl oxygens.
$[CoL_6]X_2$ [T82 - T86, T198] L = EtOAc, X = $SbCl_6$ L = Me_2CO, X = $FeCl_4$, $InCl_4$ L = $MeNO_2$, X = $FeCl_4$, $SbCl_6$ L = $PhNO_2$, X = $SbCl_6$	Halide abstraction from $CoCl_2$ by the appropriate Lewis acid in a mixture of ligand/$MeNO_2$.
Rh	
$[Rh(DMSO)_6][BF_4]_3$ [T199]	Reaction of $Rh(OH)_3$ with HBF_4 in DMSO [T199]. Method not reproducible [T200, T201].
$[Rh(H_2O)_6][ClO_4]_3$ [T202 - T204]	Dissolution of halide-free hydrated Rh(III) oxide in concentrated $HClO_4$ (*ca.* 70%), reflux at 95 °C for 48h. Evaporate under reduced pressure at 60 °C until crystals form. Kinetic study of proton exchange [T203b].
$[Rh_2(MeCN)_{10}][BF_4]_4$ [T205]	Reaction of $[Rh_2(O_2CMe)_4(MeOH)_2]$ and $[Et_3O][BF_4]$ in $MeCN/CH_2Cl_2$ under reflux. High-pressure ^1H NMR study of MeCN exchange kinetics.

Ni	
$[Ni(HMP)_4]X_2$ [T1, T87, T88] X = ClO_4, BF_4	Dehydration of hydrated salts by TEOF or DMP followed by addition of HMP.
$[NiA_4]B_2$ [T7] A = py, γ-pic, B = CF_3CO_2, CCl_3CO_2	Reaction of NiB_2 with four equivalents of A in Et_2O. Complex recrystallized from $CHCl_3$.
$[Ni(DMSO)_6]X_2$ X = ClO_4, BF_4, NO_3, I [T9 - T12, T152, T155]	Dissolution of hydrated salts in DMSO followed by removal of solvent *in vacuo*.
X = ClO_4 [T176, T206]	Dehydration of hydrated $Ni(ClO_4)_2$ by TEOF in DMSO.
X = BF_4 [T93, T94]	Electrosynthesis. Yields of 80-90% based on dissolved metal.
$[Ni(DMSO)_4][NiCl_4]$ [T8]	Dissolution of $NiCl_2$ in DMSO followed by precipitation with benzene.
$[Ni(DMAA)_6][BF_4]_2$ [T207]	Dissolution of hydrated $Ni(BF_4)_2$ in DMAA and removal of H_2O and excess DMAA with a rotary evaporator below 70 °C. Hygroscopic material recrystallized from dimethylacetamide and dried *in vacuo* over P_4O_{10} at room temperature.

$[NiL_6]X_2$	
L = DMF, DEF, DMAA, X = ClO_4, BF_4 [T99, T160, T193, T208]	Dehydration of hydrated salts by TEOF or DMP followed by addition of ligand.
L = DMF, X = $SbCl_6$ [T137]	Halide abstraction from $NiCl_2$ by $SbCl_5$ in DMF/CCl_4.
L = EtOAc, X = $SbCl_6$ L = Me_2CO, X = $FeCl_4$ L = $MeNO_2$, X = $FeCl_4$, $SbCl_6$ L = $PhNO_2$, X = $SbCl_6$ [T83 - T86]	Halide abstraction from $NiCl_2$ by the appropriate Lewis acid in a mixture of ligand/$MeNO_2$.
$[Ni(MeCN)_6][ClO_4]_2$	
[T38, T195, T209 - T211]	Dehydration of $[Ni(H_2O)_6][ClO_4]_2$ by P_4O_{10} [T38, T209, T210], DMP or TEOF in MeCN [T195, T211]. Dehydration by DMP does not result in complete removal of coordinated H_2O [T209].
[T212]	Oxidation of metallic Ni by $NOClO_4$ in MeCN.
$[Ni(MeCN)_6][NiCl_4]$ [T169]	Dissolution of $NiCl_2$ in MeCN.
$[Ni(MeCN)_n]X_2$ [T30 - T35, T213]	
n = 4, X = BCl_4, $AlCl_4$, $GaCl_4$, $InCl_4$, $TlCl_4$, $FeCl_4$ n = 6, X = 1/2$[ZnCl_4]$, 1/2$[Sn(NCS)_6]$, 1/2$[SnCl_6]$ n = 7, X = 1/2$[SnCl_6]$	Halide abstraction from $NiCl_2$ by the appropriate Lewis acid in MeCN.
$[Ni(RCN)_6][SbCl_6]_2$ [T39]	
R = Pr, Bu, Ph	Halide abstraction from $NiCl_2$ by $SbCl_5$ in RCN.

[Ni(ROH)$_6$]X$_2$ R = Me, Et, X = ClO_4, BF_4, NO_3 [T43, T79, T198, T214] R = Et, X = ClO_4, NO_3 [T215]	 Dehydration of hydrated salts by TEOF, TMOF, DMP, or DEP. Halide abstraction from $NiBr_2$ by AgX in EtOH.
[Ni(AcOH)$_6$]X$_2$ [T44] X = ClO_4, BF_4, NO_3	 Dehydration of [Ni(H_2O)$_6$]X$_2$ by acetic anhydride. Acetic acid ligands coordinate through the carbonyl oxygens.
Pd	
[PdL$_4$][BF$_4$]$_2$ [T216, T217, T218] L = DMSO, DMF, DMAA	 Halide abstraction from [$PdCl_2L_2$] by $AgBF_4$ in a solution of the ligand; substitution of MeCN in [Pd(MeCN)$_4$][BF$_4$]$_2$ with L. X-ray structure of *cis*-[Pd(*O*-DMSO)$_2$(*S*-DMSO)$_2$][BF$_4$]$_2$·DMSO has been reported [T218].
[Pd(DMSO)$_4$][NO$_3$]$_2$ [T219]	Dissolution of K$_2$[Pd(NO$_3$)$_4$] in DMSO. EXAFS study has been reported.
[Pd(MeCN)$_4$][TlCl$_4$]$_2$ [T30]	Halide abstraction from $PdCl_2$ by $TlCl_3$ in MeCN.
[Pd(MeCN)$_4$]X$_2$ X = BF_4 [T216, T217] X = ClO_4, BF_4 [T220 - T222]	 Oxidation of metallic Pd by $NOBF_4$ in MeCN. Halide abstraction from [$PdCl_2$(MeCN)$_2$] by AgX in acetonitrile.

Pt	
$[Pt(py)_4]Cl_2{\cdot}3H_2O$ [T223]	Reaction of $[PtCl_2(py)_2]$ with py in aqueous solution followed by precipitation of the complex on addition of $EtOH/Et_2O$.
$[Pt(py)_4][O_2CC_6F_5]_2$ [T224]	Halide abstraction from *trans*-$[PtCl_2(py)_2]$ by $TlO_2CC_6F_5$ in pyridine.
$[Pt(DMSO)_4][CF_3SO_3]_2$ [T225]	Reaction of $Pt(OH)_2{\cdot}xH_2O$ with HO_3SCF_3 in DMSO. X-ray structure of *cis*-$[Pt(O\text{-}DMSO)_2(S\text{-}DMSO)_2][O_3SCF_3]_2$ has been reported [T225]. X-ray absorption edge, EXAFS spectra and LAXS studies have been reported [T219].
$[Pt(H_2O)_4][ClO_4]_2$ [T226]	Generated *in situ* through halide abstraction from $K_2[PtCl_4]$ by $AgClO_4$ followed by removal of a large excess of Ag^+ by electrolysis.
$[Pt(EtCN)_4][CF_3SO_3]_2$ [T227]	Halide abstraction from *cis*-$[PtCl_2(EtCN)_2]$ by $AgSO_3CF_3$ in $CH_2Cl_2/EtCN$ followed by removal of solvent *in vacuo* and precipitation of the complex by addition of Et_2O. The compound is soluble in common organic solvents, relatively insensitive to atmospheric moisture, and exhibits good thermal stability.
$[Pt(MeCN)_4][CF_3SO_3]_2$ [T219]	Dissolution of $Pt(OH)_2{\cdot}xH_2O$ in $MeCN/CF_3SO_3H$. Generated *in situ*. X-ray absorption edge and EXAFS studies have been reported.
$[Pt(MeCN)_4]X_2$ [T228, T229] X = ClO_4, BF_4	Halide abstraction from $[PtCl_2(MeCN)_2]$ by AgX in MeCN.

References

1. L. I. Elding and Å. Oskarsson, *Inorg. Chim Acta* **130** (1987) 209.
2. B. F. G. Johnson, J. Puga, and P. R. Raithby, *Acta Crystallogr., Sect. B.* **37** (1981) 953.
3. Yu. N. Kukushkin and L. I. Danilina, *Izv. VUZ, Khim. i khim. tekhn.* **24** (1981) 139.
4. I. A. Zakharova, A. P. Kurbakova, Z. V. Belyakova, Kh. I. Gasanov, T. Kh. Kurbanov, and G. Ponticelli, *Polyhedron* **6** (1987) 1065.
5. Z. V. Belyakova, A. P. Kurbakova, I. A. Zakharova, and N. V. Gerebeleu, *Abstr. XVI All-Union Conf. on Coord. Chem.*, Krasnoyarsk, (1987) 369.
6. K. H. Pearson, W. R. Howell, Jr., P. E. Reinbold, and S. Kirschner, *Inorg. Synth.* **14** (1973) 57.
7. L. G. L. Ward, *Inorg. Synth.* **13** (1972) 154.
8. F. R. Hartley and J. A. Davies, *Chem. Rev.* **81** (1981) 79.
9. W. Levason, C. A. McAuliffe, and S. G. Murray, *J. Chem. Soc., Dalton Trans.* (1975) 1567.
10. W. Levason, C. A. McAuliffe, and S. G. Murray, *Inorg. Chim. Acta* **17** (1976) 247.
11. F. R. Hartley, S. G. Murray, W. Levason, H. E. Soutter, and C. A. McAuliffe, *Inorg. Chim. Acta* **35** (1979) 265.
12. F. A. Cotton and G. Wilkinson, *Advanced Inorganic Chemistry* (Wiley, New York, 1980) p. 801.
13. T. N. Sorrell and D. L. Jameson, *J. Am. Chem. Soc.* **105** (1983) 6013.
14. B. J. Hathaway and A. E. Underhill, *J. Chem. Soc.* (1960) 3705.
15. B. J. Hathaway, D. G. Holah, and J. D. Postlethwaite, *J. Chem. Soc.* (1961) 3215.
16. (a) R. F. Schramm and B. B. Wayland, *J Chem. Soc., Chem. Commun.* (1968) 898; (b) B. B. Wayland and R. F. Schramm, *Inorg. Chem.* **8** (1969) 971.
17. R. R. Thomas and A. Sen, *Inorg. Synth.* **28** (1990) 63.
18. M. Albin, A. C. Goldstone, A. S. Withers, and W. DeW. Horrocks, Jr., *Inorg. Chem.* **22** (1983) 3182.

19. R. R. Thomas, V. Chebolu, and A. Sen, *J. Am. Chem. Soc.* **108** (1986) 4096.

20. G. Bergerhoff, *Z. Anorg. Allg. Chem.* **327** (1964) 139.

21. H. H. Morgan, *J. Chem. Soc.* **CXXIII**, Part II, (1923) 2901.

22. R. M. Golding, K. F. Mok, and J. F. Duncan, *Inorg. Chem.* **5** (1966) 774.

23. H. W. Roesky, M. Thomas, M. Noltemeyer, and G. M. Sheldrick, *Angew. Chem.* **94** (1982) 861.

24. H. W. Roesky, M. Thomas, J. W. Bats, and H. Fuess, *J. Chem. Soc., Dalton Trans.* (1983) 1891.

25. J. J. Habeeb, F. F. Said, and D. G. Tuck, *J. Chem. Soc., Dalton Trans.* (1981) 118.

26. J. J. Habeeb, F. F. Said, and D. G. Tuck, *Inorg. Nucl. Chem. Lett.* **15** (1979) 113.

27. H. Meerwein, V. Henderich, and K. Wunderlich, *Arch. Pharm.* **291** (1958) 541.

28. J. M. Mayer and E. H. Abbott, *Inorg. Chem.* **22** (1983) 2774.

29. V. Gutmann, *Coordination Chemistry in Non-aqueous Solutions* (Springer-Verlag, New York, 1968) p. 23.

30. K. Burger, *Solvation, Ionic and Complex Formation Reactions in Non-Aqueous Solvents. Experimental Methods for Their Investigation* (Elsevier, Amsterdam, 1983) p. 38.

31. M. Sandström, I. Persson, and P. Persson, *Acta Chem. Scand.* **44** (1990) 653.

32. (a) V. V. Konshin and B. N. Chernishov, *Zh. Strukt. Khim.* **24** (1983) 137; (b) D. Richardson and T. D. Alger, *J. Phys. Chem.* **79** (1975) 1733.

33. A. J. Carty and D. G. Tuck, *J. Chem. Soc.* (1964) 6012.

34. V. P. Komarov and N. S. Shapligin, *Zh. Neorg. Khim.* **27** (1982) 2974.

35. D. Knol, N. J. Koole, and M. J. A. De Bie, *Org. Magn. Reson.* **8** (1976) 213.

36. A. E. Wickenden and R. A. Krause, *Inorg. Chem.* **4** (1965) 404.

37. S. F. Lincoln and R. J. West, *Aust. J. Chem.* **26** (1973) 255.

38. V. K. Kapur and B. B. Wayland, *J. Phys. Chem.* **77** (1973) 634.

39. P. W. N. M. Van Leeuwen and W. L. Groeneveld, *Rec. Trav. Chim.* **87** (1968) 86.

40. G. Beech and K. Miller, *J. Chem. Soc., Dalton Trans.* (1972) 801.

41. A. R. Pray, *Inorg. Synth.* **5** (1957) 153.

42. J. C. Chang, L. E. Gerdom, N. C. Baenziger, and H. M. Goff, *Inorg. Chem.* **22** (1983) 1739.

43. L. M. Jackman, R. M. Scott, R. H. Portman, and J. F. Dormish, *Inorg. Chem.* **18** (1979) 1497.

44. L. M. Jackman, J. F. Dormish, R. M. Scott, R. H. Portman, and R. D. Minard, *Inorg. Chem.* **18** (1979) 1503.

45. K. Starke, *Trans. Kentucky Acad. Sci.* **20** (1959) 41; *Chem. Abstr.* **54**: 19479h (1960).

46. K. Starke, *J. Inorg. Nucl. Chem.* **11** (1959) 77.

47. V. Imhof and R. S. Drago, *Inorg. Chem.* **4** (1965) 427.

48. P. W. N. M. Van Leeuwen, *Rec. Trav. Chim.* **86** (1967) 247.

49. N. M. Karayannis, C. Owens, L. L. Pytlewski, and M. M. Labes, *J. Inorg. Nucl. Chem.* **31** (1969) 2059.

50. A. D. Van Ingen Schenau, W. L. Groeneveld, and J. Reedijk, *Rec. Trav. Chim.* **91** (1972) 88.

51. M. J. Sisley, Y. Yano, and T. W. Swaddle, *Inorg. Chem.* **21** (1982) 1141.

52. R. J. West and S. F. Lincoln, *Aust. J. Chem.* **24** (1971) 1169.

53. R. N. Butler and M. C. R. Symons, *J. Chem. Soc., Trans. Faraday Soc.* **65** (1969) 945.

54. B. Kratochvil and R. Long, *Can. J. Chem.* **48** (1970) 1414.

55. G. D. Davis and E. C. Makin, Jr., *Separ. Purif. Methods* **1** (1972) 199.

56. J. Selbin, W. E. Bull, and L. H. Holmes, Jr., *J. Inorg. Nucl. Chem.* **16** (1961) 219.

57. G. S. Vigee and P. Ng, *J. Inorg. Nucl. Chem.* **33** (1971) 2477.

58. F. A. Cotton and R. Francis, *J. Am. Chem. Soc.* **82** (1960) 2986.

59. D. L. Pisaniello and S. F. Lincoln, *Aust. J. Chem.* **34** (1981) 1195.

60. L. I. Elding, *Inorg. Chim. Acta* **20** (1976) 65.

61. Ö. Gröning, T. Drakenberg, and L. I. Elding, *Inorg. Chem.* **21** (1982) 1820.

62. L. I. Elding and Ö. Gröning, *Inorg. Chem.* **17** (1978) 1872.

63. S. Okeya and S. Kawaguchi, *Inorg. Synth.* **20** (1980) 65.

64. J. Mink and P. L. Goggin, *Kem. Közl.* **57** (1982) 161.

65. (a) F. R. Hartley and J. A. Davies, *Rev. Inorg. Chem.* **4** (1982) 27; (b) M. Sato and S. Akabori, *Bull. Chem. Soc. Japan* **62** (1988) 3492; (c) M. Sato, M. Sekino, and S. Akabori, *J. Organomet. Chem.* **344** (1988) C31.

66. F. R. Hartley, S. G. Murray, and C. A. McAuliffe, *Inorg. Chem.* **18** (1979) 1394.

67. (a) A. De Renzi, A. Panunzi, A. Vitagliano, and G. Paiaro, *J. Chem. Soc., Chem. Commun.* (1976) 47; (b) V. Yu. Kukushkin, Å. Oskarsson, and L. I. Elding, *Inorg. Synth.* **31** (1995) in press.

68. S. Buffagni, L. M. Vallarino, and J. V. Quagliano, *Inorg. Chem.* **3** (1964) 480.

69. Z. Luz and S. Meiboom, *J. Chem. Phys.* **40** (1964) 1058.

70. R. Poupko and Z. Luz, *J. Chem. Phys.* **57** (1972) 3311.

71. J. Reedijk and W. L. Groeneveld, *Rec. Trav. Chim.* **87** (1968) 513.

72. A. P. Zuur and W. L. Groeneveld, *Rec. Trav. Chim.* **86** (1967) 1089.

73. J. Reedijk and W. L. Groeneveld, *Rec. Trav. Chim.* **86** (1967) 1103.

74. J. Reedijk, A. P. Zuur, and W. L. Groeneveld, *Rec. Trav. Chim.* **86** (1967) 1127.

75. J. Reedijk, J. B. Vervelde, and W. L. Groeneveld, *Rec. Trav. Chim.* **89** (1970) 42.

76. G. Brokaar, W. L. Groeneveld, and J. Reedijk, *Rec. Trav. Chim.* **89** (1970) 1117.

77. A. C. Baxter, J. H. Cameron, A. McAuley, F. M. McLaren, and J. M. Winfield, *J. Fluorine Chem.* **10** (1977) 289.

78. G. J. Kubas, *Inorg. Synth.* **19** (1979) 90.

79. G. J. Kubas, *Inorg. Synth.* **28** (1990) 68.

References to Table 4.2

T1. M. W. G. De Bolster and W. L. Groeneveld, *Rec. Trav. Chim.* **91** (1972) 171.

T2. J. T. Donoghue and R. S. Drago, *Inorg. Chem.* **2** (1963) 1158.

T3. A. H. Lewin, I. A. Cohen, and R. J. Michl, *J. Inorg. Nucl. Chem.* **36** (1974) 1951.

T4. S. Kitagawa, M. Munakata, and A. Higashie, *Inorg. Chim. Acta* **59** (1982) 219.

T5. K. Nilsson and Å. Oskarsson, *Acta Chem. Scand.* **A36** (1982) 605.

T6. I. Persson, J. E. Penner-Hahn, and K. O. Hodgson, *Inorg. Chem.* **32** (1993) 2497.

T7. S. Amasa, D. H. Brown, and D. W. A. Sharp, *J. Chem. Soc. A* (1969) 2892.

T8. R. S. Drago and D. Meek, *J. Phys. Chem.* **65** (1961) 1446.

T9. D. W. Meek, D. Straub, and R. S. Drago, *J. Am. Chem. Soc.* **82** (1960) 6013.

T10. J. Selbin, W. E. Bull, and L. H. Holmes, Jr., *J. Inorg. Nucl. Chem.* **16** (1961) 219.

T11. F. A. Cotton and R. Francis, *J. Am. Chem. Soc.* **82** (1960) 2986.

T12. W. Libus, R. Pastewski, and T. Sadowska, *J. Chem. Soc., Faraday Trans.* **78** (1982) 377.

T13. F. Kutek, *Collect. Czech. Chem. Commun.* **33** (1968) 1930.

T14. N. M. Karayannis, C. Owens, L. L. Pytlewski, and M. M. Labes, *J. Inorg. Nucl. Chem.* **31** (1969) 2059.

T15. B. J. Hathaway and A. E. Underhill, *J. Chem. Soc.* (1960) 3705.

T16. B. J. Hathaway, D. G. Holah, and J. D. Postlethwaite, *J. Chem. Soc.* (1961) 3215.

T17. H. H. Morgan, *J. Chem. Soc.* **CXXIII**, Part II, (1923) 2901.

T18. H. Meerwein, V. Henderich, and K. Wunderlich, *Arch. Pharm.* **291** (1958) 541.

T19. E. Heckel, *F.R.G. Patent* 1.230.025 (1966); *Chem. Abstr.* **66**: 46487e (1967).

T20. (a) G. J. Kubas, *Inorg. Synth.* **19** (1979) 90; (b) G. J. Kubas, *Inorg. Synth.* **28** (1990) 68.

T21. P. Hemmerich and C. Sigwart, *Experientia* **19** (1963) 488.

T22. P. F. Barron, J. C. Dyason, L. M. Engelhardt, P. C. Healy, and A. H. White, *Aust. J. Chem.* **38** (1985) 261.

T23. M. Linss, M. G. Weller, and U. Weser, *Inorg. Chim. Acta* **80** (1983) 201.

T24. C. Kolomyjec, J. Whelan, and B. Bosnich, *Inorg. Chem.* **22** (1983) 2343.

T25. I. Csöregh, P. Kierkegaard, and R. Norrestam, *Acta Crystallogr., Sect. B* **31** (1975) 314.

T26. D. Knol, N. J. Koole, and M. J. A. De Bie, *Org. Magn. Reson.* **8** (1976) 213.

T27. M. Kubota and D. L. Johnston, *J. Inorg. Nucl. Chem.* **29** (1967) 769.

T28. J. van Rijn, J. Reedijk, M. Dartmann, and B. Kregs, *J. Chem. Soc., Dalton Trans.* (1987) 2579.

T29. G. D. Davis and E. C. Makin, Jr., *Separ. Purif. Methods* **1** (1972) 199.

T30. J. Reedijk and W. L. Groeneveld, *Rec. Trav. Chim.* **87** (1968) 513.

T31. A. P. Zuur and W. L. Groeneveld, *Rec. Trav. Chim.* **86** (1967) 1089.

T32. J. Reedijk and W. L. Groeneveld, *Rec. Trav. Chim.* **86** (1967) 1103.

T33. J. Reedijk, A. P. Zuur, and W. L. Groeneveld, *Rec. Trav. Chim.* **86** (1967) 1127.

T34. J. Reedijk, J. B. Vervelde, and W. L. Groeneveld, *Rec. Trav. Chim.* **89** (1970) 42.

T35. G. Brokaar, W. L. Groeneveld, and J. Reedijk, *Rec. Trav. Chim.* **89** (1970) 1117.

T36. A. C. Baxter, J. H. Cameron, A. McAuley, F. M. McLaren, and J. M. Winfield, *J. Fluorine Chem.* **10** (1977) 289.

T37. R. J. West and S. F. Lincoln, *J. Chem. Soc., Dalton Trans.* (1974) 281.

T38. V. K. Kapur and B. B. Wayland, *J. Phys. Chem.* **77** (1973) 634.

T39. A. P. Zuur, A. H. L. Reintjes, and W. L. Groeneveld, *Rec. Trav. Chim.* **89** (1970) 385.

T40. C. D. Jansen-Ligthelm, W. L. Groeneveld, and J. Reedijk, *Inorg. Chim. Acta* **7** (1973) 113.

T41. R. Poupko and Z. Luz, *J. Chem. Phys.* **57** (1972) 3311.

T42. L. Helm, S. F. Lincoln, A. E. Merbach, and D. Zbinden, *Inorg. Chem.* **25** (1986) 2550.

T43. P. W. N. M. Van Leeuwen, *Rec. Trav. Chim.* **86** (1967) 247.

T44. P. W. N. M. Van Leeuwen and W. L. Groeneveld, *Rec. Trav. Chim.* **87** (1968) 86.

T45. P. L. Verheijdt, P. H. Van der Voort, W. L. Groeneveld, and W. L. Driessen, *Rec. Trav. Chim.* **91** (1972) 1201.

T46. (a) F. Hultén and I. Persson, *Acta Chem. Scand.* **A41** (1987) 87; (b) M. Ignaczak and A. Grzejdziak, *Polish J. Chem.* **63** (1989) 43.

T47. M. Ignaczak and A. Grzejdziak, *Monatsh. Chem.* **115** (1984) 943.

T48. F. Hultén and I. Persson, *Acta Chem. Scand.* **A41** (1987) 87.

T49. (a) S. Ahrland and N.-O. Björk, *Acta Chem. Scand.* **A28** (1974) 823; (b) F. Gaizer and G. Johansson, *Acta Chem. Scand.* **A42** (1988) 259.

T50. N.-O. Björk and A. Cassel, *Acta Chem. Scand.* **A30** (1976) 235.

T51. V. P. Komarov and N. S. Shapligin, *Zh. Neorg. Khim.* **27** (1982) 2974.

T52. T.-C. G. Chang and D. E. Irish, *J. Sol. Chem.* **3** (1974) 161.

T53. K. Nilsson and Å. Oskarsson, *Acta Chem. Scand.* **A38** (1984) 79.

T54. K. Nilsson and I. Persson, *Acta Chem. Scand.* **A41** (1987) 139.

T55. A. S. Morkovnik, Z. S. Morkovnik, and V. V. Bessonov, *Zh. Neorg. Khim.* **32** (1987) 3106.

T56. S. S. D. Brown and I. D. Salter, *Organomet. Synth.* **3** (1986) 315.

T57. A. E. Comyns and H. J. Lucas, *J. Am. Chem. Soc.* **76** (1954) 1019.

T58. G. A. Olah and H. W. Quinn, *J. Inorg. Nucl. Chem.* **14** (1960) 295.

T59. R. J. Puddephatt, *The Chemistry of Gold* (Elsevier, Amsterdam, 1978) p. 48.

T60. H. Willner, J. Schaebs, G. Hwang, F. Mistry, R. Jones, J. Trotter, and F. Aubke, *J. Am. Chem. Soc.* **114** (1992) 8972.

T61. (a) G. Bergerhoff, *Z. Anorg. Allg. Chem.* **327** (1964) 139; (b) S. Ahrland, K. Nilsson, J. E. Penner-Hahn, I. Persson, and A. Yuchi, *Inorg. Chem.* **28** (1989) 1833.

T62. S. F. Lincoln and M. N. Tkaczuk, *Ber. Buns. Phys. Chem.* **86** (1982) 221.

T63. M. N. Tkaczuk and S. F. Lincoln, *Ber. Buns. Phys. Chem.* **86** (1982) 147.

T64. S. F. Lincoln and M. N. Tkaczuk, *Ber. Buns. Phys. Chem.* **85** (1981) 433.

T65. P.-A. Pittet, G. Elbaze, L. Helm, and A. E. Merbach, *Inorg. Chem.* **29** (1990) 1936.

T66. M. N. Tkaczuk and S. F. Lincoln, *Aust. J. Chem.* **35** (1982) 1555.

T67. N. A. Matwiyoff and W. G. Movius, *J. Am. Chem. Soc.* **89** (1967) 6077.

T68. M. Chaudhry, I. Persson, M. Sandström, and H. Yokoyama, *unpublished results* (1995).

T69. L. Huang, F. Jian, L. Wang, and J. Lu, *Jiegou Huaxue* **4** (1985) 313; *Chem. Abstr.* **106** (1987) 42888h.

T70. D. L. Pisaniello and S. F. Lincoln, *Aust. J. Chem.* **32** (1979) 715.

T71. J. Crea, S. F. Lincoln, and R. J. West, *Aust. J. Chem.* **26** (1973) 1227.

T72. D. Pisaniello, S. F. Lincoln, and E. H. Williams, *Inorg. Chim. Acta* **31** (1978) 237.

T73. Ch. P. Rao, A. M. Rao, and C. N. R. Rao, *Inorg. Chem.* **23** (1984) 2080.

T74. M. G. B. Drew, P. P. K. Claire, and G. R. Willey, *J. Chem. Soc., Dalton Trans.* (1988) 215.

T75. N. A. Matwiyoff and H. Taube, *J. Am. Chem. Soc.* **90** (1968) 2796.

T76. S. Nakamura and S. Meiboom, *J. Am. Chem. Soc.* **89** (1967) 1765.

T77. T. D. Alger, *J. Am. Chem. Soc.* **91** (1969) 2220.

T78. G. W. Stockton and J. C. Martin, *Can. J. Chem.* **52** (1974) 744.

T79. A. D. Van Ingen Schenau, W. L. Groeneveld, and J. Reedijk, *Rec. Trav. Chim.* **91** (1972) 88.

T80. P. Sobota, *Proc. 10th Conf. Coord. Chem. Smolenice-Bratislava*, 4-7 June, 1985. S.I, s. a., p. 357.

T81. E. Giesbrecht, W. G. R. De Camargo, G. Vicentini, and M. Perrier, *J. Inorg. Nucl. Chem.* **24** (1962) 381.

T82. W. L. Driessen, W. L. Groeneveld, and F. W. Van der Wey, *Rec. Trav. Chim.* **89** (1970) 353.

T83. W. L. Driessen and W. L. Groeneveld, *Rec. Trav. Chim.* **88** (1969) 977.

T84. W. L. Driessen, L. M. Van Geldrop, and W. L. Groeneveld, *Rec. Trav. Chim.* **89** (1970) 1271.

T85. W. L. Driessen and W. L. Groeneveld, *Rec. Trav. Chim.* **88** (1969) 491.

T86. W. L. Driessen and W. L. Groeneveld, *Rec. Trav. Chim.* **88** (1969) 620.

T87. M. W. G. De Bolster and W. L. Groeneveld, *Rec. Trav. Chim.* **90** (1971) 477.

T88. J. T. Donoghue and R. S. Drago, *Inorg. Chem.* **1** (1962) 866.

T89 M. N. Tkaczuk and S. F. Lincoln, *Aust. J. Chem.* **33** (1980) 2621.

T90. L. I. Chudinova, *Zh. Neorg. Khim.* **10** (1965) 1300.

T91. R. Åkesson, M. Sandström, C. Stålhandske, and I. Persson, *Acta Chem. Scand.* **45** (1991) 165.

T92. I. Persson, *Acta Chem. Scand.* **A36** (1982) 7.

T93. J. J. Habeeb, F. F. Said, and D. G. Tuck, *J. Chem. Soc., Dalton Trans.* (1981) 118.

T94. J. J. Habeeb, F. F. Said, and D. G. Tuck, *Inorg. Nucl. Chem. Lett.* **15** (1979) 113.

T95. (a) M. Sandström, *Acta Chem. Scand.* **A32** (1978) 519; (b) M. Sandström, I. Persson, and S. Ahrland, *Acta Chem. Scand.* **A32** (1978) 607.

T96. M. Nieuwenhuyzen, H. Wen, and C. J. Wilkins, *Z. Anorg. Allg. Chem.* **615** (1992) 143.

T97. A. M. Bond, R. Colton, J. Ebner, and S. R. Ellis, *Inorg. Chem.* **28** (1989) 4509.

T98. (a) I. Persson and M. Sandström, *Acta Chem. Scand., Ser. A* **32** (1978) 95; (b) I. Persson, *personal communication.*

T99. M. W. G. De Bolster, W. L. Drissen, W. L. Groeneveld, and C. J. Van Kerkwijk, *Inorg. Chim. Acta* **7** (1973) 439.

T100. M. N. Tkaczuk and S. F. Lincoln, *Aust. J. Chem.* **32** (1979) 1915.

T101. R. N. Butler and M. C. R. Symons, *J. Chem. Soc., Trans. Faraday Soc.* **65** (1969) 945.

T102. S. A. Al-Baldawi and T. E. Gough, *Can. J. Chem.* **47** (1969) 1417.

T103. J. T. Donoghue, E. Fernandez, J. A. McMillan, and D. A. Peters, *J. Inorg. Nucl. Chem.* **31** (1969) 1431.

T104. E. Giesbrecht and L. B. Zinner, *Inorg. Nucl. Chem. Lett.* **5** (1969) 575.

T105. L. B. Zinner, G. Vicentini, and P. O. Dunstan, *J. Less-Common Metals* **112** (1985) 393.

T106. V. N. Krishnamurthy and S. Soundarajan, *J. Inorg. Nucl. Chem.* **29** (1967) 517.

T107. P. B. Bertan and S. K. Madan, *J. Inorg. Nucl. Chem.* **34** (1972) 3081.

T108. F. Kutek and F. Petru, *Collect. Czech. Chem. Commun.* **33** (1968) 296.

T109. G. Vicentini and L. S. P. Braga, *An. Acad. brasil ciênc.* **53** (1981) 453.

T110. D. L. Pisaniello and S. F. Lincoln, *Aust. J. Chem.* **34** (1981) 1195.

T111. S. F. Lincoln, A. M. Hounslow, and A. J. Jones, *Aust. J. Chem.* **35** (1982) 2393.

T112. F. Kutek, *Collect. Czech. Chem. Commun.* **32** (1967) 3767.

T113. D. L. Pisaniello and S. F. Lincoln, *Inorg. Chem.* **20** (1981) 3689.

T114. D. L. Pisaniello and S. F. Lincoln, *J. Chem. Soc., Dalton Trans.* (1980) 699.

T115. D. L. Pisaniello, S. F. Lincoln, and E. H. Williams, *J. Chem. Soc., Dalton Trans.* (1979) 1473.

T116. D. L. Pisaniello, S. F. Lincoln, and E. H. Williams, *J. Chem. Soc., Chem. Commun.* (1978) 1047.

T117. D. L. Pisaniello and S. F. Lincoln, *Inorg. Chim. Acta* **36** (1979) 85.

T118. A. P. Zuur, J. J. Van Houte, and W. L. Groeneveld, *Rec. Trav. Chim.* **87** (1968) 755.

T119. C. W. Eigenbrot, Jr. and K. N. Raymond, *Inorg. Chem.* **22** (1983) 2972.

T120. (a) C. W. Eigenbrot, Jr. and K. N. Raymond, *Inorg. Chem.* **21** (1982) 2867; (b) M. Albin, A. C. Goldstone, A. S. Withers, and W. DeW. Horrocks, Jr., *Inorg. Chem.* **22** (1983) 3182.

T121. R. R. Thomas, V. Chebolu, and A. Sen, *J. Am. Chem. Soc.* **108** (1986) 4096.

T122. K.-M. Mühlsiepen and R. Mattes, *Z. Anorg. Allg. Chem.* **506** (1983) 115.

T123. H. Fukutomi, T. Harazono, and T. Kojima, *Bull. Chem. Soc. Japan* **59** (1986) 2387.

T124. G. J. Honan, S. F. Lincoln, and E. H. Williams, *Inorg. Chem.* **17** (1978) 1855.

T125. R. P. Bowen, S. F. Lincoln, and E. H. Williams, *Inorg. Chem.* **15** (1976) 2126.

T126. S. F. Lincoln, A. Ekstrom, and G. J. Honan, *Aust. J. Chem.* **35** (1982) 2385.

T127. J. Crea, R. Digiusto, S. F. Lincoln, and E. H. Williams, *Inorg. Chem.* **16** (1977) 2825.

T128. A. M. Hounslow, S. F. Lincoln, P. A. Marshall, and E. H. Williams, *Aust. J. Chem.* **34** (1981) 2543.

T129. R. P. Bowen, G. J. Honan, S. F. Lincoln., T. M. Spotswood, and E. H. Williams, *Inorg. Chim. Acta* **33** (1979) 235.

T130. J. McB. Harrowfield, D. L. Kepert, J. M. Patric, A. H. White, and S. F. Lincoln, *J. Chem. Soc., Dalton Trans.* (1983) 393.

T131. K. Ishihara, S. Funahashi, and M. Tanaka, *Inorg. Chem.* **25** (1986) 2898.

T132. K. Ishihara, S. Funahashi, and M. Tanaka, *Inorg. Chem.* **22** (1983) 2564.

T133. J. MacB. Harrowfield, B. W. Skelton, and A. H. White, *Aust. J. Chem.* **43** (1990) 759.

T134. A. J. Carty and D. G. Tuck, *J. Chem. Soc.* (1964) 6012.

T135. W. G. Movius and N. A. Matwiyoff, *Inorg. Chem.* **6** (1967) 847.

T136. W. G. Movius and N. A. Matwiyoff, *Inorg. Chem.* **8** (1969) 925.

T137. I. Lupu, E. Petrovici, and F. Grigorescu, *2nd Nat. Congr. Chem.*, Bucharest, 7-10 Sept. (1981) Abstr. Part I, S. I. s. a., p. 268.

T138. J. Crea and S. F. Lincoln, *Inorg. Chem.* **11** (1972) 1131.

T139. D. Richardson and T. D. Alger, *J. Phys. Chem.* **79** (1975) 1733.

T140. (a) J. Glaser and G. Johansson, *Acta Chem. Scand.* **A35** (1981) 639; (b) J. Glaser and G. Johansson, *Acta Chem. Scand.* **A36** (1982) 125.

T141. J. Glaser, *Ph.D. Thesis* (KTH, Stockholm, 1981).

T142. J. Berglund, P. Werndrup, and L. I. Elding, *J. Chem. Soc., Dalton Trans.* (1994) 1435.

T143. I. Dellavia, L. Helm, and A. E. Merbach, *Inorg. Chem.* **31** (1992) 2230.

T144. I. Dellavia, L. Helm, and A. E. Merbach, *Inorg. Chem.* **31** (1992) 4151.

T145. A. R. Bassoteli, A. C. D. Angelo, E. Tfouni, and N. R. Stradiotto, *Polyhedron* **3** (1984) 627.

T146. H. Schmidt, *Z. Anorg. Allg. Chem.* **271** (1953) 305.

T147. C. H. Langford and F. N. Chung, *Can. J. Chem.* **48** (1970) 2969.

T148. L. F. Larkworthy and M. W. O'Donoghue, *Inorg. Chim. Acta* **71** (1983) 81.

T149. H. J. Seifert and T. Auel, *Z. Anorg. Allg. Chem.* **360** (1968) 50.

T150. F. A. Cotton, S. A. Duraj, L. E. Manzer, and W. J. Roth, *J. Am. Chem. Soc.* **107** (1985) 3850.

T151. H. Hartmann and H. L. Schläfer, *Z. Naturforsch.* **6a** (1951) 754.

T152. G. S. Vigee and P. Ng, *J. Inorg. Nucl. Chem.* **33** (1971) 2477.

T153. J. J. Habeeb and D. G. Tuck, *Inorg. Synth.* **19** (1979) 123.

T154. D. L. Carle and T. W. Swaddle, *Can. J. Chem.* **51** (1973) 3795.

T155. H. L. Schläfer and W. Schaffernicht, *Angew. Chem.* **72** (1960) 618.

T156. Ya. Ya. Makarov-Zemlyanskii and E. D. Temyakov, *Zh. Obsch. Khim.* **51** (1981) 2644.

T157. Ya. Ya. Makarov-Zemlyanskii, *Zh. Obsch. Khim.* **50** (1980) 954.

T158. Ya. Ya. Makarov-Zemlyanskii, *Zh. Obsch. Khim.* **47** (1977) 223.

T159. H. Zawadski, *Polish J. Chem.* **56** (1982) 647.

T160. R. S. Drago, D. W. Meek, M. D. Joesten, and L. LaRoche, *Inorg. Chem.* **2** (1963) 124.

T161. L. Chong-de and R. B. Jordan, *Inorg. Chem.* **26** (1987) 3855.

T162. P. J. Jones, A. L. Hale, W. Levason, and F. R. McCullough, *Inorg. Chem.* **22** (1983) 2642.

T163. J. M. Mayer and E. H. Abbott, *Inorg. Chem.* **22** (1983) 2774.

T164. F. A. Cotton and K. J. Wiesinger, *Inorg. Synth.* **29** (1992) 134.

T165. F. A. Cotton and K. J. Wiesinger, *Inorg. Chem.* **30** (1991) 871.

T166. C. P. Prabhakaran and C. C. Patel, *J. Inorg. Nucl. Chem.* **30** (1968) 867.

T167. R. S. Nyholm and A. Turco, *J. Chem. Soc.* (1962) 1121.

T168. M. J. Sisley, Y. Yano, and T. W. Swaddle, *Inorg. Chem.* **21** (1982) 1141.

T169. B. J. Hathaway and D. G. Holah, *J. Chem. Soc.* (1964) 2400.

T170. (a) J. C. Bryan, F. A. Cotton, L. M. Daniels, S. C. Haefner, and A. P. Sattelberger, *Inorg. Chem.* **34** (1995) 1875; (b) S. N. Bernstein and K. R. Dunbar, *Angew. Chem. Int. Ed. Engl.* **31** (1992) 1360.

T171. (a) M. W. G. De Bolster, A. Vermaas, and W. L. Groeneveld, *J. Inorg. Nucl. Chem.* **35** (1973) 83; (b) O. Baudish and W. H. Hartung, *Inorg. Synth.* **1** (1939) 184.

T172. D. M. L. Goodgame, M. Goodgame, M. A. Hitchman, and M. J. Weeks, *Inorg. Chem.* **5** (1966) 635.

T173. R. M. Golding, K. F. Mok, and J. F. Duncan, *Inorg. Chem.* **5** (1966) 774.

T174. C. D. Burbridge, D. M. L. Goodgame, and M. Goodgame, *J. Chem. Soc. A.* (1967) 349.

T175. A. Müller, H. Bögge, U. Schimanski, M. Penk, K. Nieradzik, M. Dartmann, E. Krickemeyer, J. Schimanski, C. Römer, M. Römer, H. Dornfeld, U. Wienböker, W. Hellmann, and M. Zimmermann, *Monatsh. Chem.* **120** (1989) 367.

T176. C. H. McAteer and P. Moore, *J. Chem. Soc., Dalton Trans.* (1983) 353.

T177. G. Gritzner, W. Linert, and V. Gutmann, *J. Inorg. Nucl. Chem.* **43** (1981) 1193.

T178. W. Linert, V. Gutmann, O. Baumgartner, G. Wiesinger, and H. Kirchmayr, *Inorg. Chim. Acta* **74** (1983) 123.

T179. R. J. West and S. F. Lincoln, *Aust. J. Chem.* **24** (1971) 1169.

T180. S. Pohl and W. Saak, *Z. Naturforsch.* **39b** (1984) 1236.

T181. B. J. Hathaway and D. G. Holah, *J. Chem. Soc.* (1964) 2408.

T182. B. Chaudret, F. Jalón, M. Pérez-Manrique, F. Lahoz, F. J. Plou, and R. Sánchez-Delgado, *New J. Chem.* **14** (1990) 331.

T183. (a) I. P. Evans, A. Spencer, and G. Wilkinson, *J. Chem. Soc., Dalton Trans.* (1973) 204; (b) A. R. Davies, F. W. B. Einstein, N. P. Farrel, B. R. Jones, and R. S. McMillan, *Inorg. Chem.* **17** (1978) 1965.

T184. T. Bora and M. M. Singh, *Zh. Neorg. Khim.* **20** (1975) 419.

T185. T. Bora and M. M. Singh, *J. Inorg. Nucl. Chem.* **38** (1976) 1815.

T186. P. Bernhard and A. M. Sargeson, *J. Chem. Soc., Chem. Commun.* (1985) 1516.

T187. (a) A. Ludi and R. Judd, *personal communication*; (b) R. J. Judd, R. Cao, M. Biner, T. Armbruster, H.-B. Bürgi, A. E. Merbach, and A. Ludi, *Inorg. Chem.* **34** (1995) 5080.

T188. P. Bernhard, H. Lehmann, and A. Ludi, *J. Chem. Soc., Chem. Commun.* (1981) 1216.

T189. P. Bernhard, M. Biner, and A. Ludi, *Polyhedron* **9** (1990) 1095.

T190. N. Aebischer, G. Laurenczy, A. Ludi, and A. E. Merbach, *Inorg. Chem.* **32** (1993) 2810.

T191. I. Rapaport, L. Helm, A. E. Merbach, P. Bernhard, and A. Ludi, *Inorg. Chem.* **27** (1988) 873.

T192. (a) A. N. Zhilyaev, T. A. Fomina, I. V. Kuzmenko, S. B. Katser, and I. B. Baranovskii, *Zh. Neorg. Khim.* **35** (1990) 2253; (b) D. A. Freedman, D. J. Magneson, and K. R. Mann, *Inorg. Chem.* **34** (1995) 2617.

T193. S. F. Lincoln, A. M. Hounslow, and A. N. Boffa, *Inorg. Chem.* **25** (1986) 1038.

T194. V. Gutmann, R. Beran, and W. Kerber, *Monatsch. Chem.* **103** (1972) 764.

T195. N. A. Matwiyoff and S. V. Hooker, *Inorg. Chem.* **6** (1967) 1127.

T196. R. J. West and S. F. Lincoln, *Inorg. Chem.* **11** (1972) 1688.

T197. Z. Luz and S. Meiboom, *J. Chem. Phys.* **40** (1964) 1058.

T198. W. L. Driessen and J. Reedijk, *Inorg. Synth.* **29** (1992) 11.

T199. S. Sen and M. M. Singh, *Indian J. Chem.* **11** (1973) 497.

T200. J. R. Barnes, P. L. Goggin, and R. J. Goodfellow, *J. Chem. Res. Synop.* (1979) 118.

T201. J. R. Barnes and R. J. Goodfellow, *J. Chem. Res. Synop.* (1979) 350.

T202. M. C. Read, J. Glaser, and M. Sandström, *J. Chem. Soc., Dalton Trans.* (1992) 233.

T203. (a) M. C. Read, J. Glaser, M. Sandström, and I. Toth, *Inorg. Chem.* **31** (1992) 4155; (b) I. Bányai, J. Glaser, M. C. Read, and M. Sandström, *Inorg. Chem.* **34** (1995) 2423.

T204. C. Berecki-Biedermann, *Ark. Kem.* **20** (1962) 437.

T205. (a) K. R. Dunbar and L. E. Pence, *Inorg. Synth.* **29** (1992) 182; (b) P.-A. Pittet, L. Dadci, P. Zbinden, A. Abou-Hamdan, and A. Merbach, *Inorg. Chim. Acta* **206** (1993) 135.

T206. D. Oliveira and Osorio V. K. Lakatos, *Polyhedron* **6** (1987) 1313.

T207. H. Suzuki and S. Ishiguro, *Inorg. Chem.* **31** (1992) 4178.

T208. N. A. Matwiyoff, *Inorg. Chem.* **5** (1966) 788.

T209. A. E. Wickenden and R. A. Krause, *Inorg. Chem.* **4** (1965) 404.

T210. S. F. Lincoln and R. J. West, *Aust. J. Chem.* **26** (1973) 255.

T211. P. W. N. M. Van Leeuwen and W. L. Groeneveld, *Inorg. Nucl. Chem. Lett.* **3** (1967) 145.

T212. A. D. Van Ingen Schenau, W. L. Groeneveld, and J. Reedijk, *Spectrochim. Acta* **30A** (1974) 213.

T213. T. Nortia, J. Arpalahti, and M. Karppinen, *Spectrochim. Acta* **40A** (1984) 257.

T214. V. Imhof and R. S. Drago, *Inorg. Chem.* **4** (1965) 427.

T215. S. Buffagni, L. M. Vallarino, and J. V. Quagliano, *Inorg. Chem.* **3** (1964) 480.

T216. B. B. Wayland and R. F. Schramm, *Inorg. Chem.* **8** (1969) 971.

T217. R. R. Thomas and A. Sen, *Inorg. Synth.* **28** (1990) 63.

T218. B. F. G. Johnson, J. Puga, and P. R. Raithby, *Acta Crystallogr., Sect. B* **37** (1981) 953.

T219. B. Hellquist, L. A. Bengtsson, B. Holmberg, B. Hedman, I. Persson, and L. I. Elding, *Acta Chem. Scand.* **45** (1991) 449.

T220. M. Sato and S. Akabori, *Bull. Chem. Soc. Japan* **62** (1988) 3492.

T221. M. Sato, M. Sekino, and S. Akabori, *J. Organomet. Chem.* **344** (1988) C31.

T222. F. R. Hartley and J. A. Davies, *Rev. Inorg. Chem.* **4** (1982) 27.

T223. I. I. Chernyaev, ed., *Synthesis of Complex Compounds of Platinum Group Metals* [in Russian] (Mir, Moscow, 1964) p. 79.

T224. G. B. Deacon and I. L. Grayson, *Transit. Metal Chem.* **7** (1982) 97.

T225. L. I. Elding and Å. Oskarsson, *Inorg. Chim Acta* **130** (1987) 209.

T226. L. I. Elding, *Inorg. Chim. Acta* **20** (1976) 65.

T227. V. Yu. Kukushkin, Å. Oskarsson, and L. I. Elding, *Inorg. Synth.* **31** (1995) in press.

T228. F. R. Hartley, S. G. Murray, and C. A. McAuliffe, *Inorg. Chem.* **18** (1979) 1394.

T229. A. De Renzi, A. Panunzi, A. Vitagliano, and G. Paiaro, *J. Chem. Soc., Chem. Commun.* (1976) 47.

Chapter 5. SYNTHESIS *VIA* ION EXCHANGE IN NON-AQUEOUS SOLVENTS

Ion exchange reactions are routinely employed in inorganic synthesis and such reactions typically have one common feature, *i.e.* an equilibrium is shifted to the right by the formation of a product of low solubility. Ion exchange reactions performed in aqueous media commonly result in precipitation of the desired complex from solution while a more soluble by-product, usually a salt, remains dissolved in solution. In non-aqueous media, the opposite is typically observed as the desired complex remains in solution and precipitation of the by-product occurs. While this reversal might seem insignificant, it actually allows the synthesis of complexes whose generation and isolation might otherwise prove difficult or impossible.

Although perturbation of equilibrium by precipitation of a sparingly soluble product is a major factor in ion exchange, it is, of course, only one of a number of important considerations that govern the outcome of a reaction. Further, due consideration must be given to thermodynamic and kinetic limitations.

In this chapter ion exchange reactions in non-aqueous media are examined and the discussion is divided into: *(i)* classical ligand substitution by ion exchange in non-aqueous solvents, *(ii)* synthesis of complexes containing "noncoordinating" anions as ligands, *(iii)* formation of chelate complexes by ion exchange in non-aqueous solvents, *(iv)* synthesis of complexes containing metal-metal bonds between transition metals, *(v)* synthesis of complexes containing transition metal-main group metal/metalloid bonds, and *(vi)* replacement of counterions in ionic complexes. Throughout the discussion ion exchange is considered to be the transfer of charged or formally charged groups between reactants.

5.1 Ligand substitution by ion exchange in non-aqueous solvents

One successful strategy for ligand substitution by ion exchange involves the use of halide or pseudohalide metal complexes as starting materials in reactions with silver salts that contain the anion that is to be introduced as a new ligand. Since the solubility of silver halides in nonpolar, aprotic solvents is low (see Table 3.1), ion exchange in solvents of poor donor ability[1] typically occurs according to Eq. 5.1:

$$M\text{–}X \; + \; AgY \; \rightarrow \; M\text{–}Y \; + \; AgX \qquad (5.1)$$

This methodology is illustrated by the synthesis of platinum carboxylates through the reaction of platinum bromide complexes with silver carboxylates in dichloromethane,[2] Scheme 5.1:

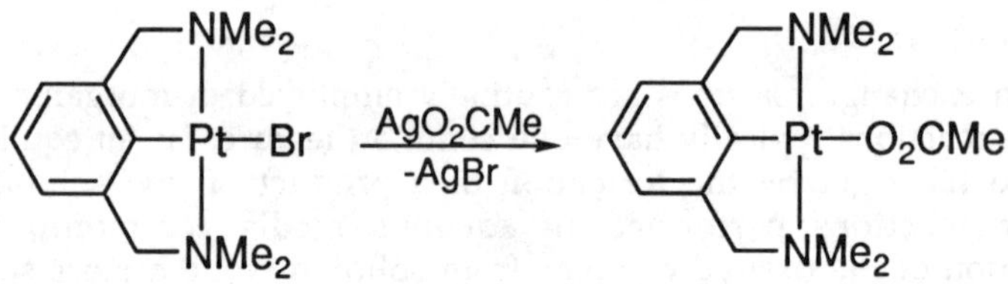

Scheme 5.1

Because carboxylate anions are hard Lewis bases,[3] they have low affinities for soft metal ions such as platinum(II). Conversely, the relatively soft bromide ion has a high affinity for this metal and yet halide abstraction by Ag^+, with concomitant formation of the sparingly soluble salt, AgBr, results in substitution of bromide by the carboxylate anion. The use of dichloromethane, a poor ligand for platinum(II), as the solvent for this reaction prevents any competitive formation of a solvento-complex, although such species may be involved as short-lived intermediates in the reaction.

A major application of ion exchange in non-aqueous solvents is for the synthesis of organometallic compounds. The common methods fall into two distinct classes. In one, the reagents are electrophilic carbon compounds, such as alkyl halides, and nucleophilic metal complexes, such as carbonylmetallates. In the other, the relative polarities of the reagents are reversed and nucleophilic carbon compounds, *e.g.* alkyllithium reagents, and electrophilic metal complexes, such as metal halides, are employed. Both classes of reaction are ion exchange processes and typically involve the formation of a salt as the low-solubility by-product.

Mononuclear metal-alkyl complexes with long-chain alkyl groups have been prepared from sodium carbonylmetallates and alkyl halides,[4] Eq. 5.2:

$$\mathrm{Na[M(CO)_2Cp]} + \mathrm{RX} \rightarrow \mathrm{[MR(CO)_2Cp]} + \mathrm{NaX} \qquad (5.2)$$

where M = Fe, Ru; R = n-C_6H_{13} to n-$C_{12}H_{25}$; X = Cl, Br

Similar reactions with α,ω-dihaloalkanes yield binuclear alkanediyl complexes,[5] Eq. 5.3:

$$2Na[Ru(CO)_2Cp] + I(CH_2)_nI \rightarrow [Cp(CO)_2Ru(CH_2)_nRu(CO)_2Cp] + 2NaI \qquad (5.3)$$

where n = 2, 5 to 10

The alkanediyl complex where n = 2 has been characterized by X-ray crystallography,[6] Figure 5.1:

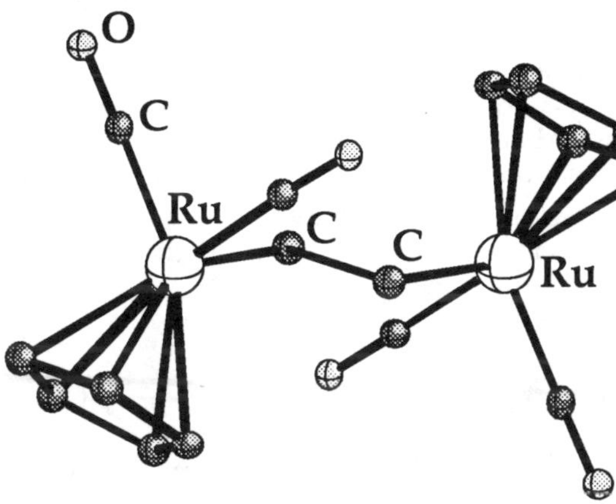

Figure 5.1

The structure of the complex where n = 5 also has been determined by X-ray crystallography.[5]

Metal acyl complexes have been prepared from sodium carbonylmetallates.[7] In the reaction of $Na[Mn(CO)_5]$ with acyl halides in dry THF, the acyl complex that is formed remains in solution and the sodium halide by-product precipitates. The authors report[8] that the reaction is driven mainly by the high nucleophilicity of the carbonylmetallate anion but also by the poor solubility of the sodium halide in THF, Eq. 5.4:

$$Na[Mn(CO)_5] + RC(O)Cl \rightarrow [Mn(C\{O\}R)(CO)_5] + NaCl \qquad (5.4)$$

The introduction of organic groups into metal complexes has been accomplished with a variety of nucleophilic reagents in ion exchange reactions in non-aqueous solvents. The common preparation of bis(cyclopentadienyl)metal complexes[9-15] may be considered as a simple ion exchange between a transition metal halide, MX_2, and a main group cyclopentadienyl, M*Cp, Eq. 5.5:

$$MX_2 + 2M^*Cp \xrightarrow{\text{solv}} [MCp_2] + 2M^*X \quad (5.5)$$

where M = Mn, Fe, Ru, Co, Ni, Hg; M* = Na, Tl, MgX; X = Cl, Br; solv = THF, benzene, 1,2-dimethoxyethane, Et_2O

The mixed-ligand cyclopentadienyl complex, $[PtCl(CO)(PPh_3)Cp]$, was synthesized by Cross and McLennan[16] by the reaction of TlCp with *cis*-$[PtCl_2(CO)(PPh_3)]$ as a suspension in benzene. Nucleophilic attack on *cis*-$[PtCl_2(CO)(PPh_3)]$ by anions typically results in replacement of the chloride ligand *trans* to the carbonyl group because of the high *trans*-effect of CO. However, in the reaction with a single equivalent of TlCp, the isolated isomer is the one in which the chloride *trans* to PPh_3 has been replaced by σ-Cp. The reaction of *cis*-$[PtCl_2(CO)(PPh_3)]$ with excess TlCp results in substitution of both chloride ligands, Scheme 5.2:

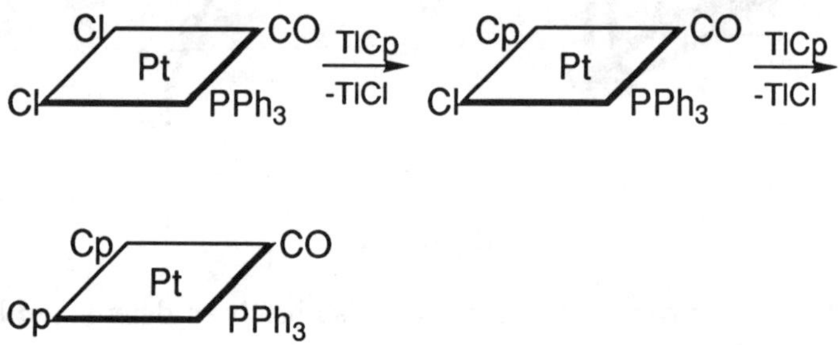

Scheme 5.2

The complex $[PtCl(CO)(PPh_3)Cp]$ may also be synthesized through the reaction of *cis*-$[PtCl_2(CO)(PPh_3)]$ with $[HgCp_2]$, Eq. 5.6:

$$cis\text{-}[PtCl_2(CO)(PPh_3)] + [HgCp_2] \rightarrow [PtCl(CO)(PPh_3)Cp] + [HgClCp] \quad (5.6)$$

The general methodology outlined above for the synthesis of platinum(II) cyclopentadienyls from platinum halide complexes and either TlCp[17-19] or $[HgCp_2]$[20] has been utilized in a number of synthetic schemes.

Similarly, Grignard reagents have been employed for the introduction of hydrocarbyl substituents into complexes *via* ion exchange. For example, the synthesis of the novel *di*-Grignard reagent $[Ti(CH_2MgBr)_2Cp_2]$ through the reaction of $[TiCl_2Cp_2]$ with $CH_2(MgBr)_2$ has been reported.[21] The titanium-containing Grignard reagent reacts with 0.5 molar equivalents of $SiCl_4$ to produce the compound shown in Figure 5.2:

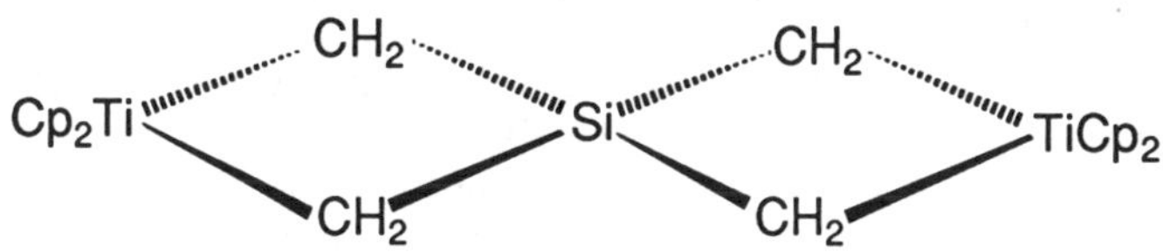

Figure 5.2

In a similar manner, $[TiCl_2Cp_2]$ has been shown to react with one equivalent of the Grignard reagent $ClMg(CH_2)_3CH{=}CH_2$ to generate the corresponding monosubstituted titanium complex,[22] Eq. 5.7:

$$[TiCl_2Cp_2] + ClMg(CH_2)_3CH{=}CH_2 \rightarrow [TiCl\{(CH_2)_3CH{=}CH_2\}Cp_2] + MgCl_2 \quad (5.7)$$

The bis(methyl) complex, $[Mo(O)_2Me_2(bipy)]$, has been synthesized through the reaction of $[Mo(O)_2Br_2(bipy)]$ with two equivalents of methylmagnesium bromide in THF.[23]

Organolithium reagents have been employed in ion exchange reactions with transition metal halide complexes. For example, [IrCl(triphos)] reacts with $LiCH_2SiMe_3$ in diethyl ether at room temperature over the course of two hours to generate $[Ir(CH_2SiMe_3)(triphos)]$ along with LiCl as an insoluble by-product.[24] Transition metal nitrates similarly react with alkyllithium reagents in nonpolar solvents and form $LiNO_3$ as a by-product. This strategy has been utilized in the synthesis of different organoplatinum(II) complexes.[25]

The range of reagents that may be used for halide abstraction reactions in *aqueous media* is quite narrow with silver salts, AgY, often the reagents of choice. In *non-aqueous media*, the range of reagents, MY, is broader because the corresponding metal halides, MX, typically show poor solubility in solvents characterized by low donor numbers,[1] D_N. Sodium carbonylmetallates, for example, react with $[PtCl_2(PhCN)_2]$ in THF to generate the sparingly soluble by-product, NaCl. Braunstein and coworkers[26-31] have utilized this methodology (see also section 5.4) in the synthesis of the trinuclear complex *trans*-$[Pt\{Mo(CO)_3Cp\}_2(PhCN)_2]$ by ion exchange between $[PtCl_2(PhCN)_2]$ and $Na[Mo(CO)_3Cp]$ in THF, Eq. 5.8:

$$[PtCl_2(PhCN)_2] + 2Na[Mo(CO)_3Cp] \rightarrow \textit{trans}\text{-}[Pt\{Mo(CO)_3Cp\}_2(PhCN)_2] + 2NaCl \quad (5.8)$$

Braunstein has stated[32] that one of the important driving forces for this reaction is the formation of NaCl but, in this specific case, the complex *trans*-$[Pt\{Mo(CO)_3Cp\}_2(PhCN)_2]$ is also only slightly soluble in THF.

The reaction described by Eq. 5.8, in which the chloride ligands of $[PtCl_2(PhCN)_2]$ are replaced while the benzonitrile ligands are unaffected, may be contrasted with the more common mode of reactivity for this complex in which the labile benzonitrile ligands undergo facile substitution, not only by neutral ligands such as tertiary phosphines, sulfoxides, *etc.*, but also by anions, such as the chloride ion,[33] Eq. 5.9:

$$[PtCl_2(PhCN)_2] + 2[Ph_3PCH_2Ph]Cl \rightarrow [Ph_3PCH_2Ph]_2[PtCl_4] + 2PhCN \quad (5.9)$$

The fundamental differences in the chemical reactions shown in Eq. 5.8 and Eq. 5.9 illustrate the importance of the formation of an insoluble by-product as a driving force in ion exchange.

The reactivity of the chloride and benzonitrile ligands of $[PtCl_2(PhCN)_2]$ under different conditions has been exploited in the synthesis of *trans*-$[Pt\{Mn(CO)_5\}_2(CO)_2]$.[29] Thus, reaction of $[PtCl_2(PhCN)_2]$ with two equivalents of $Na[Mn(CO)_5]$ in THF results in precipitation of NaCl and generation of *trans*-$[Pt\{Mn(CO)_5\}_2(PhCN)_2]$. Subsequent reaction with CO in toluene leads to substitution of the benzonitrile ligands and formation of *trans*-$[Pt\{Mn(CO)_5\}_2(CO)_2]$.

Ion exchange in the presence of easily-substituted neutral ligands is further exemplified by the reaction of $[SmI_2(THF)_2]$ with KCp in THF which produces the sparingly soluble salt KI and the soluble complex $[Sm(THF)_2Cp_2]$.[34] In a similar manner, Hadj-Bagheri and Puddephatt[35] have shown that $[PtCl_2(SMe_2)_2]$ reacts by formal anion exchange with phenyllithium in diethyl ether, leaving the normally labile Pt–SMe_2 bonds intact, Eq. 5.10:

$$[PtCl_2(SMe_2)_2] + 2PhLi \rightarrow [PtPh_2(SMe_2)_2] + 2LiCl \quad (5.10)$$

Kavanagh, Steed and Tocher[36] have compared the replacement of chloride by acetate in the bis(allyl)ruthenium(IV) complex $[\{RuCl(\mu\text{-}Cl)(\eta^3{:}\eta^3\text{-}C_{10}H_{16})\}_2]$ with silver acetate, sodium acetate, and acetic acid as potential sources of acetate ion. With the silver salt, ion exchange in acetone proceeds rapidly at room temperature. The reaction is driven by precipitation of AgCl and generates [$Ru(O_2CMe)Cl(\eta^3{:}\eta^3\text{-}C_{10}H_{16})$] in solution. The Ru(IV) complex has been characterized crystallographically, Figure 5.3:

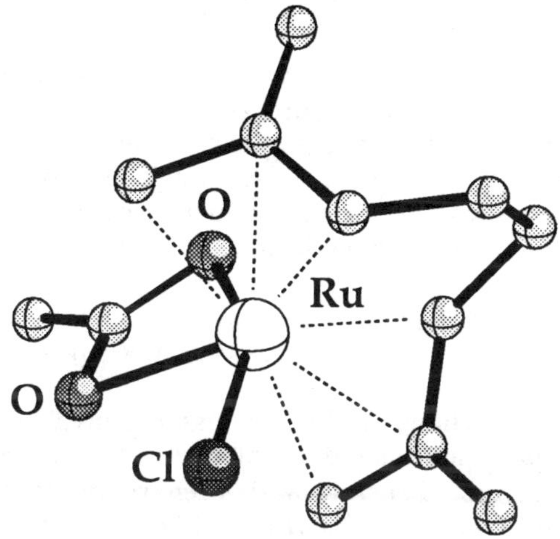

Figure 5.3

With sodium acetate as the source of the anionic chelating ligand, the precipitation of NaCl is slow and other factors may also be important in driving the reaction to completion. Attempts to generate the ruthenium(IV) complex [$Ru(O_2CMe)Cl(\eta^3{:}\eta^3\text{-}C_{10}H_{16})$] using acetic acid failed.[37]

Borkowski, Baenziger and Jordan[38] have shown that [$TiCl(CH_2Ph)Cp_2$] reacts with $NaBPh_4$ in acetonitrile to generate the solvento-complex, Scheme 5.3:

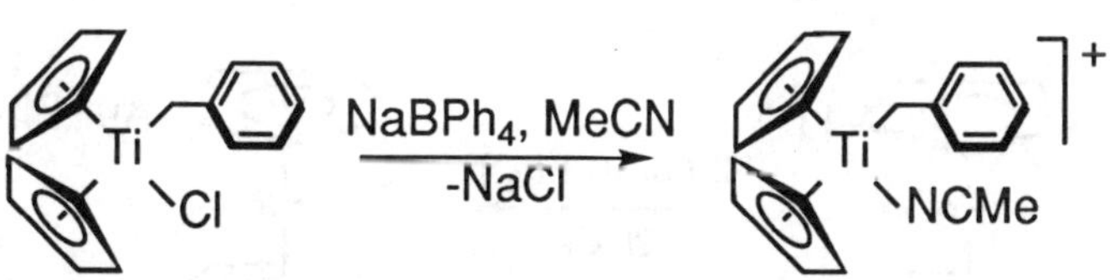

Scheme 5.3

The reaction is apparently driven by precipitation of chloride ion, formed by dissociation from titanium as NaCl.

Exchange of formally anionic groups occurs in the reaction between certain metal hydrides and organic halides. The reaction shown in Eq. 5.11 is illustrative:[39]

$$[NbH(PMe_2Ph)Cp_2] + C_4H_9Br \rightarrow [NbBr(PMe_2Ph)Cp_2] + C_4H_{10} \qquad (5.11)$$

Formation of the saturated hydrocarbon is apparently an important driving force in this reaction. Saturated hydrocarbons are also the by-products in a number of reactions between complexes containing M–H bonds and complexes containing M–C bonds. Although not strictly ion exchange reactions, processes such as that shown in Eq. 5.12[40] may be conceptually related to the ion exchange reaction described by Eq. 5.11.

$$[MoH(CO)_3Cp] + [ZrMe_2Cp_2] \rightarrow [Cp(CO)_3MoZrMeCp_2] + CH_4 \qquad (5.12)$$

Ion exchange reactions need not necessarily involve the formation and cleavage of M–X bonds and may occur at sites remote from the metal center. For example, Chivers, Lensink and Richardson[41a] have shown that treatment of $[Mo(CO)_5(PPh_2Cl)]$ with KNSO in acetonitrile results in the precipitation of KCl and the formation of $[Mo(CO)_5(PPh_2NSO)]$ in solution. Reactions of some bromoalkyl complexes with sodium iodide in acetone give high yields of the corresponding iodoalkyl complexes, Eq. 5.13:

$$CpM(CO)_2(CH_2)_4Br + NaI \rightarrow CpM(CO)_2(CH_2)_4I + NaBr \qquad (5.13)$$

where M = Fe,[41b] Ru[41c]

The auration of the metallated cyclopentadienyl ligands in $[Fe(C_5H_4Li)_2]$ is a related example of ion exchange at a site remote from the metal center,[42] Scheme 5.4:

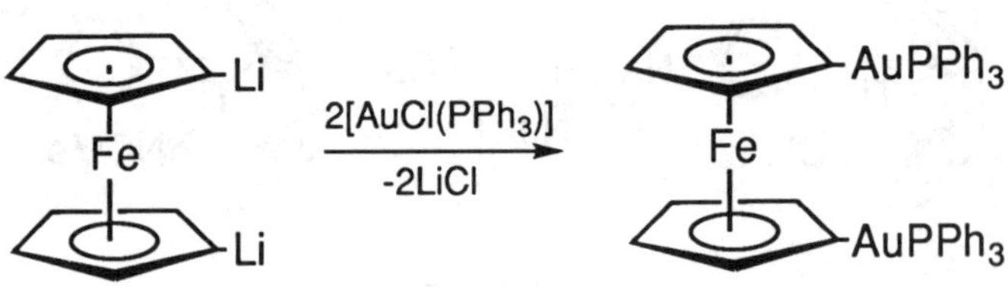

Scheme 5.4

Some closely-related ion exchange reactions of vanadium compounds also have been reported.[43]

5.2 Synthesis of complexes containing classically "noncoordinating" anions as ligands

It is well-documented that certain anions of strong protic acids, such as perchlorate (ClO_4^-), tetrafluoroborate (BF_4^-), hexafluorophosphate (PF_6^-), and trifluoromethanesulfonate (triflate, $CF_3SO_3^-$), are very poor donors that do not typically function as ligands, particularly for late transition metals. In the 1960s and early 1970s, few complexes containing these anions as ligands had been isolated and characterized and so the term "noncoordinating" entered common usage.[44] Since the early 1970s a number of synthetic methods have evolved that allow the generation of complexes where these "noncoordinating" anions function as ligands. The terminology, nonetheless, remains in common use, although sometimes a touch of irony might be implied (*e.g.* a paper entitled "*Coordination of 'Noncoordinating' Anions etc.*" [45]).

Weakly bonded anions may be displaced from certain metal centers by poor, neutral donors, including ligands such as THF, acetone, benzene, and dichloromethane. Thus, complexes containing classically noncoordinating anions as ligands function as inorganic synthons and serve as precursors to a wide range of derivatives by substitution.[46] This utility in synthesis, combined with applications in homogeneous catalysis, has led to considerable interest in the chemistry of complexes containing poor donor anions and a number of review articles have appeared.[46-49]

X-ray crystallography has played an important role in the developing chemistry of classically noncoordinating anions and, in general, structural studies reveal little perturbation in the geometric parameters associated with the anion upon coordination. This has been interpreted[50] in terms of a large ionic contribution to the metal-anion bonding.

Beck and Sünkel, in their review article,[47] discuss the relative donor abilities of anions. Based upon IR and NMR spectroscopic data for $[MX(CO)_3Cp]$ (M = Mo, W), the following order of donor abilities for X^- was suggested:[51-55]

$$FAsF_5^-, FSbF_5^-, FPF_5^- < FBF_3^- < OSO_2F^-, OSO_2CF_3^- < OClO_3^-$$

$$< OTeF_5^- \ll OReO_3^- \ll Cl^-$$

The relative donor strength of anions has been the subject of a number of studies.[45,56]

Several synthetic methods have been developed for the preparation of complexes containing weakly bonded anions derived from strong protic acids. However, ion exchange in a poorly coordinating solvent has proved to be the most productive method and examples are cited in the following paragraphs.

Hartley and his research group[57,58] have described the reactions of $[PdCl_2(dppe)]$ with $AgClO_4$ in a variety of solvent systems. Precipitation of AgCl and formation of $[Pd(solv)_2(dppe)]^{2+}$ *in situ* occurred when the solvent was THF or methanol but attempts to isolate these complexes led to displacement of the weakly coordinated solvent molecules by the perchlorate anions and formation of $[Pd(OClO_3)_2(dppe)]$. Halide abstraction from $[PdCl_2(dppe)]$ with $AgClO_4$ in the presence of benzonitrile allowed the isolation of $[Pd(NCPh)_2(dppe)][ClO_4]_2$ but an analogous reaction in the presence of acetonitrile led to isolation of a mixture of products with coordinated acetonitrile and coordinated perchlorate ligands. The complex $[Pd(OClO_3)_2(dppe)]$ could be obtained directly by halide abstraction in a dichloromethane/benzene solvent mixture.

Gowik, Klapötke and Thewal[59] have established that the reaction between $[TiF_2Cp_2]$ and two equivalents of SbF_5 in liquid sulfur dioxide produces $[Ti(FSbF_5)_2Cp_2]$, Eq. 5.14:

$$[TiF_2Cp_2] + 2SbF_5 \rightarrow [Ti(FSbF_5)_2Cp_2] \quad (5.14)$$

Beck and coworkers[51,60] have shown that the reaction of $[MH(CO)_3Cp]$ (M = Mo, W) with Ph_3CX (X = BF_4, PF_6) proceeds *via* formal ion exchange to generate $[MX(CO)_3Cp]$ where the coordination of X^- was confirmed by IR spectroscopy. Reactions of $[MX(CO)_3Cp]$[61,62] with CO, PPh_3, PCl_3, C_2H_4, MeCN, Me_2CO, THF, or H_2O lead to facile substitution of X^-.

The reaction of $[Re(Me)(NO)(PPh_3)Cp]$ with trifluoromethanesulfonic acid (triflic acid) in dichloromethane at 0 °C results in cleavage of the Re–Me bond and formation of $[Re(OSO_2CF_3)(NO)(PPh_3)Cp]$ by formal ion exchange of methyl anion for the triflate anion.[63,64] Methane was not detected in this reaction. However, in other protonation reactions of rhenium alkyls, $[Re(R)(NO)(PPh_3)Cp]$, formation of alkane, RH, has been confirmed.[63]

The reactions of $[PdCl(R)(PPh_3)_2]$ (R = 2-{6-chloropyridyl}) with AgX (X = BF_4, ClO_4, PF_6) in dichloromethane/benzene produce bridged, dimeric complexes of the type $[Pd_2R_2(\mu\text{-}X)_2(PPh_3)_2]$.[65] Addition of stoichiometric amounts of pyridine to solutions of these complexes leads to bridge cleavage and the generation of the mononuclear pyridine complexes $[PdRX(py)(PPh_3)]$ in solution. For the case where X = ClO_4, the reaction is illustrated in Scheme 5.5:

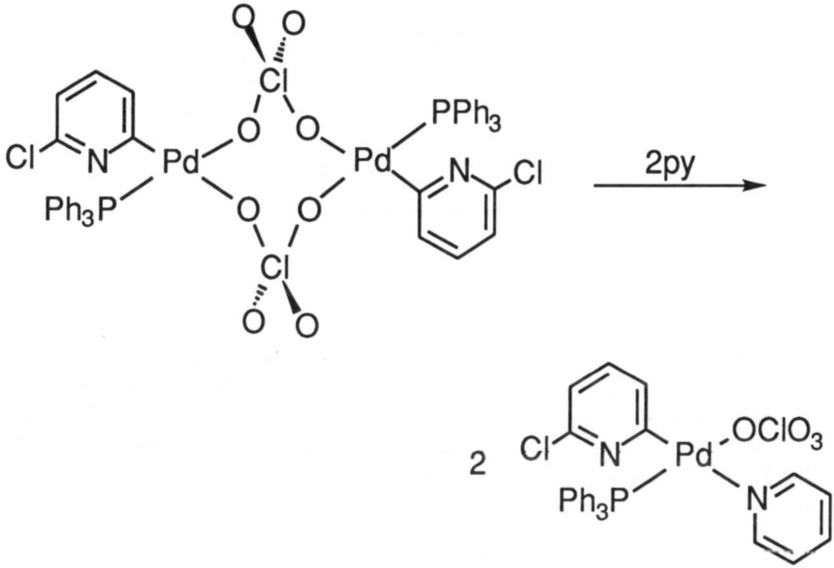

Scheme 5.5

The complex $[Pd_2R_2(\mu\text{-}ClO_4)_2(PPh_3)_2]$ has been described as the first example of a dinuclear complex where two perchlorate ligands bridge the metal centers.

5.3 Formation of chelate complexes by ion exchange in non-aqueous solvents

Replacement of two monodentate, anionic ligands by a bidentate, formally dianionic ligand results in the generation of a chelate complex. For example, the reactions of Grignard reagents, $BrMgCH_2CR_2CH_2MgBr$, with $[MCl_2Cp_2]$ (M = Ti, Zr, Hf) yield metallacycles with four-membered chelate rings,[66] *e.g.* Scheme 5.6:

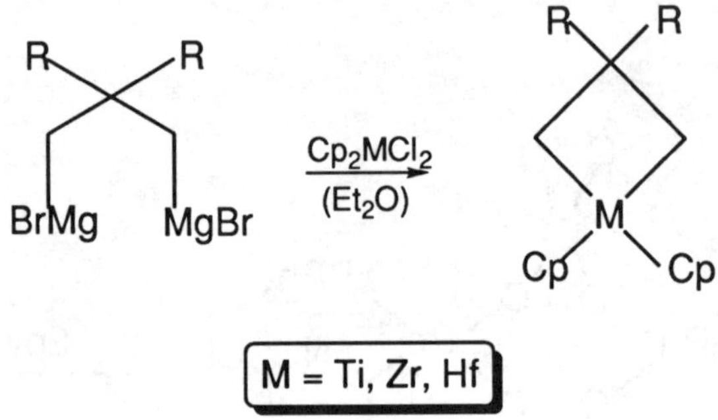

Scheme 5.6

A study[67] of the reactions of $[MoX_2Cp_2]$ (X = Cl, Br, I) with excess $Li(CH_2)_4Li$ in Et_2O shows that the corresponding molybdenacyclopentane is formed in reactions in which yields are strongly dependent upon the nature of X. Thus, yields were lowest when X = Cl and highest when X = I. The resulting molybdenacyclopentane, $[Mo\{(CH_2)_4\}Cp_2]$, has been characterized by X-ray crystallography, Figure 5.4:

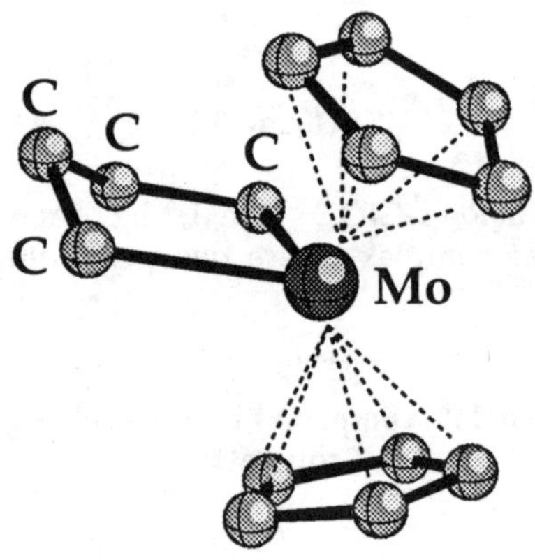

Figure 5.4

A related replacement of two anionic ligands with one bidentate ligand occurs in the generation of a palladium(II) acetylacetonate complex by removal of two bridging chloride ligands with thallium(I) and replacement by the acetylacetonate ligand,[68] Scheme 5.7:

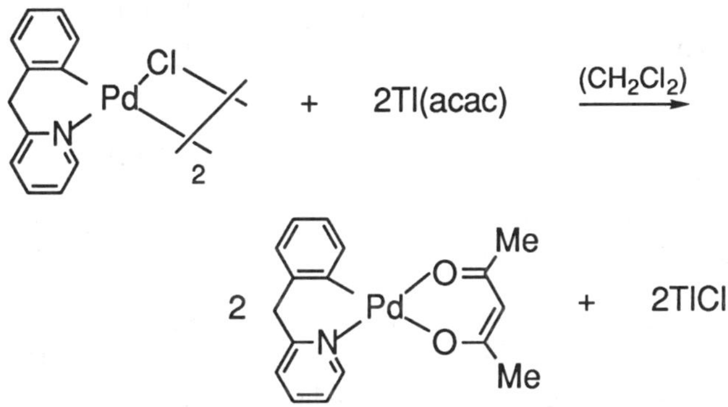

Scheme 5.7

5.4 Synthesis of complexes containing metal-metal bonds between transition metals

Several literature sources contain information on the synthesis of complexes with metal-metal bonds between transition metals or between transition metals and main group metals.[69,70] The use of ion exchange reactions in the rational synthesis of heterobimetallic compounds is widespread. Such reactions are of considerable utility because the starting materials are readily accessible and typically the reactions proceed cleanly and in high yield.[71] A number of ion exchange reactions that lead to the isolation of complexes containing metal-metal bonds have been described in *Inorganic Syntheses*.[72,73]

Carbonylmetallates and metal halide complexes can be employed as starting materials in the synthesis of compounds with M–M* bonds.[74,75] Complexes with Co–Rh bonds have been synthesized by this method, with THF as the solvent,[76] Eq. 5.15:

$$Na[Co(CO)_4] + [RhCl(CO)(PEt_3)_2] \rightarrow [(CO)_4Co\text{–}Rh(CO)(PEt_3)_2] + NaCl \qquad (5.15)$$

The precipitation of NaCl is a significant driving force in this reaction because, in the presence of a THF-soluble source of chloride such as [PPN]Cl, the $[Co(CO)_4]^-$ fragment is displaced from $[(CO)_4Co–Rh(CO)(PEt_3)_2]$, Eq. 5.16:

$$[(CO)_4Co\text{–}Rh(CO)(PEt_3)_2] + [PPN]Cl \rightarrow [PPN][Co(CO)_4] + [RhCl(CO)(PEt_3)_2] \quad (5.16)$$

The reaction of an anionic metal carbonyl and a cationic metal carbonyl in THF at -20 °C has been studied,[77-79] Eq. 5.17:

$$[Mn(CO)_6]X + Na[Co(CO)_4] \rightarrow [Mn(CO)_6][Co(CO)_4] + NaX \quad (5.17)$$

Warming to 20 °C results in elimination of CO from $[Mn(CO)_6][Co(CO)_4]$ and formation of $[(CO)_5Mn–Co(CO)_4]$.

The reaction of $[ZrCl(Me)Cp_2]$ with $K[Ru(CO)_2Cp]$ in THF at room temperature leads to the formation of the heterobimetallic compound, $[Cp_2(Me)Zr–Ru(CO)_2Cp]$.[80] The same product has been prepared from $[ZrCl_2Cp_2]$ by replacement of one chloride ligand with $[Ru(CO)_2Cp]^-$ followed by replacement of the second chloride ligand with an equivalent of methyllithium, Scheme 5.8:

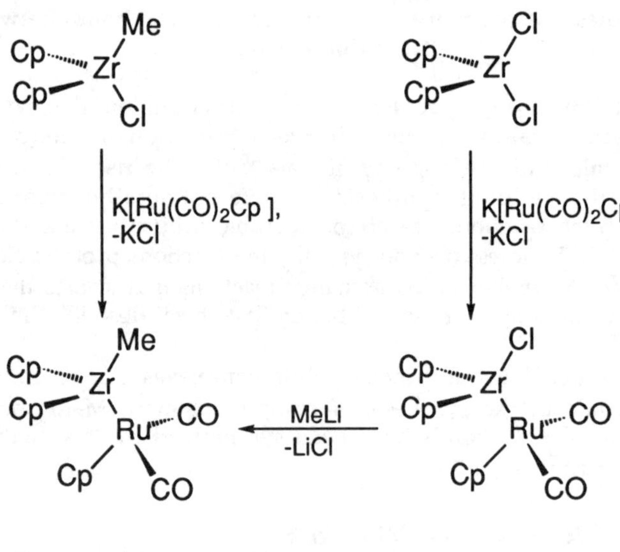

Scheme 5.8

The Zr–Fe analog, $[Cp_2(Me)Zr–Fe(CO)_2Cp]$, has been synthesized by the same method (Scheme 5.8) utilizing $K[Fe(CO)_2Cp]$.

Pfeffer and coworkers[81-83] have shown that heterobimetallic complexes containing M–M* bonds can be synthesized through the reactions of carbonylmetallate anions with cyclopalladated compounds. Reaction between the palladium(II) complex (Figure 5.5) and $Na[Mo(CO)_3Cp]$ takes place only when hexane is added to a mixture of the starting materials in THF.

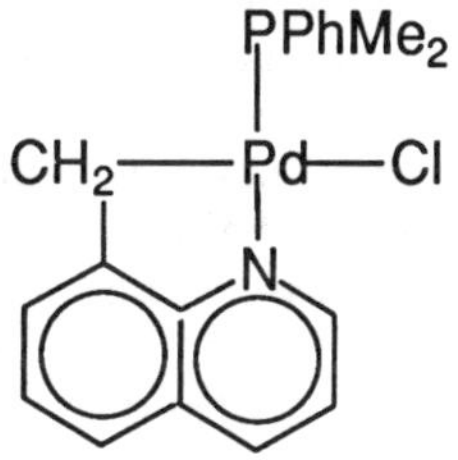

Figure 5.5

The product has been characterized by X-ray crystallography, Figure 5.6:

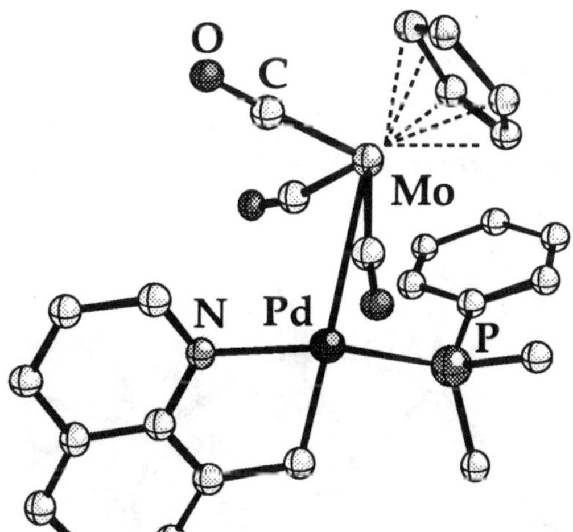

Figure 5.6

The Pd–Mo bond of the compound shown in Figure 5.6 can be cleaved easily by the addition of NaCl to a THF solution of the complex. This observation[81] suggests that the Pd–Mo interaction has a significant ionic component.

5.5 Synthesis of complexes containing transition metal-main group metal/metalloid bonds

Ion exchange methods for the synthesis of compounds that contain bonds between transition metals and main group metals or metalloids are, in general, similar to those described in the preceding section for the generation of complexes containing bonds between transition metals. A number of review articles describing the chemistry of these so-called *inorganometallic* compounds have appeared with many examples of ion exchange reactions in synthesis. Accordingly, only representative examples are cited here.

Main group metal/metalloid halides and carbonylmetallates react to generate compounds containing M–M* bonds. For example, compounds containing Mo–Sn and Mn–Ge bonds have been synthesized by such reactions in THF solution,[72, 73] Eqs. 5.18 and 5.19:

$$SnClMe_3 + Na[Mo(CO)_3Cp] \rightarrow [Me_3Sn{-}Mo(CO)_3Cp] + NaCl \qquad (5.18)$$

$$GeH_3Br + Na[Mn(CO)_5] \rightarrow [H_3Ge{-}Mn(CO)_5] + NaBr \qquad (5.19)$$

In a similar manner, the reaction between $SbCl_3$ and $Na[Mo(CO)_3Cp]$ in THF solution has been shown to produce the dimetallaantimony complex, $[SbCl\{Mo(CO)_3Cp\}_2]$, accompanied by precipitation of NaCl.[84] The antimony compound has been characterized by X-ray crystallography, Figure 5.7:

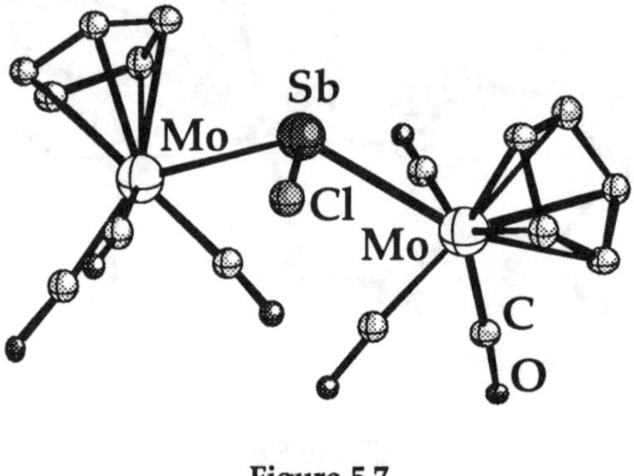

Figure 5.7

The molecule consists of a central antimony atom in a trigonal planar geometry bonded to one chlorine atom and to two "piano-stool" molybdenum fragments. It has been established that treatment of $SbCl_3$ with three equivalents of the carbonylmetallate reagent generates the trimetallaantimony compound, $[Sb\{Mo(CO)_3Cp\}_3]$.

Sodium carbonylmetallates similarly react with compounds containing P–X and As–X bonds in hydrocarbon solvents with formation of NaX as an insoluble by-product,[85-87] Eq. 5.20 and 5.21:

$$2Na[Co(CO)_4] + 2PClPh_2 \rightarrow [Co_2(\mu\text{-}PPh_2)_2(CO)_6] + 2NaCl + 2CO \quad (5.20)$$

$$Na[Mo(CO)_3Cp] + AsClMe_2 \rightarrow [Mo(AsMe_2)(CO)_3Cp] + NaCl \quad (5.21)$$

Main group metal halide fragments coordinated to transition metals are prone to attack by sodium carbonylmetallates. Firfiray, Irving and Moss[88] have shown that $[Os_3Cl(SnCl_3)(CO)_{12}]$ reacts with $Na[M(CO)_3Cp]$ (M = Mo, W) by substitution at tin, Eq. 5.22:

$$[Os_3Cl(SnCl_3)(CO)_{12}] + Na[M(CO)_3Cp] \rightarrow [Os_3Cl(SnCl_2)(CO)_{12}\{M(CO)_3Cp\}] + NaCl \quad (5.22)$$

Moss describes the driving forces behind this reaction as the high nucleophilicity of the sodium carbonylmetallate and the poor solubility of the sodium halide formed as a by-product.[89] The tungsten compound, $[Os_3Cl(SnCl_2)(CO)_{12}\{W(CO)_3Cp\}]$, has been characterized by X-ray crystallography and consists of a chain of five metal atoms, held together only by metal-metal bonds, Figure 5.8.

The opposite approach to that illustrated in Eq. 5.18 - 5.22, *i.e.* one in which the reaction involves a transition metal halide complex and a main group element anion or anion equivalent, has been widely employed. For example, compounds containing Pt–Sn, Pt–Ge, and Pt–Si bonds have been synthesized through this approach, Eq. 5.23:

$$[PtX_2L_2] + 2LiA(R_1R_2R_3) \xrightarrow{solv} [Pt\{A(R_1R_2R_3)\}_2L_2] + 2LiX \quad (5.23)$$

where L_2 = dppe; X = Cl; A = Si, Ge; R_1 = Me; $R_2 = R_3$ = Ph; solv = C_6H_6;[90,91] L = PMe_2Ph; X = Cl; A = Si; R_1 = Me; $R_2 = R_3$ = Ph; solv = THF;[90,91] L_2 = dppe; X = I; A = Ge; $R_1 = R_2 = R_3$ = Ph; solv = Et_2O or C_6H_6;[92] L = PEt_3; X = Cl; A = Ge; $R_1 = R_2 = R_3$ = Ph; solv = Et_2O;[93] L = PPh_3; X = Cl; A = Sn; $R_1 = R_2 = R_3$ = Ph; solv = THF;[94] L = PEt_3; X = Br;

A = Ge; $R_1 = R_2 = R_3$ = Ph; solv = 1,2-dimethoxyethane.[92]

Each of these reactions involves formation of a lithium salt as an insoluble by-product.

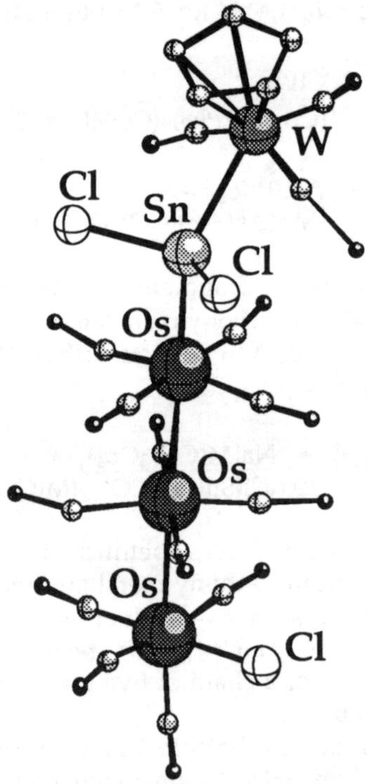

Figure 5.8

As an extension of the methodology represented by Eq. 5.23, compounds containing both Pt–Si and Pt–Ge bonds have been synthesized by ion exchange in diethyl ether,[95] Eq. 5.24:

$$[PtCl(SiMe_3)(PEt_3)_2] + LiGePh_3 \rightarrow [Pt(GePh_3)(SiMe_3)(PEt_3)_2] + LiCl \qquad (5.24)$$

The reagent $LiGePh_3$ has been employed in the synthesis of compounds containing Cu–Ge, Ag–Ge, and Au–Ge bonds,[94,96,97] Eq. 5.25:

$$[MCl(PPh_3)_n] + LiGePh_3 \xrightarrow{solv} [M(GePh_3)(PPh_3)_n] + LiCl \qquad (5.25)$$

where M = Cu, Ag, Au; n = 1 to 3; solv = ethylene glycol, THF, or Et_2O.

5.6 Replacement of counterions in ionic complexes

The replacement of one counterion by another in an ionic complex represents a convenient way to modify the solubility of the compound (see Chapter 2). Often this approach is more convenient than modification of the structure of a coordinated ligand in order to alter the solubility properties of a complex. Ion exchange thus represents a method by which solubility may be modified. Such reactions are performed in aqueous media, so that a sparingly soluble complex precipitates from solution, or in non-aqueous media, so that a sparingly soluble by-product, usually a salt, precipitates from solution.

The addition of tetra(*n*-butyl)ammonium chloride to an aqueous solution of $K[PtCl_3(MeCN)]$ results in cation exchange and precipitation of $[(n\text{-}Bu)_4N][PtCl_3(MeCN)]$ with KCl remaining in solution.[98] The addition of sodium perchlorate to aqueous $[Fe(bipy)_3]Cl_2$ results in anion exchange and precipitation of $[Fe(bipy)_3][ClO_4]_2$ with NaCl remaining in solution.

Chugaev[99] has shown that treatment of solid $K_2[PtCl_4]$ with alcoholic tri(*iso*-propyl)ammonium chloride at 60 °C results in the formation of solid KCl and an alcoholic solution of $[(i\text{-}Pr)_3NH]_2[PtCl_4]$. A series of complexes, $[AH]_2[MCl_4]$ (where A = amine; M = Pd, Pt) has been synthesized[100,101] *via* the general scheme described by Chugaev. A convenient method involves treatment of the potassium tetrachlorometallate(II), $K_2[MCl_4]$, with amine hydrochloride as a suspension in DMF. Because KCl is only sparingly soluble in DMF, ion exchange is typically complete in 10 to 15 minutes.

Combinations of ion exchange processes have been employed for conversions where a one-step method is not available. For example, to generate the cesium salt from $Na_3[IrCl_4(C_2O_4)]$ requires two steps.[102] First, tetra(*n*-pentyl)ammonium chloride is added to an aqueous solution of $Na_3[IrCl_4(C_2O_4)]$ and $[(n\text{-}Pn)_4N]_3[IrCl_4(C_2O_4)]$ is extracted into CH_2Cl_2. Second, cesium trichloroacetate, which is soluble in dichloromethane, is added to the organic fraction and the resultant precipitate of $Cs_3[IrCl_4(C_2O_4)]$ is separated by filtration.

In order to exchange sodium cations for cesium in $Na_2[Fe(CN)_5(NO)]$, a different two-step process is employed.[103] First, $Na_2[Fe(CN)_5(NO)]$ is treated with $AgNO_3$ in aqueous solution to produce the sparingly soluble silver salt, $Ag_2[Fe(CN)_5(NO)]$, and leave $NaNO_3$ in solution. Second, treatment of $Ag_2[Fe(CN)_5(NO)]$ with cesium chloride leaves silver chloride as an insoluble precipitate and $Cs_2[Fe(CN)_5(NO)]$ in solution.

In order to generate the barium salt of the titanium tris(catecholate) dianion from its ammonium salt, an acid-base reaction with barium hydroxide has been employed,[104] Eq. 5.26:

$$[NH_4]_2[Ti(C_6H_4O_2)_3] + Ba(OH)_2 \rightarrow Ba[Ti(C_6H_4O_2)_3] + 2NH_3 + 2H_2O \qquad (5.26)$$

However, in order to generate the analogous lead(II) salt a comparable method cannot be employed since lead(II) hydroxide does not exist and PbO and related oxide derivatives are insufficiently basic to react with the ammonium cation. Fortunately, the large sizes of the lead(II) cation and the $[Ti(C_6H_4O_2)_3]^{2-}$ dianion result in spontaneous precipitation of $Pb[Ti(C_6H_4O_2)_3]$ upon addition of aqueous lead nitrate to aqueous $[NH_4]_2[Ti(C_6H_4O_2)_3]$. Ammonium nitrate remains in solution.[105]

References

1. V. Gutmann, *Coordination Chemistry in Non-aqueous Solutions* (Springer-Verlag, New York, 1968) p. 23.
2. A. F. M. van der Ploeg, G. van Koten, and K. Vrieze, *Inorg. Chem.* **21** (1982) 2026.
3. R. G. Pearson, *Coord. Chem. Rev.* **100** (1990) 403.
4. A. Emeran, M. A. Gafoor, J. K. I. Goslett, Y.-H. Liao, L. Pimble, and J. R. Moss, *J. Organomet. Chem.* **405** (1991) 237.
5. K. P. Finch, J. R. Moss, and M. L. Niven, *Inorg. Chim. Acta* **166** (1989) 181.
6. M. A. Gafoor, A. T. Hutton, and J. R. Moss, *unpublished results.*
7. J. M. Andersen and J. R. Moss, *J. Organomet. Chem.* **439** (1992) C25.
8. J. M. Andersen and J. R. Moss, *unpublished results.*
9. R. B. King, ed., *Organometallic Synthesis* (Wiley, New York, 1966), vol. 1, p. 186.
10. A. N. Nesmeyanov, *Ferrocene and Related Compounds* [in Russian] (Nauka, Moscow, 1982) p. 440.
11. A. N. Nesmeyanov, R. B. Materikova, and N. S. Kochetkova, *Izv. Akad. Nauk SSSR, Ser. Khim.* (1963) 1334.
12. A. N. Nesmeyanov, R. B. Materikova, E. M. Brainina, and N. S. Kochetkova, *Izv. Akad. Nauk SSSR, Ser. Khim.* (1969) 1323.
13. A. N. Nesmeyanov, R. B. Materikova, N. S. Kochetkova, and E. V. Leonova, *Dokl. Akad. Nauk SSSR* **177** (1967) 131.
14. A. N. Nesmeyanov, R. B. Materikova, and N. S. Kochetkova, *Dokl. Akad. Nauk SSSR* **147** (1962) 113.
15. A. N. Nesmeyanov, R. B. Materikova, N. S. Kochetkova, and R. V. Lukyanova, *Izv. Akad. Nauk SSSR., Ser. Khim.* (1969) 975.
16. R. J. Cross and A. J. McLennan, *J. Chem. Soc., Dalton Trans.* (1983) 359.
17. H. C. Clark and A. Shaver, *Can. J. Chem.* **54** (1976) 2068.
18. H. Kurosawa, T. Majima, and N. Asada, *J. Am. Chem. Soc.* **102** (1980) 6996.
19. A. B. Goel, S. Goel, D. van der Veer, and H. C. Clark, *Inorg. Chim. Acta* **53** (1981) L117.

20. G. K. Anderson and R. J. Cross, *J. Chem. Soc., Dalton Trans.* (1979) 1246.

21. B. J. J. van der Heisteeg, G. Schat, O. S. Akkerman, and F. Bickelhaupt, *Organometallics* **5** (1986) 1749.

22. H. Lehmkuhl, Y.-L. Tsien, E. Jannsen, and R. Mynott, *Chem. Ber.* **116** (1983) 2426.

23. G. N. Schrauzer, L. A. Hughes, N. Strampach, F. Ross, D. Ross, and E. O. Schlemper, *Organometallics* **2** (1983) 481.

24. E. Arpac and L. Dahlenburg, *J. Organomet. Chem.* **251** (1983) 361.

25. J. K. Jawad, R. J. Puddephatt, and M. A. Staleri, *Inorg. Chem.* **21** (1982) 332.

26. P. Braunstein in *Perspectives in Coordination Chemistry*, eds. A. F. Williams, C. Floriani, and A. Merbach (VCH, Weinheim, 1992) p. 67.

27. P. Braunstein, R. Bender, and J.-M. Jud, *Inorg. Synth.* **26** (1989) 341.

28. R. Bender, P. Braunstein, J.-M. Jud, and Y. Dusausoy, *Inorg. Chem.* **23** (1984) 4489.

29. O. Bars, P. Braunstein, and J.-M. Jud, *Nouv. J. Chim.* **8** (1984) 771.

30. P. Braunstein, C. de M. de Bellefon, S.-E. Bouaoud, D. Grandjean, J.-F. Halet, and J.-Y. Saillard, *J. Am. Chem. Soc.* **113** (1991) 5282.

31. P. Braunstein, E. de Jesús, A. Dedieu, M. Lanfranchi, and A. Tiripicchio, *Inorg. Chem.* **31** (1992) 399.

32. P. Braunstein, *personal communication*.

33. Yu. N. Kukushkin, *Reactivity of Coordination Compounds* [in Russian] (Khimia, Leningrad, 1987).

34. W. J. Evans and T. A. Ulibarri, *Inorg. Synth.* **27** (1990) 155.

35. N. Hadj-Bagheri and R. J. Puddephatt, *Polyhedron* **7** (1988) 2695.

36. B. Kavanagh, J. W. Steed, and D. A. Tocher, *J. Chem. Soc., Dalton Trans.* (1993) 327.

37. D. A. Tocher, *personal communication*.

38. S. L. Borkowski, N. C. Baenziger, and R. F. Jordan, *Organometallics* **12** (1993) 486.

39. C. R. Lucas, *Inorg. Synth.* **16** (1976) 107.

40. J. A. Marsella, J. C. Huffman, K. G. Caulton, B. Longato, and J. R. Norton, *J. Am. Chem. Soc.* **104** (1982) 6360.

41. (a) T. Chivers, C. Lensink, and J. F. Richardson, *Organometallics* **5** (1986) 819; (b) H. B. Freidrich, P. A. Makhesha, J. R. Moss, and B. K. Williamson, *J. Organomet. Chem.* **384** (1990) 325; (c) H. B. Freidrich, K. P. Finch, M. A. Gafoor, and J. R. Moss, *Inorg. Chim. Acta* **206** (1993) 225.

42. R. J. Puddephatt, *The Chemistry of Gold* (Elsevier, Amsterdam, 1978) p. 102.

43. F. J. Köhler, W. A. Geike, P. Hofmann, U. Schubert, and P. Stauffert, *Chem. Ber.* **117** (1984) 904.

44. M. R. Rosenthal, *J. Chem. Educ.* **50** (1973) 331.

45. R. V. Honeychuck and W. H. Hersh, *Inorg. Chem.* **28** (1989) 2869.

46. G. A. Lawrance, *Chem. Rev.* **86** (1986) 17.

47. W. Beck and K. Sünkel, *Chem. Rev.* **88** (1988) 1405.

48. (a) S. H. Strauss, *Chem. Rev.* **93** (1993) 927; (b) N. M. N. Gowda, S. B. Naikar, and G. K. N. Reddy, *Adv. Inorg. Chem. Radiochem.* **28** (1984) 255.

49. C. J. Garner and D. Hughes, *Adv. Inorg. Chem. Radiochem.* **17** (1975) 1.

50. K. Shelly, C. A. Reed, Y. J. Lee, and W. R. Scheidt, *J. Am. Chem. Soc.* **108** (1986) 3117.

51. W. Beck and K. Schloter, *Z. Naturforsch.* **33b** (1978) 1214.

52. K. D. Abney, K. M. Long, O. P. Anderson, and S. H. Strauss, *Inorg. Chem.* **26** (1987) 2638.

53. S. H. Strauss, K. D. Abney, K. M. Long, and O. P. Anderson, *Inorg. Chem.* **23** (1984) 1994.

54. J. Heidrich, D. Loderer, and W. Beck, *J. Organomet. Chem.* **312** (1986) 329.

55. M. Appel, K. Schloter, J. Heidrich, and W. Beck, *J. Organomet. Chem.* **322** (1987) 77.

56. D. M. Branan, N. W. Hoffman, E. A. McElroy, N. Prokopuk, A. B. Salazar, M. J. Robbins, W. E. Hill, and T. R. Webb, *Inorg. Chem.* **30** (1991) 1200.

57. F. R. Hartley and J. A. Davies, *Rev. Inorg. Chem.* **4** (1982) 27.

58. J. A. Davies, F. R. Hartley, and S. G. Murray, *J. Chem. Soc., Dalton Trans.* (1980) 2246.

59. P. Gowik, T. Klapötke, and U. Thewal, *J. Organomet. Chem.* **385** (1990) 345.

60. W. Beck and H. S. Smedal, *Angew. Chem.* **78** (1966) 267.

61. K. Sünkel, K. Schloter, W. Beck, K. Ackermann, and U. Schubert, *J. Organomet. Chem.* **241** (1983) 333.

62. B. Olgemöller, H. Bauer, H. Löbermann, U. Nagel, and W. Beck, *Chem. Ber.* **115** (1982) 2271.

63. J. H. Merrifield, J. M. Fernández, W. E. Buhro, and J. A. Gladysz, *Inorg. Chem.* **23** (1984) 4022.

64. M. A. Dewey, A. M. Arif, and J. A. Gladysz, *J. Organomet. Chem.* **384** (1990) C29.

65. K. Isobe, K. Nanjo, Y. Nakamura, and S. Kawaguchi, *Chem. Lett.* (1979) 1193.

66. J. W. F. L. Seetz, G. Schat, O. S. Akkermann, and F. Bickelhaupt, *Angew. Chem.* **95** (1983) 242; *Angew. Chem. Int. Ed. Engl.* **22** (1983) 248.

67. P. Diversi, G. Ingrosso, A. Lucherini, W. Porzio, and M. Zocchi, *J. Chem. Soc., Dalton Trans.* (1983) 967.

68. K. Hiraki, Y. Fuchita, and K. Takechi, *Inorg. Chem.* **20** (1981) 4316.

69. F. A. Cotton and R. A. Walton, *Multiple Bonds Between Metal Atoms* (2nd Ed., Wiley, New York, 1993).

70. (a) C. P. Casey, *J. Organomet. Chem.* **400** (1990) 205; (b) N. S. Vyazankin, G. A. Razuvaev, and O. A. Kruglaya, *Organomet. Chem. Rev.* **3A** (1968) 323; (c) D. A. Roberts and G. L. Geoffroy in *Comprehensive Organometallic Chemistry*, eds. G. Wilkinson, F. G. A. Stone, and E. W. Abel, (Pergamon, Oxford, 1982) vol. 6, p. 763.

71. M. H. Chisholm and I. P. Rothwell, *Prog. Inorg. Chem.* **29** (1982) 1.

72. R. E. J. Bichler, M. R. Booth, H. C. Clark, and B. K. Hunter, *Inorg. Synth.* **12** (1970) 60.

73. S. R. Stobart, *Inorg. Synth.* **15** (1974) 174.

74. W. Beck, *Angew. Chem.* **103** (1991) 173; *Angew. Chem. Int. Ed. Engl.* **30** (1991) 168.

75. J. E. Ellis, *Adv. Organomet. Chem.* **31** (1990) 1.

76. D. A. Roberts, W. C. Mercer, G. L. Geoffroy, and C. G. Pierpont, *Inorg. Chem.* **25** (1986) 1439.

77. T. Kruck and M. Höfler, *Angew. Chem. Int. Ed. Engl.* **3** (1964) 701.

78. T. Kruck and M. Höfler, *Chem. Ber.* **97** (1964) 2289.

79. T. Kruck, M. Höfler, and M. Noack, *Chem. Ber.* **99** (1966) 1153.

80. C. P. Casey, R. F. Jordan, and A. L. Rheingold, *J. Am. Chem. Soc.* **105** (1983) 665.

81. M. Pfeffer, D. Grandjean, and G. Le Borgne, *Inorg. Chem.* **20** (1981) 4426.

82. J. Dehand and M. Pfeffer, *J. Organomet. Chem.* **104** (1976) 377.

83. G. Le Borgne, S. E. Bouaoud, D. Grandjean, P. Braunstein, J. Dehand, and M. Pfeffer, *J. Organomet. Chem.* **136** (1977) 375.

84. A. M. Barr, M. D. Kerlogue, N. C. Norman, P. M. Webster, and L. J. Farrugia, *Polyhedron* **8** (1989) 2495.

85. A. D. Harley, G. J. Guskey, and G. L. Geoffroy, *Organometallics* **2** (1983) 53.

86. W. Malisch, M. Luksza, and W. S. Sheldrick, *Z. Naturforsch.* **36b** (1981) 1580.

87. M. Luksza, K. Jörg, and W. Malisch, *Inorg. Chim. Acta* **85** (1984) L49.

88. D. B. Firfiray, A. Irving, and J. R. Moss, *J. Chem. Soc., Chem. Commun.* (1990) 377.

89. J. R. Moss, *personal communication.*

90. J. Chatt, C. Eaborn, S. D. Ibekwe, and P. N. Kapoor, *J. Chem. Soc. A* (1970) 1343.

91. J. Chatt, C. Eaborn, S. D. Ibekwe, and P. N. Kapoor, *J. Chem. Soc., Chem. Commun.* (1967) 869.

92. E. H. Brooks, R. J. Cross, and F. Glocking, *Inorg. Chim. Acta.* **2** (1968) 17.

93. R. J. Cross and F. Glocking, *J. Chem. Soc.* (1965) 5422.

94. M. C. Baird, *J. Inorg. Nucl. Chem.* **29** (1967) 367.

95. F. Glocking and K. A. Hooton, *J. Chem. Soc.* (1967) 1066.

96. F. Glocking, *Quart. Rev.* **20** (1966) 45.

97. F. Glocking and K. A. Hooton, *J. Chem. Soc.* (1962) 2658.

98. V. Yu. Kukushkin, E. Yu. Pankova, T. N. Fomina, and N. P. Kiseleva, *Koord. Khim.* **14** (1988) 1110.

99. L. Tchugaev [Chugaev], *Compt. Rend.* **159** (1914) 188.

100. Yu. N. Kukushkin and L. I. Danilina, *Iz. Vyssh. Uchebn. Zaved., Khim., Khim. Tekhnol.* **24** (1981) 139.

101. V. V. Sibirskaya, N. V. Vorobiov-Desiatovsky, Yu. N. Kukushkin, and E. S. Postnikova, *Zh. Neorg. Khim.* **25** (1980) 2486.

102. H. Schultz and W. Preetz, *Z. Naturforsch.* **37b** (1982) 1553.

103. M. M. Vergara and E. L. Varetti, *J. Phys. Chem. Solids* **48** (1987) 13.

104. (a) J. A. Davies and S. G. Dutremez, *J. Am. Ceram. Soc.* **73** (1990) 1429; (b) J. A. Davies and S. G. Dutremez, *Inorg. Synth.* (1995) in press.

105. J. A. Davies and S. G. Dutremez, *J. Am. Ceram. Soc.* **73** (1990) 2570.

Chapter 6. SYNTHESIS OF BRIDGED COMPLEXES AND RING-CLOSURES

In previous Chapters the substitution of an anionic ligand by a neutral solvent molecule (Chapter 3) or by a second anionic ligand (Chapter 5) has been discussed. Two other modes of reaction are observed in selected instances when an anionic ligand is removed from a metal center. Thus, in the absence of other potential ligands, two or more unsaturated fragments can combine to form a bridged product or a remote donor group already present in a coordinated ligand can participate in ring-closure to the vacant coordination site. Some synthetically useful examples of these two classes of reaction are described in the following sections.

6.1 Synthesis of bridged complexes

A synthetic method based upon the elimination of a ligand from the coordination sphere of a complex in a solvent of poor donor ability is often employed for the preparation of bridged complexes. For example, the reaction of complexes of the type $[MCl_2(dppe)]$ (M = Pt, Pd) with one equivalent of $AgClO_4$ in diethyl ether or acetone results in the precipitation of silver chloride. The solvents used in the synthesis are characterized by low values of D_N and their donor abilities are lower than that of the remaining chloride ligand and dimerization, therefore, takes place. This results in the formation of bridged products of the type $[M_2(\mu\text{-}Cl)_2(dppe)_2][ClO_4]_2$.[1] The reaction of the complex, $[Pd(CN)_2(dppe)]$, with one equivalent of silver perchlorate in acetonitrile results in the formation of a product with the empirical formula $[Pd(CN)(ClO_4)(dppe)]$.[2] It was established by X-ray crystallography[3] that the compound obtained does not possess the expected planar, tetranuclear structure with linear Pd–C≡N–Pd bridges, but is, in fact, trimeric with significantly bent bridging groups,[4] Figure 6.1:

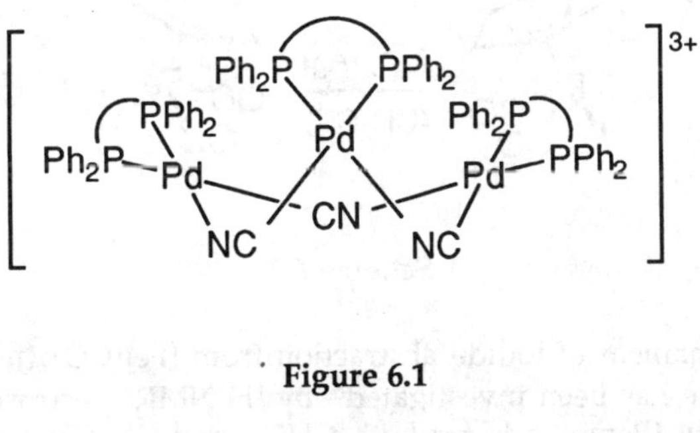

Figure 6.1

The unusual structure was explained on the basis of minimization of electrostatic repulsions between charge centers in the various possible oligomeric forms.[3,4]

The synthesis of the halide-bridged dimers, $[Pt_2(\mu\text{-}X_2)(R_3E)_4][BF_4]_2$ (R = Me, Ph; E = P, As, Sb; X = Cl, Br, I), has been described.[5,6] The reactions are complete in acetone or dichloromethane solution at room temperature within 1 hour, Scheme 6.1:

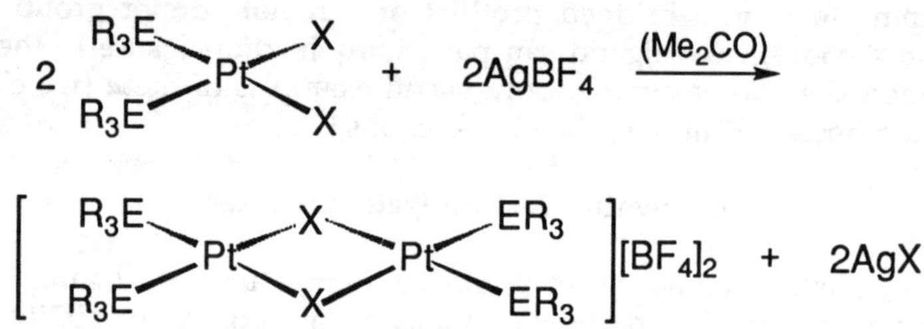

Scheme 6.1

The silyl bridge between the cyclopentadienyl rings in the *di*-iodide complex, $[Fe_2I_2(CO)_4(\mu\text{-}\{C_5H_4\}_2SiMe_2)]$, maintains the two iron centers in close proximity to one another.[7] When the complex is treated with silver tetrafluoroborate in dichloromethane, one iodide is abstracted and the other iodide is positioned to bridge the two metal centers with formation of an iodonium salt, Scheme 6.2:

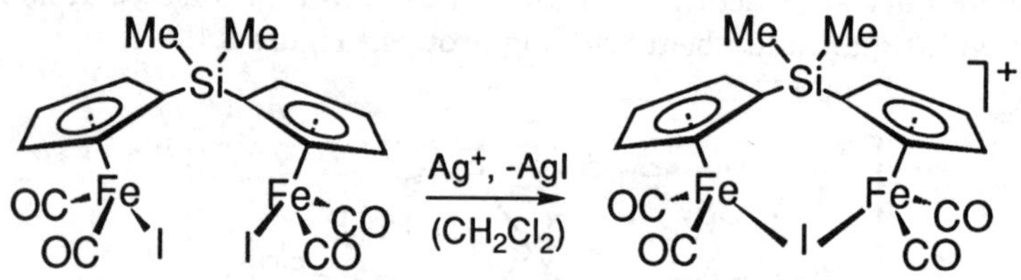

Scheme 6.2

The mechanism of iodide abstraction from $[FeI(CO)_2(\eta^5\text{-}Cp)]$ by silver tetrafluoroborate has been investigated[8] by ^1H NMR spectroscopy for $CDCl_3$ solutions and by IR methods for both $CHCl_3$ and CH_2Cl_2 solutions. It was suggested that $[FeI(CO)_2(\eta^5\text{-}Cp)]$ initially forms a silver(I) adduct, $[(\eta^5\text{-}Cp)(CO)_2Fe\text{–}I\text{–}Ag]^+$, and that this complex decomposes to precipitate AgI. In the presence of excess $[FeI(CO)_2(\eta^5\text{-}Cp)]$, the bridged compound, $[Fe_2(\mu\text{-}I)(CO)_4(\eta^5\text{-}Cp)_2][BF_4]$, is formed.

Binuclear compounds can form as a result of processes involving the cleavage of M–H bonds in complexes dissolved in poor donor solvents. For example, the reaction of the complex, $[WH(NO)_2(\eta^5\text{-}Cp)]$, with the hydride-ion acceptors, Ph_3CX (X = BF_4, PF_6), in the donor solvent acetonitrile produces a solvento-complex, $[W(NO)_2(MeCN)(\eta^5\text{-}Cp)][X]$. However, dimerization of the unsaturated intermediate formed by hydride abstraction generates a binuclear complex with the formula $[W_2H(X)(NO)_4(\eta^5\text{-}Cp)_2]$ when the reaction occurs in CH_2Cl_2.[9]

When the platinum(II) complexes, $[PtX_2L_2]$ (L = PPh_3, PEt_3, $PMePh_2$, $P(p\text{-}CNC_6H_4Me)_3$; X = Cl; L = CNMe; X = Cl, Br, I), were treated with the strong alkylating agent trimethyloxonium tetrafluoroborate, $[Me_3O][BF_4]$, in dichloromethane, the halide-bridged, dimeric species, $[Pt_2(\mu\text{-}X)_2L_4][BF_4]_2$, were formed in good yield,[10] Eq. 6.1:

$$2[PtX_2L_2] + 2[Me_3O][BF_4] \rightarrow [Pt_2(\mu\text{-}X)_2L_4][BF_4]_2 + 2Me_2O + 2MeX \quad (6.1)$$

In this case, the oxonium salt alkylates the coordinated halide ion and, because the synthesis takes place in CH_2Cl_2, the coordination site freed as a result of the alkylation is not occupied by a solvent molecule. Accordingly, dimerization takes place. The use of dichloromethane in these reactions is important for more than one reason. Possible alternatives are polar solvents (*e. g.* acetonitrile, acetone, ethanol) but these solvents react with $[Me_3O][BF_4]$. Similar to the reactions described above, the reaction of $[Pr_4N][PtCl_3(SMe_2)]$ with the methylating reagent, $MeSO_3F$, in dichloromethane produces the dimeric complex, $[Pt_2Cl_4(\mu\text{-}SMe_2)_2]$.[11] The relationship between alkylation reactions of this type and ion exchange (Chapter 5) has been noted by Belluco.[12]

The synthesis of bridged compounds through the elimination of alkylated ligands from mononuclear precursors has been demonstrated for some pseudohalide complexes. For example, the abstraction of azide ions from $[M(N_3)_2(PPh_3)_2]$ (M = Pd, Pt) was carried out with stoichiometric quantities of $[Et_3O][BF_4]$ and $(EtO)_2SO_2$ in dichloromethane solution.[13] As a result of the alkylation, an azide ion is removed from the coordination sphere of the complex in the form of ethyl azide. The poor donor solvent does not occupy the vacant coordination site and dimerization of the coordinatively unsaturated intermediate takes place, Eq. 6.2:

$$2[M(N_3)_2(PPh_3)_2] + 2[Et_3O][BF_4] \rightarrow [M_2(\mu\text{-}N_3)_2(PPh_3)_4][BF_4]_2 + 2Et_2O + 2EtN_3 \quad (6.2)$$

Azide alkylation followed by dimerization of a coordinatively unsaturated intermediate was observed[14] when the rhodium(I) complex, $[Rh(N_3)(\eta^2\text{-}CS_2)(triphos)]$ (triphos = $MeC(CH_2PPh_2)_3$), was treated with $MeOSO_2CF_3$ in dichloromethane solution. This reaction results in the formation of a bridged complex, $[(triphos)Rh(\mu\text{-}C_2S_4)Rh(triphos)]^{2+}$, that contains a tetrathiooxalate-like C_2S_4 unit. The use of boron trihalides was proposed by Druce, Lappert and Riley[15,16] for the synthesis of bridged, cationic complexes of platinum(II) through halide or pseudohalide abstraction reactions. Boron trihalides form tetrahaloborates with halides and form analogous ions with pseudohalides and, thus, remove these ligands from the coordination spheres of the complexes. When the reaction (X = Cl or F, Y = Cl; X = Cl, Y = NCO or N_3) depicted in Scheme 6.3:

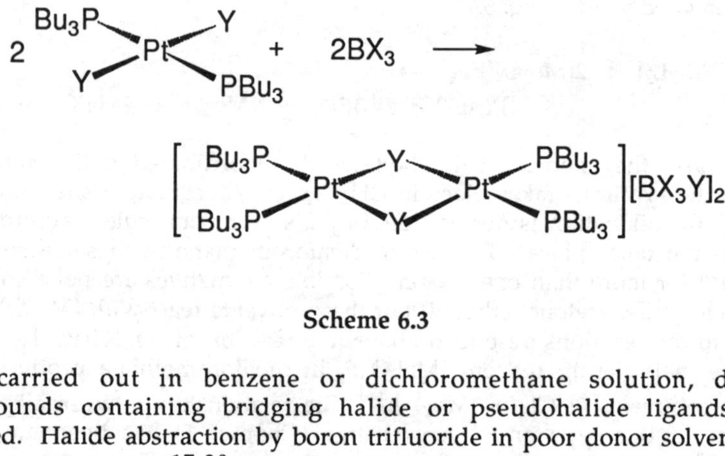

Scheme 6.3

was carried out in benzene or dichloromethane solution, dimeric compounds containing bridging halide or pseudohalide ligands were isolated. Halide abstraction by boron trifluoride in poor donor solvents was subsequently used[17-20] for the preparation of bridged complexes of platinum(II) and palladium(II) of the type $[M_2(\mu\text{-}Y)_2(PBu_3)_4][BX_3Y]_2$.

Boron trifluoride etherate, $BF_3{\cdot}OEt_2$, is a more convenient reagent for halide abstraction than gaseous boron trifluoride or trichloride. Boron trifluoride etherate is a liquid with a fairly high boiling point (126 °C[21]), is relatively stable, and is convenient to use. It has been shown[22-25] that treatment of $[Ph_3PCH_2Ph]_2[PtCl_4]$ with $BF_3{\cdot}OEt_2$ in either CH_2Cl_2 or MeCN solutions leads to the formation of different products. In the former case the bridged complex was formed, whereas, in the latter case the solvento-complex was produced, Scheme 6.4. Both of the products were characterized by X-ray diffraction methods (see Scheme 6.4).

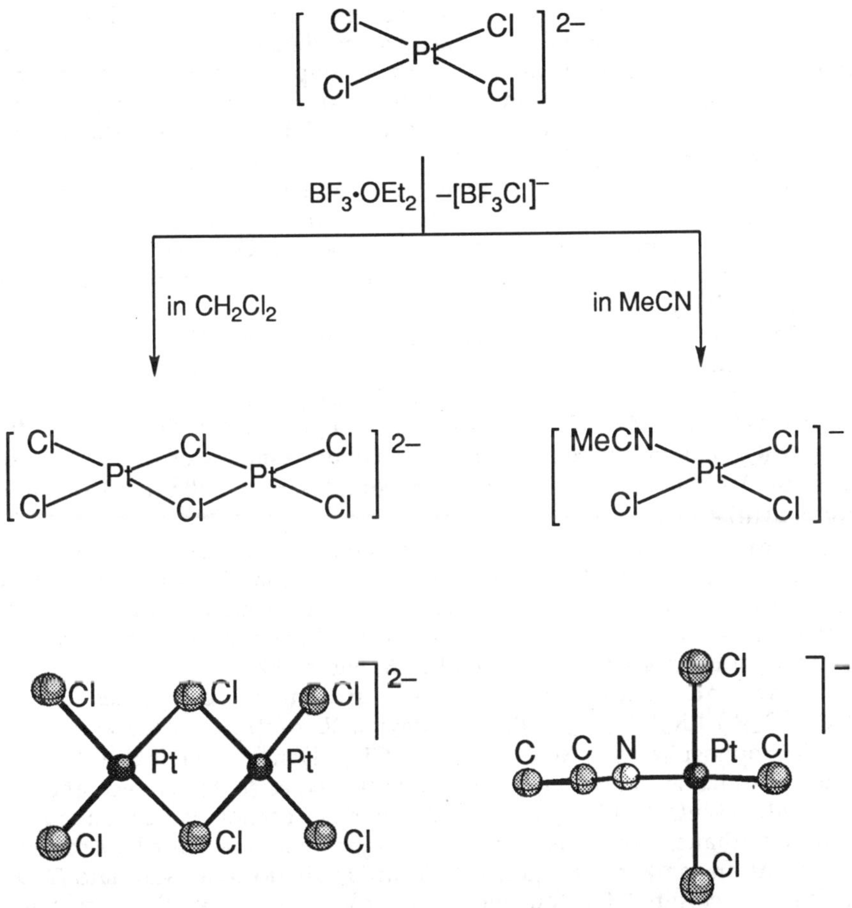

Scheme 6.4

Interestingly, and in contrast to the reactions described above, heating $K_2[PtCl_4]$ in neat $BF_3{\cdot}OEt_2$ followed by filtration and washing of the precipitate formed with 0.1 *M* HCl, yielded neither a bridged product nor a solvento-complex, but rather produced the unusual clathrate compound, $Pt_6Cl_{12}{\cdot}0.1C_2H_5Cl{\cdot}5.7H_2O$.[26-28]

6.2 Ring-closure reactions

Anion abstraction in poor donor solvents can also be used to induce ring-closure reactions of bidentate ligands coordinated to a metal center by only one of two donor atoms. For example, the reaction of *fac*-$[ReBr(CO)_3(dppm)]$ with dppm in xylene yields *cis-mer*-$[ReBr(CO)_2$(dppm-*P*)(dppm-*P,P'*)], in which one of the bis(diphenylphosphino)methane (dppm) ligands is monodentate.[29] The rhenium(I) complex reacts with $TlPF_6$ in dichloromethane solution with precipitation of TlBr, thus creating a vacant coordination site, allowing ring-closure of the uncoordinated phosphorus of the monodentate dppm ligand, and formation of $[Re(CO)_2$(dppm-*P, P'*)$_2][PF_6]$. The same general method has been used for preparation of the manganese analog, $[Mn(CO)_2$(dppm-*P, P'*)$_2][PF_6]$.[30]

In a similar manner, treatment of *cis*-$[PtCl_2(Ph_2PCH_2CH_2OMe$-*P*)$_2]$, with one equivalent of silver perchlorate in chloroform gives the cationic complex, $[PtCl(Ph_2PCH_2CH_2OMe$-*P*)$(Ph_2PCH_2CH_2OMe$-*P,O*)]$[ClO_4]$, and with two equivalents of $AgClO_4$, the dicationic complex, $[Pt(Ph_2PCH_2CH_2OMe$-*P,O*)$_2][ClO_4]_2$, is produced, Scheme 6.5.[31] Other examples of ring-closure of bidentate P~O ligands are known for complexes of palladium(II)[32] and rhodium(III).[33]

It has been established[34] that the reaction of the nortricyclic compound, **A**, with a silver salt in MeCN results in the formation of a solvento-complex containing a coordinated acetonitrile ligand. In acetone solution, halide abstraction is followed by isomerization of the ligand, ring-closure, and the formation of the norbornadiene complex, **B**, Figure 6.2.

The interaction between the sulfinato-*S* complexes *trans*-$[PtCl(SO_2R')(PR_3)_2]$ ($PR_3 = PPh_3$, PMe_2Ph; R' = alkyls, aryls) and silver hexafluorophosphate in MeCN and in CH_2Cl_2 has been investigated.[35] Halide abstraction in acetonitrile produces the expected solvento-complexes, $[Pt(SO_2R')(MeCN)(PR_3)_2][PF_6]$. When this reaction is carried out in dichloromethane, which is a poorer donor than acetonitrile, *trans-cis* isomerization and rearrangement from sulfinato-*S* to sulfinato-*O,O'*, resulting in closure of a four-membered ring, are observed, Scheme 6.6. Addition of sodium chloride to [Pt(*O,O'*-$SO_2R')(PR_3)_2][PF_6]$ regenerates the original sulfinato-*S* complex.[35]

Halide abstraction from a metal center bearing suitable organic ligands in a relatively weak donor solvent can result in cyclometallation, presumably by electrophilic attack of a positive metal center on a carbon atom that is typically part of an aromatic ring.

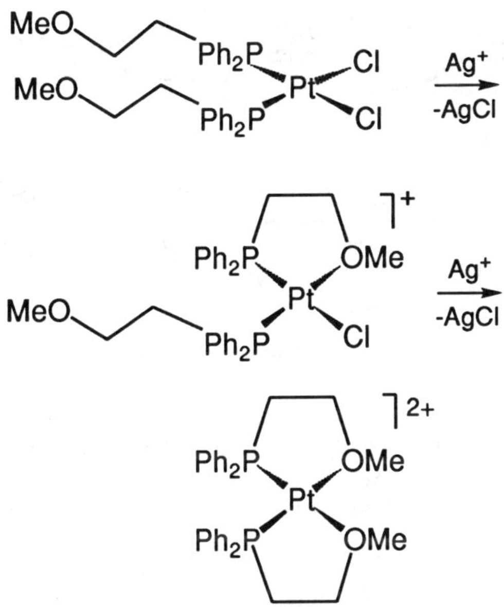

Scheme 6.5

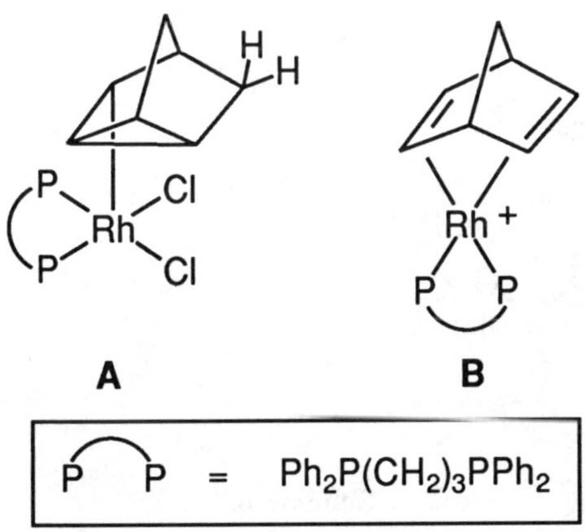

Figure 6.2

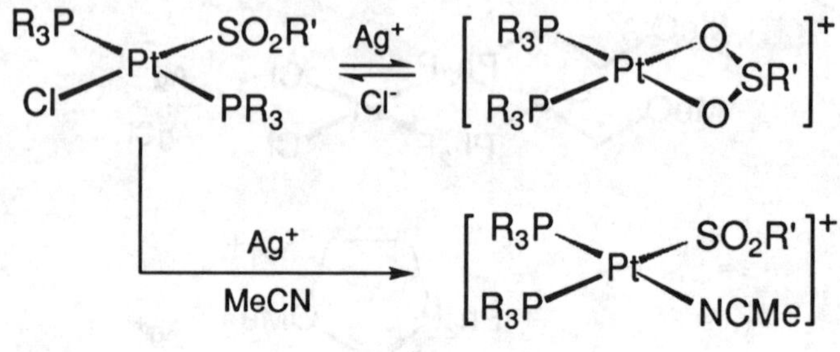

Scheme 6.6

Roundhill, Joslin and Mague[36] treated the ruthenium(II) complex, $[RuCl(PPh_3)\{P(OPh)_3\}(\eta^5\text{-}Cp)]$, with silver trifluoromethanesulfonate and piperidine in THF. The reaction occurred over 30 minutes and yielded the cyclometallated product shown in Scheme 6.7:

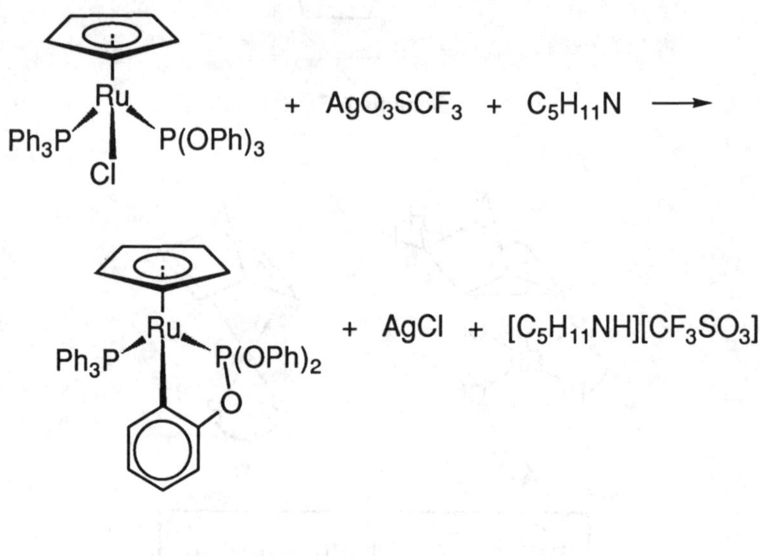

Scheme 6.7

The cyclometallation occurs at the phosphite phenyl group to form a 5-membered ring rather than at an *ortho*-position of the triarylphosphine which would lead to a more-strained, 4-membered metallocycle.

Cyclometallation reactions of *trans*-[$MI_2(NH_2CH_2Ph)_2$] (M = Pd, Pt) occur quite differently.[37] The platinum complex reacts with two equivalents of $AgBF_4$ in acetone to produce AgI. Concentration of the filtered solution followed by dissolution in H_2O/MeOH and addition of KI generates the cyclometallated product, Scheme 6.8:

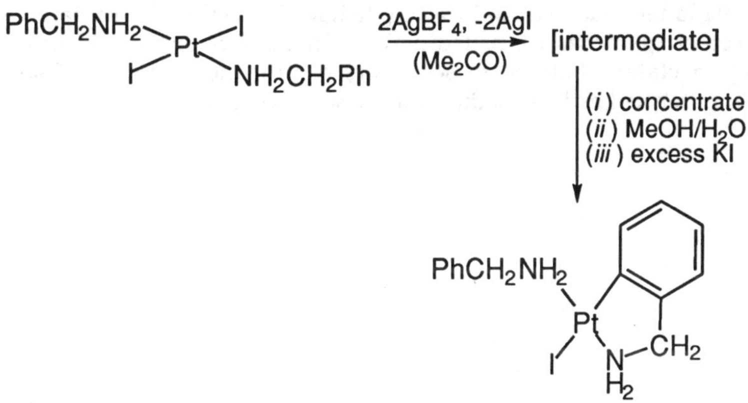

Scheme 6.8

The palladium analog reacts differently, Scheme 6.9:

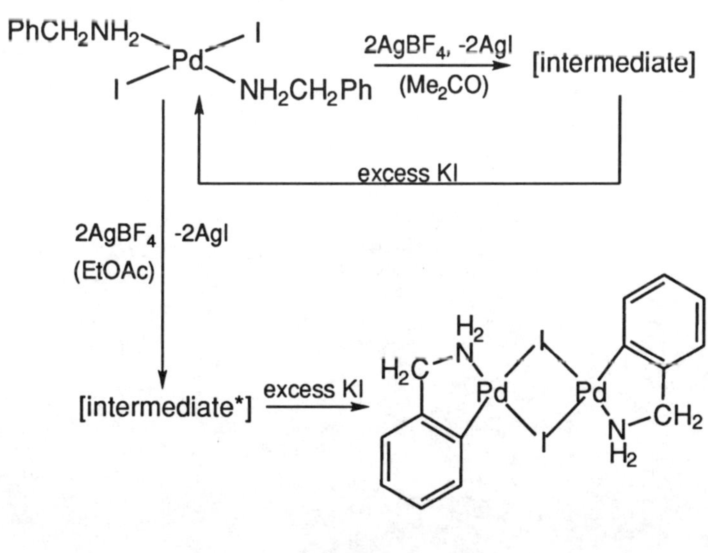

Scheme 6.9

Thus, treatment of *trans*-[$PdI_2(NH_2CH_2Ph)_2$] with two equivalents of $AgBF_4$ in acetone also generates two equivalents of AgI. However, subsequent addition of KI simply regenerates the original iodo-complex. When the reaction was carried out in ethyl acetate, formation of a dimeric, *ortho*-metallated complex occurred.

The reactions illustrated in Schemes 6.8 and 6.9 show the sensitivity of cyclometallation reactions, induced by halide abstraction, to the nature of the metal center (platinum *vs.* palladium) and the nature of the solvent (acetone *vs.* ethyl acetate). Both of these factors control the reactivity of solvento-complexes formed as intermediates in these processes.

References

1. J. A. Davies, F. R. Hartley, and S. G. Murray, *Inorg. Chem.* **19** (1980) 2299.

2. J. A. Davies, F. R. Hartley, and S. G. Murray, *Inorg. Chim. Acta.* **41** (1980) 249.

3. J. A. Davies, F. R. Hartley, S. G. Murray, and M. A. Pierce-Butler, *J. Chem. Soc., Dalton Trans.* (1983) 1305.

4. J. A. Davies and J. F. Liebman, *J. Chem. Soc., Dalton Trans.* (1983) 1793.

5. D. A. Duddell, P. L. Goggin, R. J. Goodfellow, and M. G. Norton, *J. Chem. Soc., Chem. Commun.* (1968) 879.

6. W. J. Cherwinski and H. C. Clark, *Can. J. Chem.* **47** (1969) 2665.

7. W. W. McConnell, G. O. Nelson, and M. E. Wright, *Inorg. Chem.* **22** (1983) 1689.

8. B. M. Mattson and W. A. G. Graham, *Inorg. Chem.* **20** (1981) 3186.

9. B. W. Hames and P. Legzdins, *Organometallics* **1** (1982) 116.

10. P. M. Treichel, K. P. Wagner, and W. J. Knebel, *Inorg. Chim. Acta.* **6** (1972) 674.

11. P. L. Goggin, R. J. Goodfellow, and F. J. S. Reed, *J. Chem. Soc., Dalton Trans.* (1974) 576.

12. U. Belluco, *Organometallic and Coordination Chemistry of Platinum* (Academic, New York, 1974) p. 33.

13. W. Beck, P. Kreutzer, and K. V. Werner, *Chem Ber.* **104** (1971) 528.

14. C. Bianchini, C. Meali, A. Meli, M. Sabat, and P. Zanello, *J. Am. Chem. Soc.* **109** (1987) 185.

15. P. M. Druce, M. F. Lappert, and P. N. K. Riley, *J. Chem. Soc., Chem. Commun.* (1967) 486.

16. P. M. Druce, M. F. Lappert, and P. N. K. Riley, *J. Chem. Soc., Dalton Trans.* (1972) 438.

17. H. C. Clark, K. R. Dixon, and W. J. Jacobs, *J. Am. Chem. Soc.* **90** (1968) 2259.

18. H. C. Clark and K. R. Dixon, *J. Am. Chem, Soc.* **91** (1969) 596.

19. H. C. Clark, K. R. Dixon, and W. J. Jacobs, *J. Am. Chem. Soc.* **91** (1969) 13.

20. W. P. Fehlhammer, W. Beck, and P. Pöllmann, *Chem. Ber.* **102** (1969) 3903.

21. *Aldrich Catalogue Handbook of Fine Chemicals* (Aldrich, Spain, 1994/1995) p. 207.

22. V. Yu. Kukushkin, E. Yu. Pankova, T. N. Fomina, and N. P. Kiseleva, *Koord. Khim.* **14** (1988) 1110.

23. V. Yu. Kukushkin, V. M. Tkachuk, I. A. Krol, Z. A. Starikova, B. V. Zhadanov, and N. P. Kiseleva, *Zh. Obsch. Khim.* **61** (1991) 51.

24. L. I. Elding, Å. Oskarsson, and V. Yu. Kukushkin, *Inorg. Synth.* **31** (1995) in press.

25. V. Yu. Kukushkin, I. A. Krol, Z. A. Starikova, and V. M. Tkachuk, *Koord. Khim.* **16** (1990) 1406.

26. V. Yu. Kukushkin, V. M. Tkachuk, and N. V. Vorobiov-Desiatovsky, *Inorg. Synth.* **31** (1995) in press.

27. N. V. Vorobiov-Desiatovsky, T. P. Smorodina, V. A. Paramonov, E. S. Postnikova, V. V. Sibirskaya, A. I. Marchenko, and Yu. N. Kukushkin, *Zh. Prikl. Khim.* **58** (1985) 977.

28. V. V. Sibirskaya, N. V. Vorobiov-Desiatovsky, E. S. Postnikova, and T. P. Smorodina, *Zh. Prikl. Khim.* **62** (1989) 503.

29. G. A. Carriedo, M. L. Rodriguez, S. Garcia-Granda, and A. Aguirre, *Inorg. Chim. Acta* **178** (1990) 101.

30. G. A. Carriedo, V. Riera, and J. Santamaria, *J. Organomet. Chem.* **234** (1982) 175.

31. G. K. Anderson and R. Kumar, *Inorg. Chem.* **23** (1984) 4064.

32. A. W. G. Platt and P. G. Pringle, *J. Chem. Soc., Dalton Trans* (1989) 1193.

33. P. Braunstein, D. Matt, and Y. Dusausoy, *Inorg. Chem.* **22** (1983) 2043.

34. M. F. McGuiggan and L. H. Pignolet, in: *183rd ACS National Meeting,* March 28 to April 2, 1982, Division of Inorganic Chemistry, Abstract 97.

35. M. Kubota, R. K. Rothrock, M. R. Kernan, and R. B. Haven, *Inorg. Chem.* **21** (1982) 2491.

36. F. L. Joslin, J. T. Mague, and D. M. Roundhill, *Organometallics* **10** (1991) 521.

37. A. Avshu, R. D. O'Sullivan, A. W. Parkins, N. W. Alcock, and R. M. Countryman, *J. Chem. Soc., Dalton Trans.* (1983) 1619.

Chapter 7. THE ELECTROSYNTHESIS OF COORDINATION COMPOUNDS

Electrosynthesis is one of the most interesting and yet underutilized methods of preparing coordination compounds. Electrochemical reactions make use of the universal "chemical reagent" — the electron.[1] The electrosynthesis of metal complexes can, in principle, give rise to a high selectivity, because it is possible to control electrode potentials over a wide range. Electrons can be removed from, or added to, a system without the complications associated with the presence of chemical oxidants and reductants and the by-products associated with their use. However, a new set of complications in the form of electrodes and supporting electrolytes must be dealt with.

In the late nineteenth and early twentieth centuries, following rapid developments in the electrosynthesis of organic compounds, electrochemical methods were used to prepare both organometallic and coordination compounds. In 1882, Drechsel's student, Gerdes, developed an electrochemical method for the preparation of platinum(IV) hexammine complexes that involved the anodic dissolution of a platinum electrode in ammonium carbonate solution using alternating current.[2,3] In 1908, Chugaev carried out the electrosynthesis of cobalt ammine complexes in an electrolytic cell equipped with a platinum cathode and a cobalt anode.[4] Platinum ammine complexes were similarly prepared by ac and dc electrolyses with a platinum anode. The series of experiments carried out with a cobalt anode demonstrated that the electrosynthesis of $[Co(NH_3)_6]Cl_3$, $[CoCl(NH_3)_5][SO_4]$ and $[Co(NO_2)_2(NH_3)_4][NO_3]$ was possible. Chugaev also prepared nickel dimethyl- and methylethylglyoximates electrochemically at a nickel anode in an aqueous methanol medium in the presence of ammonium acetate. He pointed out[4] that "a large quantity of a coarsely crystalline precipitate of a complex compound is obtained in a short time", and concluded that this was "a convenient method for obtaining nickel dimethylglyoximate".

Recent reviews of the electrosynthesis of metal complexes reflect an increasing level of interest in this preparative method.[5-17] In this Chapter the electrochemical generation of metal complexes is discussed with an emphasis on useful preparative methods. Other sources provide information on the application of electroanalytical techniques to coordination chemistry.[18]

7.1 Electrochemical oxidation and reduction of complexes using inert electrodes

The electrosynthesis of metal complexes commonly involves a

preliminary determination of the applied potential at which the reaction of interest takes place. Such a determination is usually made by application of a suitable electroanalytical technique, such as polarography or cyclic voltammetry. The bulk electrosynthesis of a compound at the potential determined electroanalytically may allow isolation of the expected product or may generate intermediates which are subsequently converted into the isolable product. It is noteworthy that the isolated products from an electrosynthesis may differ from the species identified electroanalytically or spectroscopically, *i.e.* the species identified in an analytical investigation may not be those isolated from a preparative experiment at a macroelectrode.[5]

An example illustrating this point was reported by Budnikov[19] who studied the electrochemical behavior of $[Ni(S_2CNEt_2)_2]$, including a preparative-scale electrochemical oxidation of the compound. It was established that a reversible one-electron process is the first step in the polarographic oxidation of the nickel(II) complex. In contrast to this result, bulk oxidative electrolysis at the potential determined in the preliminary polarographic study generated the nickel(IV) complex, $[Ni(S_2CNEt_2)_2]^{2+}$, and not the expected nickel(III) complex. A possible route for the formation of $[Ni(S_2CNEt_2)_2]^{2+}$ is a rapid disproportionation of a nickel(III) intermediate, formed during the electrosynthesis, into the original nickel(II) complex and the nickel(IV) product.

Olson and Keim[20] have observed that electrosynthesis has considerable advantages over traditional chemical methods for the preparation of complexes in low oxidation states. These advantages include, in particular, a reaction system consisting of high purity components, the selectivity of the processes, and the possibility of controlling the reactions taking place. The authors cited have studied the synthesis of the rhodium(0) complexes, $[Rh(PPh_3)_4]$ and $[Rh(PMePh_2)_4]$. To obtain $[Rh(PPh_3)_4]$,[20] the rhodium(I) complex, $[RhCl(PPh_3)_3]$, was reduced at -2.3 *V* (*vs.* Ag/0.1 *M* $AgNO_3$ in MeCN) in an acetonitrile/toluene solvent system in the presence of 0.1 *M* tetraethylammonium perchlorate and a ten-fold excess of PPh_3. The compound $[Rh(PPh_3)_4]$ precipitated from solution as the reaction proceeded with an isolated yield of 70%. The related cobalt(0) complex, $[Co\{P(OEt)_3\}_4]$, has been obtained by the reverse methodology, *i. e.* oxidation of the cobalt(-1) complex, $[Et_4N][Co\{P(OEt)_3\}_4]$, in acetonitrile.[21]

The lower oxidation states of transition metals are stabilized by ligands with good π-acceptor properties, whereas good σ-donor ligands typically stabilize higher oxidation states.[22] The electrochemical oxidation or reduction of a metal center in a complex is, therefore, often accompanied by ligand substitution. Thus, the electrochemical reduction of the ruthenium(II) complex, $[RuCl_2(PPh_3)_4]$, in acetonitrile, leads to the elimination of the two σ-donor chloride ligands and stabilization of the ruthenium(0) product by coordination of acetonitrile, which has π-acceptor properties. As a result, $[Ru(MeCN)_2(PPh_3)_4]$ is isolated.[22]

The controlled potential bulk reductive electrolysis of *cis*-$[PtCl_2(PPh_3)_2]$ at a mercury electrode in $MeCN/C_6H_6$ containing $[Et_4N][ClO_4]$ occurs by a two-electron reduction that results in expulsion of two chloride ions and generation of $[Pt(PPh_3)_2]$ equivalents, which may be trapped by acetylenes or by oxidative addition addenda,[22a,22b] Scheme 7.1:

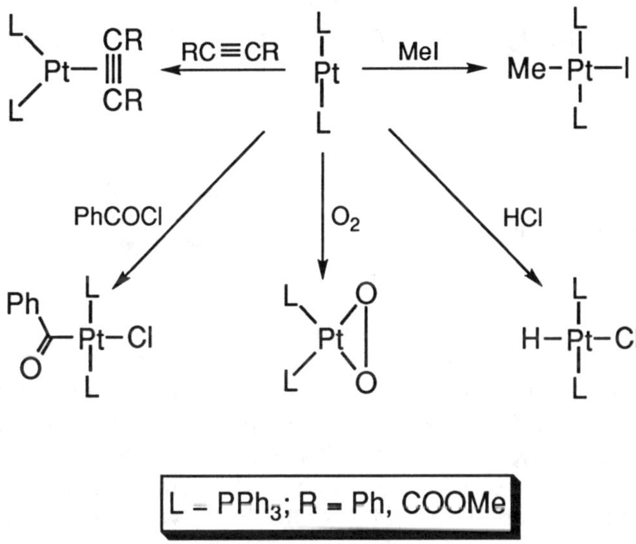

Scheme 7.1

Similar bulk reductive electrolyses of *cis*-$[PtCl_2(PEt_3)_2]$ are more complex[22b,22c] because the resultant $[Pt(PEt_3)_2]$ equivalents react with the background electrolyte by a Hoffman elimination reaction, Scheme 7.2. Accordingly, $[Pt(PEt_3)_2]$ equivalents can only be trapped effectively by reagents present in excess during the electrolysis. As the reduction of *cis*-$[PtCl_2(PEt_3)_2]$ occurs at -2.1 *V vs.* Ag/AgCl, the range of possible trapping agents is limited. Nonetheless, electrolyses in the presence of PhCl or PhCN produce *trans*-$[PtPh(X)(PEt_3)_2]$ efficiently.

The electrochemical generation and reactions of $[PtL_2]$ equivalents and related low valent, formally unsaturated intermediates, has been reviewed.[22d]

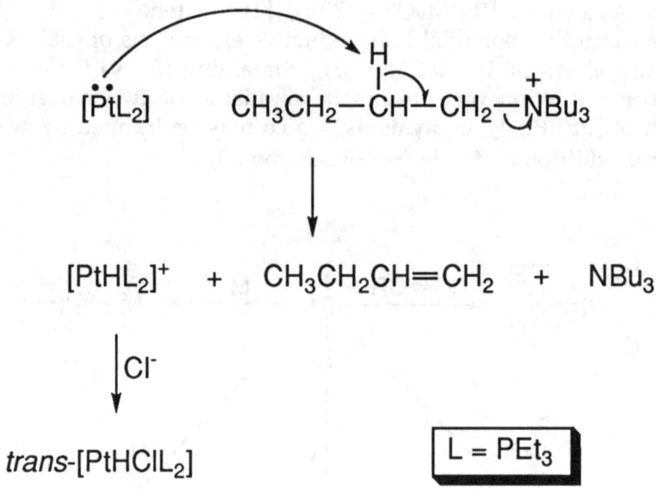

Scheme 7.2

The cathodic reduction of [Fe(acac)$_3$] in methanol in the presence of the strong π-acceptors, PPh_3 and 1,3-butadiene, yields [Fe(1,3-butadiene)$_2$(PPh$_3$)].[23] The electrochemical reduction of the d^6-complexes, [M(R)Cl$_2$(CO)(PPh$_3$)$_2$] (M = Rh, Ir; R = alkyl), leads to the elimination of two chloride ions and the formation of the d^8-complexes, [M(R)(CO)(PPh$_3$)$_2$], which are more difficult to prepare by routine chemical methods.[24]

Controlled potential electrolysis of the neutral rhodium(I) acetylide (alkynyl) complex, [Rh(CCR){N(C$_2$H$_4$PPh$_2$)$_3$}] (R = Ph, CO$_2$Et), in THF in the presence of [(*n*-Bu)$_4$N][ClO$_4$] at -0.2 *V* (*vs.* SCE) yields the rhodium(II) complex, [Rh(CCR){N(C$_2$H$_4$PPh$_2$)$_3$}][ClO$_4$], in good yield.[28]

Among the most familiar electrosynthetic reactions involving the oxidation of a metal center in a coordination compound is the conversion of alkali metal hexacyanoferrates(II) to the corresponding hexacyanoferrates(III). When the electrolysis is carried out in a divided cell equipped with diaphragm, the electrooxidation is practically quantitative. In a review of studies of the oxidation of $[Fe(CN)_6]^{4-}$ it was noted[25] that the rate of oxidation of the hexacyanoferrate(II) anion depends on a large number of factors, including the current density, the concentration of the complex in solution, the electrode material, the temperature, and the p*H* of the solution.

The electrosynthesis, in aqueous solution, of nonstoichiometrically oxidized, highly conducting complexes of platinum has been described,[26,27] Eqs. 7.1 and 7.2:

$$1.75K^+ + Pt(CN)_4^{2-} \rightarrow 0.25e^- + K_{1.75}Pt(CN)_4 \cdot 1.5\,H_2O \qquad (7.1)$$

$$2K^+ + Pt(CN)_4^{2-} + 0.3Cl^- \rightarrow 0.3e^- + K_2Pt(CN)_4Cl_{0.3} \cdot 3H_2O \qquad (7.2)$$

In addition to the metal-centered redox reactions, described above, ligand-centered redox reactions may also be employed in synthesis. Many examples of such reactions are known for ruthenium complexes of nitrogen donor ligands. Thompson and Meyer[29] found that the coordinated ammine in $[Ru(trpy)(bipy)(NH_3)]^{2+}$ is subject to oxidation in aqueous solution and that this leads to a rapid, quantitative conversion, first to the nitrite complex, $[Ru(trpy)(bipy)(NO_2)]^+$, and then to the ruthenium(II) nitrate complex, $[Ru(trpy)(bipy)(NO_3)]^+$. Electrolysis of complexes, $[Ru(bipy)_2(NH_2CH_2R)_2]^{2+}$ (NH_2CH_2R = allylamine, benzylamine, *n*-butylamine), in 0.1 *M* HCl leads to oxidative dehydrogenation of the amine ligands and generation[30] of the corresponding nitrile-containing complexes, $[Ru(bipy)_2(NCR)_2]^{2+}$. The electrolysis of $[Ru(bipy)_2(en)]^{2+}$ has been utilized to prepare the corresponding *di*-imine complex,[31] Scheme 7.3:

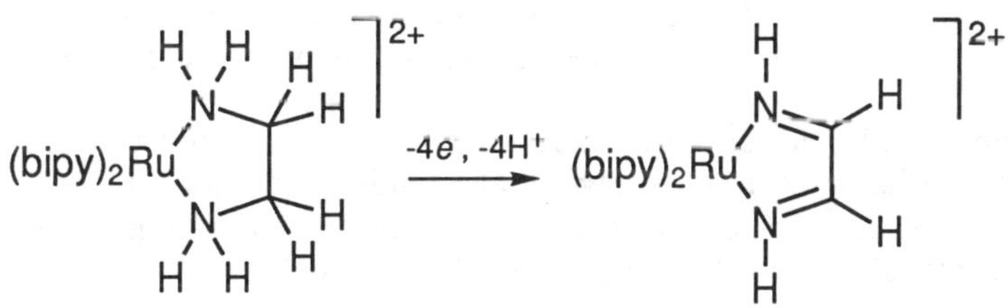

Scheme 7.3

Electrolysis of the isopropylamine complex of ruthenium(II), $[Ru(trpy)(bipy)(NH_2CHMe_2)]^{2+}$, in aqueous HCl, followed by the addition of ammonium hexafluorophosphate, results in precipitation of the imine complex, $[Ru(trpy)(bipy)(NH{=}CMe_2)]^+$, as the hexafluorophosphate salt. Further oxidation of $[Ru(trpy)(bipy)(NH{=}CMe_2)][PF_6]$ leads to the formation of the ruthenium(IV) compound, $[Ru(trpy)(bipy)(NCMe_2)][PF_6]_3$, which contains a coordinated isopropylideneamide anion, $[N{=}CMe_2]^-$. The complex was isolated as the perchlorate salt, $[Ru(trpy)(bipy)(NCMe_2)][ClO_4]_3 \cdot H_2O$, and characterized by X-ray crystallography,[32,33] Figure 7.1:

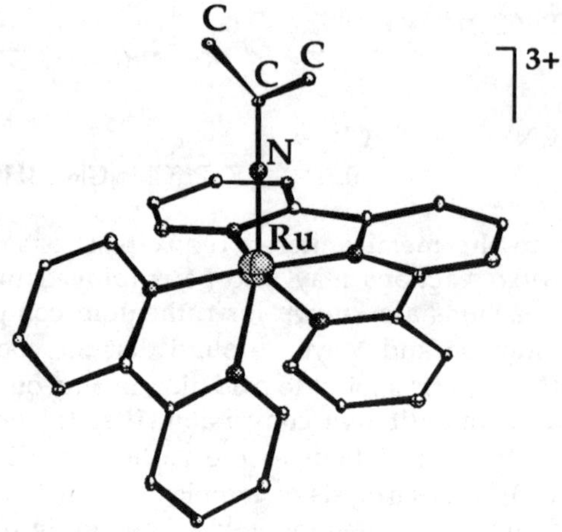

Figure 7.1

The aryldiazenate complex, $[Ru(bipy)_2(N_2C_6H_4Me)Cl]^{2+}$, in acetonitrile solution, undergoes an irreversible one-electron oxidation,[34] Eq. 7.3:

$$[Ru(bipy)_2(N_2C_6H_4Me)Cl]^{2+} + MeCN + e^- + H^+ \rightarrow [Ru(bipy)_2(MeCN)Cl]^{2+} + N_2 + C_6H_5Me \quad (7.3)$$

Although ligand oxidation is observed in the examples cited above, studies of reaction mechanisms have shown that metal-based redox transformations may be involved. For example, the oxidation of $[Ru(en)(bipy)_2]^{2+}$ involves formal conversion of ruthenium(II) to ruthenium(III), Eq. 7.4:

$$[Ru(en)(bipy)_2]^{2+} \rightarrow [Ru(en)(bipy)_2]^{3+} + e^- \quad (7.4)$$

followed by conversion of ethylenediamine to the corresponding *di*-imine with reduction of ruthenium(III) to ruthenium(II), Eq. 7.5:

$$[Ru(en)(bipy)_2]^{3+} \rightarrow [Ru(C_2H_4N_2)(bipy)_2]^{2+} + 4H^+ + 3e^- \quad (7.5)$$

7.2 Electrochemically induced reactions

Reactions involving thermal or photochemical substitution of coordinated carbon monoxide in metal carbonyl complexes are well known (see Chapter 3) and related processes may be induced electrochemically. A number of review articles[11,18] have dealt with the electrochemically-induced replacement of CO in metal carbonyls. Thus, the electrolysis of a solution of $[Co_2(CO)_6(PhCCPh)]$ and PPh_3 in acetone leads to the replacement of one carbonyl group and the formation of the substituted product, $[Co_2(CO)_5(PPh_3)(PhCCPh)]$.[35] Electrochemically-induced ligand replacement reactions of the homoleptic metal carbonyls, $[M(CO)_6]$ (M = Cr, Mo, W), have been studied.[36,37] Substitution produces the compounds $[M(CO)_{6-n}L_n]$ (L = THF, py, PMe_3, PPh_3; n = 1-3). The following reaction mechanism, which involves the intermediacy of the radical anion, $[M(CO)_5^{\bullet -}]$, was proposed,[11] Scheme 7.4:

$$[M(CO)_6] \xrightarrow[-CO]{e^-} [M(CO)_5^{\bullet -}] \xrightarrow[-e^-]{PPh_3} [M(CO)_5(PPh_3)]$$

Scheme 7.4

The replacement of carbon monoxide in the complex, $[W(CO)_4L_2]$ (L_2 = aromatic diamine), has been described.[38] The reaction takes place under mild conditions in acetonitrile upon electrochemical reduction of $[W(CO)_4L_2]$. The corresponding solvento-complexes are obtained in high yield.[38]

The thermally- and electrochemically-induced decarbonylation of a dinuclear iron carbonyl complex has been studied.[39,40] The replacement of CO upon electrolysis is selective, Scheme 7.5.

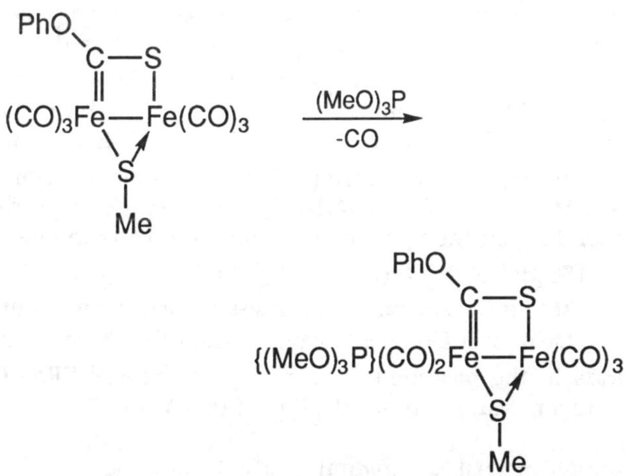

Scheme 7.5

In contrast, the thermal decarbonylation of the starting material is not selective, and the product contains $P(OMe)_3$ ligands coordinated to each iron atom.

In recent years interest has developed in the mechanisms of electrochemically-induced catalytic substitution reactions of metal carbonyls. Darchen, Mahé, and Patin[41] have shown a general mechanism for redox substitution, depicted as a so-called "square-scheme", Scheme 7.6:

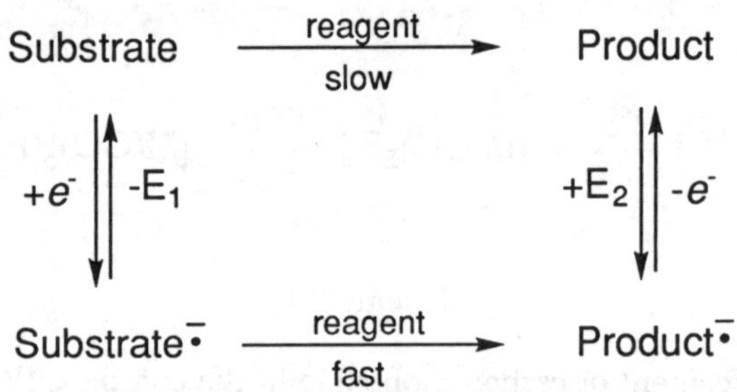

Scheme 7.6

The first reduction step is efficient when the reduced state (Substrate$^{\cdot-}$) is more reactive than the starting material (Substrate) and the potential E_2 is more negative than E_1. Indeed, if $E_1 < E_2$, the process is not catalytic.

A number of electrochemically-induced carbonyl replacement reactions have been developed that are of synthetic utility. The substitution of carbonyl ligands by trimethylphosphite in the bridged tri-iron cluster, $[Fe_3(\mu^3\text{-}S)_2(CO)_8(C_3H_2S_2)]$, is generally effected under thermal or photochemical conditions and often leads to a mixture of multiply substituted products. The selective substitution of CO ligands in the cluster, $[Fe_3(\mu^3\text{-}S)_2(CO)_8(C_3H_2S_2)]$, was accomplished by an electrochemically-induced reaction in a DMF/0.1 *M* $[(n\text{-}Bu)_4N][BF_4]$ solution. The process proceeds stepwise, depending on the applied potential, and leads to generation of the new clusters, $[Fe_3(\mu^3\text{-}S)_2(CO)_{8-n}(C_3H_2S_2)\{P(OMe)_3\}_n]$ (n = 1, 2 or 3). The authors[41] suggest that the substitution process is catalytic in nature.

Treichel, Dean and Douglas[42] have studied the thermal substitution of the CO ligands in the bicapped tri-iron cluster, $[Fe_3(\mu^3\text{-}PPh)_2(CO)_9]$. The reaction leads to a mixture of products, Eq. 7.6 (n = 1, 2 or 3):

$$[Fe_3(\mu^3\text{-}PPh)_2(CO)_9] + nP(OEt)_3 \rightarrow [Fe_3(\mu^3\text{-}PPh)_2\{P(OEt)_3\}_n(CO)_{9-n}] + nCO \qquad (7.6)$$

Ohst and Kochi[43-45] found that the tri-iron nonacarbonyl complex undergoes rapid ligand substitution under conditions which preclude the possibility of a thermal process; the stepwise substitution of the CO ligands by phosphites takes place selectively at the three separate iron centers.

Not only carbonyl ligands but also a number of other ligand types undergo substitution under electrocatalytic conditions.[46-48]

The synthesis of polymers by the electropolymerization of monomeric, vinyl-substituted complexes, Figure 7.2:

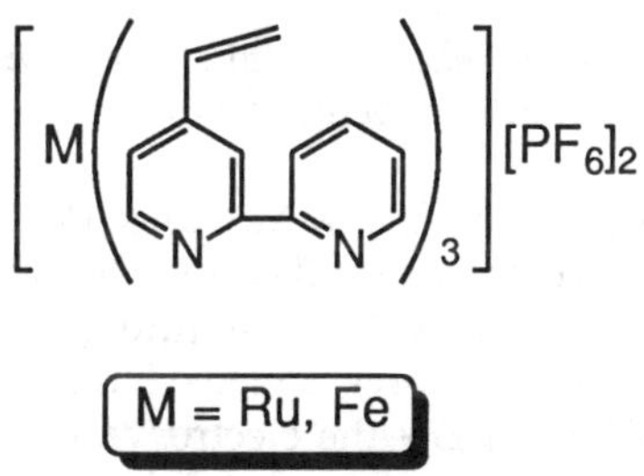

Figure 7.2

has been described.[49-51] The electrochemical polymerization of metal complexes has provided a unique method for the preparation of films containing redox centers in each repeating unit. Redox conductivity enables such a film to grow continuously since its exterior remains active for electron transfer at the film-solution interface.

Electrochemical polymerization has been examined in detail with many mixed-ligand complexes of ruthenium(II), osmium(II) and iron(II).[49-51] The ligands employed were 4-vinylpyridine, other monomers containing the pyridine skeleton, derivatives of cinnamates, stilbazole, *etc.*, in addition to unsubstituted polypyridyl ligands (2,2'-bipyridyl, phen, terpy). Polymerizations were typically carried out in MeCN/0.1 *M* $[Et_4N][ClO_4]$ by repeated cycling of a platinum electrode near the reduction potential of the ligand. Homogeneous, uniform polymer films were formed. For the majority of the complexes investigated, the first step in electrochemical polymerization is formation of a radical-anion by electron transfer into a π^*-orbital of the unsubstituted ligand, Scheme 7.7. An electron is transferred then onto the substituted ligand, either by an intramolecular mechanism or by disproportionation. Repetition of these reactions induces polymerization.

$$[Ru(bipy)_2L_2]^{2+} \underset{-e^-}{\overset{e^-}{\rightleftharpoons}} [Ru(bipy)(bipy^{\bullet -})L_2]^{+} \underset{-e^-}{\overset{e^-}{\rightleftharpoons}}$$

$$[Ru(bipy^{\bullet -})_2L_2]$$

Scheme 7.7

7.3 Electrosynthesis using a sacrificial anode

7.3.1 Preparation of anhydrous salts, metal halide complexes, and solvento-complexes

Anhydrous metal halides, $[MX_n]$ and $[MX_k]^{(k-n)-}$, mixed-ligand solvento-complexes containing halide ligands, $[MX_n(solv)_m]$, and homoleptic solvento-complexes, $[M(solv)_k]^{n+}$, are interconverted quite readily.[52] Accordingly, reactions that involve ion exchange or the coordination of solvent molecules may accompany the electrosynthesis of such compounds.

The synthesis of anhydrous metal halides is often a complex process. In thermal dehydration reactions complete removal of water is often complicated by decomposition of the compound. The reactions of metals with gaseous hydrogen halides or with halogens usually takes place at high temperatures and requires apparatus that may not be available routinely. For example, $CrBr_3$ can be obtained by the action of Br_2 on chromium metal at a temperature of 750 °C.[53] An alternative method has been proposed[54] for the preparation of this compound that involves the anodic dissolution of chromium in the presence of bromine with a platinum cathode immersed in benzene connected by a salt bridge to a chromium anode immersed in MeOH/Br_2. The electrolytic cell is represented by Eq. 7.7:

$$Pt_{(-)} \mid PhH \mid\mid MeOH, Br_2 \mid Cr_{(+)} \qquad (7.7)$$

The electrosynthesis is rapid at room temperature and gram quantities of anhydrous chromium bromide may be obtained within a few hours. The chemical yield is 93%. Additionally, molybdenum chloride, $MoCl_2$, has been prepared electrosynthetically,[55] Eq. 7.8:

$$Pt_{(-)} \mid MeCN \mid\mid Cl_2 \mid Mo_{(+)} \qquad (7.8)$$

Tuck[9] has reported that the mechanisms of these reactions involve radical and radical-chain processes.

When solvents with comparatively high donor numbers are used in electrosyntheses of this type, the reaction products may be isolated as solvates or as mixed-ligand complexes. Thus, the electrolysis of a solution of Br_2 in acetonitrile in a cell with a manganese anode and a platinum cathode in the presence of $[Et_4N][ClO_4]$ produces $MnBr_2{\cdot}MeCN$ in 75% yield.[56] The anodic dissolution of vanadium using a solvent/electrolyte system consisting of acetonitrile, bromine, and $[(n\text{-}Bu)_4N][ClO_4]$, produces the complex, $[VBr_2(MeCN)_4]Br_3$, which has been characterized crystallographically.[57]

The solvates, $ThX_4{\cdot}4MeCN$, have been obtained by the anodic dissolution of thorium in acetonitrile in the presence of X_2. When this synthesis was carried out in the presence of $[Et_4N]Cl$ or $[Et_4N]Br$, reactions that followed the electrolysis made isolation of the complexes, $[Et_4N]_2[ThX_6]$ (X = Cl, Br), possible.[58] The complexes $[Et_4N][AuCl_2]$,[59] $[Et_4N][AuBr_4]$, $[Et_4N]_2[SnBr_6]$, and related compounds[9] have been obtained in the same manner.

Electrosynthesis has been used to obtain homoleptic solvento-complexes. Typically the electrolysis is carried out in the presence of a strong acid, or its salt, with an anion which is a weaker donor than the solvent employed. For example, the electrochemical oxidation of vanadium, cobalt, nickel, zinc, and cadmium in dimethyl sulfoxide in the presence of HBF_4 has been used to synthesize hexakis(dimethyl sulfoxide) solvento-complexes of these metals.[60,61] The electrolysis of a solution of $AgClO_4$ in acetonitrile using a tin anode and a platinum cathode produces the tin(II) solvento-complex, $Sn(ClO_4)_2{\cdot}2MeCN$.[62] The electrolysis of $NaBH_4$ in liquid ammonia using a magnesium anode and a mercury cathode produces $Mg(BH_4)_2{\cdot}6NH_3$ in 98% yield.[63] The ammonia molecules are readily removed *in vacuo* at a temperature of about 100 °C to yield the borohydride, $Mg(BH_4)_2$. An electrochemical method for the preparation of $Al(BH_4)_3{\cdot}6NH_3$ in high yield by the electrolysis of $NaBH_4$ with an aluminum anode and an iron cathode has also been described.[64]

7.3.2 Electrosynthesis of complexes containing weakly acidic anions

Ligands such as 2,2'-bipyridyl and related N-donors form stable complexes with most transition metal ions. These ligand types include a wide range of azoles, Figure 7.3. Although the synthesis of cationic complexes containing these ligands is straightforward, methods for obtaining neutral complexes with a deprotonated azole ligand are often of limited applicability and give low yields of products which are often contaminated.[65] Osipov and coworkers[65-67] have proposed an electrochemical method for synthesizing chelate complexes, $[ML_2]$ (M = Co, Ni, Cu), of ligands of this type by dissolution of an anode of the appropriate metal in an alcoholic solution of azole. The complexes, $[ML_2]$, were obtained in good yields and were found to be practically free of impurities. Ethanol was first used as the solvent, but it was subsequently found that methanol was a more convenient solvent for such processes because: (*i*) its dielectric constant, which is greater than that of ethanol, appreciably increases the electrical conductivity of solutions with a given molar concentration of the ligand, (*ii*) the time required for the synthesis is reduced significantly, and (*iii*) the yields of products are increased.[65] The method requires use of an alternating current, because ac-electrolysis prevents the build-up of surface film on the electrode (a problem encountered with dc-electrolysis), which slows the electrosynthesis.

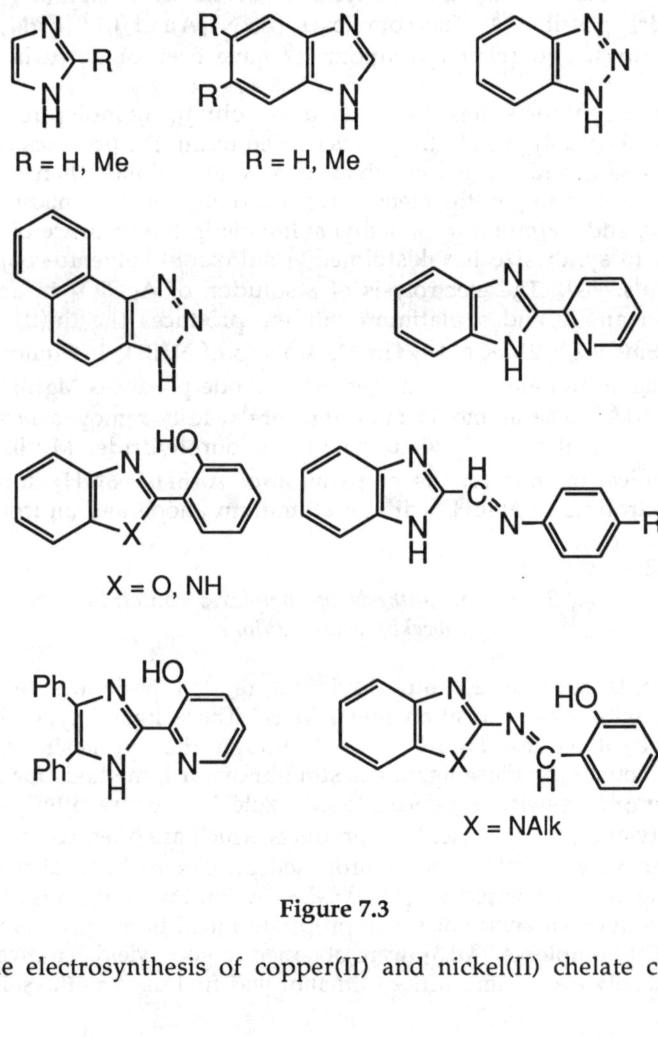

Figure 7.3

The electrosynthesis of copper(II) and nickel(II) chelate complexes,

$[ML_n]$, containing anions of a range of weak acids (HL = acetylacetone, dimethylglyoxime, 8-quinolinol, *C*-hetaryl-substituted monosaccharides, *etc.*) has been described.[68-73] The structures of the complexes are shown in Figure 7.4:

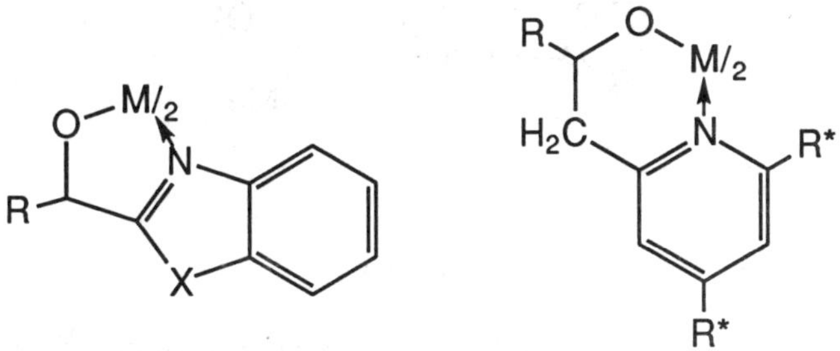

X = S, NMe; R, R* = glycoside, H, Me; M = Cu, Ni

Figure 7.4

For example, in the case of monosaccharide ligands, $[ML_n]$ complexes were synthesized with yields of 80-90% by electrolysis of methanol solutions of the ligands with a copper or nickel anode (M) and a platinum cathode, followed by precipitation of products with benzene. In contrast to the electrosynthesis, the chemical synthesis of the same copper chelates from cupric acetate in methanol proceeds with only a 30% yield and yields both $[ML_2]$ and $[M(MeCO_2)L]$. In electrosyntheses of this type[68-73] the proposed reaction mechanisms involve the anodic dissolution of the metal followed by complex formation.

Hydrogen is liberated at the cathode, Eqs. 7.9 - 7.11:

anode: $nL^- + M \rightarrow [ML_n] + ne^-$ (7.9)

cathode: $nHL - ne^- \rightarrow nL^- + n/2H_2$ (7.10)

overall: $nHL + M \rightarrow [ML_n] + n/2H_2$ (7.11)

The anodic dissolution of zinc in acetonitrile in the presence of the Schiff base **A** has been studied,[74] Figure 7.5:

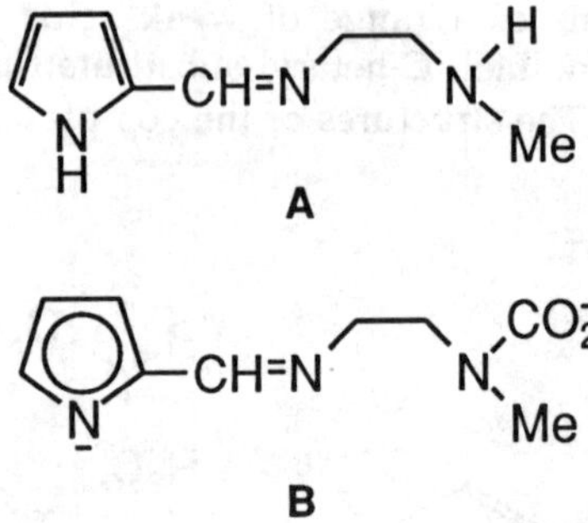

Figure 7.5

Electrosynthesis was complicated by the insertion of carbon dioxide from air into a Zn–N bond of the carbamate complex containing ligand **A** with formation of the zinc complex of the dianionic ligand **B**. Hydrogen was liberated at the platinum cathode.[74]

Electrosynthesis has been widely employed in the preparation of metal alkoxides. As early as 1906, Szilard noted that copper or lead methoxides are obtained by anodic dissolution of these metals through electrolysis of a solution of NaOMe in methanol.[75] The ethoxides and isopropoxides of gallium, scandium, yttrium, germanium, titanium, zirconium, tantalum, and niobium have similarly been synthesized by dissolution of an anode of the appropriate metal in EtOH or *i*-PrOH in the presence of the background electrolyte, $[Bu_4N]Br$.[76-78] In the patent literature it is claimed that the alkoxides of metals with atomic numbers from 14 to 82 can be obtained through electrosynthesis. A sacrificial anode is employed in an anhydrous alcohol containing 1-4 carbon atoms with a quaternary alkylammonium salt acting as the background electrolyte.[79]

A large number of metal thiolates have been obtained in high yields by dissolution of sacrificial anodes in solutions of thiols,[80-89a] Eqs. 7.12 - 7.14:

anode: $2PhS^- + M \rightarrow [M(SPh)_2] + 2e^-$ (7.12)

cathode: $2PhSH + 2e^- \rightarrow 2PhS^- + H_2$ (7.13)

overall: $2PhSH + M \rightarrow [M(SPh)_2] + H_2$ (7.14)

Compounds of the type $[M(SPh)_2]$ can also be obtained from disulfides which are reduced at the cathode, Eq. 7.15:

$$Ph_2S_2 + 2e^- \rightarrow 2PhS^- \qquad (7.15)$$

It has been observed[88] that electrosynthesis provides a simple, high-yield, one-step route to thiolato-derivatives of both main group and transition metals. Electrochemical methods for the preparation of thiolato-compounds may be adapted for the generation of selenolato ($PhSe^-$) complexes. For example, Kampf, Kumar, and Oliver have reported[89b] that $[(Ph_3P)_2Cu(\mu\text{-}SePh)_2Cu(PPh_3)_2]\cdot MeCN$ has been obtained by electrooxidation of a copper anode in a solution of diphenyldiselenide, Ph_2Se_2, and triphenylphosphine in acetonitrile/toluene. The complex has been structurally characterized as the acetonitrile solvate, Figure 7.6:

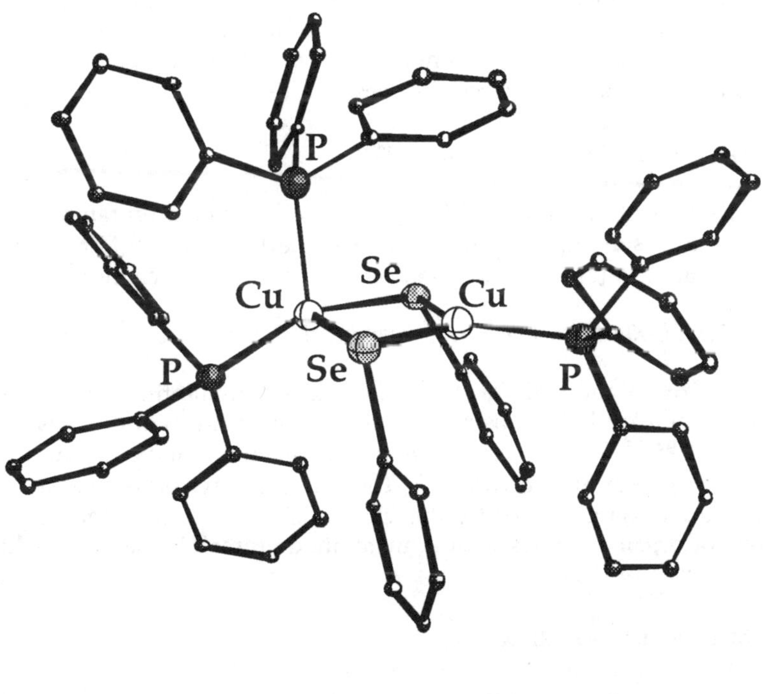

Figure 7.6

7.3.3 *Electrosynthesis of organometallic compounds*

The electrosynthesis of tetraethyllead, which has been carried out on an industrial scale, has been studied. The Grignard reagent, EtMgCl, and an excess of ethyl chloride are supplied to the electrolysis cell, which contains a copper gauze cathode and an anode consisting of lead granules.[90] During electrolysis, the anode dissolves to produce $PbEt_4$, while $MgCl^+$ is reduced at the cathode, Eq. 7.15 - 7.17:

$$\textbf{anode:}\quad Pb + 4Et^- \rightarrow PbEt_4 + 4e^- \qquad (7.15)$$

$$\textbf{cathode:}\quad 4MgCl^+ + 4e^- \rightarrow 2Mg + 2MgCl_2 \qquad (7.16)$$

$$\textbf{overall:}\quad Pb + 4EtMgCl \rightarrow PbEt_4 + 2Mg + 2MgCl_2 \qquad (7.17)$$

The metallic magnesium liberated at the cathode reacts with excess ethyl chloride to form EtMgCl, which continues the reaction. The chemical yield of tetraethyllead is *ca.* 96%. The anodic reactions involved in the formation of the metal-carbon bonds in this and related processes occur by radical mechanisms.[6,91] An electrosynthesis of $[FeCp_2]$ was reported, Eq. 7.18:

$$Fe_{(+)} + 2TlCp \rightarrow FeCp_2 + 2Tl \qquad (7.18)$$

The electrolysis of MCp (M = Li, Na, K) in tetrahydrofuran or pyridine solution at a dissolving manganese anode similarly produces $[MnCp_2]$.[93] The nickel(II) complex, $[NiCp_2]$, has also been obtained in this manner.[6]

7.3.4 *Electrochemical reactions involving oxidative addition*

In 1966, a method for the synthesis of organotin compounds by electrolysis of alkyl halides using a tin anode and a magnesium cathode was developed.[94,95] Such electrochemical reactions have been studied extensively by Tuck and coworkers.[9] These studies have been reviewed and a general mechanism proposed for this class of electrochemical reactions.[9] The reactions of organic halides with a metal anode resemble oxidative addition reactions, Eqs. 7.19 - 7.20:

$$M_{(+)} + RX \rightarrow RMX \qquad (7.19)$$

$$M_{(+)} + 2RX \rightarrow R_2MX_2 \qquad (7.20)$$

These reactions are reminiscent of those used in the preparation of Grignard reagents and of Frankland's synthesis of organozinc halides. The products, RMX and R_2MX_2, are often extremely reactive and disproportionate readily. They can be stabilized, however, by complex formation with other ligands present in solution. For example, the electrochemical oxidation of cadmium in an acetonitrile solution of the aryl halides, *o*-$O_2NC_6H_4X$ (X = Cl, Br, I), produces the intermediate, [Cd(*o*-$O_2NC_6H_4$)X], which reacts with 2,2'-bipyridyl present in the solution to generate [Cd(*o*-$O_2NC_6H_4$)X(bipy)].[96] The electrolysis of alkyl or aryl halides, RX (R = Me, Et, CF_3, C_3H_3, Ph, C_6F_5, CH_2Ph; X = Cl, Br, I, but not in all possible combinations), in acetonitrile solution at a zinc anode produces the reactive species, [ZnRX],[97] which react further with the electrolyte, [(*n*-Pr)$_4$N]X, to produce the anionic complexes, [(*n*-Pr)$_4$N][ZnRX_2].

The compounds, [M(R)XL_2] (M = Ni, Pd; R = alkyl, aryl; X = halide), are very well known and are generally prepared by reaction between [MX_2L_2] and the appropriate alkyllithium, LiR, or Grignard reagent, RMgX.[98] Habeeb and Tuck have reported a direct electrochemical synthesis.[99] For example, the compounds, [M(C_6F_5)Br(PEt_3)$_2$] (M = Ni, Pd), have been prepared, Eq. 7.21:

$$M_{(+)} + C_6F_5Br + 2PEt_3 \rightarrow [M(C_6F_5)Br(PEt_3)_2] \qquad (7.21)$$

Although applicable for cases where M = Ni, Pd, such reactions do not occur for M = Pt. Organonickel cyanides have similarly been prepared[99] *via* C–CN bond cleavage of organonitriles at a dissolving nickel electrode in the presence of PEt_3, Eq. 7.22:

$$Ni_{(+)} + RCN + 2PEt_3 \rightarrow [NiR(CN)(PEt_3)_2] \qquad (7.22)$$

The mechanisms of these electrochemical reactions involve radical intermediates.[9]

Electrochemical reactions that involve oxidative addition are possible not only for organohalides but also for other substrates. For example, the synthesis of [CdBr($SnPh_3$)(bipy)], which contains a Sn–Cd bond, can be accomplished electrochemically,[9] Eq. 7.23:

$$Cd_{(+)} + Ph_3SnBr + bipy \rightarrow [CdBr(SnPh_3)(bipy)] \qquad (7.23)$$

The anodic dissolution of zinc, cadmium and indium (M*) in the presence of the binuclear carbonyl complexes, [$Co_2(CO)_8$] and [$Mn_2(CO)_{10}$], has been investigated.[100] Reactions have been carried out in a mixture of methanol and benzene using [Et_4N][ClO_4] as the background electrolyte and result in the formation of heterobinuclear compounds containing M–M* bonds, M*$[M(CO)_n]_m$ (M = Co, Mn; n = 4, 5; m = 2, 3). Similarly, Sn$[Co(CO)_4]_4$ has been prepared[101] by controlled potential electrolysis of [$Co_2(CO)_8$] in 2-propanol at a sacrificial tin anode, Eqs. 7.24 - 7.27:

$$\textbf{anode:} \quad Sn \rightarrow Sn^{2+} + 2e^- \tag{7.24}$$

$$\textbf{cathode:} \quad [Co_2(CO)_8] + 2e^- \rightarrow 2[Co(CO)_4]^- \tag{7.25}$$

$$\textbf{overall:} \quad Sn + [Co_2(CO)_8] \rightarrow Sn^{2+} + 2[Co(CO)_4]^- \tag{7.26}$$

$$\textbf{solution:} \quad Sn^{2+} + 2[Co(CO)_4]^- + [Co_2(CO)_8] \rightarrow Sn[Co(CO)_4]_4 \tag{7.27}$$

The structure of $Sn[Co(CO)_4]_4$ was confirmed by X-ray crystallography.[101]

7.4 Electrosynthesis at a sacrificial cathode

In certain cases, the electrolysis of ketones, alkyl halides, or unsaturated organic compounds can produce organometallic compounds by the dissolution of a metal cathode. The range of metals used as cathode materials and capable of dissolving under electrolytic conditions is extremely narrow. Most studies have dealt with the synthesis of compounds of mercury, lead, and tin. It has been pointed out that the "cathodic dissolution of a metal" is intrinsically paradoxical.[102] Although electrosynthesis at a sacrificial cathode has been known since 1902, the mechanisms of the reactions taking place are still the subject of considerable discussion.[5] It is likely that the reactions take place by radical mechanisms involving the interaction of the metal cathode with organic radicals formed at the electrode surface.

The electrolysis of halogen-containing organic compounds at a sacrificial cathode commonly produces homoleptic organometallic compounds. For example, the electrolysis of aqueous solutions of iodopropionitrile at a tin, lead, or mercury cathode yields respectively $[Sn(CH_2CH_2CN)_4]$, $[Pb(CH_2CH_2CN)_4]$, or $[Hg(CH_2CH_2CN)_2]$.[103] The electrolysis of ethyl chloride or ethyl bromide at a lead cathode has been proposed as a method for the preparation of tetraethyllead.[6]

Organometallic compounds are formed during the electrochemical reduction of ketones in aqueous solution at a lead or mercury cathode. For example, $[Hg(i\text{-}Pr)_2]$ is formed by the reduction of the carbonyl group of acetone at a mercury cathode,[102] Eq. 7.28:

$$Hg + 2Me_2C{=}O + 6H^+ + 6e^- \rightarrow [Hg(CHMe_2)_2] + 2H_2O \tag{7.28}$$

There are few reports of the formation of metal complexes by the electrochemical reduction of aldehydes. Tedoradze[6] has shown that the electrolysis of benzaldehyde at a mercury electrode produces dibenzylmercury, Eq. 7.29:

$$Hg + 2PhCHO + 6H^+ + 6e^- \rightarrow [Hg(CH_2Ph)_2] + 2H_2O \qquad (7.29)$$

Attempts to carry out similar reactions with a range of different aldehydes were unsuccessful.

Organometallic compounds also may be formed by the dissolution of a metal cathode during the reduction of unsaturated organic compounds. Tomilov and Kaabak[104,105] have studied the reduction of acrylonitrile, $H_2C{=}CHCN$, at a tin cathode in aqueous alkaline solution. Electrolysis produced $[Sn(CH_2CH_2CN)_4]$. Other reactions that take place at sacrificial cathodes have been reviewed.[6,9] Despite the long history of reactions that involve sacrificial cathodes, current ideas about the mechanisms of these reactions do not explain all aspects of the chemistry involved and have little predictive power.

7.5 Practical considerations in synthetic electrochemistry

A wide range of solvents and supporting electrolytes has been used for electrosynthesis and each of them has its own characteristics. Laube and Schmulbach[5] have observed that the choice of a particular combination of solvent and supporting electrolyte is primarily empirical and is made only after consideration of such factors as the chemical inertness towards the solutes or intermediates involved in the electrochemical reaction, ease of purification and handling, the potential range available, the solubilities of the solutes, and the reference electrodes available. Some useful data concerning solvents and supporting electrolytes commonly used in electrosynthesis are summarized in Tables 7.1 and 7.2.

Table 7.1[10]

Physicochemical properties of common solvents used in electrosynthesis

Solvent	mp	bp	Dielect-ric const.	Applied potential, *V vs.* SCE		
	°C			Pt(-)	Hg(-)	Pt(+)
THF	-80	+66	7.6	-3.2[a]	-2.95[d]	+1.6[a]
DME	-58	+82	3.5		-2.95[d]	
Diglyme		+159			-2.95[d]	
pyridine	-42	+115	12.3	-2.2	-2.4	+3.3[b]
DMF	-61	+153	38	-2.8[b]	-2.95[d]	
acetonitrile	-45	+82	38	-2.7[c]	-2.8[b]	+3.5[e]
DMSO	+18	+189	47	-2.3	-2.8[b]	
Propylene carbonate	-49	+242	64	-1.9	-2.9[b]	
sulfolane	+28	+285	44			
HMP	+7	+235	30	-3.3[a]	-2.95	+1.0[a]

Background electrolytes: (*a*) $LiClO_4$, (*b*) $[Et_4N][ClO_4]$, (*c*) $[Bu_4N][ClO_4]$, (*d*) $[Bu_4N]Br$, (*e*) $[Et_4N][BF_4]$.

Table 7.2[10]

Limiting potentials for background electrolytes in THF and DMSO, *V* (*vs.* SCE)

Electrolyte	$Hg_{(-)}$ / THF	$Pt_{(-)}$ / DMSO	$Pt_{(+)}$ / DMSO
$[Bu_4N]Br$	-2.95	-2.4	+1.45
$[Bu_4N][ClO_4]$	-2.94	-2.4	+2.10
$[Bu_4N]Cl$	-2.88		
$[Bu_4N][BF_4]$	-2.87		
$[Bu_4N][BPh_4]$	-2.89		
$[Et_4N]Br$		-2.3	
$[Me_4N]Cl$		-2.4	+1.52
LiCl		-2.68	+1.52
$NaBPh_4$	-1.68		
$NaClO_4$		-2.08	+2.10
$NaBF_4$	-1.7		
$LiClO_4$	-1.95	-2.60	+2.10
$KClO_4$		-2.33	+2.10

References

1. Yu. P. Kitaev, T. V. Troepolskaya, and G. K. Budnikov, *Intermediate Products in Electrochemical Reactions* [in Russian] (Nauka, Moscow, 1982) p. 216.
2. B. Gerdes, *J. pr. Chem.* **26** (1882) 257.
3. A. A. Grinberg, *The Physical Chemistry of Complex Compounds, Selected Works* [in Russian] (Nauka, Leningrad, 1972) p. 131.
4. V. M. Zamyatkina, Yu. N. Kukushkin, and A. A. Makarenya, *Lev Aleksandrovich Chugaev* [in Russian] (Nauka, Moscow, 1973) p. 102.
5. B. L. Laube and C. D. Schmulbach, *Progr. Inorg. Chem.* **14** (1971) 65.
6. G. A. Tedoradze, *J. Organomet. Chem.* **88** (1975) 1.
7. H. Lehmkuhl, in *Organic Electrochemistry*, M. M. Baizer, ed., (Dekker, New York, 1973) p. 621.
8. D. De Montauzon, R. Poilblanc, P. Lemoine, and M. Gross, *Electrochim. Acta* **23** (1978) 1247.
9. D. G. Tuck, *Pure Appl. Chem.* **51** (1979) 2005.
10. H. Lehmkuhl, *Synthesis* (1973) 377.
11. M. O. Albers and N. J. Coville, *Coord. Chem. Rev.* **53** (1984) 227.
12. C. Oldham and D. G. Tuck, *J. Chem. Ed.* **59** (1982) 420.
13. A. D. Garnovsky, Yu. I. Ryabukhin, and A. S. Kuzharov, *Koord. Khim.* **10** (1984) 1011.
14. M. C. Chakravorty and G. V. B. Subrahmanyam, *Coord. Chem. Rev.* **135/136** (1994) 65.
15. A. P. Tomilov, I. N. Chernikh, and Yu. M. Kargin, *Electrochemistry of Elementoorganic Compounds* [in Russian] (Nauka, Moscow, 1985).
16. (a) Z. Nagy, *Electrochemical Synthesis of Inorganic Compounds* (Plenum, New York, 1985); (b) M. Ya. Fioshin, and M. G. Smirnova, *Electrosynthesis of Oxidants and Reductants* [in Russian] (Khimia, Leningrad, 1981).
17. D. Astruc, *Angew. Chem. Int. Ed. Engl.* **27** (1988) 643.
18. P. H. Rieger, *Electrochemistry* (Chapman and Hall, London, 1993) 2nd Ed.
19. G. K. Budnikov, *Electrochemical Reactions of Metal Chelates in Organic and Mixed Solvents* [in Russian] (KGU, Kazan, 1980) p. 304.

20. D. C. Olson and W. Keim, *Inorg. Chem.* **8** (1969) 2028.

21. G. Schiavon, S. Zecchin, G. Zotti, and G. Pilloni, *Inorg. Chim. Acta.* **20** (1976) L1.

22. (a) J. A. Davies, C. T. Eagle, D. E. Otis, and U. Venkataraman, *Organometallics* **5** (1986) 1246; (b) J. A. Davies, C. T. Eagle, D. E. Otis, and U. Venkataraman, *Organometallics* **8** (1989) 1080; (c) J. A. Davies and C. T. Eagle, *Organometallics* **5** (1986) 2149; (d) J. A. Davies and R. J. Staples in *Selective Hydrocarbon Activation: Principles and Progress*, eds. J. A. Davies, P. L. Watson, J. F. Liebman, and A. Greenberg (VCH, New York, 1990) p. 379.

23. W. Schäfer, J. Kerrinnes, and U. Langbein, *Z. Anorg. Allg. Chem.* **406** (1974) 101.

24. S. Zecchin, G. Schiavon, and G. Pilloni, *J. Organomet. Chem.* **110** (1976) C45.

25. I. V. Tananaev, G. B. Seifer, Yu. Ya. Kharitonov, V. G. Kuznetsov, and A. P. Korolkov, *The Chemistry of Hexacyanoferrates(II)* [in Russian] (Nauka, Moscow, 1971) p. 13.

26. J. S. Miller, *Inorg. Synth.* **19** (1979) 13.

27. J. S. Miller, *Science* **193** (1976) 189.

28. C. Bianchini, A. Meli, M. Peruzzini, A. Vacca, F. Laschi, P. Zanello, and F. M. Ottaviani, *Organometallics* **9** (1990) 360.

29. M. S. Thompson and T. J. Meyer, *J. Am. Chem. Soc.* **103** (1981) 5577.

30. F. R. Keene, D. J. Salmon, and T. J. Meyer, *J. Am. Chem. Soc.* **98** (1976) 1884.

31. G. M. Brown, T. R. Weaver, F. R. Keene, and T. J. Meyer, *Inorg. Chem.* **15** (1976) 190.

32. P. A. Adcock, F. R. Keene, R. S. Smythe, and M. R. Snow, *Inorg. Chem.* **23** (1984) 2336.

33. P. A. Adcock and F. R. Keene, *J. Am. Chem. Soc.* **103** (1981) 6494.

34. W. L. Bowden, W. F. Little, and T. J. Meyer, *J. Am. Chem. Soc.* **99** (1977) 4340.

35. G. J. Bezems, P. H. Rieger, and S. Visco, *J. Chem. Soc., Chem. Commun.* (1981) 265.

36. J. Grobe and H. W. Zimmermann, *Z. Naturforsch.* **36b** (1981) 301.

37. J. Grobe and H. W. Zimmermann, *Z. Naturforsch.* **36b** (1981) 482.

38. A. B. Bocarsly and D. E. Bisson, *ACS Division of Inorg. Chem.*, Winter Newsletter, January (1984) Abstr. 81.

39. A. Darchen, El. K. Lhadi, and H. Patin, *J. Organomet. Chem.* **259** (1983) 189.

40. El. K. Lhadi, H. Patin, A. Benoit, J. Y. Le Marouille, and A. Darchen, *J. Organomet. Chem.* **259** (1983) 321.

41. A. Darchen, C. Mahé, and H. Patin, *J. Chem. Soc., Chem. Commun.* (1982) 243.

42. P. M. Treichel, W. K. Dean, and W. M. Douglas, *Inorg. Chem.* **11** (1972) 1609.

43. H. H. Ohst and J. K. Kochi, *J. Chem. Soc., Chem. Commun.* (1986) 121.

44. H. H. Ohst and J. K. Kochi, *J. Am. Chem. Soc.* **108** (1986) 2897.

45. H. H. Ohst and J. K. Kochi, *Inorg. Chem.* **25** (1986) 2066.

46. J. W. Hershberger, R. J. Klinger, and J. K. Kochi, *J. Am. Chem. Soc.* **104** (1982) 3034.

47. J. W. Hershberger and J. K. Kochi, *J. Chem. Soc., Chem. Commun.* (1982) 212.

48. J. W. Hershberger, R. J. Klinger, and J. K. Kochi, *J. Am. Chem. Soc.* **105** (1983) 61.

49. (a) P. Denisevich, H. D. Abruña, C. R. Leidner, T. J. Meyer, and R. W. Murray, *Inorg. Chem.* **21** (1982) 2153; (b) T. Ikeda, R. Schmehl, P. Denisevich, K. Willman, and R. W. Murray, *J. Am. Chem. Soc.* **104** (1982) 2683; (c) J. M. Calvert, R. H. Schmehl, B. P. Sullivan, J. F. Facci, T. J. Meyer, and R. W. Murray, *Inorg. Chem.* **22** (1983) 2151.

50. C. R. Leidner, B. P. Sullivan, R. A. Reed, B. A. White, M. T. Crimmins, R. W. Murray, and T. J. Meyer, *Inorg. Chem.* **26** (1987) 882.

51. H. D. Abruña, P. Denisevich, M. Umana, T. J. Meyer, and R. W. Murray, *J. Am. Chem. Soc.* **103** (1981) 1.

52. B. D. Berezin, *Sov. J. Coord. Chem. (Engl. Transl.)* **17** (1991) 324.

53. R. J. Sime and N. W. Gregory, *J. Am. Chem. Soc.* **82** (1960) 93.

54. J. J. Habeeb and D. G. Tuck, *Inorg. Synth.* **19** (1979) 123.

55. J. J. Habeeb, C. Oldham, and D. G. Tuck, *J. Inorg. Nucl. Chem.* **43** (1981) 2087.

56. J. J. Habeeb, L. Neilson, and D. G. Tuck, *Inorg. Chem.* **17** (1978) 306.

57. F. A. Cotton, G. E. Lewis, and W. Schwotzer, *Inorg. Chem.* **25** (1986) 3528.

58. N. Kumar and D. G. Tuck, *Can. J. Chem.* **60** (1982) 2579.

59. A. D. Goolsby and D. T. Sawyer, *Anal. Chem.* **40** (1968) 1978.

60. J. J. Habeeb, F. F. Said, and D. G. Tuck, *Inorg. Nucl. Chem. Lett.* **15** (1979) 113.

61. J. J. Habeeb, F. F. Said, and D. G. Tuck, *J. Chem. Soc., Dalton Trans.* (1981) 118.

62. H. Schmidt, *Z. Anorg. Allg. Chem.* **271** (1953) 305.

63. G. F. Huff, A. D. McElroy, and R. M. Adams, *US Patent* 2855353 (1958); *Chem. Abstr.* **53** (1959) 1965a.

64. J. Plesek, S. Hermanek, and V. Gregor, *Czech. Patent* 116107 (1965); *Chem. Abstr.* **65** (1966) 3354b.

65. N. N. Bogdashev, A. D. Garnovskii, O. A. Osipov, V. P. Grigorev, and N. M. Gontmakher, *Zh. Obsch. Khim.* **46** (1976) 675.

66. O. A. Osipov, N. N. Bogdashev, V. P. Grigorev, A. D. Garnovskii, and N. M. Gontmakher, *USSR Patent* 485115 (1975).

67. O. A. Osipov, N. N. Bogdashev, V. P. Grigorev, and A. D. Garnovskii, *USSR Patent* 4016474 (1972).

68. H. Lehmkuhl and W. Eisenbach, *Lieb. Ann. Chem.* (1975) 672.

69. J. J. Habeeb, D. G. Tuck, and F. H. Walters, *J. Coord. Chem.* **8** (1978) 27.

70. L. Bustos, J. H. Green, J. L. Hencher, M. A. Khan, and D. G. Tuck, *Can. J. Chem.* **61** (1983) 2141.

71. Yu. A. Zhdanov, O. A. Osipov, N. M. Gontmakher, V. P. Grigorev, A. D. Garnovskii, O. N. Nechaeva, N. N. Bogdashev, Yu. E. Alekseev, V. G. Alekseeva, and T. A. Yusman, *Dokl. Akad. Nauk SSSR* **227** (1976) 882.

72. Yu. A. Zhdanov, O. A. Osipov, V. P. Grigorev, A. D. Garnovskii, Yu. E. Alekseev, N. M. Gontmakher, V. G. Alekseeva, T. A. Yusman, V. G. Zaletov, N. N. Bogdashev, and O. N. Nechaeva, *Koord. Khim.* **3** (1977) 3.

73. Yu. A. Zhdanov, O. A. Osipov, V. P. Grigorev, A. D. Garnovskii, Yu. E. Alekseev, V. G. Alekseeva, N. M. Gontmakher, P. A. Petrov, V. G. Zaliotov, V. N. Fomina, T. A. Yusman, O. N. Nechaeva, and V. N. Mirny, *Carbohyd. Res.* (1974) 1.

74. J. Castro, A. Castiñeiras, M. L. Duran, J. A. Garcia-Vazquez, A. Macias, J. Romero, and A. Sousa, *Z. Anorg. Allg. Chem.* **586** (1990) 203.

75. B. Szilard, *Z. Elektrochem.* **12** (1906) 393.

76. E. P. Turevskaya and N. Ya. Turova, *Koord. Khim.* **9** (1983) 1064.

77. V. A. Shreider, E. P. Turevskaya, N. I. Kozlova, and N. Ya. Turova, *Izv. Akad. Nauk SSSR, Ser. Khim.* (1981) 1687.

78. V. A. Shreider, E. P. Turevskaya, N. I. Kozlova, and N. Ya. Turova, *Inorg. Chim. Acta* **53** (1981) L73.

79. Monsanto Chemicals Ltd., *Fr. Addn.* 2, 091, 229 (1972); *Chem. Abstr.* **77** (1972) 96297r.

80. F. F. Said and D. G. Tuck, *Inorg. Chim. Acta* **59** (1982) 1.

81. J. L. Hencher, M. A. Khan, F. F. Said, and D. G. Tuck, *Inorg. Nucl. Chem. Lett.* **17** (1981) 287.

82. J. L. Hencher, M. A. Khan, F. F. Said, and D. G. Tuck, *Polyhedron* **4** (1985) 1263.

83. J. L. Hencher, M. A. Khan, F. F. Said, R. Sieler, and D. G. Tuck, *Inorg. Chem.* **21** (1982) 2787.

84. R. K. Chadha, R. Kumar, and D. G. Tuck, *Can. J. Chem.* **65** (1987) 1336.

85. M. A. Khan, R. Kumar, and D. G. Tuck, *Polyhedron* **7** (1988) 49.

86. R. K. Chadha, R. Kumar, J. R. Lopez-Grado, and D. G. Tuck, *Can. J. Chem.* **65** (1988) 2151.

87. J. H. Green, R. Kumar, N. Seudeal, and D. G. Tuck, *Inorg. Chem.* **28** (1989) 123.

88. T. A. Annan, R. Kumar, and D. G. Tuck, *Inorg. Chem.* **29** (1990) 2475.

89. (a) R. Kumar and D. G. Tuck, *Inorg. Chem.* **29** (1990) 1444; (b) J. Kampf, R. Kumar, and J. P. Oliver, *Inorg. Chem.* **31** (1992) 3626.

90. *Chem. Eng. News* **72** (1965) November 8.

91. R. Galli, *J. Electroanal. Chem.* **22** (1969) 75.

92. S. Valcher, *Corsi Semin. Chim.* (1968) 37; *Chem. Abstr.* **72** (1970) 17834j.

93. A. P. Giraitis, T. H. Pearson, and R. C. Pinkerton, *US Patent* 2960450; *Chem. Abstr.* **55** (1961) 8430h.

94. K. N. Korotaevskii, E. N. Lysenko, Z. S. Smolyan, L. M. Monastyrskii, and L. V. Armenskaya, *Zh. Obsch. Khim.* **36** (1966) 167.

95. L. V. Armenskaya, K. N. Korotaevskii, E. N. Lysenko, L. M. Monastyrskii, Z. S. Smolyan, and G. A. Razuvaev, *USSR Patent* 184453 (1966).

96. F. F. Said and D. G. Tuck, *J. Organomet. Chem.* **224** (1982) 121.

97. J. J. Habeeb, A. Osman, and D. G. Tuck, *J. Organomet. Chem.* **185** (1980) 117.

98. M. H. L. Green, *Organometallic Compounds. The Transition Elements*, Vol. II (Methuen, London, 1968) p. 244.

99. J. J. Habeeb and D. G. Tuck, *J. Organomet. Chem.* **139** (1977) C17.

100. J. J. Habeeb, D. G. Tuck, and S. Zhandire, *Can. J. Chem.* **57** (1979) 2196.

101. A. Cabrera, H. Samain, A. Mortreux, F. Petit, and A. J. Welch, *Organometallics* **9** (1990) 959.

102. A. P. Tomilov, S. G. Mairanovskii, M. Ya. Fioshin, and V. A. Smirnov, *The Electrochemistry of Organic Compounds* [in Russian] (Khimiya, Leningrad, 1968) p. 481.

103. A. P. Tomilov, Yu. D. Smirnov, and S. L. Varshavskii, *Zh. Obsch. Khim.* **35** (1965) 391.

104. A. P. Tomilov and L. V. Kaabak, *Zh. Prikl. Khim.* **32** (1959) 2600.

105. L. V. Kaabak and A. P. Tomilov. *Zh. Obsch. Khim.* **33** (1963) 2808.

Chapter 8. NON-TRADITIONAL OXIDANTS IN PREPARATIVE COORDINATION CHEMISTRY

Oxidants are important reagents in preparative coordination chemistry and the halogens, oxygen, hydrogen peroxide, concentrated nitric acid, permanganates, bromates, and persulphates all play useful roles. The choice of an oxidant depends on the chemistry to be explored and the specific oxidant characteristics necessary for the reactions involved. For example, to oxidize the metal center in a complex, without affecting the ligand, a particular oxidant may be selected. Alternatively, to oxidize the ligand without affecting the metal, a different oxidant may be required. The products of the reduction of oxidants are frequently ligands themselves and this is sometimes a decisive factor in the choice of the oxidant. The larger the number of oxidants available to the synthetic chemist, the more options there are available for the investigation of specific problems.

Over the last two decades the range of oxidants used in coordination chemistry has expanded to include many that are compatible with non-aqueous solvents. The potential of a redox couple is related to the degree of solvation of the oxidant and its reduction products. For example, the ease with which chlorine is reduced in polar solvents is associated with the difference in solvation energies between two chloride ions and a Cl_2 molecule, which tends to shift the equilibrium to favor formation of the more highly solvated chloride ions. In poorly solvating solvents, the difference between the solvation energies of Cl_2 and chloride ions decreases and this results in a decrease in the oxidation potential, *i.e.* chlorine is a weaker oxidant in a poorly solvating solvent. Thus, attempts[1] to obtain platinum(IV) complexes of hydroxylamine by oxidation of $[PtX_2(NH_2OH)_2]$ (X = Cl or Br) with halogens in aqueous solution leads to decomposition of coordinated hydroxylamine and formation of halo- and hydroxo-complexes of platinum(IV). In carbon tetrachloride, platinum(IV) complexes with coordinated hydroxylamine are readily obtained.

In recent years a number of non-traditional oxidants have been employed in preparative coordination chemistry. The term *new oxidants* would not be altogether accurate because individual instances of their use have been known for some time. The oxidants discussed in this Chapter often are specific for the oxidation of particular ligands or ligand types in coordination compounds. Accordingly, these "ligand-specific" reagents are of interest to synthetic chemists because the opportunity exists to direct an oxidation process to a specific site in a complex.

8.1 Nitrosonium, thionitrosonium and nitronium salts

Nitrosonium salts, NOX, are one-electron oxidants and the standard potential[2,3] of the NO^+/NO couple in aqueous solution is 1.45 *V* (1.23 *V vs.* Ag/AgCl electrode). When nitrosonium salts are employed with weak donor, non-aqueous solvents, the differences in solvation energies between NO^+ and NO diminish and the oxidizing ability of NO^+ increases.[4] Indeed, the potential of the NO^+/NO couple in nitromethane is 1.58 *V vs.* Ag/AgCl.[5]

A number of methods for the preparation of nitrosonium reagents are known. Nitrosyl chloride, NOCl, is formed through the reaction of molecular chlorine and NO, by treatment of $NOHSO_4$ or $NaNO_2$ with HCl, as well as by reaction of KCl and NO_2.[6] Nitrosyl bromide, NOBr, is usually obtained from Br_2 and NO.[6] Nitrosyl fluoride, NOF, is produced in a similar manner.[6] Among the different procedures[7] for the preparation of $NOBF_4$, the reaction of N_2O_3 with a mixture of HBF_4 and H_2SO_4 or aqueous HBF_4 is commonly employed and $NOClO_4$ is obtained analogously,[7,8] Eq. 8.1:

$$N_2O_3 + 2HX \rightarrow 2NOX + H_2O \tag{8.1}$$

where X = BF_4, ClO_4

Mocella, Okamoto and Barefield[9] advise that the use of nitrosonium salts in synthesis requires careful selection and handling of solvents. Some solvents (for example, acetonitrile, nitromethane, dichloromethane, ethyl acetate, benzene, hexane, and carbon tetrachloride), when carefully purified, do not react with NOX salts. However, common solvents like acetone, pyridine, ethers, alcohols, and water are reactive towards nitrosonium salts, even when the solvents have been carefully purified. Reactions of NOX salts are best carried out in an oxygen-free atmosphere, so that the production of nitrate ion is avoided. Nitrate ion is generated in a stepwise manner, Eqs. 8.2 - 8.4:

$$NO\ (g) + 1/2O_2\ (g) \rightarrow NO_2\ (g) \tag{8.2}$$

$$2NO_2\ (g) \rightarrow N_2O_4\ (aq) \tag{8.3}$$

$$N_2O_4\ (aq) \rightarrow NO^+ + NO_3^- \tag{8.4}$$

In fact, in spite of its relatively weak donor properties, nitrate ion can significantly interfere with some reactions.

Nitrosonium salts containing weak donor anions are employed for the oxidation of metals in the preparation of both metal salts and homoleptic solvento-complexes. Thus, one of the first syntheses in which a nitrosonium salt was employed as the oxidant was reported by Seel[10] in 1950, Eq. 8.5:

$$\text{Na} + \text{NOSbCl}_6 \xrightarrow{\text{SO}_2} \text{Na[SbCl}_6] + \text{NO} \quad (8.5)$$

More recently the oxidation of metals by nitrosonium salts in acetonitrile has been used to synthesize homoleptic solvento-complexes. Thus, the complexes, $[Cu(MeCN)_4]X$ ($X = BF_4$, ClO_4),[11,12] have been obtained by oxidation of metallic copper in acetonitrile with NOX; the perchlorate, $NOClO_4$, in MeCN oxidizes metallic gold to form $[Au(MeCN)_4][ClO_4]$,[13,14] Eq. 8.6:

$$\text{Au} + \text{NOClO}_4 + 4\text{MeCN} \rightarrow [\text{Au(MeCN)}_4][\text{ClO}_4] + \text{NO} \quad (8.6)$$

The solvated cations, $[Ag(MeCN)_4]^+$,[14,15] $[Eu(MeCN)_n]^{3+}$,[16] $[Ni(MeCN)_6]^{2+}$,[17] and $[Pd(MeCN)_4]^{2+}$ [18-20] have been synthesized by metal oxidation with NOX salts.

In addition to the oxidation of metals, the oxidation of metal ions in complexes by nitrosonium salts is synthetically useful. The oxidation of $[Re_2Cl_4(PMe_2Ph)_4]$ by nitrosonium hexafluorophosphate in acetonitrile at room temperature has been reported.[21] Addition of one equivalent of $NOPF_6$ to $[Re_2Cl_4(PMe_2Ph)_4]$ led to the formation of $[Re_2Cl_4(PMe_2Ph)_4][PF_6]$ in 73% yield. Reaction of $[Re_2Cl_4(PMe_2Ph)_4]$ with 3.5 equivalents of $NOPF_6$ produced $[Re_2Cl_4(PMe_2Ph)_4][PF_6]_2$ in 90% yield. Cyclic voltammograms of the starting material and the two products gave rise to identical $E_{1/2}$ values (+0.85 and -0.26 *V vs.* SCE). All three compounds were characterized by X-ray diffraction.[21]

The one-electron oxidation, by $NOPF_6$ in acetonitrile, of a copper(I) chelate has been described,[22] Scheme 8.1. The reaction proceeds rapidly and yields the copper(II) product. Ring-closure and an increase in coordination number accompany the redox process. In another example, reaction of the platinum(0) complex, $[Pt(C_2H_4)(PPh_3)_2]$, with $NOBF_4$ in acetonitrile leads to oxidation of the metal center with concurrent substitution of the coordinated ethylene by solvent molecules,[23] Eq. 8.7:

$$[\text{Pt(C}_2\text{H}_4)(\text{PPh}_3)_2] + 2\text{NOBF}_4 + 2\text{MeCN} \rightarrow [\text{Pt(MeCN)}_2(\text{PPh}_3)_2][\text{BF}_4]_2 + \text{C}_2\text{H}_4 + 2\text{NO} \quad (8.7)$$

Oxidation with nitrosyl chloride of the complex, $[MoL(CO)_2(NO)]$ (L = tris(3,5-dimethylpyrazolyl)borate), is accompanied by substitution of the coordinated carbonyl groups by chloride ions,[24] Eq. 8.8:

$$[MoL(CO)_2(NO)] + 2NOCl \xrightarrow{CH_2Cl_2} [MoCl_2L(NO)] + 2CO + 2NO \qquad (8.8)$$

Further examples of nitrosyl-promoted oxidations have been cited in several reviews.[25]

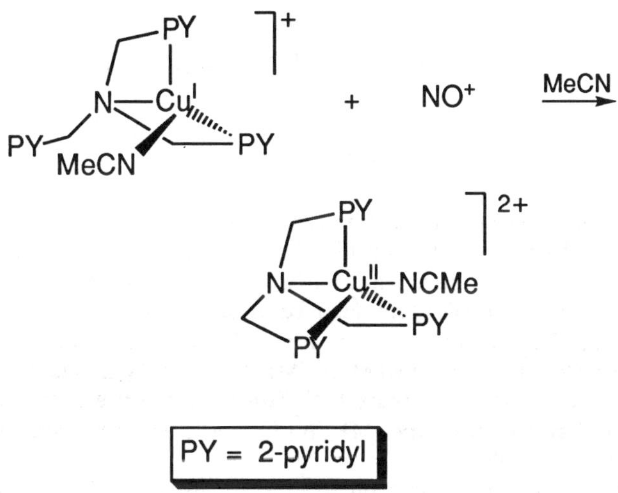

Scheme 8.1

In the previous examples the nitrosyl group did not participate in substitution reactions with the metal complexes. However, it has been established that the NO^+ cation can react to give either oxidation products or the corresponding metal nitrosyl derivatives. Both processes may take place simultaneously. The products[26] obtained in reactions of $[Mo(CO)_6]$ with NO^+ in acetonitrile depend on the experimental conditions. Thus, under a nitrogen atmosphere, $[Mo(CO)_6]$ and $NOPF_6$ react in acetonitrile solution to give *cis*-$[Mo(NO)_2(MeCN)_5][PF_6]_2$, but passage of a vigorous stream of nitrogen gas through the solution gives $[Mo(NO)(MeCN)_4][PF_6]_2$.[26] Further, when the reaction between $[Cr(CO)_4L]$ (L = dppm, dppe) and $NOBF_4$ in CH_2Cl_2 was carried out in a closed, nitrogen-filled vessel, the product was $[Cr(CO)_3(NO)L]^+$. On the other hand, if the reaction was carried out with a nitrogen stream passing through the mixture to remove the NO gas evolved, the oxidation product, $[Cr(CO)_4L]^+$, was formed.[26] These studies led to the proposal of a stepwise mechanism,[27] Scheme 8.2:

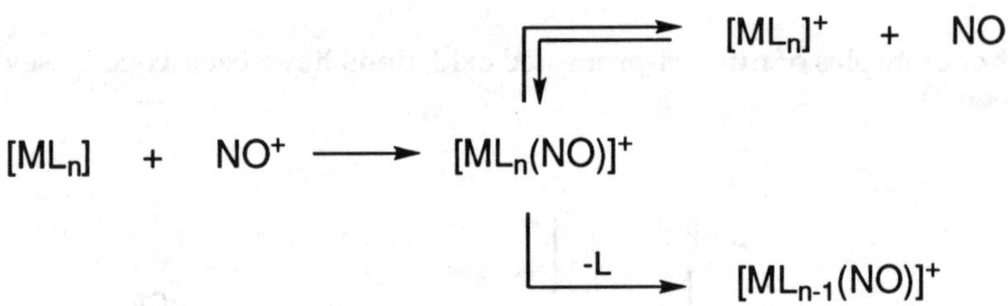

Scheme 8.2

The mechanism illustrates how the manipulation of reaction conditions can cause a shift in the equilibrium and lead to formation of either the oxidation product, $[ML_n]^+$, or the substituted nitrosyl complex, $[ML_{n-1}(NO)]^+$. The substituted nitrosyl complex can also be regarded as an oxidation product since the coordinated nitrosyl can be formally considered as an NO^- anion. Changing the properties of L and M, in $[ML_n]$, markedly affects the position of the equilibrium shown in Scheme 8.2. The ability of the complex, $[ML_n]$, to form intermediates of the type $[ML_n(NO)]^+$ with an increased coordination number is also important.

Reaction of the complex, $[Et_4N][V(CO)_6]$, with $NOBF_4$ in dichloromethane at -40 °C results in the formation of the substituted complex $[V(CO)_5(NO)]$.[28] The infrared spectra of the products obtained by the interaction of $[V(CO)_6]^-$ with NO^+ and the reaction[29] of $[V(CO)_6]$ with NO are identical. The following reaction mechanism has been proposed, Scheme 8.3:

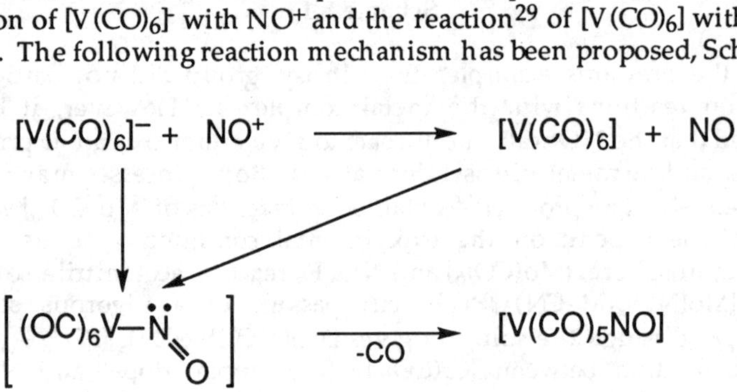

Scheme 8.3

This mechanism does not conflict with the mechanism shown in Scheme 8.1 and, in fact, extends it. The following examples illustrate the dual reactivity of nitrosonium salts. When the isonitrile complex of tungsten, $[W(CNPh)_6]$, is mixed with $NOPF_6$ in tetrahydrofuran, two products are formed,[30] Eq. 8.9:

$$2[W(CNPh)_6] + 2NOPF_6 \rightarrow [W(NO)(CNPh)_5][PF_6] + [W(CNPh)_7][PF_6] + NO \quad (8.9)$$

When the ruthenium complex, $[RuBr(CO)(PEt_3)(\eta^5\text{-}C_5Ph_5)]$, and $NOPF_6$ were reacted in dichloromethane and the mixture cooled rapidly to -196 °C, the electron spin resonance spectrum of $[RuBr(CO)(PEt_3)(\eta^5\text{-}C_5Ph_5)]^+$ was readily detected. In contrast, at room temperature, the same reaction gives a mixture of products from which $[RuBr(PEt_3)(NO)(\eta^5\text{-}C_5Ph_5)][PF_6]$ was isolated in 44% yield.[31]

In a related example, the reaction of $[Mo(CO)_6]$ in dichloromethane solution with $NOPF_6$ has been described,[32] Eq. 8.10:

$$[Mo(CO)_6] + 2NOPF_6 \rightarrow cis\text{-}[Mo(NO)_2(PF_6)_2] + 6CO \quad (8.10)$$

The product, *cis*-$[Mo(NO)_2(PF_6)_2]$, has a polymeric structure with bridging hexafluorophosphate ligands,[32] Figure 8.1:

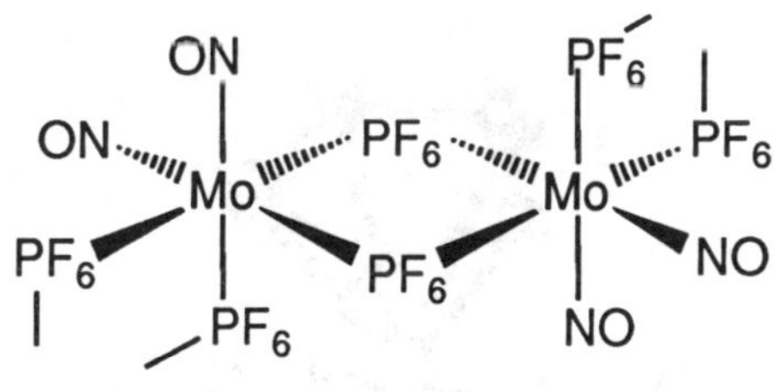

Figure 8.1

The corresponding solvates, $[Mo(NO)_2(solv)_4][PF_6]_2$, are formed from similar reactions employing MeCN[20,26] or $MeNO_2$ as solvents.[32]

Additionally, nitrosonium salts exhibit dual reactivity in reactions with cluster compounds.[33] For example, the oxidation of $[Os_4H_3(CO)_{12}]^-$ by NOX salts (X = BPh_4, PF_6) in acetonitrile gives rise to the clusters, $[Os_4H_3(CO)_{12}(MeCN)_2]X$, in 35% yield while the substitution of THF or dichloromethane for acetonitrile results in the formation of the nitrosyl cluster, $[Os_4H_3(CO)_{12}(NO)]$.[33]

In addition to oxidation of metals or metal ions in coordination compounds, the NO^+ ion is used for oxidation of coordinated ligands. The ability of nitrosonium salts to oxidize coordinated azide ion has been applied extensively in preparative coordination chemistry since the reaction was discovered in 1968,[34] Eq. 8.11:

$$[CoN_3(NH_3)_5][ClO_4]_2 + NOClO_4 + solv \rightarrow [Co(NH_3)_5(solv)][ClO_4]_3 + N_2O + N_2 \quad (8.11)$$

In these reactions the solvent may be tetramethylene sulfone,[34] triethyl phosphate,[35] or acetonitrile.[36] The gaseous reaction products are lost from the reaction medium and the vacant coordination site is occupied by a solvent molecule (see Chapter 3).

The first example of the insertion of nitrosonium ion into a metal-carbon bond was reported in 1986,[37] Eq. 8.12:

$$[CrMe(NO)_2(\eta^5\text{-}Cp)] + NOPF_6 \xrightarrow{CH_2Cl_2} [Cr(NO)_2\{N(CH_2)OH\}(\eta^5\text{-}Cp)][PF_6] \quad (8.12)$$

The structure of the organometallic chromium-formaldoxime product, $[Cr(NO)_2\{N(CH_2)OH\}(\eta^5\text{-}Cp)][PF_6]$, has been confirmed by single-crystal X-ray diffraction, Figure 8.2:

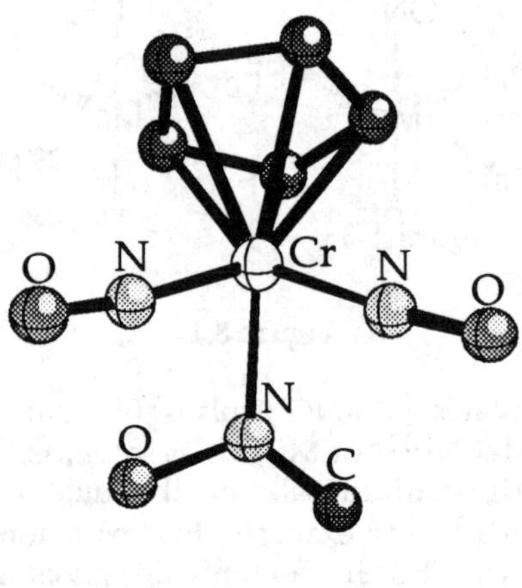

Figure 8.2

The mechanistic pathways for the insertion of NO^+ into metal-carbon bonds have been described in the literature,[37] Scheme 8.4:

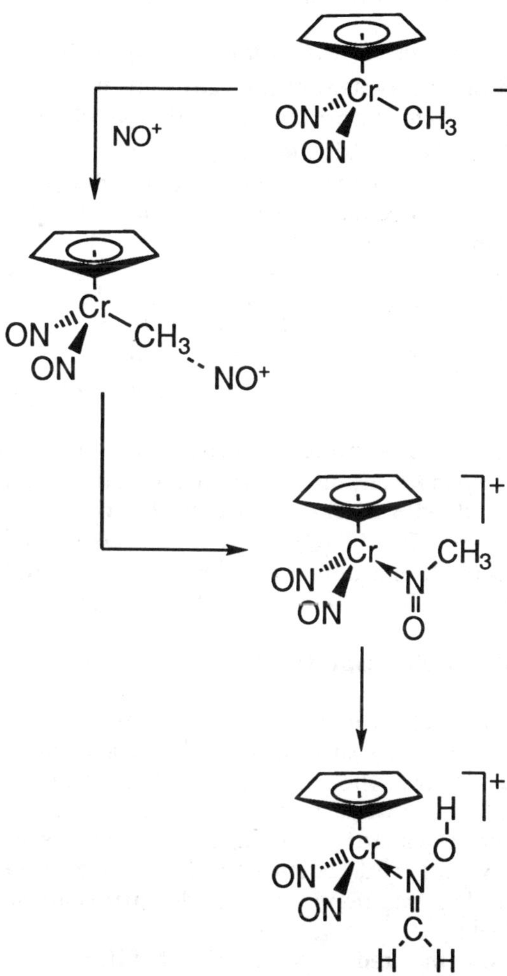

Scheme 8.4

A related example of the insertion of NO^+ into a metal-carbon bond occurs in the reaction of the cobalt complex, $[\{CpCo\}_3\{\mu_3\text{-}\eta^1\text{-}CR\}\{\mu_3\text{-}\eta^1\text{-}CR^*\}]$, and $NOBF_4$ in $MeNO_2/CH_2Cl_2$,[38] Eq. 8.13:

$$[\{CpCo\}_3\{\mu_3\text{-}\eta^1\text{-}CR\}\{\mu_3\text{-}\eta^1\text{-}CR^*\}] + NOBF_4 \rightarrow [\{CpCo\}_3\{\mu_3\text{-}\eta^1\text{-}CR\}\{\mu_3\text{-}\eta^1\text{-}C(R^*)=N(O)\}][BF_4] \quad (8.13)$$

In dichloromethane reaction of the ruthenium thiolate complex, $[Ru(SPh)(CO)_2(\eta^5\text{-}Cp)]$, with one equivalent of the nitrosonium salt, $NOPF_6$, results in oxidation of the thiolate ligand and precipitation of the dimer, $[\{Ru(CO)_2(\eta^5\text{-}Cp)\}_2(\mu\text{-}PhSSPh)][PF_6]$.[39] Oxidation by NO^+ occurs also in the reaction[40] between the nickel(II) phosphine complexes, $[NiX_2(PPh_3)_2]$ (X = Cl, Br, NO_3), and NOCl in a mixture of benzene and cyclohexane. The reactions result in formation of the dimers, $[NiXCl(OPPh_3)]_2$.

Examples of the oxidation of metal complexes by thionitrosonium salts, NSX, are uncommon but some useful reactions have been reported. Thus, the reaction between $[ReBr(CO)_5]$ and $NSSbF_6$ leads to formation of $[Re(CO)_5(NS)][SbF_6]$ in low yield.[41] In addition, interaction between $[Cr(CO)_3(C_6H_6)]$ and $NSPF_6$ results in formation[42] of the solvento-complex, $[Cr(MeCN)_5(NS)]^{2+}$.

Oxidation of coordination compounds by nitronium salts, NO_2X, is also known.[43] For instance, the reaction of nitronium tetrafluoroborate, NO_2BF_4, with $[Pt(PPh_3)_4]$ and with Vaska's compound, $[IrCl(CO)(PPh_3)_2]$, results in formation of $[Pt(NO_2)_2(PPh_3)_2]$ and $[IrCl(CO)(NO)(PPh_3)_2]^+$, respectively.[44] In most cases the interaction of coordination compounds with nitrosonium or with nitronium salts leads to identical products.[44,45]

8.2 Aryldiazonium salts

The structural and synthetic chemistry of aryldiazenato-complexes have been described in the literature[46] and similarities between the nitrosyl and aryldiazenato-complexes of transition metals have been noted. These similarities are particularly evident when resonance structures of linear (*a*), bent (*b*), and bridged (*c*) complexes are examined, Scheme 8.5. Analogous reactivity patterns for nitrosyl and aryldiazenato-complexes are also observed and this extends[27] to redox reactions in which nitrosonium and aryldiazonium salts participate.

No literature data on the redox potentials of nitrosonium and diazonium ions in non-aqueous solvents under comparable conditions are available.[27] However, it follows from available potentials that NO^+ (+1.58 *V vs.* Ag/AgCl, nitromethane)[47] is a stronger oxidant than ArN_2^+ (+0.295 *V vs.* SCE, sulfolane)[48] although the potentials quoted are clearly not directly comparable. A comparative study has been made[27] of the interaction of the complexes, $[Cr(CO)_2L(\eta^6\text{-}C_6Me_6)]$ (L = PPh_3, $PMePh_2$, PMe_2Ph, $P(OPh)_3$, $P(OMe)_3$), with $NOPF_6$ and PhN_2PF_6 in toluene/methanol and dichloromethane, respectively. It has been established that nitrosonium salts react to produce only the products of CO or L substitution, *i.e.* the complexes $[Cr(CO)(NO)L(\eta^6\text{-}C_6Me_6)][PF_6]$ and $[Cr(CO)_2(NO)(\eta^6\text{-}C_6Me_6)][PF_6]$. However, interaction with a diazonium salt leads to formation of the one-electron oxidation product, $[Cr(CO)_2L(\eta^6\text{-}C_6Me_6)][PF_6]$, in addition to the complexes $[Cr(CO)(N_2Ph)L(\eta^6\text{-}C_6Me_6)][PF_6]$ and $[Cr(CO)_2(N_2Ph)(\eta^6\text{-}C_6Me_6)][PF_6]$, which result from substitution.

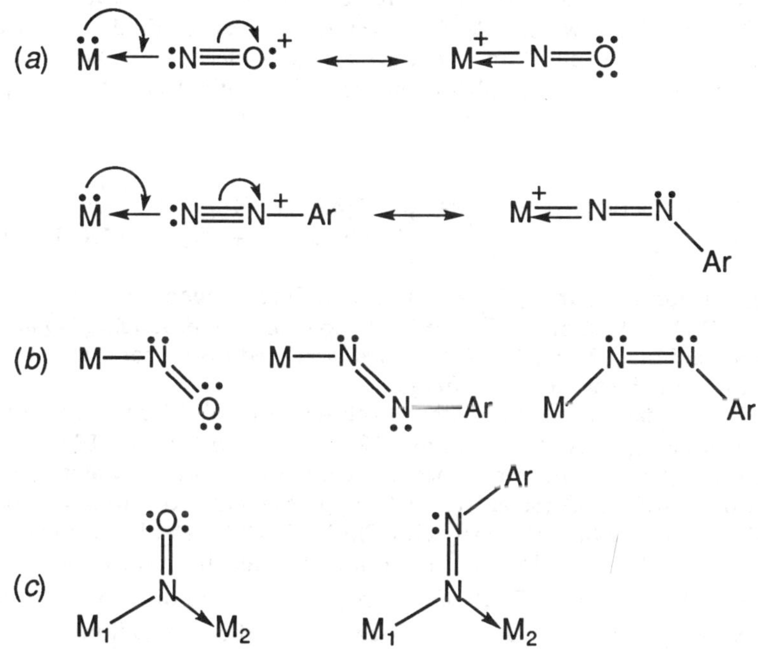

Scheme 8.5

To account for this finding, a mechanism describing the interaction of carbonyl complexes with nitrosonium salts was proposed[27] (see Section 8.1). On the basis of similarity between the reactivities of NOX and ArN_2X, the scheme was extended to include reactions of complexes of the type $[ML_n]$ with ArN_2X. Despite the more negative potential of the ArN_2^+/ArN_2 couple compared with that of the NO^+/NO couple, diazonium salts are more reactive one-electron oxidants. This greater reactivity may be rationalized in several ways: steric hindrance of the ArN_2^+ ligand decreases the likelihood of formation of the intermediate, $[ML_n(ArN_2)]^+$, and the formation of the oxidation product, $[ML_n]^+$, is favored by the irreversible formation of N_2 and biaryl, Eq. 8.14:

$$2ArN_2^+ + 2e^- \rightarrow 2N_2 + Ar{-}Ar \quad (8.14)$$

The use of aryldiazonium salts as oxidants in synthetic coordination chemistry is illustrated in the reactions of metal hydrides with aryldiazonium cations, which give aryldiazene or arylhydrazido complexes as products. For example, the reactivity of the cobalt(I) hydride complex, $[CoH\{P(OEt)_2Ph\}_4]$, towards p-$MeC_6H_4N_2BF_4$ has been studied.[33] Instead of insertion or substitution reactions, one-electron oxidation of $[CoH\{P(OEt)_2Ph\}_4]$ to the cobalt(II) complex, $[CoH\{P(OEt)_2Ph\}_4]^+$, takes place, Eq. 8.15:

$$[CoH\{P(OEt)_2Ph\}_4] + p\text{-}MeC_6H_4N_2BF_4 + H^+ + e^- \xrightarrow{CH_2Cl_2} [CoH\{P(OEt)_2Ph\}_4][BF_4] + N_2 + MeC_6H_5 \quad (8.15)$$

Nitrogen is formed during this reaction, as well as toluene. It is possible that the p-$MeC_6H_4N_2^+$ cation oxidizes cobalt(I) to cobalt(II), producing N_2 and the unstable radical, p-$MeC_6H_4^\bullet$, which abstracts hydrogen from the solvent, resulting in the formation of toluene.

The oxidation of Schiff-base complexes of cobalt(II) by aryldiazonium salts has been reported.[49] Reactions of [Co(Salen)] with p-$XC_6H_4N_2BF_4$ (X = H, OMe, NO_2), in chloroform or acetone containing traces of water, gave the cobalt(III) complex, $[Co(Salen)(H_2O)][BF_4]$, as the product. Reaction of the cobalt(I) complex, Na[Co(Salen)], with PhN_2BF_4 in tetrahydrofuran/acetone, followed by the addition of pyridine, results in formation of the σ-arylcobalt(III) complex, [Co(Ph)(Salen)py]. Removal of dinitrogen from the corresponding aryldiazenato-complexes, with the formation of σ-aryl derivatives,[50,51] and their use as arylating agents,[52,53] has been described.

Diazonium salts in acetonitrile oxidize metallic copper or silver with formation of the corresponding homoleptic solvento-complexes.[54] Oxidation of coordinated ligands by diazonium salts is also known. Thus, the interaction of $[Rh(CO)(PPh_3)(\eta^5\text{-}Cp)]$ with an equimolar quantity of p-$FC_6H_4N_2BF_4$ in dichloromethane leads to the formation of the fulvalene complex, $[Rh_2(CO)_2(PPh_3)_2(\eta^5{:}\eta'^5\text{-}C_{10}H_8)]^{2+}$, formed by dimerization of two $[Rh(CO)(PPh_3)(\eta^5\text{-}Cp)]^+$ radical cations,[55] Scheme 8.6:

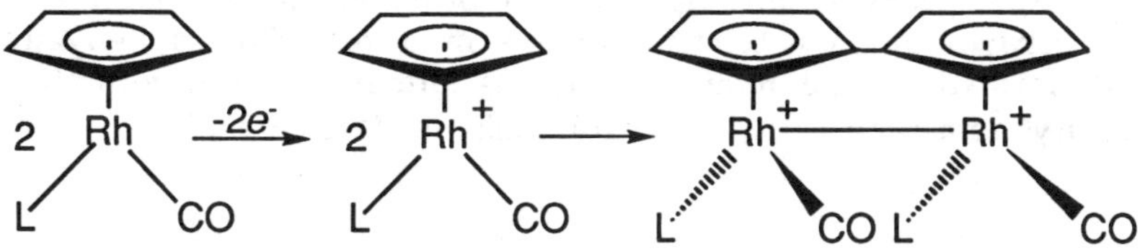

Scheme 8.6

8.3 Triphenylcarbenium, tropylium, and other carbenium salts

In 1958, Dauben and Honnen[56] studied the reaction between the cycloheptatriene complex, $[Mo(CO)_3(\eta\text{-}C_7H_8)]$, and triphenylcarbenium tetrafluoroborate, Ph_3CBF_4, in dichloromethane. They found that these compounds react immediately to form $[Mo(CO)_3(\eta\text{-}C_7H_7)][BF_4]$. This reaction was perhaps the first in which a triphenylcarbenium (trityl) salt was used as an acceptor for hydride ion in organometallic chemistry. The widespread use of Ph_3CX in preparative coordination chemistry stems from four fundamental considerations. First, the trityl cation is a one-electron oxidant. Second, this cation is capable of abstracting hydride ion from certain organic ligands. Third, triphenylcarbenium salts act as acceptors for coordinated hydrides. Fourth, the trityl cation can alkylate metal complexes.

The syntheses of triphenylmethyl perchlorate and triphenylmethyl tetrafluoroborate have been reported.[57] These reagents are obtained by reaction of triphenylmethanol with 71% $HClO_4$ or 48% HBF_4 in the presence of acetic or propionic anhydrides (introduced to remove water), Eqs. 8.16 and 8.17:

$$Ph_3COH + H^+ \rightarrow Ph_3C^+ + H_2O \qquad (8.16)$$

$$(RCO)_2O + H_2O \rightarrow 2RCOOH \qquad (8.17)$$

Yields are 76% and 92% for the perchlorate and the tetrafluoroborate, respectively. The tetrafluoroborate, Ph_3CBF_4, has advantages over the perchlorate, Ph_3CClO_4, because of its greater stability and lack of light sensitivity.[57]

It has been shown[54] that Ph_3CBF_4 can function as a one-electron oxidant and is able to oxidize metallic copper with formation of the homoleptic solvento-complex, $[Cu(MeCN)_4][BF_4]$. Yamazaki[58] has reported that the phosphine and phosphite complexes of palladium(II), $[Pd(PPh_3)_4]X_2$ (X = BF_4, PF_6), and $[Pd\{P(OPh)_3\}_4][PF_6]_2$, have been synthesized in 60% and

50% yields, respectively, through oxidation of [Pd(PPh$_3$)$_4$] and [Pd{P(OPh)$_3$}$_4$] by the trityl cation in a benzene/acetone mixture. The triphenylmethyl radical, formed in the reduction process, dimerizes with formation of 1-diphenylmethylene-4-trityl-2,5-cyclohexadiene,[59] Scheme 8.7:

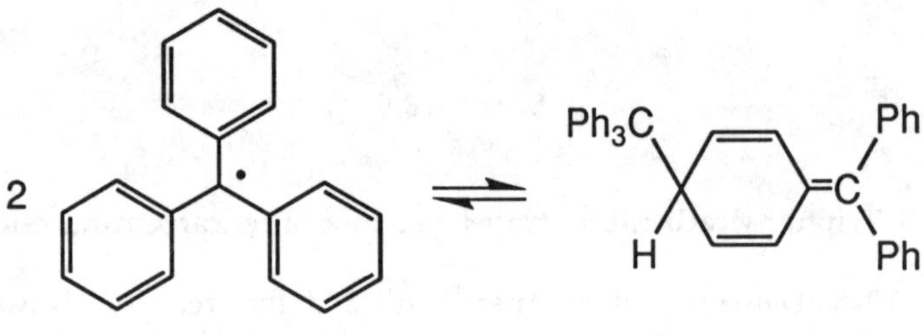

Scheme 8.7

This organic by-product was isolated in good yields during the syntheses of [Pd(PPh$_3$)$_4$]X$_2$ and [Pd{P(OPh)$_3$}$_4$][PF$_6$]$_2$.

The one-electron oxidation of the complex, [W(NO)$_2${P(OPh)$_3$}(η^5-Cp)], with triphenylcarbenium tetrafluoroborate[60] in CH$_2$Cl$_2$ under air led to the isolation of [W(NO)$_2${P(OPh)$_3$}(η^5-Cp)]$^+$, a compound described previously by Bachmann.[61] In addition, Ph$_3$COOCPh$_3$ is formed in 13% yield. Presumably Ph$_3$COOCPh$_3$ is formed[60] from Ph$_3$C$^\bullet$ during work-up of the reaction mixture in air. The triphenylcarbenium cation acts as a one-electron oxidant towards cobalt(II) hydrido-complexes of the type [CoHL$_4$] (L = P(OEt)$_2$Ph, P(OMe)$_2$Ph, P(OPh)$_3$) and does not cleave the Co–H bonds,[62] which are known to be relatively inert in comparison to most transition metal-hydrogen bonds.[62] Through the reactions of [CoHL$_4$] with Ph$_3$CX (X = PF$_6$, BF$_4$), compounds of the type [CoHL$_4$][X] are formed.

The solvent dependence of the reactivity of the triphenylcarbenium ion has been demonstrated[63] in a study of the reaction of Ph$_3$CPF$_6$ with the polyhydride, [Re$_2$H$_8$(PPh$_3$)$_4$]. The abstraction of a hydride ion from the rhenium center, in acetonitrile solution, results in the formation of a coordinatively unsaturated intermediate, stabilized as the solvento-complex, [Re$_2$H$_7$(PPh$_3$)$_4$(MeCN)][PF$_6$]. When the reaction is carried out in CH$_2$Cl$_2$, Ph$_3$CPF$_6$ functions as a one-electron oxidant leading to the formation of [Re$_2$H$_8$(PPh$_3$)$_4$][PF$_6$].

In addition to the triphenylmethyl salts, other sources of carbenium ions are available for use in one-electron oxidation reactions. Thus, the complex, [W(NO)$_2${P(OMe)$_3$}(η^5-Cp)], is oxidized to the corresponding monocation by treatment with $MeOSO_2F$,[60] Eq. 8.18:

$$[W(NO)_2\{P(OMe)_3\}(\eta^5\text{-}Cp)] + MeOSO_2F \xrightarrow{CH_2Cl_2} [W(NO)_2\{P(OMe)_3\}(\eta^5\text{-}Cp)][OSO_2F] + 1/2C_2H_6 \qquad (8.18)$$

The ethane evolved was detected by gas chromatography.

The one-electron oxidation of *trans*-$[MoCl_2(dppe)_2]$ by the alkylating agents, $[Et_3O][BF_4]$ or $MeOSO_2F$, which are capable of acting as oxidizing agents, in dichloromethane leads to the formation of the molybdenum(III) derivatives, *trans*-$[MoCl_2(dppe)_2][X]$ (X = BF_4, OSO_2F).[64] It has also been established that triethyloxonium tetrafluoroborate, $[Et_3O][BF_4]$, oxidizes $[ReCl(CO)_3(PPh_3)_2]$ to $[ReCl(CO)_3(PPh_3)_2][BF_4]$. Although the reduction products were not identified in the study,[64] by analogy with other results[60] it is likely that the alkyl cations are reduced as a result of the reaction.

A brief survey of the applications of triphenylcarbenium salts in organometallic chemistry has been published,[65] and there are many examples that serve to demonstrate the utility of these salts in synthesis.

On treating $[MoH(CO)_3Cp^*]$ with Ph_3CPF_6 in CH_2Cl_2 in the presence of carbon monoxide, triphenylphosphite, or phosphines (PPh_3, $PMePh_2$ or PMe_3), the cationic complexes, $[Mo(CO)_3LCp^*][PF_6]$, are formed.[66] Hydride abstraction from organic ligands by trityl hexafluorophosphate in CH_2Cl_2 similarly has been carried out,[67] Scheme 8.8:

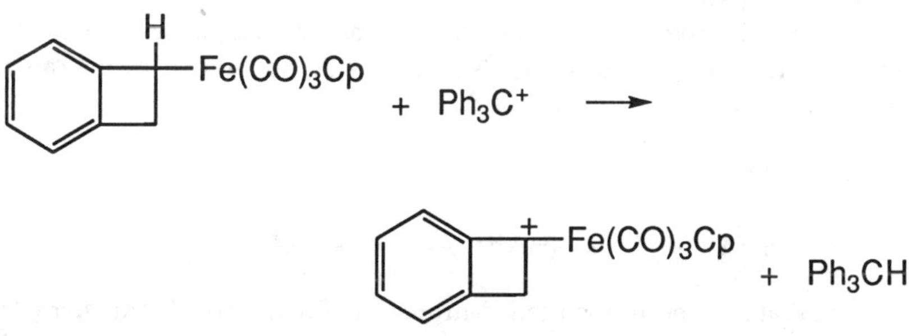

Scheme 8.8

An analogous carbene complex has been isolated from the oxidation of a [bis(organothio)methyl]iron complex[68] by Ph_3CBF_4 in CH_2Cl_2, Scheme 8.9:

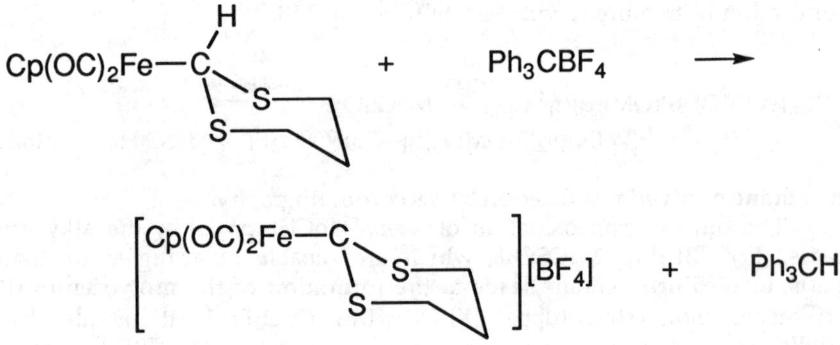

Scheme 8.9

A μ-methylidyne complex of iron[69] is formed from $[(\eta^5\text{-Cp})_2(CO)_2Fe_2\text{-}(\mu\text{-}CO)(\mu\text{-}CH_2)]$ by treatment with Ph_3CPF_6, Eq. 8.19:

$$[(\eta^5\text{-Cp})_2(CO)_2Fe_2(\mu\text{-}CO)(\mu\text{-}CH_2)] + Ph_3CPF_6 \xrightarrow{CH_2Cl_2} [(\eta^5\text{-Cp})_2(CO)_2Fe_2(\mu\text{-}CO)(\mu\text{-}CH)][PF_6] + Ph_3CH \quad (8.19)$$

The starting material can be regenerated by treating the μ-methylidyne complex with $Li[HBEt_3]$.

The mechanism of hydride transfer has proved to be controversial. However, it has been demonstrated[70] that hydride transfer reactions can take place in either one or two steps, Eqs. 8.20 - 8.21:

$$B^+ + H\text{–}A \rightarrow BH + A^+ \quad (8.20)$$

$$B^+ + H{:}A \rightarrow B^\bullet + [HA]^{+\bullet} \rightarrow BH + A^+ \quad (8.21)$$

But, in several studies, it has been demonstrated that hydride transfer occurs by a mechanism involving a one-electron transfer. The extent to which this mechanism is operative is not yet known.[70]

The ability of the trityl cation to interact with a coordinated hydride is seen in reactions of the complex $[RuH_2(PPh_3)_4]$.[71a] Treatment of this complex with a stoichiometric amount of trityl hexafluorophosphate yielded $[RuH(\eta^6\text{-}PhPPh_2)(PPh_3)_3][PF_6]$.[71a] Abstraction of the hydride ligand in $[Ph_4As][CrH(CO)_5]$ led to the formation of a bridged complex,[71b] Eq. 8.22:

$$2[Ph_4As][CrH(CO)_5] + Ph_3CBF_4 \xrightarrow{THF} [Ph_4As][(CO)_5Cr(\mu\text{-}H)Cr(CO)_5] + Ph_3CH + [Ph_4As][BF_4] \quad (8.22)$$

The reaction between Ph_3CX (X = BF_4, PF_6) and the hydride complex, $[MH(NO)_2(\eta^5\text{-}Cp)]$ (M = Mo, W), exhibits a significant solvent dependence.[72] Thus, the solvento-complexes, $[M(NO)_2(MeCN)(\eta^5\text{-}Cp)]X$, are formed in acetonitrile, while the dimers, $[M_2H(NO)_4(\eta^5\text{-}Cp)_2]$, are produced in dichloromethane, a very poor donor. The abstraction of hydride from the complexes, $[MnH(CO)_3\{P(OMe)_3\}_2]$[73] and $[MH(CO)_3Cp]$ (M = Mo, W),[74] by Ph_3CPF_6 in CH_2Cl_2 leads to the formation of $[Mn(FPF_5)(CO)_3\{P(OMe)_3\}_2]$ and $[M(FPF_5)(CO)_3Cp]$, respectively. These compounds are formed because the solvent cannot compete with the PF_6^- anion as a ligand (see Chapter 5). The reaction of $[ReH(CO)(NO)(\eta^5\text{-}Cp)]$ with triphenylcarbenium tetrafluoroborate in CH_2Cl_2 at -78 °C yields $[Re(CO)(Ph_3CH)(NO)(\eta^5\text{-}Cp)][PF_6]$ as the product,[75] Eq. 8.23:

$$[ReH(CO)(NO)(\eta^5\text{-}Cp)] + Ph_3CPF_6 \rightarrow [Re(CO)(Ph_3CH)(NO)(\eta^5\text{-}Cp)][PF_6] \quad (8.23)$$

This product is described as containing an uncommon η^2-arene ligand, Figure 8.3:

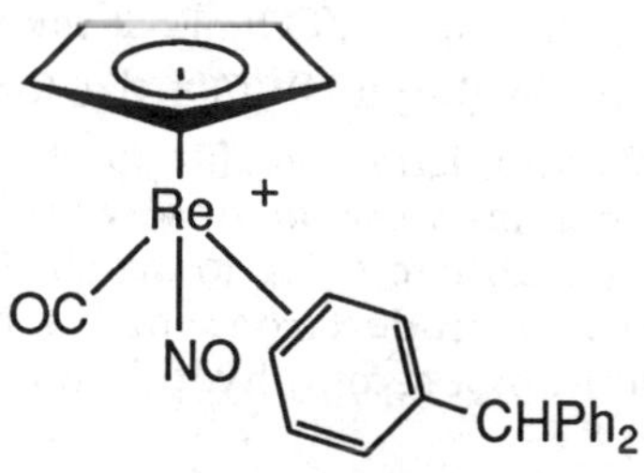

Figure 8.3

In addition to trityl cations, other carbenium ions can be used as hydride acceptors in preparative coordination chemistry. Although the tropylium cation is a much less-active hydride acceptor than the triphenylcarbenium cation,[70] oxidation of hydride complexes with this cation can be achieved. Thus, the interaction of tropylium tetrafluoroborate,

$C_7H_7BF_4$, with the complex, [ReH(CO)(NO)(η^5-Cp)], leads to hydride abstraction and formation of a 1,2-η^2-C_7H_8 complex in which cycloheptatriene coordinates to the metal center,[75,76] Scheme 8.10:

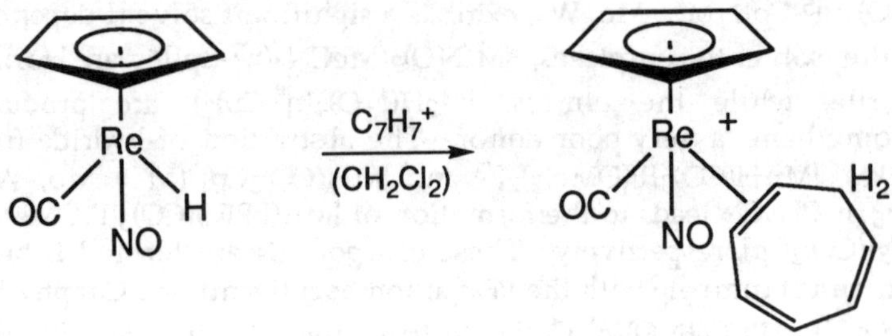

Scheme 8.10

Similarly, hydride abstraction from [OsH(CO)$_2$(η^5-Cp)] by tropylium tetrafluoroborate in dichloromethane at 25 °C for six days gives the complex, [Os(CO)$_2$(1,2-η^2-C_7H_8)(η^5-Cp)][BF_4], in 80% yield.[77]

The reaction of [Re_3(μ-H)$_4$(CO)$_{10}$]$^-$ with tropylium tetrafluoroborate results in H^- abstraction with the formation of C_7H_8.[78] From this study it was concluded that hydride abstraction leads to the production of the putative unsaturated intermediate "[Re_3(μ-H)$_3$(CO)$_{10}$]". If the reaction proceeds in the presence of carbon monoxide, [Re_3(μ-H)$_3$(CO)$_{12}$] is formed but, in MeCN, the reaction leads to the solvento-derivative, [Re_3(μ-H)$_3$(CO)$_{10}$(MeCN)$_2$]. When the reaction is performed in the weak donor solvent CH_2Cl_2, coordination of the newly-formed cycloheptatriene, C_7H_8, to the cluster might be anticipated. However, the only isolated product contained a cycloheptadienyl ligand, C_7H_9, which is the product of a second hydride transfer from the cluster to the cycloheptatriene, Eq. 8.24:

$$[Re_3(\mu\text{-}H)_4(CO)_{10}]^- + C_7H_7^+ \rightarrow \text{"}Re_3(\mu\text{-}H)_3(CO)_{10}\text{"} + C_7H_8 \rightarrow [Re_3(\mu\text{-}H)(\mu_3\text{-}H)(CO)_{10}(\eta^5\text{-}C_7H_9)] \quad (8.24)$$

[Re_3(μ-H)(μ_3-H)(CO)$_{10}$(η^5-C_7H_9)] has been structurally characterized. Related studies of hydride abstraction by tropylium cation have been reported.[79]

In conclusion, it is important to note that the trityl cation can alkylate coordinated ligands. For example, treatment of the phenylthiomethyl complex, [Fe(CH_2SPh)$(CO)_2$(η^5-Cp*)], with Ph_3CPF_6 in dichloromethane leads to the formation of a mixture of the alkylated iron(II) complex, [Fe{CH_2(SPh)CPh_3}$(CO)_2$(η^5-Cp*)][PF_6], and the product of hydride abstraction, [Fe(=CHSPh)$(CO)_2$(η^5-Cp*)][PF_6],[80] Scheme 8.11:

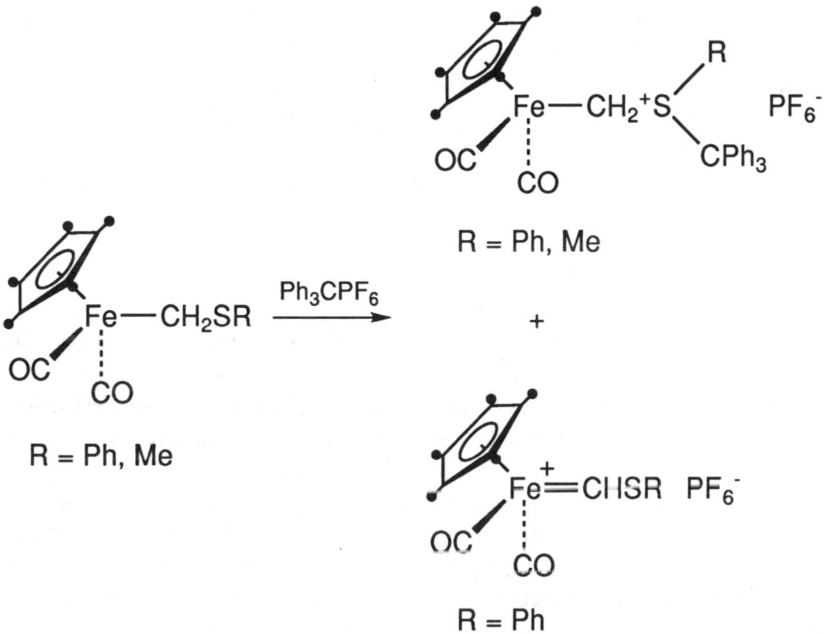

Scheme 8.11

Interaction of [Fe(CH_2SMe)$(CO)_2$(η^5-Cp*)] and Ph_3CPF_6 results only in formation of the alkylated product, [Fe{CH_2(SMe)CPh_3}$(CO)_2$(η^5-Cp*)][PF_6]. The generation of this single product has been explained on the basis of the stability of the sulphonium salt and the increased nucleophilicity of sulfur upon substitution of Ph for Me. Additional studies of alkylation by trityl cation have been described.[81]

8.4 Protic acids

In non-aqueous solvents the proton can function as an effective one-electron oxidant. Thus, the interaction between metallic titanium or chromium and HBF_4 in acetonitrile has been studied.[82] The reactions lead to the solvento-complexes, $[Ti(MeCN)_6][BF_4]_3$ and $[Cr(MeCN)_6][BF_4]_3$, in good yield. The reaction between $[Mo(CO)_6]$ and neat trifluoromethanesulfonic acid also has been studied.[83] When a suspension of the reagents is refluxed for four hours, $Mo(O_3SCF_3)_3$ is isolated. It was noted in this study that interaction of $[Mo(CO)_6]$ and 100% formic acid results in oxidation of the metal, elimination of CO ligands, and formation of $[Mo_2(O_2CH)_4]$, which contains the Mo_2^{4+} unit. The complex $[Rh(PhC{\equiv}CPh)(PPh_3)(\eta^5\text{-}Cp)]$ reacts with an excess of trifluoroacetic acid at 40 °C by displacement of the acetylene and formation of the oxidation product, $[Rh(O_2CCF_3)(PPh_3)(\eta^5\text{-}Cp)]$.[84] The reaction of the dimer, $[Cr(NO)_2(\eta^5\text{-}Cp)]_2$, with two moles of $HBF_4{\cdot}OMe_2$ in dichloromethane leads to oxidative cleavage with formation of the unstable monomeric compound, $[Cr(BF_4)(NO)_2(\eta^5\text{-}Cp)]$, which contains a coordinated BF_4^- ligand, Eq. 8.25:

$$[Cr(NO)_2(\eta^5\text{-}Cp)]_2 + 2HBF_4{\cdot}OMe_2 \xrightarrow{CH_2Cl_2} 2[Cr(BF_4)(NO)_2(\eta^5\text{-}Cp)] + H_2 + 2OMe_2 \quad (8.25)$$

The tetrafluoroborate complex, $[Cr(BF_4)(NO)_2(\eta^5\text{-}Cp)]$, is converted into the more stable salt, $[Cr(NO)_2(MeCN)(\eta^5\text{-}Cp)][BF_4]$, on treatment with MeCN.[85] Similarly, the interaction of the isoelectronic complex, $[Fe(CO)_2(\eta^5\text{-}Cp)]_2$, with *p*-toluenesulphonic acid in acetonitrile leads to oxidative cleavage of the Fe–Fe bond and formation of the acetonitrile solvento-complex, $[Fe(CO)_2(MeCN)(\eta^5\text{-}Cp)][O_3SC_6H_4Me]$.[86] For this reaction it was reported that the protic acid was responsible for oxidation of the complex. However, the evolution of molecular hydrogen was not observed.

Together with triphenylcarbenium and silver salts, protic acids are used in hydride abstraction reactions. It has been shown[87] that $[MoH_4(PMePh_2)_4]$ reacts with HBF_4 or HPF_6 in tetrahydrofuran to form the cation, $[MoH_3(PMePh_2)_3]^+$. Reactions with weaker acids, CF_3CO_2H or *p*-$MeC_6H_4SO_3H$, in toluene, lead to the formation of the molybdenum *bis*-hydrides, $[MoH_2(CF_3CO_2)_2(PMePh_2)_3]$ and $[MoH_2(p\text{-}MeC_6H_4SO_3)_2(PMePh_2)_3]$, respectively. The amount of hydrogen evolved in these reactions was determined manometrically to be *ca*. 2 mole equivalents. The monohydride complex, $[WH(NO)_2(\eta^5\text{-}Cp)]$, reacts with anhydrous *p*-toluenesulphonic acid in THF to yield $[W(OSO_2C_6H_4Me)(NO)_2(\eta^5\text{-}Cp)]$.[88] The dihydride complex, $[RhH_2(PPr_3)(\eta^5\text{-}Cp)]$, reacts rapidly with trifluoroacetic acid in methanol in the presence of NH_4PF_6 to produce hydrogen and the complex, $[(\eta^5\text{-}Cp)(PPr_3)Rh(\mu\text{-}H)_3Rh(PPr_3)(\eta^5\text{-}Cp)][PF_6]$.[85] Tungsten, rhenium, osmium, and iridium polyhydrides react with $HBF_4{\cdot}OEt_2$ in acetonitrile.[89] The starting materials and products of these reactions are presented below, Table 8.1:

Table 8.1

Polyhydride reactions with tetrafluoroboric acid etherate

$[WH_6(PMe_2Ph)_3]$	→	$[WH_2(MeCN)_3(PMe_2Ph)_3]^{2+}$
$[ReH_5(PMe_2Ph)_3]$	→	$[ReH(MeCN)_3(PMe_2Ph)_3]^{2+}$
$[ReH_7(PPh_3)_2]$	→	$[ReH(MeCN)_4(PPh_3)_2]^{2+}$
$[OsH_4(PMe_2Ph)_3]$	→	$[Os(MeCN)_3(PMe_2Ph)_3]^{2+}$
$[IrH_5(PPh_3)_2]$	→	$[IrH_2(MeCN)_2(PPh_3)_2]^{+}$

Because the metal centers in both $[WH_6(PMe_2Ph)_3]$ and $[ReH_7(PPh_3)_2]$ have d^0 electronic configurations, protonation of these metal ions is unlikely.[89] The more likely sites of protonation are the M–H bonds to give either linear or bent $[M\cdots H\cdots H]^+$ intermediates. On the other hand, complexes such as $[OsH_4(PMe_2Ph)_3]$, $[ReH_5(PMe_2Ph)_3]$, and $[IrH_5(PPh_3)_2]$ are perhaps more likely to be protonated at the metal centers. The protonated intermediates are relatively unstable and decompose at room temperature with evolution of hydrogen and, at that point, the vacant coordination sites are occupied by solvent molecules. It was noted in this study that dicationic complexes are frequently produced in these reactions. However, the reason for this occurance is unexplained as yet.

The interaction[90] of $[Mn(CO)_5]^-$ with trifluoromethanesulphonic acid in THF solution yields $[MnH(CO)_5]$ which has a pK_a of 7.[91] Consequently, the hydrogen in $[MnH(CO)_5]$ has relatively weak hydridic character. Nevertheless, the interaction of the complex, $[PPN][Mn(CO)_5]$, with neat CF_3SO_3H in the absence of solvent leads to the formation of $[Mn(CF_3SO_3)(CO)_5]$ and the evolution of hydrogen, presumably by protonation of $[MnH(CO)_5]$,[92] Eq. 8.26:

$$[MnH(CO)_5] + CF_3SO_3H \rightarrow [Mn(CF_3SO_3)(CO)_5] + H_2 \tag{8.26}$$

Thus, when the nature of the reaction medium is altered in this manner, the hydride ligand, which does not exhibit significant hydridic character in other media, is formally oxidized.

8.5 Silver, mercury and other metal salts

The importance of silver salts as reagents for halide abstraction was discussed in Chapter 3. However, the salts, AgX, also are used as one-electron oxidants in syntheses performed in non-aqueous media. The standard potential, E°, of the Ag^+/Ag couple in aqueous solution is 0.799 *V*,[93] while in the less-solvating solvent, dimethylformamide, the potential is 1.0 *V*.[94] The redox potential of the Ag^+/Ag couple is not easily determined in weak donor solvents[95] such as CH_2Cl_2 in which $AgBF_4$ or $AgPF_6$, for example, are only poorly soluble. In order to estimate the oxidizing ability of Ag^+ in non-aqueous solvents, reactions between $AgBF_4$ and $N(C_6H_4Br\text{-}p)_3$, NPh_3, and $[FeCp_2]$ were carried out in CH_2Cl_2, C_6H_5Me, MeCN, and THF.[95] The results are summarized in Table 8.2:

Table 8.2

Oxidation of substrates by $AgBF_4$ in non-aqueous solvents

Substrate/ Solvent	**CH_2Cl_2**	**C_6H_5Me**	**MeCN**	**THF**
$N(C_6H_4Br\text{-}p)_3$ (E_p = 1.16 *V*)	+	no oxid.	no oxid.	no oxid.
NPh_3 (E_p = 0.88 *V*)	+	no oxid.	no oxid.	no oxid.
$[FeCp_2]$ (E_p = 0.38 *V*)	+	+	+	+

In each of the solvents ferrocene is oxidized to ferrocenium ion. However, amines are oxidized to the corresponding $[NR_3]^+$ cations only when a suspension of $AgBF_4$ in dichloromethane is employed. Thus, $AgBF_4$ was found to be a more powerful oxidant in CH_2Cl_2 than it is in C_6H_5Me, MeCN, or THF, in agreement with the solvating abilities of these solvents.

Another important advantage of the application of silver salts as oxidants in synthesis is the ease of separation from the reaction mixture of the metallic silver formed on reduction. As a rule, reactions with Ag^+ as the oxidant are rapid, selective, and do not require an excess of the reagent. Among commercially available silver salts, $AgCF_3SO_3$ is one of the most useful. It is stable, relatively soluble in a number of organic solvents, and non-hygroscopic.

The ability of the Ag^+ ion to act as a one-electron oxidant for both metals and complexed metal ions is well known. For example, as early as 1923, Morgan[96] suggested that the complex, $[Cu(MeCN)_6][NO_3]$, could be prepared by oxidizing metallic copper with $AgNO_3$ in acetonitrile. The widespread use of silver salts as oxidants, however, began more recently and many examples are reported in the literature. Thus, the isonitrile complex of chromium, $[Cr(CNPh)_6]$, is rapidly oxidized in acetone by one equivalent of $AgPF_6$ with formation of $[Cr(CNPh)_6][PF_6]$.[97] When this complex is allowed to interact with a second equivalent of $AgPF_6$, or on interaction of the starting material directly with two equivalents of silver hexafluorophosphate, $[Cr(CNPh)_6][PF_6]_2$ is obtained. Other chromium(I) and chromium(II) complexes have been prepared analogously, *e.g.* the complexes, $[CrL_3][PF_6]$ and $[CrL_3][PF_6]_2$, where L is depicted in Figure 8.4, have been reported:[98]

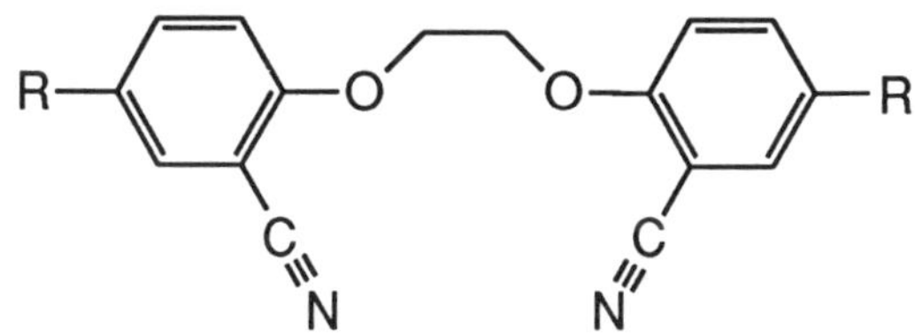

Figure 8.4

It has been reported[99] that $[Sm(THF)_2(\eta^5\text{-}Cp^*)_2]$ reacts readily with a slight excess of silver tetraphenylborate to form metallic silver and the one-electron oxidation product, Eq. 8.27:

$$[Sm(THF)_2(\eta^5\text{-}Cp^*)_2] + AgBPh_4 \rightarrow [Sm(THF)_2(\eta^5\text{-}Cp^*)_2][BPh_4] + Ag \qquad (8.27)$$

The samarium(III) complex was identified by spectroscopic and X-ray crystallographic methods.

Although many metal complexes are oxidized by loss of an electron formally associated with the metal center, chemistry based on redox processes involving ligands is also well-documented. For example, the one-electron oxidation of $[Fe(SPh)(CO)_2(\eta^5\text{-}Cp^*)]$ by silver hexafluorophosphate[100] yields the dimer, $[(\eta^5\text{-}Cp^*)(CO)_2Fe(\mu\text{-}PhSSPh)Fe(CO)_2(\eta^5\text{-}Cp^*)][PF_6]_2$, with a bridging diphenyl disulfide ligand, Scheme 8.12. In similar, but substituted, compounds, $[Fe(SPh)(CO)L(\eta^5\text{-}C_5R_5)]$, the metal center is oxidized, Eq. 8.28.

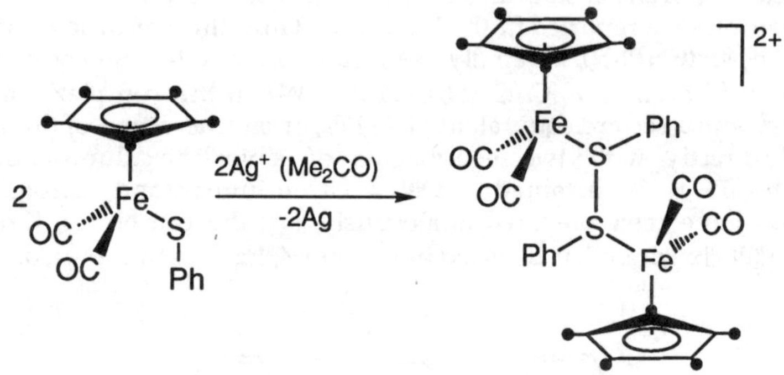

Scheme 8.12

$$[Fe(SPh)(CO)L(\eta^5\text{-}C_5R_5)] \ + \ AgX \ \rightarrow \ [Fe(SPh)(CO)L(\eta^5\text{-}C_5R_5)]X \ + \ Ag \qquad (8.28)$$

where R = Me, H; L = PMe_3, PMe_2Ph, $PMePh_2$

Generally, for oxidations, as demonstrated in the case of the complexes, $[Fe(SPh)(CO)L(\eta^5\text{-}C_5R_5)]$, it is assumed that electron loss occurs from the HOMO of the compound. However, it has been suggested[100] that the HOMO in $[Fe(SPh)(CO)_2(\eta^5\text{-}Cp^*)]$ is the sulfur lone pair and not a metal-centered orbital. The loss of one electron from this orbital leads to the formation of a radical species with the unpaired electron primarily localized on the ligand. The dimeric product is formed from coupling of two such species.

The reaction of $[Os(C_2Ph)_2(PMe_3)_4]$ and silver hexafluorophosphate in dichloromethane, where $[Os(\eta^3\text{-}PhC{\cdots}C{\cdots}C{\cdots}CPh)(PMe_3)_4][PF_6]$ was obtained, has been reported.[101] The product contains a new η^3-PhC⋯C⋯C⋯CPh ligand, formed by oxidative coupling of two acetylide (alkynyl, $-C_2Ph$) ligands.

Oxidation of one alkyl ligand in $[ZrR_2Cp_2]$ (R = Me, CH_2Ph), with silver tetraphenylborate in acetonitrile or tetrahydrofuran leads to the formation of the corresponding solvento-complexes,[102] Eq. 8.29:

$$[ZrR_2Cp_2] \ + \ AgBPh_4 \ + \ solv \ \rightarrow \ [Zr(solv)(R)Cp_2][BPh_4] \ + \ Ag \ + \ 1/2R{-}R \qquad (8.29)$$

Liberation of ethane during the reaction producing $[Zr(MeCN)(Me)Cp_2][BPh_4]$ was monitored by NMR. It was suggested that oxidation of the methyl ligand by Ag^+ proceeds through intermediate formation of a $AgCH_3$ compound.

The possibility of hydride ligand oxidation in the complexes,

$[Mo(H)_4L_4]$ (L = PMe_2Ph, $PMePh_2$), by silver tetrafluoroborate in acetonitrile has been demonstrated,[103] Eq. 8.30:

$$[Mo(H)_4L_4] + 2AgBF_4 + 2MeCN \rightarrow [Mo(H)_2(MeCN)_2L_4][BF_4]_2 + H_2 + 2Ag \quad (8.30)$$

The reactions were complete within several minutes and products were obtained in good yields.

Mercury salts, in addition to silver salts, can be used as oxidants in reactions performed in non-aqueous solvents. Like the Ag^+ ion, the Hg^{2+} ion is an effective oxidant. In both cases the reduction products, metallic silver and mercury, can be separated easily from reaction mixtures. The oxidation of lanthanides, such as Sm, Er, and Yb,[104] as well as Sc and Y,[105] by $HgCl_2$ in THF yields tetrahydrofuran complexes of the corresponding trichlorides, Eq. 8.31:

$$2Ln + 3HgCl_2 + 6THF \rightarrow 2LnCl_3{\cdot}3THF + 3Hg \quad (8.31)$$

Reaction of metallic lanthanum with $HgBr_2$ in THF results in the formation of $LaBr_3$ and cleavage of THF. The complex, $[La(H)Br(OCH{=}CH_2)]{\cdot}2THF$, is ultimately formed.[106]

The complex $[Cr(O_2CCF_3)(CO)_5]^-$ is formed upon treatment of $[Cr_2(CO)_{10}]^{2-}$ with $Hg(O_2CCF_3)_2$ as a result of oxidative substitution by the mercury salt,[107] Eq. 8.32:

$$[Cr_2(CO)_{10}]^{2-} + Hg(O_2CCF_3)_2 \rightarrow 2[Cr(O_2CCF_3)(CO)_5]^- + Hg \quad (8.32)$$

Many examples of the oxidation of organometallic compounds by silver and mercury salts have been described in the literature.[108]

Metal ions other than Ag^+ and Hg^{2+} serve as useful oxidants in preparative coordination chemistry. For example, it is well known that coordination of MeCN stabilizes copper(I) complexes, which are otherwise prone to disproportionation to copper(0) and copper(II). Accordingly, in acetonitrile, copper(II) ions are found to be oxidants capable of oxidizing, for example, tetrathiafulvalene[109] or 2,2'-bithiophene,[110] with concomitant formation of copper(I).

It has been shown[111,112] that the electrochemical oxidation of $[Ru_6C(CO)_{16}]^{2-}$ at a platinum electrode occurs at +0.5 *V* (*vs.* Ag/AgCl) and, on the same potential scale, the $[FeCp_2]^+/[FeCp_2]$ and $FeCl_3/FeCl_2$ couples occur at +0.53 *V*. This allows the mild chemical oxidation of $[Ru_6C(CO)_{16}]^{2-}$ by these reagents. It has been established that, on treatment of $[Ru_6C(CO)_{16}]^{2-}$

with two equivalents of either $[FeCp_2][BF_4]$ or $FeCl_3$ in the presence of alkynes, RC≡CR, the clusters, $[Ru_6C(CO)_{15}(\mu^3\text{-}\eta^2\text{-}RC\equiv CR)]$ (R = H, Me, Et, Ph), are formed. It is also known that oxidation of $[Os_5(CO)_{15}]^{2-}$ in THF with $FeCl_3$ in the presence of carbon monoxide yields $[Os_5(CO)_{16}]$ and $[Os_5Cl(CO)_{15}]^-$.[113] In the absence of CO, the reaction produces mainly $[Os_5Cl(CO)_{15}]^-$. It is believed that the initial step in the process is a two-electron oxidation of $[Os_5(CO)_{15}]^{2-}$ by $FeCl_3$ giving the coordinatively unsaturated intermediate "$[Os_5(CO)_{15}]$" which then reacts with CO and/or Cl^-. The chloride ions may be generated *in situ* by reduction of $FeCl_3$ to $FeCl_2$.

8.6 Non-metal halides

Coordination chemistry frequently exploits preparative methods and reagents developed originally for organic synthesis. For example, phosphorus and sulfur halides — typical inorganic compounds used commonly in organic chemistry — are now included in the group of reagents used for the synthesis of inorganic and coordination compounds. In this section, the use of non-metal halides as oxidants for coordination compounds, along with their ability to deoxygenate (*i.e.* reduce) coordinated ligands, is discussed.

8.6.1 Phosphorus halides

By the latter part of the 19th century, Schutzenberger and Fontaine[114] had already noted that PCl_5 was a strong oxidant. They demonstrated that phosphorus pentachloride interacts with platinum black to give a bridged homobinuclear platinum(II) complex, Eq. 8.33:

$$Pt + PCl_5 \rightarrow 1/2[Pt_2(\mu\text{-}Cl)_2Cl_2(PCl_3)_2] \qquad (8.33)$$

This synthesis of $[Pt_2(\mu\text{-}Cl)_2Cl_2(PCl_3)_2]$ was reproduced, confirmed, and described in detail by Arbuzov and Zoroastrova.[115] It has also been shown that phosphorus pentabromide, PBr_5, similarly reacts with platinum black to form $[Pt_2(\mu\text{-}Br)_2Br_2(PBr_3)_2]$.[116]

Phosphorus pentachloride not only reacts with metallic platinum but also oxidizes platinum(II) complexes to the platinum(IV) analogs.[117] Indeed, PCl_5 will oxidatively chlorinate platinum(II) complexes so effectively that it compares favorably with molecular chlorine, the reagent most commonly used for this purpose. In general, PCl_5 is most useful when (*i*) the coordinated ligands in a complex exhibit fairly strong reducing properties and the routine oxidation with chlorine proceeds without selectivity, *i.e.* when oxidation of both the metal center and the coordinated ligands takes place upon treatment with chlorine; (*ii*) precise amounts of a chlorinating agent are required; and (*iii*) the starting materials have low solubilities and high temperatures are normally necessary to achieve reasonable rates.

Oxidative chlorination of anionic platinum(II) complexes with PCl_5 has been described,[117] Eqs. 8.34 and 8.35:

$$[Ph_3PCH_2Ph][PtCl_3(py)] + PCl_5 \rightarrow [Ph_3PCH_2Ph][PtCl_5(py)] + PCl_3 \quad (8.34)$$

$$[Ph_3PCH_2Ph]_2[PtCl_4] + PCl_5 \rightarrow [Ph_3PCH_2Ph]_2[PtCl_6] + PCl_3 \quad (8.35)$$

Similar reactions of neutral complexes of platinum(II) are known, Eq. 8.36:

$$[PtCl_2L_2] + PCl_5 \xrightarrow{\text{solv}} [PtCl_4L_2] + PCl_3 \quad (8.36)$$

where L = Me_2S, py; solv = MeCN or $MeNO_2$

The oxidation of cationic platinum(II) complexes also can be accomplished with phosphorus pentachloride, Eq. 8.37:

$$[Pt(py)_4]Cl_2{\cdot}3H_2O + 4PCl_5 \xrightarrow{\text{MeCN}} [PtCl_4(py)_2] + PCl_3 + 2py{\cdot}HCl + 3OPCl_3 + 4HCl \quad (8.37)$$

Chlorination is accompanied by elimination of two pyridine ligands.

One of the advantages of PCl_5, as compared to chlorine, is that PCl_5 is a milder oxidant. For example, the coordinated hydroxylamine ligands in $[Pt(NH_2OH)_4][OH]_2$ exhibit reducing properties. On chlorination of this compound with molecular chlorine in water, both the metal center and the coordinated ligands are subject to oxidation and a mixture of products is formed. The reactivity with chlorine can be inhibited by use of non-aqueous solvents (in particular, $CHCl_3$ or CCl_4[1]) but the high reactivity of Cl_2 nonetheless results in low yield and impurities in the product. The use of PCl_5 in nitromethane for this reaction yields $[PtCl_4(NH_2OH)_2]$ with high purity and yield.[118] The elimination of two hydroxylamine ligands and replacement by chloride is similar to the process observed in the reaction of $[Pt(py)_4]Cl_2{\cdot}3H_2O$ with PCl_5 (*vide supra*). Selectivity in the oxidation of platinum(II) complexes can be improved by the use of oxidants that are milder than phosphorus pentachloride. Derivatives of $SbCl_5$ — the adducts $PhC(O)Cl{\cdot}SbCl_5$ and $[Et_3NCH_2Ph][SbCl_6]$ — are useful for this purpose.[119] These adducts oxidize platinum(II) amine complexes having low redox potentials but are unable to oxidize, for example, *cis*- and *trans*-$[PtCl_2(Me_2S)_2]$, which are characterized by a relatively high redox potential.

Phosphorus pentachloride is known to be an effective reagent for the deoxygenation of selected substrates due primarily to the high strength of the P=O bond in phosphoranes.[120] Thus, PCl_5 is able not only to oxidize platinum(II) complexes but is able also to react with oxygen-containing ligands. In particular, phosphorus pentachloride readily reacts with isomeric complexes, *cis*- and *trans*-$[PtCl_2(NH_2C_2H_4OH)_2]$, with formation of *cis*- and *trans*-$[PtCl_4(NH_2C_2H_4Cl)_2]$, respectively.[121] In this reaction, PCl_5 functions as an oxidant and as a reagent for OH^-/Cl^- substitution. PCl_5 also transforms *cis*- and *trans*-$[PtCl_4(NH_2C_2H_4OH)_2]$ into the corresponding 2-chloroethylamine derivatives,[121] Scheme 8.13:

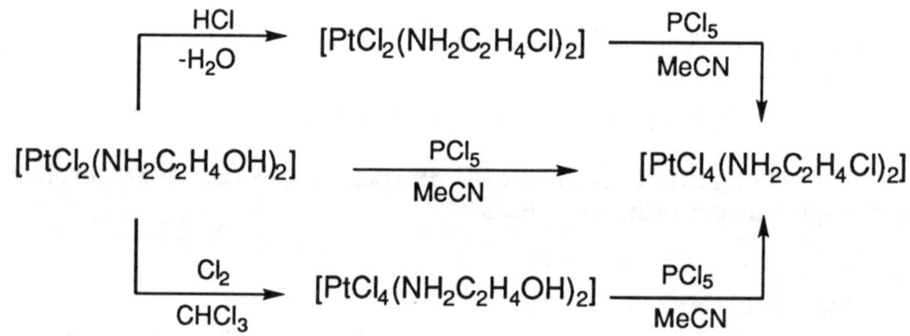

Scheme 8.13

The reaction of PCl_5 with $[PtCl_2(NH_2CH_2CO_2H)_2]$ in chloroform proceeds in a similar manner. Thus, mixing PCl_5 with $[PtCl_2(NH_2CH_2CO_2H)_2]$ results in oxidation of platinum(II) to platinum(IV) and substitution of OH^- by Cl^-. The product, $[PtCl_4(NH_2CH_2COCl)_2]$, was isolated as a solid in high yield,[122] Scheme 8.14.

It has been shown[123] that PCl_5 deoxygenates metal nitro complexes. For example, $K_2[Pt(NO_2)_4]$, $[Pt(NO_2)_2(en)]$, and the isomeric complexes, *cis*- and *trans*-$[Pt(NO_2)_2(NH_3)_2]$, react with PCl_5 in acetonitrile, or PCl_5 in a melt, to form platinum(IV) chloride derivatives. The reaction of $K_2[Pt(NO_2)_4]$ with PCl_5 illustrates the two pathways by which the reaction occurs. The first is the oxidative chlorination of platinum(II) (ligands are omitted for clarity), Eq. 8.38. The second is deoxygenation of the nitro groups, Eq. 8.39. The replacement of NO_2^- by Cl^- accompanies deoxygenation.

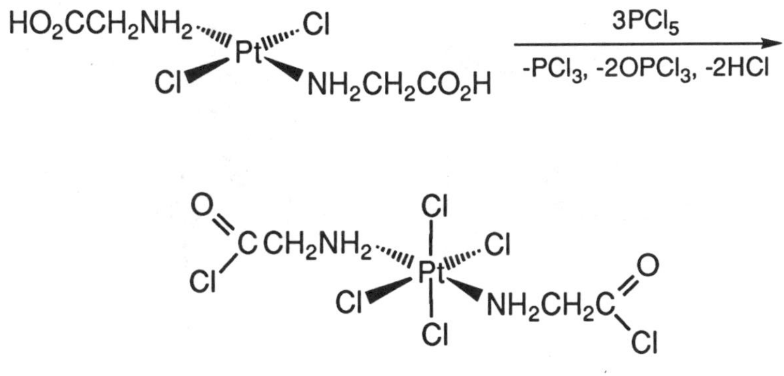

Scheme 8.14

$$[Pt(II)] + PCl_5 \rightarrow [Cl\text{–}Pt(IV)\text{–}Cl] + PCl_3 \quad (8.38)$$

$$NO_2^- + PCl_5 \rightarrow Cl^- + NOCl + OPCl_3 \quad (8.39)$$

The reaction between PCl_5 and platinum(II) complexes containing coordinated dimethyl sulfoxide is complicated. It has been shown[124] that interaction of $K[PtCl_3(S\text{-}Me_2S{=}O)]$ or *cis*-$[PtCl_2(S\text{-}Me_2S{=}O)_2]$ with phosphorus pentachloride in acetonitrile, followed by treatment of the reaction mixture with 2-propanol, removal of solvent *in vacuo*, and addition of $[Et_4N]Cl$ to an aqueous solution of the residue, leads, in both cases, to isolation of the platinum(IV) dimethyl sulfide complex, $[Et_4N][PtCl_5(Me_2S)]$, as solid. The structure of $[Et_4N][PtCl_5(Me_2S)]$ has been confirmed by X-ray crystallography,[124] Figure 8.5. When water is used instead of 2-propanol, only the products of oxidative chlorination, $[Et_4N][PtCl_5(S\text{-}Me_2S{=}O)]$ (confirmed by X-ray crystallography,[124] see Figure 8.5), and $[PtCl_4(S\text{-}Me_2S{=}O)_2]$, are formed. Hence, the interaction of platinum(II) dimethyl sulfoxide complexes with PCl_5, followed by addition of 2-propanol, results in deoxygenation (reduction) of the *S*-coordinated Me_2SO and oxidation of platinum(II) to platinum(IV). Presumably the 2-propanol is oxidized to acetone.

Phosphorus pentachloride also has been used as an oxidant for molybdenum and tungsten carbonyl complexes.[125] For example, the reaction of $[W(CO)_3(Me)Cp^*]$ with PCl_5 in dichloromethane[125] leads to formation of the dimer, $[Cp^*WCl_4]_2$, in a yield of 90%. Organophosphorus halides, R_3PX_2, have been used as mild oxidants in inorganic syntheses. The reagents are synthesized by the direct reactions of tertiary phosphines with halogens.

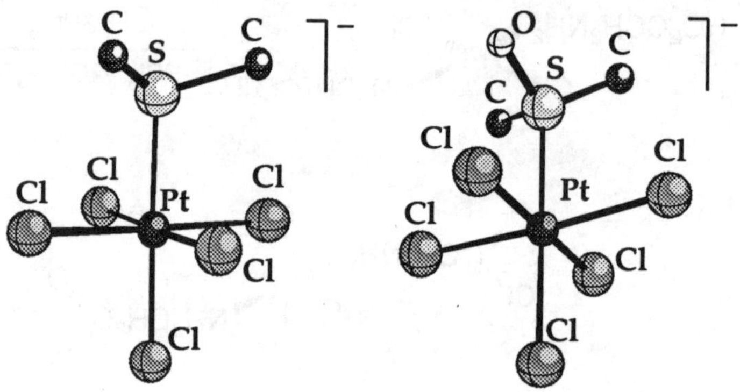

Figure 8.5

For example, PPh_3 is oxidized by I_2 or Br_2 in dry diethyl ether to produce Ph_3PX_2 (X = I, Br). Although $^{31}P\{^1H\}$ NMR data for Ph_3PI_2 indicate an ionic structure, $[Ph_3PI]I$, in $CDCl_3$ solution, X-ray crystallography shows that both of these materials (X = Br, I) are molecular in the solid state, with structures of the type $Ph_3P–X–X$.[126] The molecular structure of Ph_3PI_2 is depicted below, Figure 8.6:

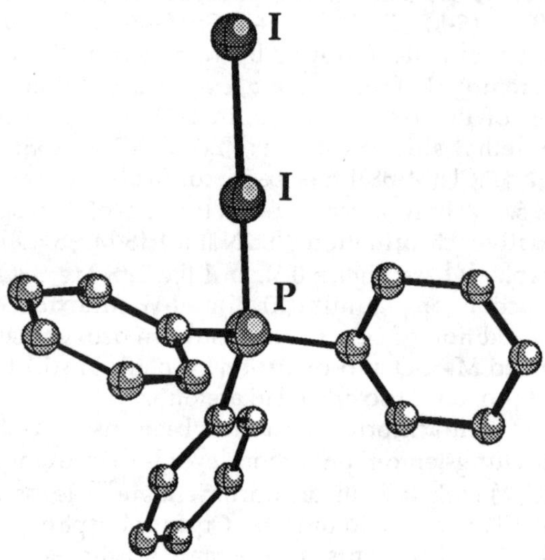

Figure 8.6

The utility of R_3PX_2 compounds as oxidants has been demonstrated in the preparation of zirconium fulvalene complexes.[127] The oxidation of $[Zr(PMe_3)(\eta^1{:}\eta^5\text{-}C_5H_4)Cp]_2$ to the zirconium(III) fulvalene derivative cannot be carried out by treatment with Cl_2 or Br_2. Interaction with these reagents leads to $[ZrX_3Cp]$ as a product. With R_3PX_2 (R = *n*-Bu, Me; X = Cl, Br) the reaction proceeds *via* initial formation of the unstable intermediate, $[ZrX(PMe_3)(\eta^1{:}\eta^5\text{-}C_5H_4)Cp]_2$, followed by thermal conversion to the fulvalene complex, Scheme 8.15:

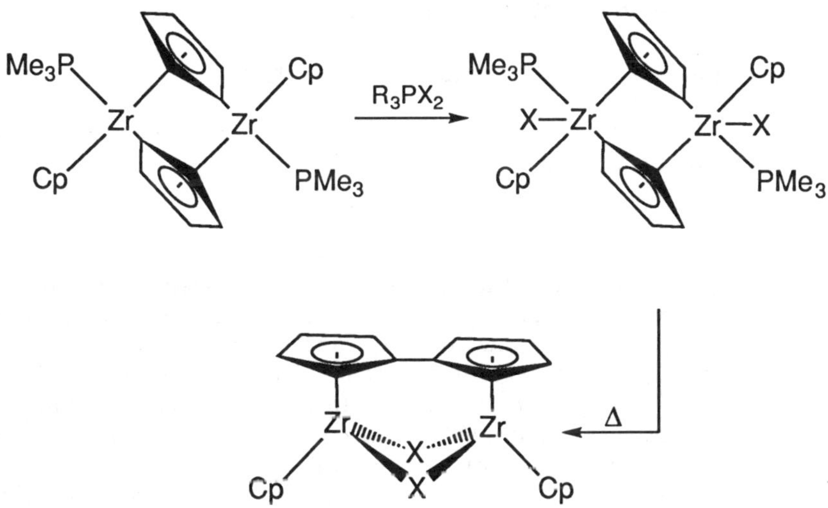

Scheme 8.15

Unusual reactions have been reported between coarse-grained, unactivated metal powders[128] and R_3PX_2 that produce metal phosphine complexes. For example, nickel powder reacts with two equivalents of Me_3PI_2 to give the nickel(III) complex, $[NiI_3(PMe_3)_2]$, Eq. 8.40:

$$Ni + 2Me_3PI_2 \rightarrow [NiI_3(PMe_3)_2] + 1/2I_2 \tag{8.40}$$

The composition and structure of the product appear to be dependent on the nature of both the R and X substituents[128] in R_3PX_2. Thus, for example, treatment of cobalt powder with Me_3PI_2 in dry diethyl ether results in the formation of $[CoI_3(PMe_3)_2]$. In contrast, reaction of metallic cobalt with

$(n\text{-Bu})_3PI_2$ yields $[\{(n\text{-Bu})_3PI\}_2(\mu\text{-I})][CoI_3(\mu\text{-I})\{(n\text{-Bu}_3)PI\}]$. Both complexes have been structurally characterized.

8.6.2 $SOCl_2$ and SO_2Cl_2 as oxidants

Thionyl chloride, $SOCl_2$, used primarily in organic chemistry, has no established history as a reagent in inorganic and coordination chemistry, except in reactions with Grignard reagents, and published reviews[129] contain few references to the use of thionyl chloride in synthesis. However, thionyl chloride has properties that make it a versatile oxidant and it can serve as a chlorinating agent in the following manner, Eq. 8.41:

$$[\text{substrate}] + SOCl_2 \rightarrow [\text{substrate}Cl_2] + [SO] \qquad (8.41)$$

The sulfur(II) oxide liberated is extremely unstable and decomposes[129,130] by disproportionation, Eq. 8.42:

$$2[SO] \rightarrow S + SO_2 \qquad (8.42)$$

$SOCl_2$ acts as an oxidant toward metals. Thus, moisture-free, boiling thionyl chloride reacts slowly with sodium, magnesium, mercury, aluminum, tin, bismuth, and iron to convert these metals into the corresponding metal chlorides; sulfur and sulfur(IV) oxide are by-products of these reactions.[130] The preparation of $[PdCl_2(py)_2]$ from metallic palladium is described in the patent literature.[131] The complex is formed by the reaction of palladium black with a mixture of pyridine and thionyl chloride in *o*-dichlorobenzene. The synthesis is carried out in an autoclave under oxygen pressure at 150 °C and $[PdCl_2(py)_2]$ is formed in 50% yield. Although this process may seem unduly complicated, it serves as an illustration of the potential applicability of $SOCl_2$ in the synthesis of coordination compounds.

Thionyl chloride is able to oxidize the metal center in a number of complexes. For example, the reaction of iron pentacarbonyl, $[Fe(CO)_5]$, with $SOCl_2$ is extremely vigorous and, in addition to $FeCl_2$ and CO, yields $[FeCl_2(CO)_3]$, CO_2, S, and SO_2 as products.[132] The palladium(0) complex, $[Pd(PPh_3)_4]$, reacts with an excess of $SOCl_2$ to form the palladium(II) complex, $[PdCl_2(PPh_3)_2]$. Triphenylphosphine, liberated in the synthesis, is oxidized by thionyl chloride with the formation of both Ph_3PO and Ph_3PS.[133]

The oxidative chlorination of the cerium(III) compound, $CeCl_3(H_2O)_7$, in $SOCl_2$ in the presence of tetraethylammonium chloride has been investigated.[134] At room temperature slow dehydration occurs. Upon boiling, however, the yellow thionyl chloride solution gradually becomes orange and then red with the simultaneous disappearance of the starting material and formation of the cerium(IV) salt, $[Et_4N]_2[CeCl_6]$. It was noted that bubbling chlorine gas through a solution of cerium(III) chloride and tetra(n-butyl)ammonium chloride in thionyl chloride at 20 °C does not yield an oxidation product in a reasonable time period. However, refluxing a thionyl chloride solution of the same reagents for 80 minutes gives rise to quantitative oxidation, producing $[CeCl_6]^{2-}$.

The uranium complex, $[(n\text{-Bu})_4N][UCl_6]$, is obtained by oxidation of $[(n\text{-Bu})_4N]_2[UCl_6]$ with $SOCl_2$,[135] Eq. 8.43:

$$4[(n\text{-Bu})_4N]_2[UCl_6] + 2SOCl_2 \rightarrow 4[(n\text{-Bu})_4N][UCl_6] + 4[(n\text{-Bu})_4N]Cl + SO_2 + S \quad (8.43)$$

The starting material is dissolved in thionyl chloride and stirred at room temperature. Oxidation to $[(n\text{-Bu})_4N][UCl_6]$ is complete within 2-3 hours. In addition to the uranium(V) compound, SO_2 and S were identified as products. No H_2S, SO_3, HCl, or Cl_2 were found.

Schmidbaur and Jandik[136] have reported the oxidative chlorination of a gold(I) dimethylphosphonium-*bis*-methylide dimer. The reaction, in toluene, yields the tetrachlorodigold(III) derivative, Scheme 8.16:

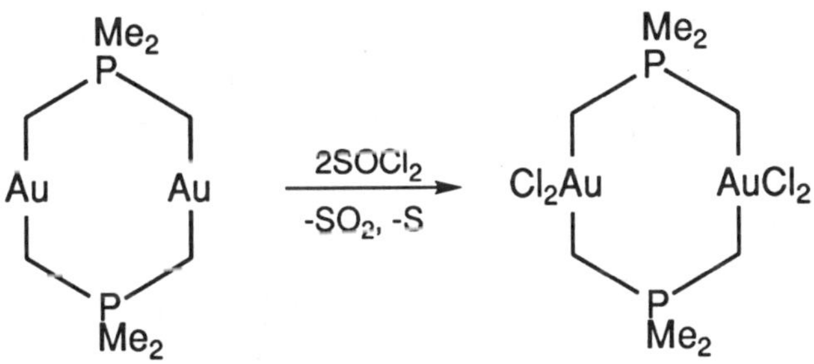

Scheme 8.16

The sulfur-containing by-products, sulfur and SO_2, were not identified and their formation in this reaction is unconfirmed.

The vanadium(I) complex, $[V(CO)_4Cp]$, when treated with thionyl chloride at -25 °C in CCl_4, is oxidized to $[VCl_3Cp]$ in 65% yield.[137] The same method has been applied to the synthesis of $[VCl_3Cp^*]$, but the yield is very low.[137] The reaction of $[Pb_2Et_6]$ with $SOCl_2$ produces $[PbClEt_3]$ in 86% yield.[138] The oxidation of $[W(\mu\text{-H})(CO)_2(\eta^5\text{-}C_5R_5)]_2$ (R = H, Me) by thionyl chloride in toluene at -78 °C yields $[WCl_2(CO)_2(\eta^5\text{-}C_5R_5)]_2$, which contains a tungsten-tungsten bond,[139] Scheme 8.17:

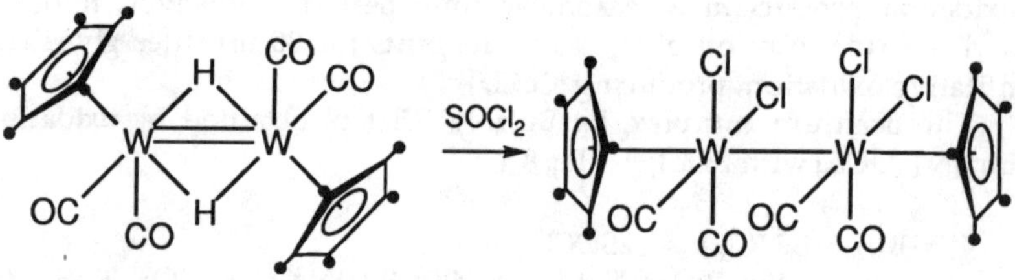

Scheme 8.17

Thionyl chloride can not only oxidize metal ions in complexes, it can also oxidize coordinated ligands. Thus, the phthalocyaninatocobalt(II) complex, $[Pc^{2-}Co^{II}Cl_2]$, is oxidized by thionyl chloride in two steps.[140] In the first step, the metal ion is oxidized to give the cobalt(III) derivative, $[Pc^{2-}Co^{III}Cl_2]$, and, in the second step, the complex, $[Pc^{1-}Co^{III}Cl_2]$, is formed with a radical ligand.

Chelated α-amino acids in *bis*(1,2-ethanediamine)cobalt(III) complexes may be oxidized to the α-imino acid derivatives with thionyl chloride in DMF accompanied by formation of elemental sulfur,[141] Scheme 8.18:

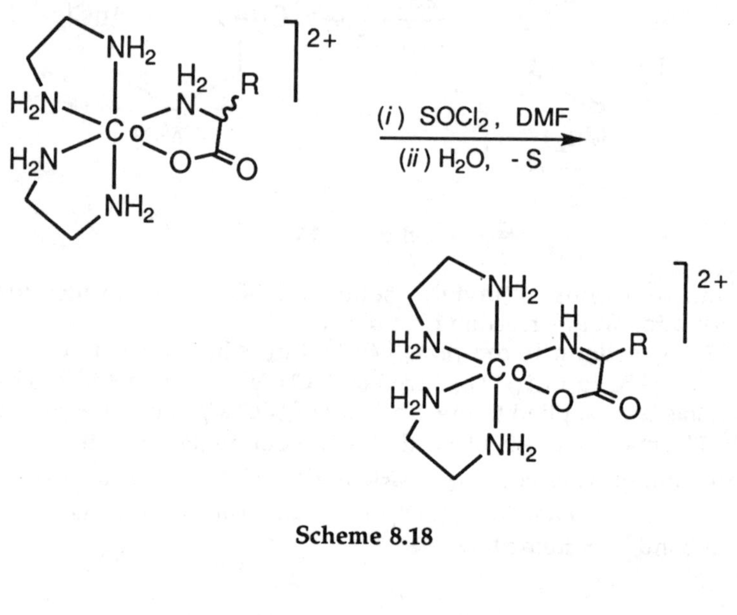

Scheme 8.18

Pendant functional groups in these complexes are modified by $SOCl_2$ in the normal fashion but such modification may result in further intramolecular reactions. The glycinate- and *N*-methylglycinate cobalt(III) complexes, $[Co(N(R)HCH_2COO)(en)_2]^{2+}$, do not produce the imines in reactions with thionyl chloride but, instead, yield the 2-thiooxamato complexes, $[Co(N(R)C(S)COO)(en)_2]^{+}$, and sulfate. However, if the glycinato complex, dimethylformamide, and $SOCl_2$ are premixed, the *N*-formyloxamato complex, $[Co(N(CHO)C(O)COO)(en)_2]^{+}$, is the major product (formed with liberation of elemental sulfur) demonstrating the ability of this mixture of reagents[141] to promote both formylation and oxidation reactions in this case, Scheme 8.19:

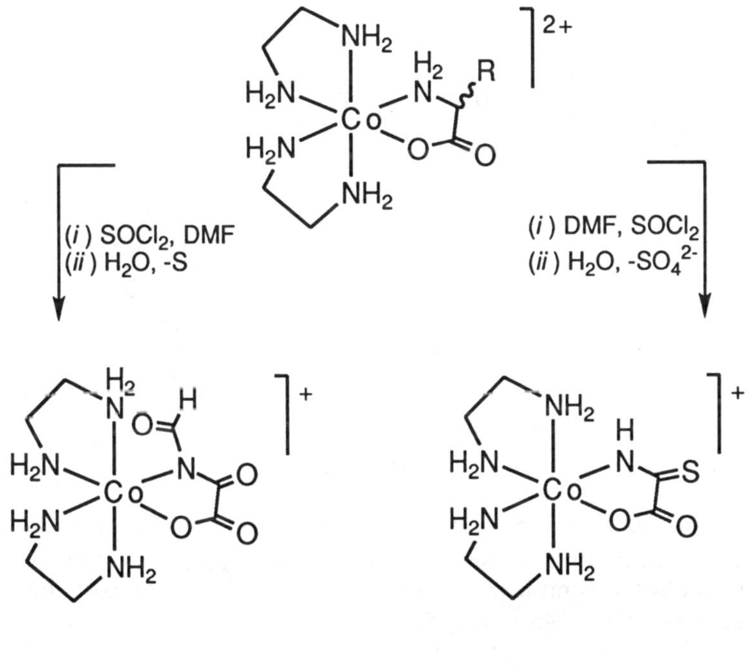

Scheme 8.19

Further, thionyl chloride can take part in oxidative addition reactions. For example, the reaction of $SOCl_2$ with Vaska's compound in benzene produces the iridium(III) complex, $[IrCl_2(SOCl)(CO)(PPh_3)_2]$,[133,142] Eq. 8.44:

$$[IrCl(CO)(PPh_3)_2] + SOCl_2 \rightarrow [IrCl_2(SOCl)(CO)(PPh_3)_2] \quad (8.44)$$

The reaction of the cluster, $[(t\text{-Bu})NH_3][Fe_3(CO)_9(\mu_3\text{-SBu})]$, with thionyl chloride takes place by two pathways,[143] Scheme 8.20:

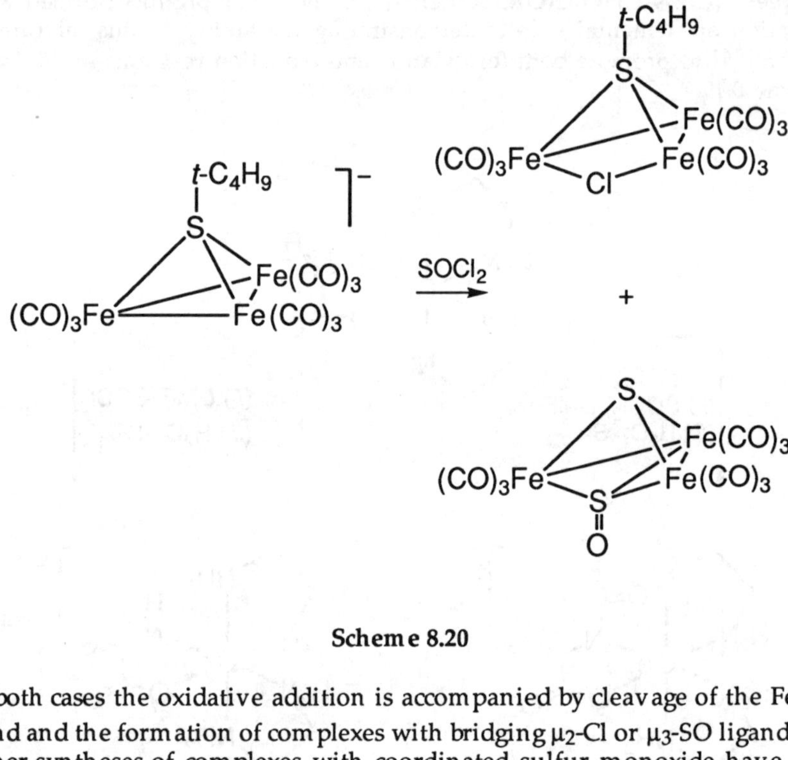

Scheme 8.20

In both cases the oxidative addition is accompanied by cleavage of the Fe–Fe bond and the formation of complexes with bridging μ_2-Cl or μ_3-SO ligands.[143] Other syntheses of complexes with coordinated sulfur monoxide have also been described,[144] Eq. 8.45 and Eq. 8.46:

$$[FeH_2(CO)_2\{P(OPh)_3\}_2] + SOCl_2 + 2NEt_3 \rightarrow [FeSO(CO)_2\{P(OPh)_3\}_2] + 2[Et_3NH]Cl \quad (8.45)$$

$$2[MnH(CO)_4(PPh_3)] + SOCl_2 + 2NEt_3 \rightarrow [Mn(CO)_4(PPh_3)]_2(\mu\text{-SO}) + 2[Et_3NH]Cl \quad (8.46)$$

Although these reactions can be classified as metathetical exchange processes, they are also formal redox reactions.[145]

Oxidations involving sulfuryl chloride, SO_2Cl_2, are not common and information is available on only a few examples of such reactions. SO_2Cl_2 is a fairly mild chlorinating agent.[146] For example, the reaction of sulfuryl chloride with the iron complex, $[Fe_2(CO)_4(\mu\text{-dienyl})_2]$ (dienyl = Cp, MeC_5H_4, C_9H_7), is instantaneous and yields a mixture of $[FeCl(CO)_2(\mu\text{-dienyl})]$ and $[Fe(CO)_3(\mu\text{-dienyl})][FeCl_4]$. Overall, the process is a two-electron oxidative cleavage of the Fe–Fe bond in the starting material, $[Fe_2(CO)_4(\mu\text{-dienyl})_2]$, that gives mononuclear complexes as products. In a related reaction, $[M_2(CO)_{10}]$ (M = Mn, Re) gives $[MCl(CO)_5]$ in high yield as the sole product upon treatment with sulfuryl chloride. These results have been confirmed by Davis, Durrant, and Rowland.[147] It has been established that treatment of $[M(PPh_3)_4]$ (M = Pt, Pd) with SO_2Cl_2 yields $[MCl_2(PPh_3)_2]$.[147] Sulfuryl chloride has been used for the oxidation of antimony(III) compounds, $MeSb(OR)_2$ (R = Me, Et).[148] The reactions proceed in dichloromethane at -60 °C and yield dimeric antimony(V) alkoxy-bridged compounds of the type $[MeSbCl_2(OR)(\mu\text{-OR})]_2$.

Thionyl chloride and sulfuryl chloride are useful oxidants but the number of reactions studied is small and so the scope and limitations are unknown.

8.6.3 Iodobenzene dichloride

Iodobenzene dichloride, $PhICl_2$, is a very useful reagent for oxidative addition of chlorine to metal complexes. Lucas and Kennedy[149a] have shown that $PhICl_2$ is obtained *via* molecular chlorine oxidation of iodobenzene in dry chloroform. The solid product is relatively stable if it is kept cold and protected from moisture. X-ray crystal structures of $PhICl_2$[149b] and its analog, $(2,4,6\text{-}i\text{-}Pr_3C_6H_2)ICl_2$,[149c] have been reported.

The reactivity of iodobenzene dichloride as an oxidant parallels that of Cl_2. But, unlike molecular chlorine, $PhICl_2$ is a solid and may be weighed easily for use in stoichiometric reactions. Cotton, Price, and Vidyasagar[150] have called $PhICl_2$ *a solid chlorine equivalent*. Iodobenzene, a by-product of oxidations with $PhICl_2$, is a poor ligand toward most complexes, although carbon-iodine oxidative addition reactions of this substrate are known. It is a liquid that is soluble in common organic solvents and usually it can be separated easily from the reaction product.

Representative examples of the use of iodobenzene dichloride in oxidative chlorination reactions are compiled in Table 8.3.

Table 8.3

Oxidative chlorination by iodobenzene dichloride

Starting material	Product(s)	Solvent
$[Re_2Cl_4(PMe_3)_4]$[150]	$[Re_2Cl_5(PMe_3)_3]$	CH_2Cl_2
$[Rh_2Cl_2(CO)_2(dppm)_2]$[151]	$[Rh_2Cl_4(CO)_2(dppm)_2]$ $[Rh_2Cl_5(dppm)_2]$ $[Rh_2Cl_6(dppm)_2]$	CH_2Cl_2
$[M(CO)_6]^-$ [152] M = Nb, Ta	$[M_2Cl_3(CO)_8]^-$	
$[Au(o\text{-}C_6H_4NO_2)(PPh_3)]$[153]	$[Au(o\text{-}C_6H_4NO_2)Cl_2(PPh_3)]$	CH_2Cl_2
$[Au_2Pt\{CH_2P(S)Ph_2\}_4]$[154]	$[Au_2PtCl_2\{CH_2P(S)Ph_2\}_4]$	CH_2Cl_2
$HgL_2(AuCl)_2$[155] L = $CH_2P(S)Ph_2$	$HgAuL_2AuCl_4$	CH_2Cl_2
$[M(CNR)_3(CO)_3]$ + CNR[156] M = Mo, W; R = alkyl, aryl	$[MCl(CNR)_6]^+$	EtOH

8.6.4 *Other non-metal halides*

Among other non-metal halides, the higher halides of the Group V elements are useful reagents for oxidative halogenation. Thus, reaction of the molybdenum complex, $[Mo(\mu\text{-}C_8Me_8)Cp_2]$, with the mild oxidants, $[Pr_4N][SbCl_6]$ or $[PhCO][SbCl_6]$, in dichloromethane leads to the formation of $[Mo(\mu\text{-}C_8Me_8)(\mu\text{-}Cl)Cp_2][SbCl_4]$ in quantitative yield.[157] The authors[157] suggest that the product is obtained as a result of oxidation followed by halogenation of the cation so formed.

In order to obtain complexes with a coordinated SO_2 ligand, AsF_5 in sulfur dioxide is widely used.[158] Thus, the reaction of metallic zinc or cadmium with an excess of arsenic pentafluoride in liquid SO_2 leads to the formation of homoleptic SO_2-complexes, Eq. 8.47:

$$M + 3AsF_5 \xrightarrow{SO_2} [M(SO_2)_2][AsF_6]_2 + AsF_3 \qquad (8.47)$$

The complexes, $[M(AsF_6)_2(SO_2)_2]_n$, have been obtained by the interaction of the metal, M (M = Mn, Fe, Co, Ni, Cu, Zn, or Mg), and AsF_5 in a 1:5 molar ratio in liquid SO_2.[158]

Reaction of $[NbCl_2Cp_2]$ with SbF_5 in liquid SO_2 yields the oxidation product, $[NbCl_2Cp_2][SbF_6]$, and SbF_3.[159] In the same report, oxidation of $[VCl_2Cp_2]$ by AsF_5 with the formation of $[VCl_2Cp_2][AsF_6]$ was described.

Oxalyl chloride and oxalyl bromide are also non-traditional oxidants that have been usefully employed in inorganic and coordination chemistry. It has been established[160] that $(COCl)_2$, in the gas phase, can oxidize many transition metals including the platinum group metals ruthenium, rhodium, iridium and platinum. In particular, co-condensation of rhodium or platinum with oxalyl chloride leads to formation of $[Rh(\mu\text{-}Cl)(CO)_2]_2$ and *cis*-$[PtCl_2(CO)_2]$, respectively.

A novel synthesis of metal-carbyne complexes that involves the treatment of metal-acyl derivatives with $(COX)_2$ in dichloromethane has been described.[161] Deoxygenation of $[Me_4N][M(COR)(CO)_5]$ (M = Cr, Mo, W) with oxalyl halides results in generation of $[X(CO)_4M(CR)]$ (R = Me, Ph) in good yield,[161] Scheme 8.21:

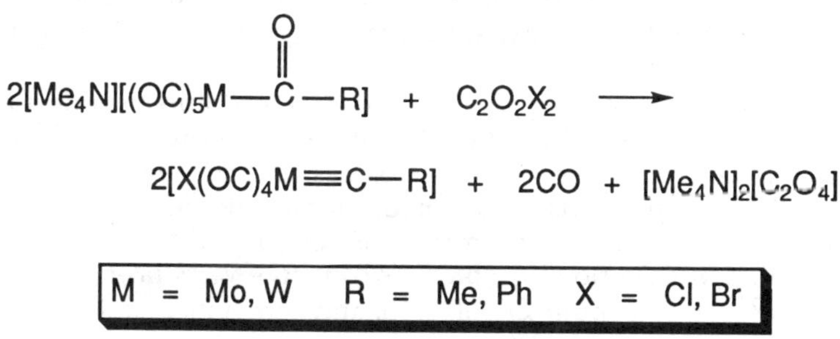

Scheme 8.21

In this reaction phosgene can be used as an oxidant instead of oxalyl chloride.

Organic 1,2-dihalides can oxidize certain metals. For example, the slow addition of samarium or ytterbium to a solution of 1,2-diiodoethane in THF leads to the formation of SmI_2 and YbI_2 in quantitative yields,[162] Eq. 8.48:

$$M + ICH_2CH_2I \rightarrow MI_2 + CH_2{=}CH_2 \qquad (8.48)$$

The reaction is rapid at room temperature. Karraker[163] has reported the interaction of metallic plutonium with a solution of diiodoethane in THF.

The adduct $PuI_3(C_2H_4)\cdot 4THF$ was isolated and identified.

Organic dihalides also can oxidize metal centers in complexes. For example, on treatment of Vaska's compound or its rhodium analog, $[MCl(CO)(PPh_3)_2]$ (M = Ir, Rh), with 1,1,2,2-tetrachloroethane, oxidation occurs,[164] Eq. 8.49:

$$[MCl(CO)(PPh_3)_2] + Cl_2CH{-}CHCl_2 \rightarrow [MCl_3(CO)(PPh_3)_2] + cis\text{-}ClCH{=}CHCl \quad (8.49)$$

8.7 Aminoxides and other oxygen donors

Carbon dioxide is a poorer ligand than carbon monoxide. Accordingly, one strategy for the elimination of CO from carbonyl complexes is oxidation of a carbonyl ligand to CO_2 followed by replacement by a stronger electron donor, often a solvent molecule. Oxygen donors are used widely for the transformation of coordinated CO into CO_2. Trimethylamine oxide, Me_3NO (abbreviated as TMAO), and trimethylamine oxide dihydrate, $Me_3NO{\cdot}2H_2O$, are often used for this purpose[165] as are other compounds that contain N–O or N=O bonds, (nitroarenes, $ArNO_2$,[166] nitrosoarenes, ArNO,[167] oximes, RR*N=OH,[168] [PPN][NO_2],[169] and N_2O[170]), I=O bonds (iodosobenzene, PhI=O[171]), S=O bonds (dimethyl sulfoxide, $Me_2S{=}O$[172a]) or Te=O bonds ($(p\text{-}MeOC_6H_4)_2Te{=}O$[172b]).

In 1959 Hieber and Lipp[173] showed that pyridine-*N*-oxide interacts with iron pentacarbonyl, $[Fe(CO)_5]$, at high temperature through a redox reaction that produces pyridine and iron oxides. In 1965, Kmeicik[174] reported the reduction of nitroaryls with carbon monoxide in the presence of $[Fe(CO)_5]$. Aromatic azo- and amino-compounds were isolated and the transformation of CO into CO_2 noted. About five years later the ability of nitriloxides[175] and a number of aliphatic, aromatic, and heterocyclic compounds with N–O bonds[176] to oxygenate carbonyl ligands in cobalt and iron complexes was reported. It was suggested[176] that the oxidation of carbonyl ligands proceeds *via* nucleophilic attack by the oxygen atom donor on the carbonyl carbon atom which, as a result of coordination, has electrophilic character.

The reaction of TMAO with metal carbonyls has been utilized in preparative coordination chemistry only since the mid-1970's, after the publication of work by Shvo and Hazum.[177] It was suggested that the reactions of amine oxides with substituted metal carbonyls could be used to decompose such complexes and isolate the free organic ligands. Thus, when the complexes, $[Fe(CO)_3L]$ (L = C_8H_8, norbornadiene[177]), in benzene, react with an excess of Me_3NO or $Me_3NO{\cdot}2H_2O$, complete decarbonylation of the complexes takes place along with decomposition. The free ligands, L, were isolated in yields of 75-95%. Oxidative decarbonylation has been shown to be of significant utility in synthesis. Thus, [Fe(CO)$_5$], in the presence of diene ligands and TMAO, is rapidly converted under mild conditions to diene tricarbonyl complexes, Eq. 8.50:

$$[Fe(CO)_5] + \text{diene} + Me_3NO \rightarrow [Fe(CO)_3(\text{diene})] + Me_3N + CO_2 + CO \quad (8.50)$$

The mechanism of oxygenation of carbonyl ligands by TMAO has been described.[176,178] Hogeveen, Elzinga, Eekhof, and Kellogg have suggested a possible mechanism for oxidative decarbonylation.[178] On interaction of [Fe(CO)$_5$] with a two-fold excess of TMAO in tetrahydrofuran at -30 °C, the evolution of CO_2 takes place and the complex, [Fe(CO)$_4$(NMe$_3$)], is isolated in a yield of 45%. Based on this observation a mechanism involving nucleophilic attack by TMAO on the carbonyl carbon atom was suggested, Scheme 8.22:

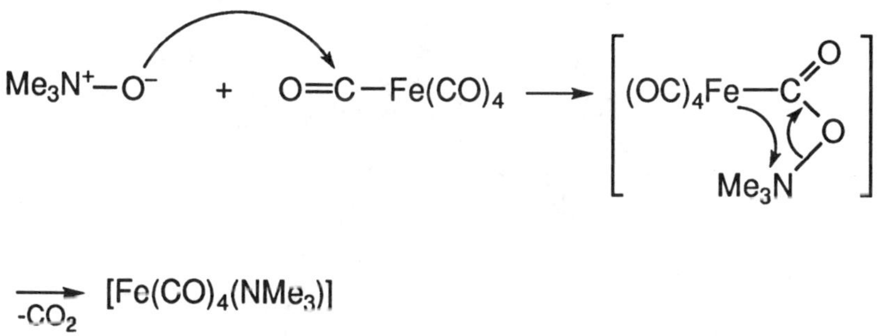

Scheme 8.22

Trimethylamine, in the unstable [Fe(CO)$_4$(NMe$_3$)] complex, is replaced easily by other ligands present in the reaction medium.[179] Many intermediate complexes with coordinated NMe$_3$ ligands have been isolated and characterized.[178,180] For example, the reaction of [Os$_3$(μ^3-S)(CO)$_{10}$] with two equivalents of Me$_3$NO·2H$_2$O in dichloromethane, which leads to the formation of [Os$_3$(μ-H)(μ-OH)(μ^3-S)(CO)$_8$(NMe$_3$)] containing a coordinated NMe$_3$ ligand has been reported.[180d] This cluster was characterized by X-ray crystallography, Figure 8.7. The bridging hydride was not located crystallographically but was identified by ^{1}H NMR spectroscopy. The reaction results in the loss of two carbonyl groups from the cluster and coordination of one NMe$_3$ ligand and the equivalent of one water molecule. The water is present in the form of μ-OH and μ-H ligands.

1,3-dipolar cycloaddition reactions[181] between low-valent metal carbonyls and aryl nitrile *N*-oxides, ArCNO, have been described, Scheme 8.23. Thermal elimination of CO_2 and stabilization by η^2-coordination of the resultant nitrile occurs. For example, for [Ir{η^2-C(-C_6H_4Cl-*p*)NO(C=O)}(CO)-(η^5-Cp*)], the reaction proceeds as follows, Scheme 8.24.

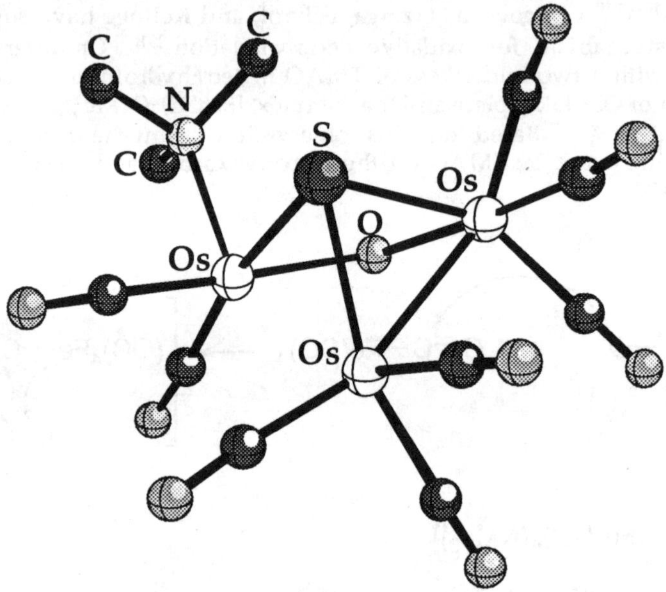

Figure 8.7

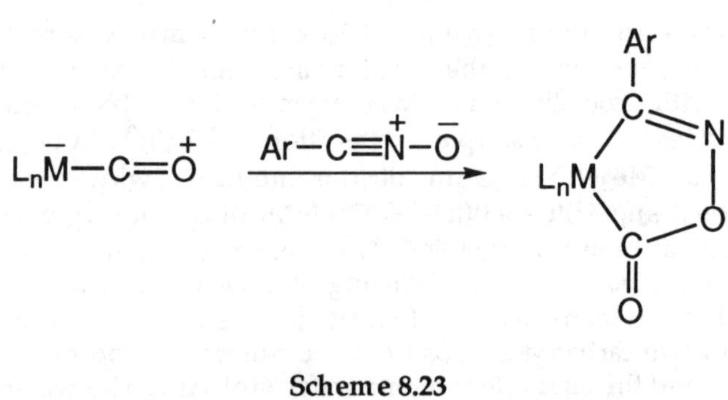

Scheme 8.23

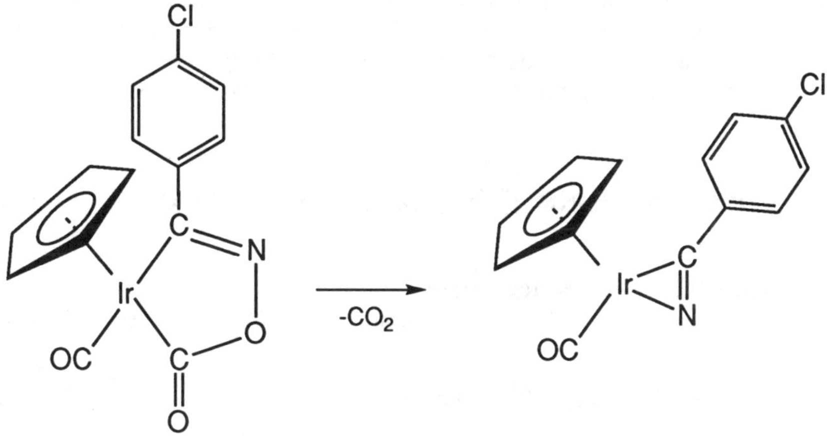

Scheme 8.24

In this case the reduction product formed from the reagent remains coordinated to the metal center. The product has been characterized by X-ray crystallography, Figure 8.8:

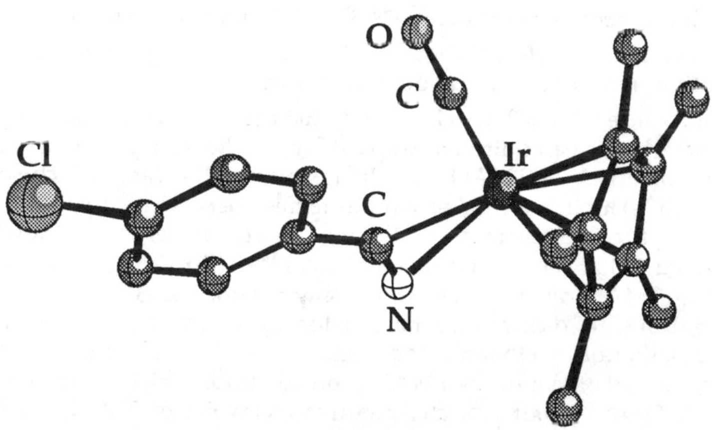

Figure 8.8

IR and UV spectroscopy have been employed to study the kinetics of the reaction of chromium, molybdenum and tungsten carbonyls with TMAO in various solvents, both in the presence and absence of triphenylphosphine (oxidative decarbonylation of $[M(CO)_6]$ by TMAO also has been studied[182]). It was found[178f] that the reactions are first-order in $[M(CO)_6]$ and TMAO but zero-order in triphenylphosphine. The rate of reaction was shown to decrease with a decrease in the dielectric constant of the solvent. The authors proposed the following mechanism for the nucleophilic oxygenation, Eqs. 8.51 - 8.53:

$$[M(CO)_6] + Me_3NO \xrightarrow{slow} [(OC)_5MC(O)\cdots O\cdots NMe_3] \qquad (8.51)$$

$$[(OC)_5MC(O)\cdots O\cdots NMe_3] \xrightarrow{fast} "(OC)_5M" + NMe_3 + CO_2 \qquad (8.52)$$

$$"(OC)_5M" + PPh_3 \xrightarrow{fast} [M(CO)_5(PPh_3)] \qquad (8.53)$$

Trimethylamine oxide is a strong nucleophile with respect to the carbonyl groups in $[M(CO)_6]$ complexes. The relative activity of nucleophiles towards these complexes decreases in the order:

$$MeLi > Me_3NO > PhCH_2MgBr >> N_3^- > NCO^- > NCS^- > Cl^- > Br^- > I^-$$

Several features of the reactions of metal carbonyl compounds with TMAO have been elucidated:[165,180,183] (*i*) Substituted and unsubstituted neutral metal carbonyls with ν(CO) > 2000 cm^{-1} participate in nucleophilic oxygenation reactions. If cationic compounds are considered then the threshold value of ν(CO) may be below 2000 cm^{-1} as a result of the positive charge which increases the electrophilicity of the complex and facilitates nucleophilic attack of TMAO. (*ii*) In heterometallic carbonyl clusters, CO ligands can exhibit varying electrophilicities, depending upon the metal center. In these cases there is the possibility of selective oxidation of coordinated CO at one of the metal centers.[184] (*iii*) Oxygenation reactions involving TMAO can be carried out using a wide range of solvents. For example, benzene, dichloromethane, chloroform, tetrahydrofuran, acetone, 2-methoxyethanol, methanol, and acetonitrile have all been used. (*iv*) Oxidative decarbonylation by TMAO proceeds under fairly mild conditions, usually at 25-70 °C, but typically requires an excess of TMAO. The mild conditions allow the synthesis of complexes not normally accessible under more extreme thermal conditions. Trimethylamine is a poor ligand, characterized by high volatility. It can be removed easily from reaction mixtures. (*v*) The products of the reaction depend on the donor ability of the solvent (*vide infra*).

If oxidative decarbonylation is carried out in the presence of donor solvents, the reaction can lead to the formation of solvento-complexes. For example, the removal of a carbonyl group from the triosmium cluster, $[Os_3(CO)_{12}]$, with trimethylamine oxide in acetonitrile leads to the formation of $[Os_3(CO)_{11}(MeCN)]$. The reaction of $[Os_3(CO)_{11}(MeCN)]$ with a second equivalent of Me_3NO gives a high yield of $[Os_3(CO)_{10}(MeCN)_2]$.[185] The reaction does not proceed further and the *tris*(acetonitrile) complex is not formed. This behavior can be rationalized in terms of the influence of the acetonitrile ligands, which are relatively good electron donors, on the electrophilicity of the carbonyl groups, resulting in reduced susceptibility to nucleophilic attack.

In the absence of a donor ligand (or a donor solvent) the coordinatively unsaturated species formed by loss of CO can react with the starting material to form products that contain metal-metal bonds.[186] Thus, refluxing $Me_3NO{\cdot}2H_2O$ with $[Rh(CO)_2(\eta^5\text{-}Cp)]$ in benzene leads to the formation of a trimer as the major product, Figure 8.9:

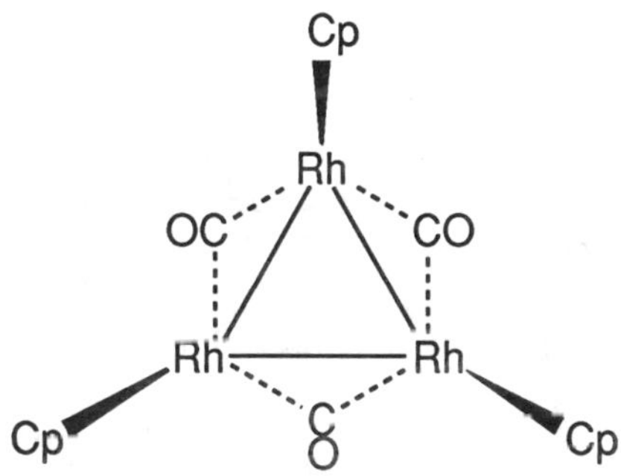

Figure 8.9

The decarbonylation of $[Rh_2(CO)_2(CF_3C{\equiv}CCF_3)(\eta^5\text{-}Cp)_2]$ by TMAO in dry acetone is accompanied by a 90° twist of the coordinated alkyne and a shift of the remaining, intact carbonyl group from a terminal to a bridging position,[187] Scheme 8.25. The structures of both the starting material and the product were determined by X-ray crystallography,[188] Figure 8.10.

The reaction of $[Ru_2(CO)_6(DAB)]$ (DAB = 1,4-disubstituted 1,4-diaza-1,3-butadiene, RN=C(H)C(H)=NR) with Me_3NO in a mixture of weak donor solvents (dichloromethane and toluene) leads to the formation of $[Ru_2(CO)_5(DAB)]$.[189] The binding of the DAB ligand is changed from the $6e^-$ (σ-N, μ^2-N', η^2-C=N') to the $8e^-$ (σ-N, σ-N', η^2-C=N, η^2-C'=N') mode.

Nucleophilic ligand oxygenation by organic oxygen donors is not restricted to carbonyl chemistry. Ligands that acquire electrophilic character

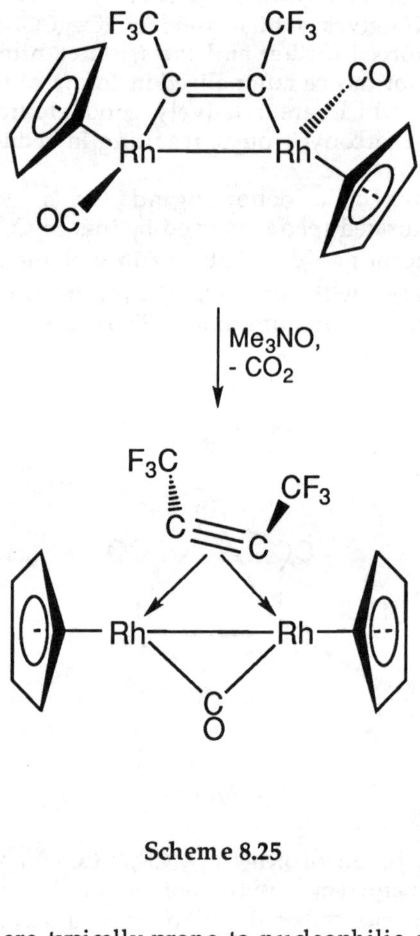

Scheme 8.25

upon coordination are typically prone to nucleophilic oxygenation. Thus, isocyanides are known to resemble carbon monoxide in many of their properties[190] and trimethylamine oxide has been shown to oxidize the isocyanide ligands, RNC (R = Me, Bu, Pr, Cy), in the rhodium complexes, $[Rh_2(CO)(CNR)(CF_3C{\equiv}CCF_3)(\eta^5\text{-}Cp)_2]$.[191a] The oxygen is transferred from Me_3NO to the isocyanide with formation of the corresponding isocyanate, RNCO. Subsequent intramolecular rearrangement and evolution of CO results in formation of the product, which has been characterized by X-ray crystallography, Figure 8.11.

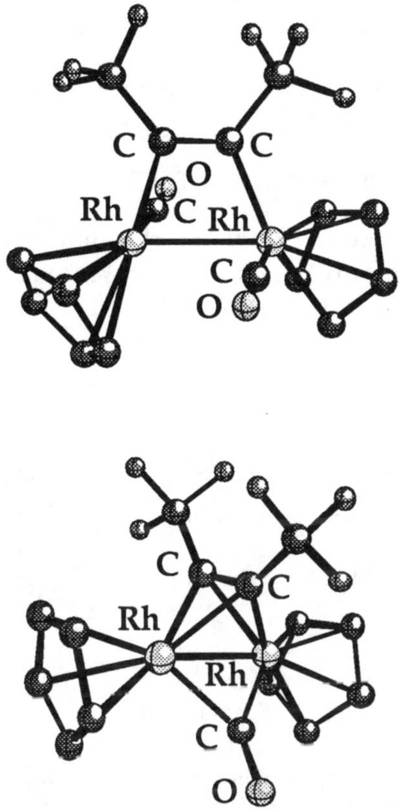

Figure 8.10

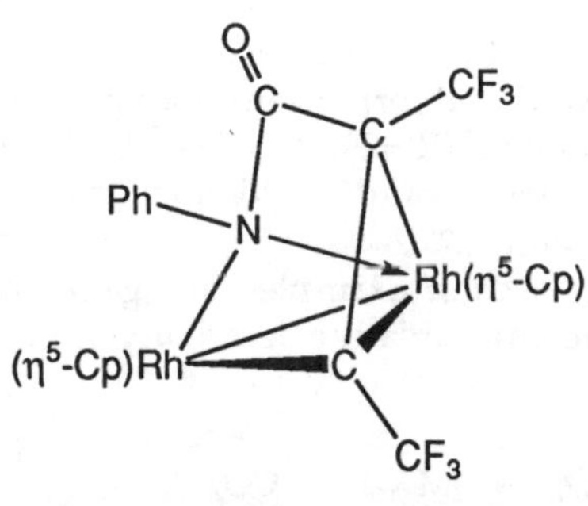

Figure 8.11

The kinetics and mechanism of oxygen atom transfer from Me_3NO to the benzyl isocyanide complex, $[Fe(CNCH_2Ph)_6]Br_2$, has been studied by Basolo and coworkers.[191b]

Organic oxygen donors also have been widely used for the oxidation of coordinated carbenes,[192] Eq. 8.54:

$$[L_nM{=}CR_2] + [O] \rightarrow [ML_n] + O{=}CR_2 \tag{8.54}$$

The interaction of the electrophilic rhenium methylidene complex, $[Re(=CH_2)(NO)(PPh_3)(\eta^5\text{-}Cp)][PF_6]$, with iodosobenzene leads to formation of the formaldehyde complex,[193] Scheme 8.26:

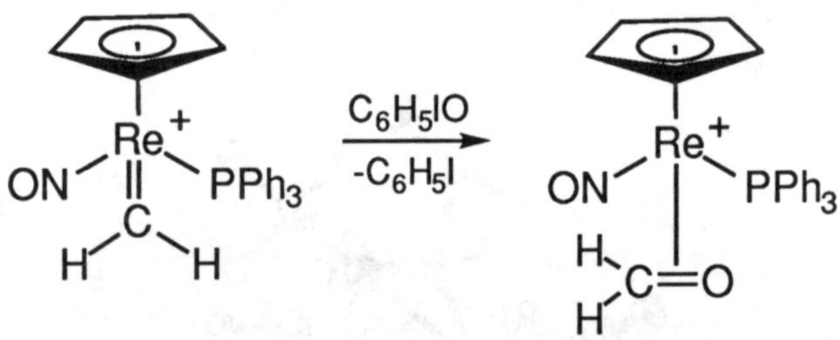

Scheme 8.26

The vinylidene complex, $[Fe(C{=}CHMe)(dppe)(\eta^5\text{-}Cp)][BF_4]$, reacts with excess iodosobenzene in acetonitrile to form the oxidation product, $[Fe_2(\mu\text{-}C_4Me_2)(dppe)_2(\eta^5\text{-}Cp)_2][BF_4]_2$.[194] The X-ray crystal structure shows that this homobinuclear complex contains two "$[Fe(dppe)(\eta^5\text{-}Cp)]$" fragments bound by a bridging "divinylidene" ligand, *i.e.* [{Fe}=C=C(Me)–C(Me)=C={Fe}], where {Fe} = $[Fe(dppe)(\eta^5\text{-}Cp)]$. The organic bridging ligand is formed by rupture of the C_β–H bond and formation of a new C–C bond. The tungsten complexes, $[W(CO)_5\{C(R)Ph\}]$ (R = H, Ph, OMe), in dichloromethane at -15 °C, react with $Me_2S{=}O$ by ligand oxidation to form O=C(R)Ph and $[W(CO)_5(SMe_2)]$,[195] Scheme 8.27.

Throughout this section examples of ligand oxygenation have been considered, for example, the oxidative decarbonylation of metal carbonyls, Eq. 8.55:

$$[MX_n(CO)] + [O] \rightarrow [MX_n] + CO_2 \tag{8.55}$$

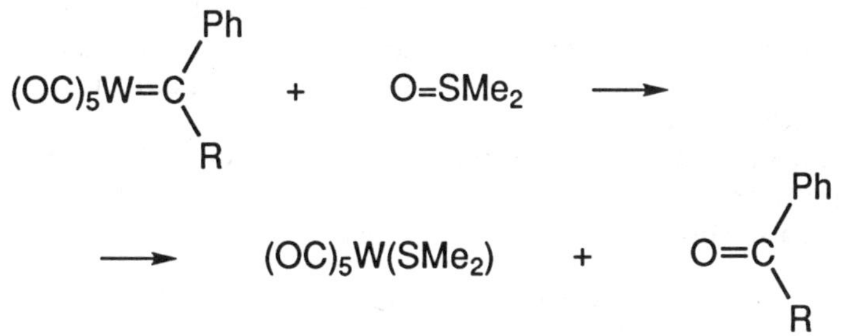

Scheme 8.27

In addition, the possibility of metal oxygenation by oxygen donors must be considered,[196] Eq. 8.56:

$$[MX_n] + [O] \rightarrow [O{=}MX_n] \quad (8.56)$$

It has been reported that dioxiranes[197] react with metal carbonyls by oxygen transfer to both the carbonyl ligands and the metal center, Eq. 8.57:

$$[MX_n(CO)] + 2R^1R^2CO_2 \rightarrow [O{=}MX_n] + CO_2 + 2R^1R^2CO \quad (8.57)$$

Thus, oxidation of the rhenium complex, $[Re(CO)_3Cp^*]$, with four equivalents of dimethyldioxirane (determined by spectroscopic titration) in anhydrous acetone solution yields $[ReO_3Cp^*]$. Both IR and GC analyses of the liberated gas indicated it to be a mixture of carbon monoxide and carbon dioxide. The first step of the reaction is likely to be the loss of a single carbonyl group by oxidative elimination, Eq. 8.58:

$$[Re(CO)_3Cp^*] + Me_2CO_2 \rightarrow [Re(CO)_2Cp^*] + CO_2 + Me_2CO \quad (8.58)$$

The resulting intermediate, $[Re(CO)_2Cp^*]$, is then rapidly converted to the trioxide, Eq. 8.59:

$$[Re(CO)_2Cp^*] + 3Me_2CO_2 \rightarrow [ReO_3Cp^*] + 2CO + 3Me_2CO \quad (8.59)$$

The oxidation of $[Et_4N][MoTp(CO)_3]$ (Tp = hydridotris(1-pyrazolyl)-borate) to the corresponding oxometallate also has been studied.[97] The generation of oxometallic complexes in this manner illustrates the utility of dioxiranes as selective, yet highly active, oxygen-atom donors for the oxidative decarbonylation of metal carbonyls, particularly in non-aqueous solvents such as acetone.

Nucleophilic oxygenation reactions employing oxides of phosphine[198] or phosphido[199] ligands, coordinated SO_2,[200] and also both organoboranes and organoaluminates,[201] have been reported.

8.8 Other selected oxidants

Benzoquinones exhibit redox behavior at easily accessible potentials and coordinate to metal ions. Catechols, products of the two-electron reduction of *o*-benzoquinones, are also good ligands, giving rise to catecholate complexes.[202] *o*-Benzoquinones have been used to obtain catecholate complexes *via* oxidative addition.[203] For example, the reaction of *trans*-$[RhCl(CO)(PPh_3)_2]$ with tetrachloro-1,2-benzoquinone results in the conversion of the four-coordinate rhodium(I) complex to a six-coordinate rhodium(III) catecholate complex, Scheme 8.28:

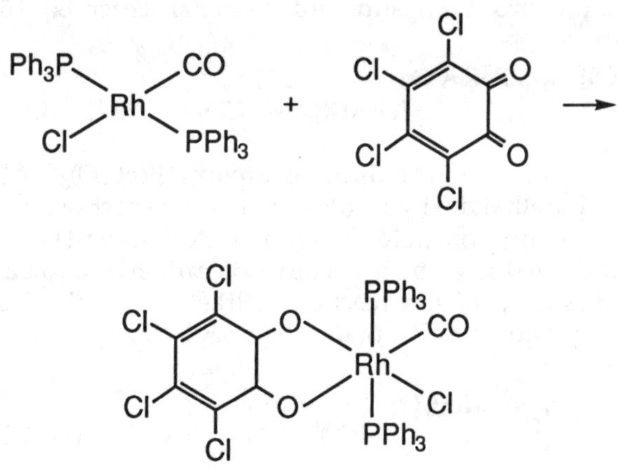

Scheme 8.28

Complexes of iridium(0), ruthenium(0), palladium(0) and platinum(0) are oxidized by tetrachloro-1,2-benzoquinone or tetrabromo-1,2-benzoquinone.[203]

A mixture of $AlCl_3$ and CH_2Cl_2 has been used as an oxidant[204] but the mechanism and nature of the by-products are not known. In dichloromethane $AlCl_3$ will oxidize $[Mo(CO)_2(bipy)_2]$. The dimer,

$[Mo(CO)_2(bipy)_2]_2[BF_4]_2$, is isolated after the addition of $NaBF_4$.[205]

N-halosuccinimides (chloro-, bromo-, and iodo-) are useful reagents for oxidations in organic chemistry and for the oxidative halogenation of metal complexes. For example, the reaction of the rhenium hydride, *trans-(H,Br)*-[Re(Br)(H)(CO)(PMe_3)Cp*], with *N*-bromosuccinimide in THF at 0 °C yields *trans*-[$ReBr_2$(CO)(PMe_3)Cp*].[206]

Tetracyanoethylene (TCNE) and tetracyanoquinodimethane (TCNQ), Figure 8.12:

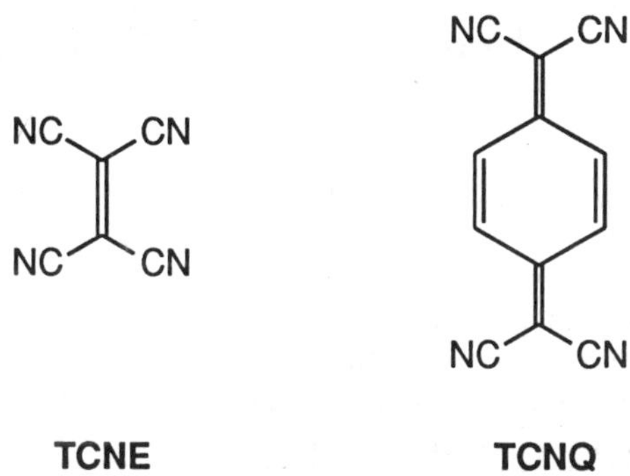

Figure 8.12

are known to be strong electron acceptors[207] that form stable radical-anions, $TCNQ^{\bullet-}$ and $TCNE^{\bullet-}$, by one-electron reduction. The half-wave potentials measured by polarographic reduction of these compounds in acetonitrile (0.127 *V* for TCNQ and 0.152 *V* for TCNE) show these materials to be mild oxidants in this solvent.[207] The physical properties and chemistry of TCNE and TCNQ, as well as applications of these compounds as reagents in organic and organometallic chemistry, have been reviewed[208] and only selected examples of their reaction chemistry are presented here. It has been established[209] that the tricobalt clusters, Figure 8.13:

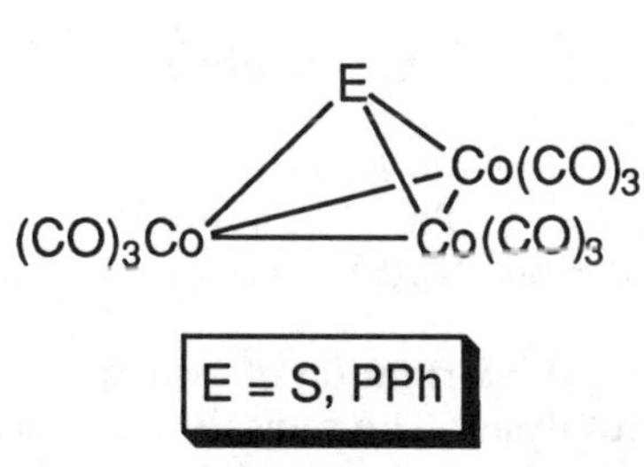

Figure 8.13

react with TCNQ in ethanol, acetone, dichloroethane, tetrahydrofuran or dichloromethane by one-electron oxidation to yield $[ECo_3(CO)_9]^+$.

TCNE oxidizes the complex, Figure 8.14:

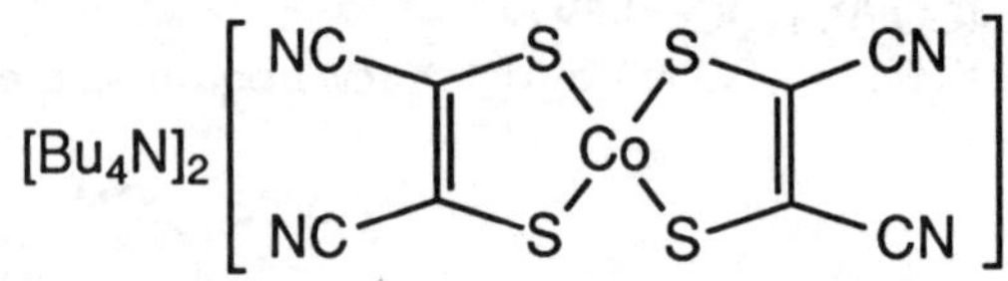

Figure 8.14

with formation of the dimer, $[Bu_4N]_2[Co(C_4N_2S_2)_2]_2$.[210] Octamethylferrocene, $[Fe(\eta^5\text{-}C_5Me_4H)_2]$, has an oxidation potential of 0.07 *V* (*vs.* SCE) and is readily oxidized by TCNE or TCNQ in acetonitrile. The reaction results in formation of the 1:1 iron(III) charge-transfer complexes $[Fe(\eta^5\text{-}C_5Me_4H)_2]^{\cdot+}[TCNE]^{\cdot-}$ and $[Fe(\eta^5\text{-}C_5Me_4H)_2]^{\cdot+}[TCNQ]^{\cdot-}$. Both salts have been structurally characterized.[211] The oxidation of $[Fe(\eta^5\text{-}C_5Me_4H)_2]$ also can be accomplished with organic electron-acceptors such as *n*-hexacyanobutadiene, $n\text{-}C_4(CN)_6$, tris(dicyanomethylene)cyclopropane, $C_6(CN)_6$, 2,3-dichloro-5,6-dicyanobenzoquinone, and perfluoro-7,7,8,8-tetracyano-*p*-quinodimethane.

Both TCNE and TCNQ are poor ligands for most metals. However, upon electron uptake by back-donation or by complete reduction, they bind to metal fragments *via* the C=C π-system or by the nitrogen atoms of the C≡N groups and a wide variety of di-, tri-, and tetranuclear metal complexes are formed.[212-214]

Organic disulphides can oxidize metal carbonyls[215] by a photochemical redox reaction, Eq. 8.60:

$$[M(CO)_nCp]_2 + S_2R_2 \xrightarrow{h\nu} 2[M(SR)(CO)_nCp] \qquad (8.60)$$

where M = Fe, R = Me, n = 2;
M = Ru, R = Me, Ph, CH_2Ph, n = 2;
M = W; R = $Me_2NC(S)$; n = 3.

The mechanisms involve the formation of extremely reactive metal carbonyl radicals as a result of photolysis. The radicals are oxidized in a dark reaction by the organic disulphide, accompanied by formation of thiolate complexes.[215]

The preparation of tris(benzenethiolato)indium(III), $[In(SPh)_3]$, by the oxidation of metallic indium with diphenyl disulfide in toluene, has been described and the selenium analog, $[In(SePh)_3]$, has been obtained similarly,[216] Eq. 8.61:

$$In + 3/2Ph_2Se_2 \rightarrow [In(SePh)_3] \quad (8.61)$$

2,2,6,6-Tetramethylpiperidine-1-oxyl (TEMPO) has been employed routinely as a radical trap in the measurement of metal-carbon bond strengths in macrocyclic complexes in solution. Leidner, Seyler and Fanwick[217a] have reported that this reagent can oxidize a coordinated methyl group to a carbonyl. Thus, addition of excess TEMPO to a benzene solution of [RuMe(OEP)] (where H_2OEP = octaethylporphyrin) yields [Ru(OEP)(CO)-(TEMPO)], which has been isolated and characterized by X-ray crystallography. The mechanism of this conversion is not known and this work[217a] is the first report of the transformation of M–Me to M–CO.

Chan and Leung[217b] have reported a novel, high-yielding and convenient preparation of the metal-metal bonded dimers, $[M_2(OEP)_2]$ (M = Rh, Ir), by the oxidation of [MH(OEP)] with TEMPO in benzene solution, Eq. 8.62:

$$2[MH(OEP)] + 2TEMPO \rightarrow [M_2(OEP)_2] + 2TEMPOH \quad (8.62)$$

8.9 "Modification" of water-soluble oxidants

"Modification" of traditional oxidants in order to impart hydrophobic properties and so allow their use in non-aqueous solvents is of interest in the development of broadly applicable reagents. For example, the pyridinium analog of potassium dichromate, $[pyH]_2[Cr_2O_7]$, has been synthesized.[218] This compound is freely soluble in DMF, diethylacetamide, and dimethyl sulfoxide, and is sparingly soluble in dichloromethane and chloroform. Similarly, a "hydrophobic" analog of hydrogen peroxide, $[Me_4N]_2O_2$, has been used as an oxidant for metal complexes dissolved in acetonitrile.[219]

The modification of potassium permanganate, $KMnO_4$, has been carried out in the same manner. Although this reagent is commonly used as an oxidant in aqueous solutions,[220] a serious disadvantage that accompanies its use is its thermodynamic instability in aqueous media, Eq. 8.63:

$$4MnO_4^- + 2H_2O \rightarrow 4MnO_2 + 3O_2 + 4OH^- \quad (8.63)$$

This process is catalyzed by acids, bases, and even MnO_2.[221] Because manganese dioxide is a product of $KMnO_4$ reduction by oxidizable substrates, catalytic decomposition of $KMnO_4$ by the pathway shown in Eq. 8.62 generally occurs. Accordingly, potassium permanganate is typically used in large excess and yields are often low, frequently due to over-oxidation. An additional disadvantage is the *pH*-dependence of oxidation reactions in aqueous solutions. Reactions must be carried out at *pH* 4.5 - 5.5 in order to minimize decomposition and these aqueous, acidic conditions preclude extension of the method to complexes of strongly basic ligands which protonate under such conditions.

Sala and Sargent[221] have synthesized tetra(*n*-butyl)ammonium permanganate, $[(n\text{-Bu})_4N][MnO_4]$, in high yield by a metathetical exchange reaction between $KMnO_4$ and $[(n\text{-Bu})_4N]Br$ in aqueous solution. The crystalline precipitate of $[(n\text{-Bu})_4N][MnO_4]$ is soluble in DMF, acetone, pyridine, chloroform and dichloromethane and may be stored as a solid for prolonged periods without decomposition. However, there have been reports of the detonation of quarternary ammonium permanganates during drying at elevated temperatures and care in handling is indicated.[222] A systematic study[222] has shown that salts containing unsaturated organic groups in the cation ($Et_3NCH_2Ph^+$, $Me_3NCH_2Ph^+$, Ph_3NMe^+, $Ph_3PC_2H_4PPh_3^{2+}$ and Ph_4P^+) will explode at 80 °C or above. In contrast, although $(n\text{-Bu})_4N^+$ and Et_4N^+ salts will also decompose at such temperatures, the decomposition is not reported to be explosive.

Bock and Jaculi[223] have reported that, in DMF, the half-wave potential of the quasi-reversible MnO_4^-/MnO_4^{2-} couple is -0.73 *V* (*vs.* SCE). This represents a shift of more that 1 *V* relative to the value in aqueous solution (+0.32 *V vs.* SCE)! Thus, in organic solvents, MnO_4^- is a much milder oxidant than it is in aqueous solution. The salt, $[(n\text{-Bu})_4N][MnO_4]$, has been used[224] as an oxidant for the synthesis of coordination compounds in non-aqueous solvents. For example, $Mn(OAc)_2{\cdot}4H_2O$ reacts with stoichiometric amounts of $[(n\text{-Bu})_4N][MnO_4]$ in acetonitrile, ethanol, acetone, or DMF in the presence of picolinic acid (HL) with the formation of $[MnL_3]{\cdot}H_2O$. With salicylic acid ($salH_2$) and pyridine, the product, after recrystallization from ethanol/petrol, is $[Mn(EtOH)_4][Mn_2(sal)_4(py)_2]$. The reaction between $Mn(O_2CPh)_2$, $[(alkyl)_4N][MnO_4]$, and 2-hydroxymethyl-pyridine (pyOH) in acetonitrile in the presence of triethylamine leads to the formation of $[(alkyl)_4N][Mn_4O_2(O_2CPh)_7(pyO)_2]$ in high yield. The complex has been shown to contain a butterfly-type $(Mn^{3+})_4O_2$ core.[224]

It has been established[225] that the reaction of $MnCl_2$ with 2,2'-bipyridyl and *p*-$MeC_6H_4SO_3H$ in DMF, followed by addition of a solution of $[Et_4N][MnO_4]$ in DMF and then $[Et_4N][MCl_4]$ (M = Mn, Cd), leads to the formation of the manganese(IV) clusters, $[Mn_3O_4Cl_2(bipy)_4][MCl_4]{\cdot}3DMF$.

The modification of water-soluble perruthenate salts into more hydrophobic analogs has been described.[226] In non-aqueous media, salts of $[RuO_4]^-$ such as $[(n\text{-Bu})_4N][RuO_4]$,[227] $[(n\text{-Pr})_4N][RuO_4]$,[228] $[Ph_4P][RuO_4]$, and $[PPN][RuO_4]$[229] are remarkably mild oxidants. Stoichiometric oxidation of primary alcohols to aldehydes and secondary alcohols to ketones without competing double-bond attack can be effected.

The modification of the traditional oxidant, $K_3[Fe(CN)_6]$, has been reported by Mascharak[230] who prepared the hydrophobic analog, $[Et_4N]_3[Fe(CN)_6]$. Cyclic voltammetry showed a regular shift in $E_{1/2}$ to more positive values upon gradual addition of a more-polar solvent to a less-polar one, Table 8.4:

Table 8.4

Half-wave potentials ($E_{1/2}$) for the reduction of $[Et_4N]_3[Fe(CN)_6]$ in solvent **A** containing x% (v/v) solvent **B** and 0.1 *M* $[(n\text{-}Bu)_4N][ClO_4]$ (TBAP) or 0.1 *M* $[Et_4N][ClO_4]$ (TEAP)[230]

x	**A = MeCN, B = H_2O (TBAP)**	**A = MeCN, B = MeOH (TBAP)**	**A = DMF, B = MeOH (TEAP)**
0	-0.85	-0.85	-0.90
1.5	-0.55	-0.71	-0.84
3	-0.43	-0.66	-0.78
6	-0.31	-0.61	-0.71
9	-0.23	-0.57	-0.67
12	-0.18	-0.53	-0.63
25	-0.08	-0.46	-0.53
100	+0.15 (TEAP)	-0.30	-0.30

Although $[Et_4N]_3[Fe(CN)_6]$ has been described as a mild oxidant[230] in organic solvents, the potential can be shifted to more positive values by addition of water or methanol, thus indicating the possibility of use as a "tunable oxidant". Indeed, $[Et_4N][Fe(SPh)_4]$ is stable in acetonitrile in the presence of $[Et_4N]_3[Fe(CN)_6]$, but is converted into $[Fe(SPh)_4]^-$ upon addition of water (6%).

Finally, it should be noted that many oxidants can be used in both aqueous and non-aqueous media. For example, applications of PbO_2 in a mixture of MeCN/CF_3CO_2H or CH_2Cl_2/CF_3CO_2H,[231] of $[FeCp_2]^+$ in MeCN[232] or tetrahydrofuran,[233] of $Pb(OAc)_4$ in dichloromethane,[234] and of $TlCl_3$ in dichloromethane,[235] *etc.* are known.

References

1. A. I. Stetsenko, V. I. Nefedov, T. G. Abzaeva, and Ya. V. Salyn, *Izv. Akad. Nauk SSSR, Ser. Khim.* (1974) 530.

2. T. Mussini, *Chim. Ind. (Milan)* **50** (1968) 783; *Chem. Abstr.* **69** (1968) 73417.

3. G. Schmid and U. Neumann, *Z. Phys. Chem.* **154** (1967) 150.

4. V. Gutmann, *Coordination Chemistry in Non-aqueous Solution* (Springer-Verlag, New York, 1968) p. 31.

5. G. Cauquis and D. Serve, *Compt. Rend.* **C 266** (1968) 1591.

6. C. T. Ratcliffe and J. M. Shreeve, *Inorg. Synth.* **11** (1968) 194.

7. (a) F. Seel, W. Birnkraut, and D. Werner, *Chem. Ber.* **95** (1962) 1264; (b) R. N. Scott and D. F. Shriver, *Inorg. Chem.* **5** (1966) 158; (c) E. Wilke-Dörfurt and G. Balz, *Z. Anorg. Chem.* **159** (1927) 197; (d) G. Balz and E. Mailänder, *Z. Anorg. Chem.* **217** (1934) 161; (e) G. Brauer, *Handbook of Preparative Inorganic Chemistry*, Second Edition, Vol. 1 (Academic Press, New York, 1963) p. 320; (f) A. S. Morkovnik, Z. S. Morkovnik, and V. V. Bessonov, *Zh. Neorg. Khim.* **32** (1987) 3106.

8. (a) V. Ya. Rosolovskii and E. S. Rumyantsev, *Zh. Neorg. Khim.* **8** (1963) 1326; (b) K. A. Hofmann and G. A. Zedwitz, *Ber.* **42** (1909) 2031; (c) *French Patent* 1471198 (1967); *Chem. Abstr.* **67** (1967) 75006.

9. M. T. Mocella, M. S. Okamoto, and E. K. Barefield, *Synth. React. Inorg. Met.-Org. Chem.* **4** (1974) 69.

10. F. Seel, *Z. Anorg. Chem.* **261** (1950) 75.

11. B. J. Hathaway, D. G. Holah, and J. D. Postlethwaite, *J. Chem. Soc.* (1961) 3215.

12. B. J. Hathaway and A. E. Underhill, *J. Chem. Soc.* (1960) 3705.

13. G. Bergerhoff, *Z. Anorg. Chem.* **327** (1964) 139.

14. K. Nilsson, *Structural and Thermodynamic Studies of Some Copper(I), Silver(I), and Gold(I) Complexes in Acetonitrile Solution*, Ph.D. Thesis, University of Lund, Sweden, 1986.

15. K. Nilsson and Å. Oskarsson, *Acta Chem. Scand.* **A39** (1984) 79.

16. M. Albin, A. C. Goldstone, A. S. Withers, and W. DeW. Horrocks, *Inorg. Chem.* **22** (1983) 3182.

17. A. D. Van Ingen Schenau, W. L. Groeneveld, and J. Reedijk, *Spectrochim. Acta.* **30A** (1974) 213.

18. B. B. Wayland and R. F. Schramm, *Inorg. Chem.* **8** (1969) 971.

19. R. F. Schramm and B. B. Wayland, *J. Chem. Soc., Chem. Commun.* (1968) 898.

20. R. R. Thomas and A. Sen, *Inorg. Synth.* **28** (1990) 63.

21. F. A. Cotton, K. R. Dunbar, L. R. Falvello, M. Tomas, and R. A. Walton, *J. Am. Chem. Soc.* **105** (1983) 4950.

22. P. P. Paul, Z. Tycklár, A. Farooq, K. D. Karlin, S. Liu, and J. Zubieta, *J. Am. Chem. Soc.* **112** (1990) 2430.

23. D. A. Phillips, M. Kubota, and J. Thomas, *Inorg. Chem.* **15** (1976) 118.

24. A. S. Drane and J. A. McCleverty, *Polyhedron* **2** (1983) 53.

25. (a) K. G. Caulton, *Coord. Chem. Rev.* **14** (1975) 317; (b) D. M. P. Mingos and D. J. Shermann, *Adv. Inorg. Chem.* **34** (1989) 293.

26. S. Clamp, N. G. Connelly, G. E. Taylor, and T. S. Louttit, *J. Chem. Soc., Dalton Trans.* (1980) 2162.

27. (a) P. K. Ashford, P. K. Baker, N. G. Connelly, R. L. Kelly, and V. A. Woodley, *J. Chem. Soc., Dalton Trans.* (1982) 477; (b) N. G. Connelly, Z. Demidowicz, and R. L. Kelly, *J. Chem. Soc., Dalton Trans.* (1975) 2335; (c) N. G. Connelly and G. A. Johnson, *J. Organomet. Chem.* **77** (1974) 341.

28. K. L. Fjare and J. E. Ellis, *J. Am. Chem. Soc.* **105** (1983) 2303.

29. W. Heiber, J. Peterhans, and E. Winter, *Chem. Ber.* **94** (1961) 2572.

30. D. D. Klendworth, W. W. Welters, and R. A. Walton, *J. Organomet. Chem.* **213** (1981) C13.

31. N. G. Connelly, A. C. Loyns, I. Manners, D. L. Mercer, K. E. Richardson, and P. H. Rieger, *J. Chem. Soc., Dalton Trans.* (1990) 2451.

32. P. Legzdins and J. C. Oxley, *Inorg. Chem.* **23** (1984) 1053.

33. (a) G. Albertin, P. Amendola, S. Antoniutti, and E. Bordignon, *J. Chem. Soc., Dalton Trans.* (1990) 2979; (b) B. F. G. Johnson, J. Lewis, W. J. H. Nelson, J. Puga, P. R. Raithby, and K. H. Whitmire, *J. Chem. Soc., Dalton Trans.* (1983) 1339; (c) B. F. G. Johnson, P. R. Raithby, and C. Zuccaro, *J. Chem. Soc., Dalton Trans.* (1980) 99; (d) B. F. G. Johnson, J. Lewis, W. J. H. Nelson, J. Puga, P. R. Raithby, M. Schröder, and K. H. Whitmore, *J. Chem. Soc., Chem. Commun.* (1982) 610.

34. J. K. Hurst and H. Taube, *J. Am. Chem. Soc.* **90** (1968) 1174.

35. R. B. Jordan, A. M. Sargeson, and H. Taube, *Inorg. Chem.* **5** (1966) 1091.

36. (a) W. R. Ellis and W. L. Pursell, *Inorg. Chem.* **21** (1982) 834; (b) R. J. Balahura, *Can. J. Chem.* **52** (1974) 1762.

37. (a) P. Legzdins, B. Wassink, F. W. B. Einstein, and A. C. Willis, *J. Am. Chem. Soc.* **108** (1986) 317; (b) P. Legzdins, G. B. Richter-Addo, B. Wassink, F. W. B. Einstein, R. H. Jones, and A. C. Willis, *J. Am. Chem. Soc.* **111** (1989) 2097.

38. A. Goldhaber, K. P. C. Vollhardt, E. C. Walborsky, and M. Wolfgruber, *J. Am. Chem. Soc.* **108** (1986) 516.

39. P. M. Treichel, M. S. Schmidt, and R. A. Crane, *Inorg. Chem.* **30** (1991) 379.

40. (a) K. P. Sarma and R. K. Poddar, *Transit. Metal Chem.* **8** (1983) 225; (b) K. P. Sarma and R. K. Poddar, *Transit. Metal Chem.* **8** (1983) 31.

41. R. Mews and C.-H. Liu, *Angew. Chem. Int. Ed. Engl.* **22** (1983) 162; *Angew. Chem.* **95** (1983) 156.

42. M. Herberhold and L. Haumaier, *Z. Naturforch.* **35b** (1980) 1277.

43. Yu. V. Guk, M. A. Ilyushin, E. L. Golod, and B. V. Gidaspov, *Uspekhi Khimii* **52** (1983) 499; *Russ. Chem. Rev. (Engl. Transl.)* **52** (1983) 284.

44. J. Reed, *J. Inorg. Nucl. Chem.* **38** (1976) 2239.

45. H. N. Rabinowitz, K. D. Karlin, and S. J. Lippard, *J. Am. Chem. Soc.* **99** (1977) 1420.

46. D. Sutton, *Chem. Soc. Rev.* **4** (1975) 443.

47. R. M. Elofson and F. F. Gadallah, *J. Org. Chem.* **34** (1969) 854.

48. G. Cauquis and D. Serve, *Compt. Rend.* **C266** (1968) 1591.

49. N. Farrell, M. N. De Oliveira Bastos, and A. A. Neves, *Polyhedron* **2** (1983) 1243.

50. A. W. B. Garner and M. J. Mays, *J. Organomet. Chem.* **67** (1974) 153.

51. K. R. Laing, S. D. Robinson, and M. F. Uttley, *J. Chem. Soc., Dalton Trans.* (1973) 2713.

52. N. Farrell and D. Sutton, *Can. J. Chem.* **55** (1977) 360.

53. A. N. Nesmeyanov, Y. A. Chapovskii, I. V. Polovyanyuk, and L. G. Makarova, *J. Organomet. Chem.* **7** (1967) 329.

54. H. Meerwein, V. Henderich, and K. Wunderlich, *Arch. Pharm.* **291** (1958) 541.

55. N. G. Connelly, A. R. Lucy, J. D. Payne, A. M. R. Galas, and W. E. Geiger, *J. Chem. Soc., Dalton Trans.* (1983) 1879.

56. H. P. Dauben and L. R. Honnen, *J. Am. Chem. Soc.* **80** (1958) 5570.

57. (a) L. Fieser and M. Fieser, *Reagents for Organic Syntheses* (Wiley-Interscience, New York, 1970 - 1978) Vols. I - VII; (b) H. J. Dauben, L. R. Honnen, and K. M. Harmon, *J. Org. Chem.* **25** (1960) 1442.

58. S. Yamazaki, *Inorg. Chem.* **21** (1982) 1638.

59. (a) H. Lankamp, W. Th. Nauta, and C. MacLean, *Tetrahedron Lett.* (1968) 249. (b) D. V. Ioffe and L. S. Efros, *Reactivities of Organic Compounds* [in Russian] **9** (1972) 1165; (c) W. D. Hounshell, D. A. Dougherty, J. P. Hummel, and K. Mislow, *J. Am. Chem. Soc.* **99** (1977) 1916; (d) J. M. McBride, *Tetrahedron* **30** (1974) 2009.

60. Y. S. Yu, R. A. Jacobson, and R. J. Angelici, *Inorg. Chem.* **21** (1982) 3106.

61. W. E. Bachmann, *J. Am. Chem. Soc.* **53** (1931) 2758.

62. J. R. Sanders, *J. Chem. Soc., Dalton Trans.* (1973) 748.

63. J. D. Allison and R. A. Walton, *J. Chem. Soc., Chem. Commun.* (1983) 401.

64. C. Eaborn, N. Farrell, J. L. Murphy, and A. Pidcock, *J. Chem. Soc., Dalton Trans.* (1976) 58.

65. E. W. Abel and S. P. Tyfield, *Adv. Organomet. Chem.* **8** (1970) 126.

66. A. Asdar, C. Lapinte, and L. Toupet, *Organometallics* **8** (1989) 2708.

67. A. Sanders, T. Bauch, C. V. Magatti, C. Lorenc, and W. P. Giering, *J. Organomet. Chem.* **107** (1976) 359.

68. F. B. McCormick, R. J. Angelici, R. A. Pickering, R. E. Wagner, and R. A. Jacobson, *Inorg. Chem.* **20** (1981) 4108.

69. (a) C. P. Casey, P. J. Fagan, and W. H. Miles, *J. Am. Chem. Soc.* **104** (1982) 1134; (b) C. P. Casey, P. J. Fagan, W. H. Miles, and S. R. Marder, *J. Mol. Catal.* **21** (1983) 178.

70. D. N. Kursanov, Z. N. Parnes, M. I. Kalinkin, and N. M. Loim, *Ionic Hydrogenation* [in Russian] (Khimia, Moscow, 1979).

71. (a) J. R. Sanders, *J. Chem. Soc., Dalton Trans.* (1973) 743; (b) M. Y. Darensbourg and J. C. Deaton, *Inorg. Chem.* **20** (1981) 1644.

72. B. W. Hames and P. Legzdins, *Organometallics* **1** (1982) 116.

73. H. Berke and G. Weiler, *Z. Naturforsch.* **39b** (1984) 431.

74. W. Beck and K. Schloter, *Z. Naturforsch.* **33b** (1978) 1214.

75. J. R. Sweet and W. A. G. Graham, *Organometallics* **2** (1983) 135.

76. J. R. Sweet and W. A. G. Graham, *J. Organomet. Chem.* **217** (1981) C37.

77. J. K. Hoyano, C. J. May, and W. A. G. Graham, *Inorg. Chem.* **21** (1982) 3095.

78. T. Beringhelli, G. Ciani, G. D'Alfonso, P. Romiti, A. Sironi and M. Freni, *Inorg. Chem.* **23** (1984) 2849.

79. (a) T. Beringhelli, G. D'Alfonso, and M. De Angelis, *J. Organomet. Chem.* **322** (1987) C21; (b) G. A. Moehring, P. E. Fanwick, and R. A. Walton, *Inorg. Chem.* **26** (1987) 1861; (c) G. A. Moehring and R. A. Walton, *J. Chem. Soc., Dalton Trans.* (1987) 715.

80. V. Guerchais, J.-Y. Thépot, and C. Lapinte, *J. Chem. Soc., Chem. Commun.* (1989) 1554.

81. (a) H. P. Dauben and D. J. Bertelli, *J. Am. Chem Soc.* **83** (1961) 497; (b) F. M. Chaudhari and P. L. Pauson, *J. Organomet. Chem.* **5** (1966) 73; (c) G. S. Bodner, D. E. Smith, W. G. Hatton, P. C. Heah, S. Georgiou, A. L. Rheingold, S. J. Geib, J. P. Hutchinson, and J. A. Gladysz, *J. Am. Chem. Soc.* **109** (1987) 7688.

82. J. J. Habeeb, F. F. Said, and D. G. Tuck, *J. Chem. Soc., Dalton Trans.* (1981) 118.

83. J. M. Mayer and E. H. Abbott, *Inorg. Chem.* **22** (1983) 2774.

84. H. Werner and J. Wolf, *Angew. Chem.* **94** (1982) 308.

85. P. Legzdins, D. T. Martin, C. R. Nurse, and B. Wassink, *Organometallics* **2** (1983) 1238.

86. B. Callan and A. R. Manning, *J. Organomet. Chem.* **252** (1983) C81.

87. E. Carmona-Guzman and G. Wilkinson, *J. Chem Soc., Dalton Trans.* (1977) 1716.

88. P. Legzdins and D. T. Martin, *Inorg. Chem.* **18** (1979) 1250.

89. R. H. Crabtree, G. G. Hlatky, C. P. Parnell, B. E. Segmüller, and R. J. Uriarte, *Inorg. Chem.* **23** (1984) 354.

90. J. A. Gladysz, W. Tam, G. M. Williams, D. Lee Johnson, and D. W. Parker, *Inorg. Chem.* **18** (1979) 1163.

91. R. Schunn in *Transition Metal Hydrides*, ed. E. L. Muetterties (Marcell Dekker, New York, 1971) p. 239.

92. W. C. Trogler, *J. Am. Chem. Soc.* **101** (1979) 6459.

93. *Standard Potentials in Aqueous Solution*, eds. A. J. Bard, R. Parsons, and J. Jordan (Marcell Dekker, New York, 1985) p. 307.

94. D. Astruc, *Angew. Chem. Int. Ed. Engl.* **27** (1988) 643.

95. P. K. Baker, K. Broadley, N. G. Connelly, B. A. Kelly, M. D. Kitchen, and P. Woodward, *J. Chem. Soc., Dalton Trans.* (1980) 1710.

96. H. H. Morgan, *J. Chem. Soc.* **CXXIII**, Part II, (1923) 2901.

97. P. M. Treichel and G. J. Essenmacher, *Inorg. Chem.* **15** (1976) 146.

98. D. T. Plummer and R. J. Angelici, *Inorg. Chem.* **22** (1983) 4063.

99. W. J. Evans, T. A. Ulibarri, L. R. Chamberlain, J. W. Ziller, and D. Alvarez, Jr., *Organometallics* **9** (1990) 2124.

100. (a) P. M. Treichel, L. D. Rosenhein, and M. S. Schmidt, *Inorg. Chem.* **22** (1983) 3960; (b) P. M. Treichel and L. D. Rosenhein, *Inorg. Chem.* **23** (1984) 4018.

101. J. Gotzig, H. Otto, and H. Werner, *J. Organomet. Chem.* **287** (1985) 247.

102. (a) R. F. Jordan, C. S. Bajgur, R. Willett, and B. Scott, *J. Am. Chem. Soc.* **108** (1986) 7410; (b) R. F. Jordan, W. E. Dasher, and S. F. Echols, *J. Am. Chem. Soc.* **108** (1986) 1718; (c) R. F. Jordan, C. S. Bajgur, W. E. Dasher, and A. L. Rheingold, *Organometallics* **6** (1987) 1041.

103. L. F. Rhodes, J. D. Zubkowski, K. Folting, J. C. Huffman, and K. G. Caulton, *Inorg. Chem.* **21** (1982) 4185.

104. O. V. Kravchenko, S. E. Kravchenko, V. D. Makhaev, V. B. Polyakova, G. V. Slobodenchuk, and K. N. Semenenko, *Koord. Khim.* **8** (1982) 1356.

105. G. B. Deacon and A. J. Koplick, *Inorg. Nucl. Chem. Lett.* **15** (1979) 263.

106. K.-H. Thiele, K. Unverhau, M. Geitner, and K. Jacob, *Z. Anorg. Allg. Chem.* **548** (1987) 175.

107. W. J. Schleintz, Y. Lavender, W. Welcman, R. B. King, and J. K. Ruff, *J. Organomet. Chem.* **33** (1971) 357.

108. O. A. Reutov, I. P. Beletskaya, G. A. Artamkina, and A. N. Kashin, *Reactions of Organometallic Compounds as Redox Processes* [in Russian] (Nauka, Moscow, 1981) p. 155.

109. M. B. Inoue, M. Inoue, Q. Fernando, and K. W. Nebesny, *Inorg. Chem.* **25** (1986) 3976.

110. M. Inoue, A. R. Mendivil-Sol, H. Grijalva, and M. B. Inoue, *Inorg. Chim. Acta*, **161** (1989) 179.

111. S. R. Drake, B. F. G. Johnson, J. Lewis, G. Conole, and M. McPartlin, *J. Chem. Soc., Dalton Trans.* (1990) 995.

112. C.-M. T. Hayward and J. R. Shapley, *Inorg. Chem.* **21** (1982) 3816.

113. R. Khattar, B. F. G. Johnson, J. Lewis, P. R. Raithby, and M. J. Rosales, *J. Chem. Soc., Dalton Trans.* (1990) 2167.

114. (a) M. Schutzenberger and Fontaine, *Bull. Soc. Chim.* **17** (1872) 482; (b) M. M. Schutzenberger and Fontaine, *Bull. Soc. Chim.* **18** (1872) 101; (c) M. M. Schutzenberger and Fontaine, *Bull. Soc. Chim.* **18** (1872) 148.

115. A. E. Arbuzov and V. M. Zoroastrova, *Izv. AN SSSR, Ser. Khim.* (1952) 818.

116. A. Rosenheim and W. Levy, *Z. Anorg. Chem.* **43** (1905) 334.

117. V. Yu. Kukushkin and N. P. Kiseleva, *Koord. Khim.* **14** (1988) 693.

118. N. V. Vorobiov-Desiatovskii, V. Yu. Kukushkin, Yu. N. Lukin, K. M. Patrabansh, and N. L. Morozov, *Zh. Obsch. Khim.* **60** (1990) 226.

119. V. Yu. Kukushkin and V. M. Tkachuk, *Koord. Khim.* **15** (1989) 136.

120. (a) D. E. C. Corbridge, *Phosphorus: An Outline of its Chemistry, Biochemistry and Technology*, Fourth Edition (Elsevier, Amsterdam, 1990); (b) J. I. G. Cadogan, *Quart. Rev.* **22** (1968) 222.

121. V. Yu. Kukushkin, S. V. Yakovlev, and V. B. Ukraintsev, *Koord. Khim.* **14** (1988) 969.

122. W. Beck and B. Purucker, *Chem. Ber.* **107** (1974) 3476.

123. (a) V. Yu. Kukushkin and V. M. Tkachuk, *Zh. Neorg. Khim.* **32** (1987) 3118; (b) V. Yu. Kukushkin, V. M. Tkachuk, and V. B. Lebedev, *Zh. Neorg. Khim.* **34** (1989) 2355; (c) V. Yu. Kukushkin, V. M. Tkachuk, A. M. Popov, and M. B. Egorova, *Zh. Neorg. Khim.* **34** (1989) 1439.

124. (a) V. Yu. Kukushkin, E. Yu. Pankova, S. A. Simanova, S. S. Sotman, V. S. Fundamensky, I. G. Zenkevich, and I. A. Poliakova, *Zh. Obsch. Khim.* **60** (1990) 587; (b) V. Yu. Kukushkin, V. K. Belsky, E. A. Aleksandrova, E. Yu. Pankova, V. E. Konovalov, V. N. Yakovlev, and A. I. Moiseev, *Zh. Obsch. Khim.* **61** (1991) 318; (c) V. Yu. Kukushkin, *Uspekhi Khim.* (*Russ. Chem. Rev.*) **59** (1990) 1453.

125. (a) R. C. Murray, L. Blum, A. H. Liu, and R. R. Schrock, *Organometallics* **4** (1985) 953; (b) A. H. Liu, R. C. Murray, J. C. Dewan, B. D. Santarsiero, and R. R. Schrock, *J. Am. Chem. Soc.* **109** (1987) 4282; (c) J. F. Buzinkai and R. R. Schrock, *Inorg. Chem.* **28** (1989) 2837; (d) J. W. Faller and Y. Ma, *J. Organomet. Chem.* **340** (1988) 59; (e) C. G. De Azevedo, A. R. Dias, M. H. Garcia, P. W. Jolly, C. C. Romao, and M. C. Shohoji, *XXVI Int. Conf. Coord. Chem.* Book of Abstracts (Porto, 1988) p. B70; (f) M. Scheer, A. Kolbe, E. Herrmann, V. P. Fedin, V. E. Fedorov, V. N. Ikorski, and M. A. Fedotov, *Z. Anorg. Allg. Chem.* **567** (1988) 111.

126. (a) S. M. Godfrey, D. G. Kelly, C. A. McAuliffe, A. G. Mackie, R. G. Pritchard, and S. M. Watson, *J. Chem. Soc., Chem. Commun.* (1991) 1163; (b) N. Bricklebank, S. M. Godfrey, A. G. Mackie, C. A. McAuliffe, and R. G. Pritchard, *J. Chem. Soc., Chem. Commun.* (1992) 355; (c) N. Bricklebank, S. M. Godfrey, A. G. Mackie, C. A. McAuliffe, P. G. Pritchard, and P. J. Kobryn, *J. Chem. Soc., Dalton Trans.* (1993) 101; (d) N. Bricklebank, S. M. Godfrey, H. P. Lane, C. A. McAuliffe, R. G. Pritchard, and J.-M. Moreno, *J. Chem. Soc., Dalton Trans.* (1995) 421.

127. Y. Wielstra, S. Gambarotta, A. L. Spek, and W. J. J. Smeets, *Organometallics* **9** (1990) 2142.

128. (a) S. M. Godfrey, D. G. Kelly, A. G. Mackie, P. P. Mac Rory, C. A. McAuliffe, R. G. Pritchard, and S. M. Watson, *J. Chem. Soc., Chem. Commun.* (1991) 1447; (b) C. A. McAuliffe, S. M. Godfrey, A. G. Mackie, and R. G. Pritchard, *J. Chem. Soc., Chem. Commun.* (1992) 483; (c) N. Bricklebank, S. M. Godfrey, C. A. McAuliffe, A. G. Mackie, and R. G. Pritchard, *J. Chem. Soc., Chem. Commun.* (1992) 944; (d) C. A. McAuliffe, S. M. Godfrey, A. G. Mackie, and R. G. Pritchard, *Angew. Chem. Int. Ed. Engl.* **31** (1992) 919; *Angew. Chem.* **104** (1992) 932; (e) S. M. Godfrey, C. A. McAuliffe, and R. G. Pritchard, *J. Chem. Soc., Dalton Trans.* (1993) 371.

129. (a) J. S. Pizey, *Synthetic Reagents* (Ellis Horwood, New York, 1974) p. 321; (b) K. Oka, *Synthesis* (1981) 661.

130. (a) P. W. Schenk and H. Triebel, *Z. Anorg. Allg. Chem.* **229** (1936) 305; (b) R. A. Hubbard and W. F. Luder, *J. Am. Chem. Soc.* **73** (1951) 1327.

131. (a) T. Yamahara, T. Deguchi, and S. Inokuma, *Japan. Kokai* 75,111,075 (1975); *Chem. Abstr.* **84** (1976) 121654j; (b) T. Yamahara, T. Deguchi, and S. Nakamura, *Ger. Offen.* 2,416,683 (1974); *Chem. Abstr.* **82** (1975) 98736k.

132. M. A. Englin and M. D. Vorobiov, *Zh. Obsch. Khim.* **42** (1972) 146.

133. G. Schmid and G. Ritter, *Z. Anorg. Allg. Chem.* **415** (1975) 97.

134. J. Barry, J. G. H. Du Preez, E. Els, H. E. Rohwer, and P. J. Wright, *Inorg. Chim. Acta* **53** (1981) L17.

135. W. H. Beattie, W. B. Maier, and R. F. Holland, *J. Inorg. Nucl. Chem.* **40** (1978) 1885.

136. H. Schmidbaur and P. Jandik, *Inorg. Chim. Acta* **74** (1983) 97.

137. (a) K-H. Thiele and L. Oswald, *Z. Anorg. Allg. Chem.* **423** (1976) 231; (b) F. Bottomley, J. Darkwa, L. Sutin, and P. S. White, *Organometallics* **5** (1986) 2165.

138. U. Belluco, L. Cattalini, A. Peloso, and G. Tagliavini, *Ric. Sci. Rend. Sez. A.* **3** (1963) 1107; *Chem. Abstr.* **61** (1964) 677f.

139. H. G. Alt, T. Frister, E. E. Trapl, and H. E. Engelhardt, *J. Organomet. Chem.* **362** (1989) 125.

140. (a) J. F. Myers, A. B. P. Lever, and G. W. Rayner Canham, *Proc. Int. Conf. Coord. Chem. 16th.* (1974) 15; *Chem. Abstr.* **85** (1976) 55930p; (b) J. F. Myers, G. W. Rayner Canham, and A. B. P. Lever, *Inorg. Chem.* **14** (1975) 461; (c) M. Hanack and R. Fay, *Rec. Trav. Chim.* **105** (1986) 427.

141. (a) A. Hammershøi, R. M. Hartshorn, and A. M. Sargeson, *Inorg. Chem.* **29** (1990) 4525; (b) A. Hammershøi, R. M. Hartshorn, and A. M. Sargeson, *J. Chem. Soc., Dalton Trans.* (1991) 621; (c) R. M. Hartshorn, A. C. Willis, and A. M. Sargeson, *J. Chem. Soc., Chem. Commun.* (1988) 1269; (d) L. Grøndahl, A. Hammershøi, R. M. Hartshorn, and A. M. Sargeson, *Inorg. Chem.* **30** (1991) 1800; (e) L. Grøndahl, A. Hammershøi, R. M. Hartshorn, and A. M. Sargeson, *personnal communication.*

142. (a) R. A. Vanderpool and H. B. Abrahamson, *Inorg. Chem.* **24** (1985) 2985; (b) S. J. Markham, Y. L. Chung, G. D. Branum, and D. M. Blake, *J. Organomet. Chem.* **107** (1976) 121.

143. (a) A. Winter, L. Zsolnai, and G. Huttner, *J. Organomet. Chem.* **234** (1982) 337; (b) A. Winter, L. Zsolnai, and G. Huttner, *J. Organomet. Chem.* **232** (1982) 47.

144. (a) I. P. Lorenz, J. Messelhäuser, W. Hiller, and M. Conrad, *J. Organomet. Chem.* **316** (1986) 121; (b) I. P. Lorenz, W. Hiller, and M. Conrad, *Z. Naturforsch.* **40b** (1985) 1383; (c) I. P. Lorenz and J. Messelhäuser, *24th Int. Conf. Coord. Chem.*, Book of Abstracts (1986) 394.

145. R. Bartosiewicz, V. Mecznikowska-Stolarcik, and B. Opszondec, *Methods for Reducing Organic Compounds* [in Russian] (Inostr. Lit., Moscow, 1960) p. 43.

146. (a) A. R. Manning, G. McNally, D. Cunningham, P. McArdle, and J. M. Simmie, *J. Organomet. Chem.* **338** (1988) 383; (b) A. R. Manning, G. McNally, R. Davis, and C. C. Rowland, *J. Organomet. Chem.* **259** (1983) C15.

147. R. Davis, J. L. A. Durrant, and C. C. Rowland, *J. Organomet. Chem.* **315** (1986) 119.

148. M. Wieber, J. Walz, and C. Burschka, *Z. Anorg. Allg. Chem.* **585** (1990) 65.

149. (a) H. J. Lucas and E. R. Kennedy, *Org. Synth.* **3** (1955) 482; (b) E. M. Archer and T. G. D. van Schalkwyk, *Acta Cryst.* **6** (1953) 88; (c) A. K. Mishra, M. M. Olmstead, J. J. Ellison, and P. P. Power, *Inorg. Chem.* **34** (1995) 3210.

150. F. A. Cotton, A. C. Price, and K. Vidyasagar, *Inorg. Chem.* **29** (1990) 5143.

151. F. A. Cotton, C. T. Eagle, and A. C. Price, *Inorg. Chem.* **27** (1988) 4362.

152. F. Calderazzo, M. Castellani, G. Pampaloni, and P. F. Zanazzi, *J. Chem. Soc., Dalton Trans.* (1985) 1989.

153. J. Vicente, A. Arcas, M. Mora, X. Solans, and M. Font-Altaba, *J. Organomet. Chem.* **309** (1986) 369.

154. H. H. Murray, D. A. Briggs, G. Garzùn, R. G. Raptis, L. C. Porter, and J. P. Fackler, Jr., *Organometallics* **6** (1987) 1992.

155. S. Wang and J. P. Fackler, Jr., *Organometallics* **9** (1990) 111.

156. C. M. Giandomenico, L. H. Hanau, and S. J. Lippard, *Organometallics* **1** (1982) 142.

157. S. G. Bott, N. G. Connelly, M. Green, N. C. Norman, A. Guy Orpen, J. F. Paxton, and C. J. Schaverien, *J. Chem. Soc., Chem. Commun.* (1983) 378.

158. (a) R. Hoppenheit, W. Isenberg, and R. Mews, *Z. Naturforsch.* **37b** (1982) 1116; (b) H. W. Roesky, M. Thomas, M. Noltemeyer, and G. M. Sheldrick, *Angew. Chem.* **94** (1982) 861; (c) H. W. Roesky, M. Thomas, J. W. Bats, and H. Fuess, *J. Chem. Soc., Dalton Trans.* (1983) 1891.

159. P. Gowik, T. Klapötke, and J. Pickardt, *Organometallics* **8** (1989) 2953.

160. P. R. Brown, F. G. N. Cloke, and M. L. H. Green, *Polyhedron* **4** (1985) 869.

161. (a) A. Mayr and G. A. McDermott, *J. Am. Chem. Soc.* **108** (1986) 548; (b) A. Mayr, G. A. McDermott, and A. M. Dorries, *Organometallics* **4** (1985) 608.

162. J. L. Namy, P. Girard, H. B. Kagan, and P. E. Caro, *J. Nouv. Chim.* **5** (1981) 479.

163. D. G. Karraker, *Plutonium Chem. Symp. 184th Meet. Am. Chem. Soc.* (1984) 41.

164. I. Al-Uajjar, *Inorg. Chim. Acta* **127** (1987) L47.

165. (a) T.-Y. Luh, *Coord. Chem. Rev.* **60** (1984) 255; (b) M. O. Albers and N. J. Coville, *Coord. Chem. Rev.* **53** (1984) 227; (c) R. D. Adams in *The Chemistry of Metal Cluster Complexes*, eds. D. F. Shriver, H. D. Kaesz, and R. D. Adams (VCH, New York, 1990) p. 128.

166. A. Basu, S. Bhaduri, H. Khwaja, P. G. Jones, K. Meyer-Bäse, and G. M. Sheldrick, *J. Chem. Soc., Dalton Trans.* (1986) 2501.

167. (a) J. A. Smieja and W. L. Gladfelter, *Inorg. Chem.* **25** (1986) 2667; (b) F. Porta, M. Pizzotti, and S. Cenini, *J. Organomet. Chem.* **222** (1981) 279.

168. (a) H. Alper and J. T. Edward, *Can. J. Chem.* **48** (1970) 1543; (b) J. Charalambous, L. I. B. Haines, J. S. Morgan, D. S. Peat, M. J. M. Campbell, and J. Bailey, *Polyhedron* **6** (1987) 1027; (c) A. Dondoni and G. Barbaro, *J. Chem. Soc., Chem. Commun.* 1975, 761; (d) M. Nitta, I. Sasaki, H. Miyano, and T. Koboyashi, *Bull. Chem. Soc. Japan* **57** (1984) 3357; (e) M. Langenbahn, H. Stoeckli-Evans, and G. Süss-Fink, *Helv. Chim. Acta* **74** (1991) 549; (f) K. Kaneda, K. Doken, and T. Imanaka, *Chem. Lett.* (1988) 285; (g) K. Kaneda, K. Fujita, T. Takemoto, and T. Imanaka, *Bull. Chem. Soc. Japan* **64** (1991) 602; (h) M. Akazome, Y. Tsuji, and Y. Watanabe, *Chem. Lett.* (1990) 635.

169. (a) R. E. Stevens and W. L. Gladfelter, *Inorg. Chem.* **22** (1983) 2034; (b) R. E. Stevens, P. C. C. Liu, and W. L. Gladfelter, *J. Organomet. Chem.* **287** (1985) 133; (c) K. Henrick, B. F. G. Johnson, J. Lewis, J. Mace, M. McPartlin, and J. Morris, *J. Chem. Soc., Chem. Commun.* (1985) 1617; (d) E. Delgado, J. C. Jeffery, N. D. Simmons, and F. G. A. Stone, *J. Chem. Soc., Dalton Trans.* (1986) 869; (e) D. R. Mantell and W. L. Gladfelter, *Organomet. Chem.* **4** (1988) 56; (f) R. E. Stevens, D. E. Fjare, and W. L. Gladfelter, *J. Organomet. Chem.* **347** (1988) 373.

170. (a) D. J. Berg, C. J. Burns, R. A. Andersen, and A. Zalkin, *Organometallics* **8** (1989) 1865; (b) G. A. Vaughan, P. B. Rupert, and G. L. Hillhouse, *J. Am. Chem. Soc.* **109** (1987) 5538; (c) C. Barrientos, C. K. Ghosh, W. A. G. Graham, and M. J. Thomas, *J. Organomet. Chem.* **394** (1990) C31.

171. (a) P. V. Bonnesen, P. K. L. Yau, and W. H. Hersch, *Organometallics* **6** (1987) 1587; (b) M. G. Richmond and J. K. Kochi, *Inorg. Chem.* **26** (1987) 541.

172. (a) V. Yu. Kukushkin, *Coord. Chem. Rev.* **139** (1995) 375; (b) M. Xue, Y.-C. Gao, J.-K. Shen, Q.-Z. Shi, and F. Basolo, *Inorg. Chim. Acta* **207** (1993) 207.

173. W. Hieber and A. Lipp, *Chem. Ber.* **92** (1959) 2085.

174. J. E. Kmeicik, *J. Org. Chem.* **30** (1965) 2014.

175. W. Beck, A. Mielert, and E. Schier, Z. *Naturforsch.* **24b** (1969) 936.

176. H. Alper and J. T. Edward, *Can. J. Chem.* **48** (1970) 1543.

177. (a) Y. Shvo and E. Hazum, *J. Chem. Soc., Chem. Commun.* (1974) 336; (b) Y. Shvo and E. Hazum, *J. Chem. Soc., Chem. Commun.* (1975) 829.

178. (a) J. Elzinga and H. Hogeveen, *J. Chem. Soc., Chem. Commun.* (1977) 705; (b) J. H. Eekhof, H. Hogeveen, and R. M. Kellogg, *J. Chem. Soc., Chem. Commun.* (1976) 657; (c) Y.-L. Shi, Y.-C. Gao, Q.-Z. Shi, D. L. Kershner, and F. Basolo, *Organometallics* **6** (1987) 1528; (d) J.-K. Shen, Y.-C. Gao, Q.-Z. Shi, and F. Basolo, *Inorg. Chem.* **27** (1988) 4236; (e) J. Shen, Y. Gao, Q. Shi, and F. Basolo, *Chem. J. Chin. Univ.* **9** (1988) 1035; (f) J.-K. Shen, Y.-C. Gao, Q.-Z. Shi, and F. Basolo, *Coord. Chem. Rev.* **128** (1993) 69.

179. J. A. Clucas, R. H. Dawson, P. A. Dolby, M. M. Harding, K. Pearson, and A. K. Smith, *J. Organomet. Chem.* **311** (1986) 153.

180. (a) B. F. G. Johnson, J. Lewis, and T. I. Odiaka, *J. Organomet. Chem.* **307** (1986) 61; (b) B. F. G. Johnson, J. Lewis, P. R. Raithby, and C. Zuccaro, *J. Chem. Soc., Chem. Commun.* (1979) 916; (c) R. D. Adams, J. E. Babin, and H.-S. Kim, *Inorg. Chem.* **25** (1986) 4319; (d) R. D. Adams, J. E. Babin, and H. S. Kim, *Inorg. Chem.* **25** (1986) 1122; (e) V. Yu. Kukushkin, M. B. Egorova, and A. M. Popov, *Koord. Khim.* **13** (1987) 1507; (f) D. J. Blumer, K. W. Barnett, and T. L. Brown, *J. Organomet. Chem.* **173** (1979) 71; (g) U. Koelle, *J. Organomet. Chem.* **133** (1977) 53.

181. (a) P. A. Chetcuti and M. F. Hawthorne, *J. Am. Chem. Soc.* **109** (1987) 9422; (b) P. A. Chetcuti, J. A. Walker, C. B. Knobler, and M. F. Hawthorne, *Organometallics* **7** (1988) 641; (c) P. A. Chetcuti, C. B. Knobler, and M. F. Hawthorne, *Organometallics* **7** (1988) 650.

182. (a) T. S. Andy Hor, *J. Organomet. Chem.* **319** (1987) 213; (b) T. S. Andy Hor, *Inorg. Chim. Acta* **128** (1987) L3; (c) T. S. Andy Hor and S. R. B. Rus, *J. Organomet. Chem.* **348** (1988) 343; (d) T. S. Andy Hor, *J. Organomet. Chem.* **340** (1988) 51; (e) T. S. Andy Hor, H. S. O. Chan, Y.-P. Leong, and M.-M. Tan, *J. Organomet. Chem.* **373** (1989) 221; (f) T. S. A. Hor and L.-T. Phang, *Polyhedron* **9** (1990) 2305.

183. V. Yu. Kukushkin, M. E. Dushina, V. K. Krilov, and A. V. Iretskii, *Koord. Khim.* **13** (1987) 650.

184. (a) P. Johnston, G. J. Hutchings, and N. J. Coville, *Inorg. Chim. Acta* **117** (1986) L11; (b) S. Guesmi, P. H. Dixneuf, N. J. Taylor, and A. J. Carty, *J. Organomet. Chem.* **328** (1987) 193; (c) V. A. Maksakov, V. A. Ershova, and I. V. Bragina, *Izv. AN SSSR, Ser. Khim.* (1985) 2829; (d) P. Johnston, G. J. Hutchings, and N. J. Coville, *Inorg. Chim. Acta* **117** (1986) L11.

185. (a) B. F. G. Johnson, J. Lewis, and D. A. Pippard, *J. Chem. Soc., Dalton Trans.* (1981) 407; (b) B. F. G. Johnson, J. Lewis, and D. A. Pippard, *J. Organomet. Chem.* **145** (1978) C4.

186. R. J. Lawson and J. R. Shapley, *J. Am. Chem. Soc.* **98** (1976) 7433.

187. (a) R. S. Dickson, A. P. Oppenheim, and G. N. Pain, *J. Organomet. Chem.* **224** (1982) 377; (b) R. S. Dickson and G. N. Pain, *J. Chem. Soc., Chem. Commun.* (1979) 277; (c) P. A. Corrigan, R. S. Dickson, S. H. Johnson, G. N. Pain, and M. Yeoh, *Austral. J. Chem.* **35** (1982) 2203.

188. (a) R. S. Dickson, S. H. Johnson, H. P. Kirsch, and D. J. Lloyd, *Acta Cryst., Sect. B* **33** (1977) 2057; (b) R. S. Dickson, G. N. Pain, and M. F. MacKay, *Acta Cryst., Sect. B* **35** (1979) 2321.

189. J. Keijsper, L. Polm, G. van Koten, K. Vrieze, G. Abbel, and C. H. Stam, *Inorg. Chem.* **23** (1984) 2142.

190. Yu. N. Kukushkin, *The Chemistry of the Coordination Compounds* [in Russian] (Khimia, Leningrad, 1985) p. 456.

191. (a) R. S. Dickson, R. J. Nesbit, H. Pateras, J. M. Patrick, and A. H. White, *J. Organomet. Chem.* **265** (1984) C25; (b) J.-K. Shen, W.-B. Wang, Y.-C. Gao, Q.-Z. Shi, E. J. Roskamp, and F. Basolo, *Gazz. Chim. Ital.* **122** (1992) 225.

192. (a) C. M. Lukehart and J. V. Zeile, *J. Organomet. Chem.* **97** (1975) 421; (b) F. Seitz, H. Fisher, J. Riede, and J. Vogel, *Organometallics* **5** (1986) 2187; (c) A. G. M. Bazzeett and M. A. Sturgess, *Tetrahedron Lett.* **27** (1986) 3811.

193. (a) W. E. Buhro, A. T. Patton, C. E. Strouse, J. A. Gladysz, F. B. McCormick, and M. C. Etter, *J. Am. Chem. Soc.* **105** (1983) 1056; (b) W. E. Buhro, S. Georgiou, J. M. Fernandez, A. T. Patton, C. E. Strouse, and J. A. Gladysz, *Organometallics* **5** (1986) 956.

194. R. S. Iyer and J. P. Selegue, *J. Am. Chem. Soc.* **109** (1987) 910.

195. H. Fisher, J. Schmid, and S. Zeuner, *Chem. Ber.* **120** (1987) 583.

196. (a) J. T. Anhaus, T. P. Kee, M. H. Schofield, and R. R. Schrock, *J. Am. Chem. Soc.* **112** (1990) 1642; (b) M. A. Edelman, P. B. Hitchcock, and M. F. Lappert, *J. Chem. Soc., Chem. Commun.* (1990) 1116; (c) F. Bottomley, D. F. Drummond, D. E. Paez, and P. S. White, *J. Chem. Soc., Chem. Commun.* (1986) 1752; (d) C. M. Dicken, T. C. Woon, and T. C. Bruice, *J. Am. Chem. Soc.* **108** (1986) 1636; (e) K. Shin and H. M. Goff, *J. Am. Chem. Soc.* **109** (1987) 3140; (f) F. Bottomley, D. E. Paez, and P. S. White, *J. Am. Chem. Soc.* **107** (1985) 7226.

197. S. Wolowiec and J. K. Kochi, *Inorg. Chem.* **30** (1991) 1215.

198. (a) W. Beck, M. Keubler, E. Leidl, U. Nagel, M. Schaal, S. Cenini, P. Del Buttero, E. Licandro, S. Maiorana, and A. Chiesi Villa, *J. Chem. Soc., Chem. Commun.* (1981) 446; (b) V. K. Krilov, M. E. Dushina, V. Yu. Kukushkin, and A. V. Iretskii, *Koord. Khim.* **14** (1988) 383; (c) L. K. Peterson and S. Huang, *Inorg. Chim. Acta* **203** (1993) 87.

199. W. E. Buhro, B. D. Zwick, S. Georgiou, J. P. Hutchinson, and J. A. Gladysz, *J. Am. Chem. Soc.* **110** (1988) 2427.

200. M. F. Hallam and D. M. P. Mingos, *J. Organomet. Chem.* **315** (1986) C35.

201. (a) R. Köster and Y. Morita, *Angew. Chem. Int. Ed. Engl.* **5** (1966) 580; (b) A. G. Davies and B. P. Roberts, *J. Chem. Soc., Ser. C* (1968) 1474; (c) G. W. Kabalka and H. C. Hedgecock, *J. Org. Chem.* **40** (1975) 1776; (d) G. Kabalka and S. W. Slayden, *J. Organomet. Chem.* **125** (1977) 273; (e) G. W. Kabalka and R. J. Newton, *J. Organomet. Chem.* **156** (1978) 65.

202. (a) C. G. Pierpont and C. W. Lange, *Prog. Inorg. Chem.* **41** (1994) 331; (b) C. G. Pierpont and R. M. Buchanan, *Coord. Chem. Rev.* **38** (1981) 45.

203. A. L. Balch, V. J. Catalano, and M. M. Olmstead, *Inorg. Chem.* **29** (1990) 1638 and references therein.

204. (a) H. Bock, W. Kaim, H. Nöth, and A. Semkov, *J. Am. Chem. Soc.* **102** (1980) 4421; (b) H. Bock and W. Kaim, *Chem. Ber.* **111** (1978) 3552; (c) H. Bock, G. Brähler, U. Henkel, R. Schlecker, and D. Seebach, *Chem. Ber.* **113** (1980) 289.

205. J. A. Connor, E. J. James, C. Overton, and N. El Murr, *J. Organomet. Chem.* **218** (1981) C31.

206. A. H. Klahn-Oliva, R. D. Singer, J. M. Aramini, and D. Sutton, *Inorg. Chem.* **28** (1989) 4217.

207. (a) J. B. Torrance, *Acc. Chem. Res.* **12** (1979) 79; (b) L. R. Melby, R. J. Harder, W. R. Hertler, W. Mahler, R. E. Benson, and W. E. Mochel, *J. Am. Chem. Soc.* **84** (1962) 3374.

208. (a) W. Kaim and M. Moscherosch, *Coord. Chem. Rev.* **129** (1994) 157; (b) A. J. Fatiadi, *Synthesis* (1986) 249; (c) A. J. Fatiadi, *Synthesis* (1987) 959; (d) J. K. Kochi, *Organometallic Mechanisms and Catalysis* (Academic Press, New York, 1978) Chapters 15 and 16; (e) H. Endres, in *Extended Linear Chain Compounds*, ed. J. S. Miller (Plenum Press, New York, 1983) Vol. 3.

209. U. Honrath and H. Vahrenkamp, *Z. Naturforsch.* **39b** (1984) 555.

210. A. Vlcek and A. A. Vlcek, *Inorg. Chim. Acta* **64** (1982) L273.

211. J. S. Miller, D. T. Glatzhofer, D. M. O'Hare, W. M. Reiff, A. Chakraborty, and A. J. Epstein, *Inorg. Chem.* **28** (1989) 2930.

212. (a) A. L. Crumbliss and F. Basolo, *Inorg. Chem.* **10** (1971) 1676; (b) W. Beck, R. Schlodder, and K. H. Lechler, *J. Organomet. Chem.* **54** (1973) 303; (c) F. G. Moers and J. P. Langford, *J. Inorg. Nucl. Chem.* **39** (1977) 591; (d) W. Sacher, U. Nagel, and W. Beck, *Chem. Ber.* **120** (1987) 895.

213. S. L. Amer, T. P. Dasgupta, and P. M. Henry, *Inorg. Chem.* **22** (1983) 1970.

214. R. Gross-Lannert, W. Kaim, and B. Olbrich-Deussner, *Inorg. Chem.* **29** (1990) 5046.

215. (a) R. B. King and M. B. Bisnette, *Inorg. Chem.* **4** (1965) 482; (b) S. D. Killops and S. A. R. Knox, *J. Chem. Soc., Dalton Trans.* (1978) 1260; (c) H. B. Abrahamson, M. L. Freeman, M. Bilayet Hossain, and D. van der Helm, *Inorg. Chem.* **23** (1984) 2286.

216. R. Kumar, H. E. Mabrouk, and D. G. Tuck, *Inorg. Synth.* **29** (1992) 15.

217. (a) J. W. Seyler, P. E. Fanwick, and C. R. Leidner, *Inorg. Chem.* **31** (1992) 3699; (b) K. S. Chan and Y.-B. Leung, *Inorg. Chem.* **33** (1994) 3187.

218. E. J. Corey and G. Schmidt, *Tetrahedron Lett.* (1979) 399.

219. M.-C. Lim and D. T. Sawyer, *Inorg. Chem.* **21** (1982) 2839.

220. A. J. Fatiadi, *Synthesis* (1987) 85.

221. T. Sala and M. V. Sargent, *J. Chem. Soc., Chem. Commun.* (1978) 253.

222. (a) J. A. Morris and D. C. Mills, *Chem. Ber.* **14** (1978) 326; (b) M. Jager, J. Lutolf, and M. W. Meyer, *Angew. Chem. Int. Ed. Engl.* **18** (1979) 786; (c) H.-J. Schmidt and H. J. Schafer, *Angew. Chem. Int. Ed. Engl.* **18** (1979) 787; (d) B. P. Leddy, M. A. McKervey, and P. McSweeney, *Tetrahedron Lett.* **21** (1980) 2261.

223. H. Bock and D. Jaculi, *Angew. Chem.* **96** (1984) 298; *Angew. Chem. Int. Ed. Engl.* **23** (1984) 305.

224. (a) J. B. Vincent, K. Folting, J. C. Huffman, and G. Christou, *Inorg. Chem.* **25** (1986) 996; (b) E. Bouwman, G. Christou, E. Lobkovsky, and J. C. Huffman, *J. Inorg. Biochem.* **43** (1991) 377.

225. N. Auger, J.-J. Girerd, M. Corbella, A. Gleizes, and J.-L. Zimmermann, *J. Am. Chem. Soc.* **112** (1990) 448.

226. W. P. Griffith, *Chem. Soc. Rev.* **21** (1992) 179.

227. A. C. Dengel, W. P. Griffith, and R. A. Hudson, *Transit. Met. Chem.* **10** (1985) 98.

228. W. P. Griffith and S. V. Ley, *Aldrichim. Acta* **23** (1990) 13.

229. (a) A. C. Dengel and W. P. Griffith, *Inorg. Chem.* **30** (1991) 869; (b) A. C. Dengel, J. F. Gibson, and W. P. Griffith, *J. Chem. Soc., Dalton Trans.* (1991) 2799.

230. P. K. Mascharak, *Inorg. Chem.* **25** (1986) 245.

231. (a) R. H. Magnuson, *Inorg. Chem.* **23** (1984) 387; (b) B. J. Kennedy and G. A. Heath, *Inorg. Chim. Acta* **187** (1991) 149.

232. (a) D. N. Hendrickson, Y. S. Sohn, and H. B. Gray, *Inorg. Chem.* **10** (1971) 1559; (b) M. J. Abrams, A. Davison, A. G. Jones, and C. E. Costello, *Inorg. Chim. Acta* **77** (1983) L235.

233. (a) J. L. Robbins, N. Edelstein, B. Spencer, and J. C. Smart, *J. Am. Chem. Soc.* **104** (1982) 1882; (b) P. J. Krusic, J. San Filippo, B. Hutchinson, R. L. Hance, and L. M. Daniels, *J. Am. Chem. Soc.* **103** (1981) 2129; (c) P. J. Krusic, *J. Am. Chem. Soc.* **103** (1981) 2131.

234. A. Debrov, M. P. Mahajan, M. N. Bhattacharjcc, and M. K. Chaudhuri, *J. Chem. Soc., Dalton Trans.* (1983) 2531.

235. (a) V. P. Dyadchenko, *Uspekhi Khimii* **51** (1982) 467; (b) R. J. Puddephatt, *The Chemistry of Gold* (Elsevier, New York, 1978) p. 115.

Chapter 9. REDUCTANTS IN PREPARATIVE COORDINATION CHEMISTRY

Chemical reduction is used widely in the synthesis of coordination compounds because the oxides and halides of metals in high oxidation states are common starting materials. However, despite the large number of reductants that are available, only a limited group is used typically in preparative coordination chemistry.

For example, gaseous hydrogen, a common reductant employed in catalytic hydrogenation reactions in organic chemistry, is often not the reductant of choice in synthetic coordination chemistry. It is known that hydrogen is capable of undergoing oxidative addition reactions with a number of transition metal complexes, including those of the noble metals. In such reactions hydrogen acts as a formal oxidant. At the technical level, measurement of small volumes of hydrogen for syntheses on the milligram scale may be problematic.

Complex hydrides, particularly those of boron and aluminum, often are employed as reductants in coordination chemistry. The use of boron and aluminum hydrides in this area is the subject of Chapter 10. The reduction of complexes at electrodes is a promising direction in the synthesis of coordination compounds. Electrosynthesis is characterized by high reaction selectivity because the electrode potential is controllable. The use of this method in preparative coordination chemistry is the subject of Chapter 7. The present Chapter is devoted to a discussion of other reductants useful as reagents in synthetic coordination chemistry.

9.1 Reduction by metals and amalgams

In preparative organic chemistry the alkali metals, as well as some less active metals such as zinc, magnesium, iron, tin and others, comprise a widely used group of reductants. These reductants have been used in the synthesis of organic compounds for more than 140 years.[1] Reduction by active metals is being employed increasingly in preparative coordination and organometallic chemistry. The reducing ability of metals is evaluated on the basis of the reduction potentials of the corresponding metal cations. Selected standard reduction potentials in aqueous solution at 25 °C appear in Table 9.1. Reactions with active metals are typically carried out heterogeneously.

In order to lower the activity of alkali metals, amalgams are used. The potentials of common amalgams in water are approximately 1*V* higher than the potentials of the corresponding alkali metals.[2,3] These differences are attributed to the significant exothermic effect accompanying the formation of amalgams. As reductants, amalgams have a number of advantages over alkali metals. For example, in the presence of mercury the reductant becomes mobile, facilitating stirring and heat transfer and providing better contact between the reductant and the oxidant. Sodium amalgam, containing less than 1.25% sodium, retains its liquid or semi-liquid state due to the low percentage of alkali metal present.[4] The most widely used solvent for work with Na/Hg in non-aqueous media is tetrahydrofuran, but diethyl ether, dimethoxyethane, benzene, and toluene are utilized also.

Table 9.1

Standard reduction potentials[1]

Half reaction (acidic solution)	Standard reduction potential, E°,*V*
$Li^{+}(aq) + e^{-} \rightleftharpoons Li(s)$	-3.045
$K^{+}(aq) + e^{-} \rightleftharpoons K(s)$	-2.925
$Na^{+}(aq) + e^{-} \rightleftharpoons Na(s)$	-2.714
$Al^{3+}(aq) + 3e^{-} \rightleftharpoons Al(s)$	-1.66
$Zn^{2+}(aq) + 2e^{-} \rightleftharpoons Zn(s)$	-0.763
$Fe^{2+}(aq) + 2e^{-} \rightleftharpoons Fe(s)$	-0.44
$Sn^{2+}(aq) + 2e^{-} \rightleftharpoons Sn(s)$	-0.14

In anionic carbonyl complexes the metal center is often found in a negative oxidation state and this is reflected in the standard potentials of metal carbonyls. Thus, the standard potentials of the following couples are -0.68, -0.40, and -0.74 *V*, respectively,[5] Eqs. 9.1 - 9.3:

$$[Mn_2(CO)_{10}] + 2e^{-} \rightleftharpoons 2[Mn(CO)_5]^{-} \quad (9.1)$$

$$[Co_2(CO)_8] + 2e^{-} \rightleftharpoons 2[Co(CO)_4]^{-} \quad (9.2)$$

$$[Fe_3(CO)_{12}] + 6e^{-} \rightleftharpoons 3[Fe(CO)_4]^{2-} \quad (9.3)$$

Syntheses of the carbonyl anions are most often carried out with alkali metals or their amalgams. For example, the complex, $Na[Fe(CO)_2(C_5H_4Me)]$, is obtained as a result of reduction of the dimer, $[Fe(CO)_2(C_5H_4Me)]_2$, by sodium amalgam in THF.[6] The reactions of $[M(CO)_3Cp]$ (M = Nb, Ta) with metallic sodium in liquid ammonia give $Na_2[M(CO)_3Cp]$.[7] Loubster and Lotz[8] carried out the reduction of $[Co(CO)_3(PPh_3)]_2$ with Na/Hg in tetrahydrofuran. The reaction produced $Na[Co(CO)_3(PPh_3)]$.

Sodium amalgam is used widely to reduce homobinuclear carbonyl or nitrosyl organometallic complexes. Such reduction results in the formation of anionic compounds. For example, $[Co_2(CO)_8]$ is reduced in tetrahydrofuran with sodium amalgam producing $Na[Co(CO)_4]$.[9] Several investigators[9-11] have carried out reactions of this type, Eq. 9.4:

$$[ML_nCp']_2 + 2Na/Hg \rightarrow 2Na[ML_nCp'] \quad (9.4)$$

Cp' = Cp, M = Cr, Mo, W, L = CO, n = 3;[9]
Cp' = $C_5H_4C_2H_4OH$, M = Mo, L = CO, n = 3;[10]
Cp' = Cp, M = Co, L = NO, n = 1.[11]

Methods of reduction by dissolving metals, as well as by amalgams, are used to vary the oxidation state of the metal center in complexes containing anionic ligands (usually halide ions). For example, the platinum(0) complex, $[Pt(PPhBu_2)_2]$, was obtained when the platinum(II) complex, $[PtCl_2(PPhBu_2)_2]$, was reduced with sodium amalgam in THF.[12] It is known that complexes of metals in lower oxidation states are stabilized by ligands with π-acceptor properties.[13] Accordingly, syntheses in which reduction of the metal center occurs often are carried out in the presence of such ligands. For example, the reduction of $[MoCl_3\{N(CH_2CH_2PPh_2)_3\}]$ with a sodium dispersion in tetrahydrofuran has been carried out in a nitrogen atmosphere and produced *cis*-$[Mo(N_2)_2\{N(CH_2CH_2PPh_2)_3\}]$.[14]

The reaction of $[MCl_2Cp_2]$ (M = Zr, Hf) with Mg/Hg in THF in the presence of CO yields $[M(CO)_2Cp_2]$.[15] Further, the reduction of $[TcCl_4\{P(p\text{-tol})_3\}_2]$ with zinc powder in degassed refluxing acetonitrile in the presence of $P(p\text{-tol})_3$ gives the air-sensitive technetium(III) solvento-complex, $[TcCl_3(MeCN)\{P(p\text{-tol})_3\}_2]$.[16] Its structure is shown in Figure 9.1 (only the carbon atoms of *p*-tolyl rings bonded directly to phosphorus are shown for clarity). Technetium(III) is stabilized by the MeCN and $P(p\text{-tol})_3$ ligands, which are π-acceptors.

The reduction of metal complexes is accompanied often by dimerization and/or by formation of compounds with metal-metal bonds. For example, the reduction of $[FeBr(CO)_2(\eta^5\text{-}C_5(p\text{-tol})_5)]$ with zinc powder in THF at room temperature[17] yields *trans*-$[Fe(CO)_2(\eta^5\text{-}C_5(p\text{-tol})_5)]_2$, Scheme 9.1. Reduction of cyclopentadienyl complexes of zirconium and hafnium, $[MCl_2Cp_2]$, by sodium amalgam in THF produces the dimers, $[M(\mu\text{-}Cl)Cp_2]_2$.[18]

The metal-metal bonded hafnium(III) dimer, $[Hf_2Cl_6(PEt_3)_4]$, has been synthesized[19] by treatment of a toluene suspension of $HfCl_4$ with sodium-potassium alloy in the presence of triethylphosphine, Eq. 9.5.

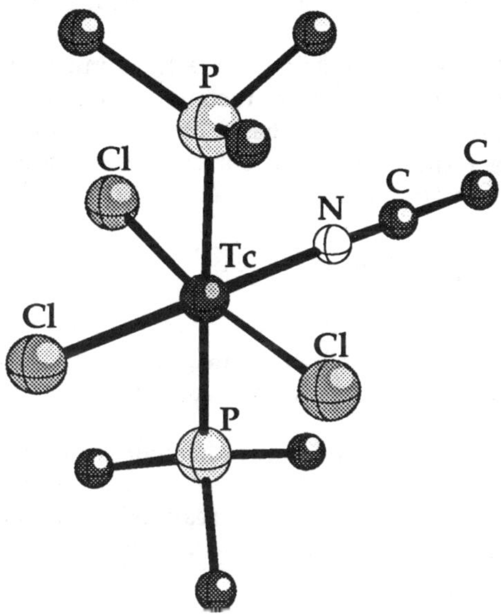

Figure 9.1

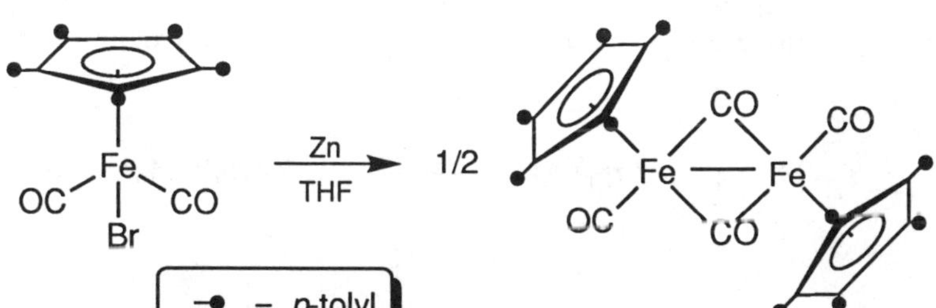

Scheme 9.1

$$2HfCl_4 + 2Na/K + 4PEt_3 \rightarrow [Hf_2Cl_6(PEt_3)_4] + 2(Na/K)Cl \qquad (9.5)$$

Sodium-potassium alloy is the preferred reductant[19,20] because $HfCl_4$ is reduced slowly, or not at all, by other common reductants such as sodium or sodium amalgam.

The possibility of transmetalation reactions must be taken into account when active metals are used as reductants. For example, Coltrain and Jackels[21] have described the synthesis of zinc complexes of macrocyclic ligands, L, by the reduction of the corresponding copper compounds by metallic zinc, which is accompanied by transmetalation, Eq. 9.6:

$$[CuL]^{2+} + Zn \rightarrow [ZnL]^{2+} + Cu \qquad (9.6)$$

Zinc amalgam in acetonitrile is an effective reductant in reactions of this kind.[22] Complexes of zinc with ethylenediamine, as well as with nitrogen-containing macrocyclic ligands, have been obtained by this method.

9.2 Reduction by organic radical-anions and anions

Both metallic sodium and lithium react with aromatic hydrocarbons to form relatively stable radical-anions, Scheme 9.2:

$$M + ArH \rightleftharpoons ArH^{\bar{\bullet}}M^+$$

Scheme 9.2

Such radical-anions form when the aromatic acceptor has sufficiently low-lying, unfilled π-orbitals. In the case of polycyclic aromatics (with low-lying LUMOs), reactions with alkali metals can result in the formation of dianions as a result of the addition of a second electron to the radical-anion, Scheme 9.3:

$$M + ArH^{\bar{\bullet}}M^+ \rightleftharpoons (ArH)^{2-}M_2^+$$

Scheme 9.3

The nature of the solvent has a significant effect on the equilibria that lead to radical anions. The reactions usually proceed rapidly to completion in dimethyl ether, dimethoxyethane, tetrahydrofuran, or liquid ammonia. It is believed[23,24] that, in these solvents, the energies of solvation of the alkali metal cations that are formed are large.

In the last decade reactions involving reduction by aromatic radical-anions have been used increasingly in preparative organometallic chemistry. Syntheses with $ArH^{\bullet -}M^+$ have much in common with reactions involving reduction by amalgams. In a number of syntheses, amalgams and aromatic radical-anions are used interchangeably as reactants. However, the radical-anions have some advantages over amalgams as reductants in synthetic coordination chemistry. For example, because $ArH^{\bullet -}M^+$ has high solubility in the solvents employed, the reduction takes place in a homogeneous manner. Variation in the reduction potential can be achieved by using different aromatic compounds, and, finally, there is little danger of transmetalation in syntheses involving $ArH^{\bullet -}M^+$.

Organouranium compounds have been prepared by reactions that involve the use of aromatic radical-anions as reductants. For example, the reduction[25] of bis(cyclooctatetraene)uranium, $[U(cot)_2]$, in THF, by a stoichiometric quantity of $Li^+Np^{\bullet -}$ yields $Li[U(cot)_2]\cdot THF$. The reaction[26] of the naphthalenide of lithium or sodium with the carbonyl-containing compounds, $[M(CO)_5(NMe_3)]$ (M = Mo, W), in THF, yields $[M(CO)_5]^{2-}$. Attempts to precipitate $[W(CO)_5]^{2-}$ by addition of a solution of [PPN]Cl in methanol to the reaction medium resulted in formation of the hydride complex, $[PPN][WH(CO)_5]$.[27] Darensbourg and Slater have shown that the alcohol is the source of hydrogen.[27] It has been established[28] that the anionic complex, $[Mo(CO)_2(bipy)_2]^-$, can be obtained by reduction of $[Mo(CO)_2(bipy)_2]$ with sodium naphthalenide or sodium amalgam in dimethoxyethane. The synthesis of high purity $Na_2[M_2(CO)_{10}]$ (M = Cr, Mo, W) by reduction of $[M(CO)_6]$ with excess sodium in the presence of catalytic amounts of 2,2'-bipyridyl has been reported.[29] It was postulated that the reduction includes a step in which an aromatic radical-anion is formed, Eq. 9.7:

$$Na + bipy \rightleftharpoons Na^+bipy^{\bullet -} \tag{9.7}$$

which then reacts with the carbonyl complex, Eq. 9.8:

$$[M(CO)_6] + Na^+bipy^{\bullet -} \rightarrow 1/2Na_2[M_2(CO)_{10}] + CO + bipy \tag{9.8}$$

Reduction of the zirconium complex, $[ZrCl_2Q_2]$ ($Q = \eta\text{-}C_5H_3(SiMe_3)_2$), with $Li_2^+(cot)^{2-}$ in ether under a carbon monoxide atmosphere yields $[ZrQ_2(CO)_2]$.[30] The reduction of $[MX_2Cp_2^*]$ (M = Sn, Ge; X = Cl, Br) by $Li_2^+(cot)^{2-}$ or $K_2^+(cot)^{2-}$ in a 1:1 molar ratio in tetrahydrofuran solution results in the formation of $[MCp_2^*]$.[31] In these reactions, cyclooctatetraene remains in solution. [MCp_2*] is also obtained with sodium naphthalenide as the reductant;[31] however, the problems associated with separation of the reaction products from naphthalene make it difficult to isolate the pure compounds.

The reaction of the iron(III) complex, [FeCl(tpp)], with two mole-equivalents of sodium anthracenide in tetrahydrofuran results in the formation of [$Na(THF)_3$][Fe(tpp)].[32] If three mole-equivalents of the reductant are used, the iron(0) complex, [$Na(THF)_3$]$_2$[Fe(tpp)], is formed.[32] It has been shown[33] that the reduction of [$Rh(dppe)_2$][BF_4] in THF, benzene, or toluene by sodium naphthalenide takes place in two discrete one-electron steps. [$Rh(dppe)_2$] forms first and then undergoes reduction to [$Rh(dppe)_2$]$^-$. Both products have been characterized in solution by spectroscopic methods.[33] The anion, [$Rh(dppe)_2$]$^-$, reacts with acetonitrile or methanol to form the known hydride complex, [RhH(dppe)].

The properties of the intercalation compound, potassium graphite, KC_8, have been detailed in several review articles.[34,35] The bonding in potassium graphite is described in terms of the limiting structure, $K^+C_8^-$, and it is believed that the anion forms as a result of the transfer of an electron from the alkali metal to the conduction band of graphite. Novikov and Volpin[35] have noted a similarity between aromatic radical-anions and alkali metal-graphite intercalation compounds. Their observation was based on inspection of reduction potentials of aromatic hydrocarbons relative to biphenyl, Table 9.2:

Table 9.2

Reduction potentials of various hydrocarbons relative to biphenyl[35]

Compound	E,*V*	Compound	E,*V*
biphenyl	0.00	dibenz [*a,h*]-anthracene	0.69
naphthalene	0.09	coronene	0.70
phenanthrene	0.17	perylene	1.12
chrysene	0.38	tetracene	1.28
pyrene	0.60		

From these data it follows that reduction is facilitated as the number of condensed rings is increased and that the reduction potential of the graphite compound, with essentially an infinite number of aromatic rings in a plane, should be the highest among the fused-ring aromatic hydrocarbons.

Accordingly, the intercalation compound, KC_8, is a relatively mild reductant. KC_8 is prepared from finely powdered graphite and potassium, mixed and heated in an evacuated vessel at 350 °C for 4-5 h.[36a] Few examples of the use of KC_8 as a reductant in synthetic coordination chemistry have been described. However, the synthesis of a nickel(0) phosphite complex by reduction of a nickel(II) precursor with KC_8 has been reported.[36a] The reaction takes place in dry benzene, Eq. 9.9:

$$[NiX_2\{P(OR)_3\}_2] + 2P(OR)_3 + 2KC_8 \rightarrow [Ni\{P(OR)_3\}_4] + 2KX + \text{graphite} \tag{9.9}$$

X = Cl, Br, I; R = Me, Et, Pr, Bu

Reduction of $[NiX_2\{P(OR)_3\}_2]$ was accomplished to some extent by heating the complex with finely dispersed potassium in toluene or xylene.[36a] The reduced product was not obtained as a pure compound by this method.

Ungurenasu and Palie[36b] have reported a highly selective route to anionic binuclear complexes in greater than 99% yield by reduction of metal carbonyls with KC_8. The reactions were carried out in dry, deoxygenated THF at 25 °C, Eqs. 9.10 and 9.11:

$$2[M(CO)_6] + 2KC_8 \rightarrow K_2[M_2(CO)_{10}] + 2CO + \text{graphite} \tag{9.10}$$

M = Cr, Mo, W

$$2[Fe(CO)_5] + 2KC_8 \rightarrow K_2[Fe_2(CO)_8] + 2CO + \text{graphite} \tag{9.11}$$

Among the other anionic organic reductants, both sodium carbide and Grignard reagents are useful. For example, a suspension of sodium carbide in THF is an effective reductant for carbonyl complexes.[38] The reducing ability of Na_2C_2 is somewhat lower than that of metallic sodium, but it is easier to work with sodium carbide than with metallic sodium because the carbide is relatively stable toward atmospheric oxygen and moisture. Sodium carbide and its oxidation product, carbon, are insoluble in THF and, thus, are easily separated from the reaction medium. Manning and Trogler[38] have used Na_2C_2 in THF to reduce manganese(I) to manganese(0), Eq. 9.12:

$$2[MnBr(CO)_5] + Na_2C_2 \rightarrow [Mn_2(CO)_{10}] + 2NaBr + 2C \tag{9.12}$$

It was also found[38] that $[Co_4(CO)_{12}]$ is reduced by sodium carbide to the cluster, $[Co_6(CO)_{14}]^{4-}$.

In 1940 it was shown that Grignard reagents function as reductants in syntheses of carbonyl complexes from metal salts.[39] For example, the synthesis of chromium hexacarbonyl is described by Eq. 9.13:

$$2CrCl_3 + 6PhMgBr + 12CO \rightarrow 2[Cr(CO)_6] + 3Ph\text{–}Ph + 3MgCl_2 + 3MgBr_2 \quad (9.13)$$

Wilkinson and coworkers[40,41] have established that homoleptic ethane-like ruthenium(III) complexes, $[Ru_2R_6]$ (R = CH_2SiMe_3, CH_2Bu), are formed by the reaction of $[Ru_2(\mu\text{-}O_2CMe)_4]Cl$ with an excess of the appropriate Grignard reagent, RMgCl, in THF at low temperature.[40] The formation of Ru(III) products from $[Ru_2(\mu\text{-}O_2CMe)_4]Cl$ (formally in the III/II oxidation states) is explained by disproportionation because yields of $[Ru_2R_6]$ are below 50%, Eq. 9.14:

$$[Ru_2^{II,III}]^{5+} \rightarrow [Ru_2^{II,II}]^{4+} + [Ru_2^{III,III}]^{6+} \quad (9.14)$$

At the same time, the reaction of $[Ru_2(\mu\text{-}O_2CMe)_4]Cl$ and Me_3SiCH_2MgCl under similar conditions, but with one equivalent of the Grignard reagent, was reported to give the reduction product, $[Ru_2(\mu\text{-}O_2CMe)_4(THF)_2]$. Me_3SiCH_2MgCl acts simply as a soluble one-electron reductant in this reaction. The oxidation products were not identified.[40,41]

Like the Grignard reagents, alkyl- or aryllithium compounds are used as reductants in syntheses of coordination compounds. For example, Thiele and Krüger[42] found that reaction of the zirconium(IV) complex, $[Zr(CH_2Ph)_2(\eta^5\text{-}Cp)_2]$, with benzyllithium in diethyl ether at -78 °C yields a zirconium(III) compound, $[Zr(CH_2Ph)_2(\eta^5\text{-}Cp)]\cdot OEt_2$. The reduction of $[ZrPh_2(\eta^5\text{-}Cp)_2]$ by phenyllithium under similar conditions gives the zirconium(II) derivative, $[ZrPhCp]\cdot 3OEt_2$.[42] These authors also have established that treatment of $[ZrR_4]$ (R = Me, CH_2Ph) with butyllithium yields $Li_2[ZrR_4]$.[43]

Lehmkuhl and coworkers[44] have described the reaction between $[(\eta^5\text{-}Cp^*)Co(\mu\text{-}Cl)_2Co(\eta^5\text{-}Cp^*)]$ and two equivalents of phenyllithium, which leads to reduction of the starting material and formation of biphenyl followed by complexation. The trinuclear complex, $[\{(\eta^5\text{-}Cp^*)Co\}_3(\mu_3,\eta^{12}\text{-}C_{12}H_{10})]$, is formed. The reduction of *fac*-$[ReCl_3(PMe_2Ph)_3]$ by potassium 2,4-dimethylpentadienide in THF has been reported by Bleeke and Moore.[45] The reaction results in the formation of $[(\eta^5\text{-}2,4\text{-}Me_2pd)Re(PMe_2Ph)_3]$, which has been structurally characterized, Figure 9.2. In this reaction, 2,4-dimethylpentadienide serves as both the alkylating agent and the reductant.

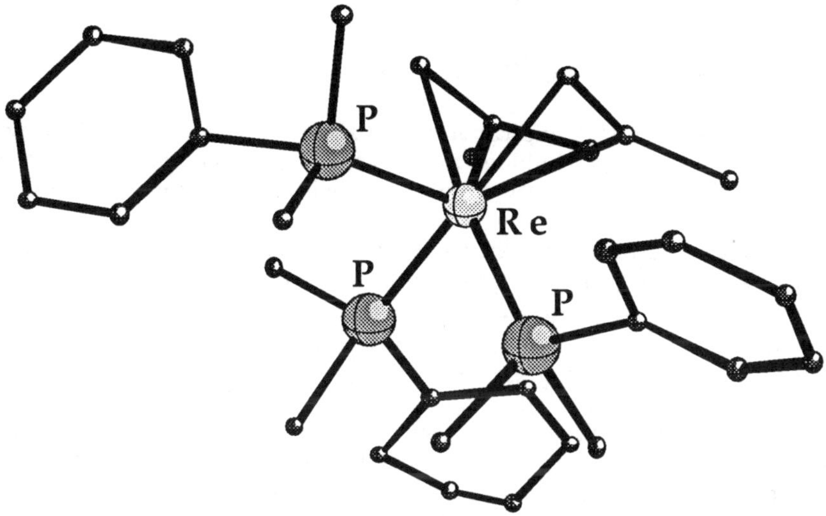

Figure 9.2

9.3 Carbon monoxide and other carbon-containing reductants

In this section reduction reactions in which carbon monoxide, carboxylic acids and their salts, alcohols, and hydroquinones participate are described. Carbon monoxide acts as a reductant and/or a carbonylating agent. For example, the palladium(II) nitrite complex, $[Pd(NO_2)_2(PPh_3)_2]$, in dichloromethane, is reduced and carbonylated by carbon monoxide.[46] The reaction between $[Pd(NO_2)_2(PPh_3)_2]$ and CO at ambient temperature in oxygen-free dichloromethane proceeds with formation of the palladium(0) cluster, $[Pd_4(CO)_5(PPh_3)_4]$. Both N_2O and CO were detected as by-products. The authors[46] believe that the initial reaction of CO and $[Pd(NO_2)_2(PPh_3)_2]$ may give a palladium nitrosyl intermediate (see below) which subsequently reacts to form N_2O and $[Pd_4(CO)_5(PPh_3)_4]$.

When ruthenium(III) chloride is heated in an atmosphere of carbon monoxide at 65 atmospheres, $[Ru_3(CO)_{12}]$ forms.[47] Reduction of hydrated iridium trichloride by CO has been reported by Balch and coworkers.[48] A mixture of the iridium salt and LiCl in 2-methoxyethanol was reacted with CO under pressure at 170 °C forming $Li[IrCl_2(CO)_2]$. When a toluene solution of 1,3-bis(diphenylphosphino)propane was added to the reaction medium,

$[Ir_2Cl_2(CO)_2(\mu\text{-dppp})_2]$ was formed.

Osmium pentacarbonyl, $[Os(CO)_5]$, can be obtained[5] by reducing OsO_4 with carbon monoxide at 300 °C, Eq. 9.15:

$$OsO_4 + 9CO \rightarrow [Os(CO)_5] + 4CO_2 \quad (9.15)$$

Calderazzo and Poli[49] have converted $[NH_4][ReO_4]$ into $[Re_2(CO)_{10}]$ with carbon monoxide. Metal sulfides as well as oxides can be reduced by carbon monoxide. For example, cobalt sulfide reacts with CO to form $[Co_2(CO)_8]$ and COS.[5]

Dimethylformamide may be used as a source of CO because DMF decomposes into CO and $HNMe_2$ upon thermolysis. Cole-Hamilton[50] has described the reaction between $RuCl_3{\cdot}3H_2O$ and two equivalents of 2,2'-bipyridyl in dimethylformamide. After several days of reflux, followed by evaporation to dryness, large crystals of $[RuCl(CO)(bipy)_2]Cl$ were obtained.

Carbon monoxide also can act as a deoxygenating reagent for coordinated ligands. Processes in which nitrate, nitrite, and nitrosyl ligands are reduced are documented. Chatt and Booth[51] performed one of the first reported oxygen atom abstractions from a coordinated ligand. Their syntheses involved abstraction of one or two oxygen atoms from coordinated nitrite or nitrate ligands with formation of nitrosyl derivatives, and evolution of carbon dioxide, Eqs. 9.16 and 9.17:

$$[Ni(NO_3)_2(PR_3)_2] + 2CO \rightarrow [Ni(NO)(NO_3)(PR_3)_2] + 2CO_2 \quad (9.16)$$

R = Et, Ph

$$[Ni(NO_2)_2(PEt_3)_2] + CO \rightarrow [Ni(NO)(NO_2)(PEt_3)_2] + CO_2 \quad (9.17)$$

Further, Feltham and coworkers[52] have prepared $[Ni(NO_2)_2(PMe_3)_2]$, reacted it with carbon monoxide, and characterized the resulting nitronitrosyl compound, $[Ni(NO)(NO_2)(PMe_3)_2]$, by X-ray diffraction, Figure 9.3. Deoxygenation reactions of this general type now are used routinely for both nickel(II)[52-56] and platinum(II)[57] complexes of the type $[M(NO_2)XL_2]$ (where M = Ni, Pt, X = Cl, Br, NO_2, and L = phosphines, $OPPh_3$, or py).

Further, the results of kinetic studies of these reactions have been discussed in the literature. The processes[52,54] appear to be associative and involve the formation of five-coordinate intermediates. The nature of the phosphine ligand in $[Ni(NO_2)_2(PR_3)_3]$ and the stereochemistry of the complexes have some influence on the rates of reactions. The proposed mechanism is depicted in Scheme 9.4.

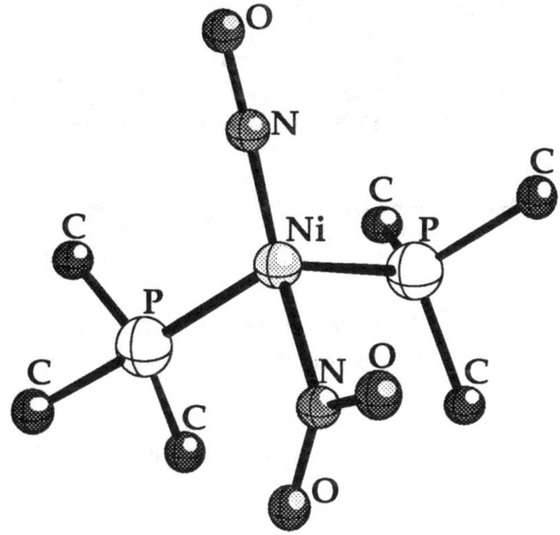

Figure 9.3

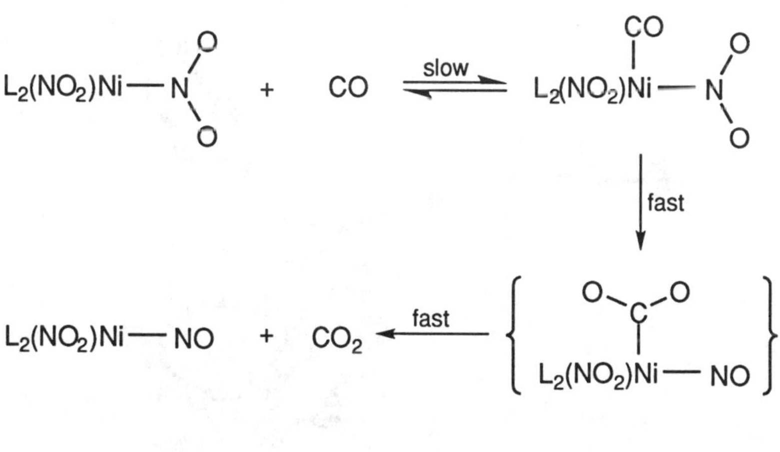

Scheme 9.4

A study with ^{18}O-labeled $[Ni(NO_2)_2(PEt_3)_2]$ showed the $-NO_2$ group to be the source of oxygen for CO_2 generation.[55]

Bos and coworkers[58,59] have described the reduction of the nitrato ligand in $[Au(NO_3)(PPh_3)]$ in dichloromethane solution by carbon monoxide. An unusual fulminato compound, $[Au(CNO)(PPh_3)]$, is formed. It has been shown that the fulminate ligand is generated by a complicated process in which the first step is the reduction of $[Au(NO_3)(PPh_3)]$, *via* $[Au(NO_2)(PPh_3)_3]$, to $[Au(NO)(PPh_3)]$ by carbon monoxide, Eq. 9.18:

$$[Au(NO_3)(PPh_3)] \xrightarrow[-CO_2]{CO} [Au(NO_2)(PPh_3)] \xrightarrow[-CO_2]{CO} [Au(NO)(PPh_3)] \quad (9.18)$$

$[Au(NO)(PPh_3)]$ acts as a chloride scavenger that exchanges NO for Cl with dichloromethane. After rearrangement, formoxime is obtained, Eqs. 9.19 and 9.20:

$$[Au(NO)(PPh_3)] + CH_2Cl_2 \rightarrow [AuCl(PPh_3)] + H_2C(Cl)NO \quad (9.19)$$

$$H_2C(Cl)NO \rightarrow HC(Cl){=}NOH \quad (9.20)$$

Dehydrohalogenation of HC(Cl)=NOH leads to the formation of HCNO which reacts with the starting material to yield the fulminato compound, $[Au(CNO)(PPh_3)]$, which has been structurally characterized,[58,59] Figure 9.4:

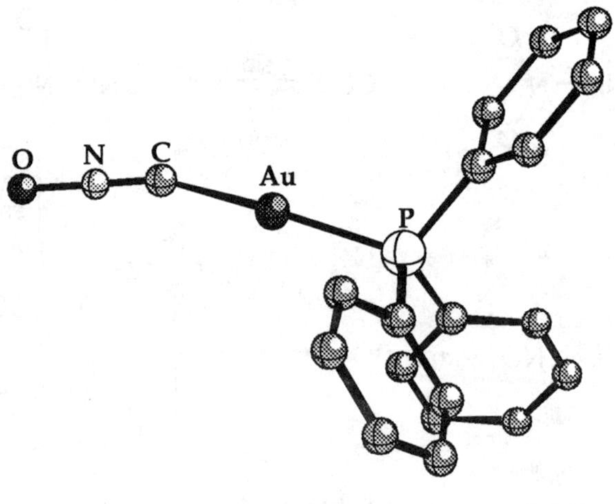

Figure 9.4

Fjare and Gladfelter[60] have investigated deoxygenation of the NO ligand in the nitrosyl cluster, $[FeRu_3(CO)_{12}(NO)]^-$, by CO, where the nitrido cluster,

$[FeRu_3N(CO)_{12}]^-$, is formed, Eq. 9.21:

$$[FeRu_3(CO)_{12}(NO)]^- + CO \rightarrow [FeRu_3N(CO)_{12}]^- + CO_2 \qquad (9.21)$$

The CO_2 eliminated in the reaction has been identified by IR spectroscopy and ^{13}C NMR spectroscopy. Although CO is required to balance equation 9.21, the nitrido cluster forms even in the absence of CO. This leads to the conclusion that the deoxygenation process is intramolecular in nature.

Deoxygenation of the nitrosobenzene complex of rhodium(I), [RhCl(cod)(PhNO)], has been studied by Vasapollo and coworkers.[61,62] The reaction occurs in ethanol under an atmosphere of carbon monoxide when [RhCl(cod)(PhNO)] is heated. Aniline and CO_2 have been identified among the reaction products. The formation of azoxybenzene has been noted where free nitrosobenzene was present in the reaction medium. The reactions are described by Eqs. 9.22 - 9.24:

$$[M]\text{–}N(=O)Ph + CO \rightarrow [M]\text{–}N\text{–}Ph + CO_2 \qquad (9.22)$$

$$[M]\text{–}N\text{–}Ph + PhN{=}O \rightarrow PhN{=}N(O)Ph + [M] \qquad (9.23)$$

$$[M] + PhN{=}O \rightarrow [M]\text{–}N(=O)Ph \qquad (9.24)$$

The formation of aniline is attributed to reaction between the nitrene species and ethanol.

These examples show the broad applicability of ligand deoxygenation of the [M–N–O] functionality with carbon monoxide in synthesis. Ligands deoxygenated in this manner include nitrate, nitrite, nitrosyl, and NOR.

Carboxylic acids and their salts, as well as alcohols and alkoxides, are common ligands and this has limited their use as reductants in synthesis. Potassium hexachloroplatinate(IV), for example, is reduced by ammonium oxalate.[63] Che and coworkers[64] have shown that the reaction of aqueous solutions of ascorbic acid with the complex, $[RuO_2(NH_3)_4]Cl_2$, in the presence of NaX (X = Cl, I, CNS) yields the complexes, $[RuX_2(NH_3)_4]X$. The reduction of a mixture of hydrated ruthenium(III) chloride and ruthenium(IV) chloride by glucose or ascorbic acid in water in the presence of 2,2'-bipyridyl results in the formation of $[Ru(bipy)_3]Cl_2{\cdot}7H_2O$.[65] Additionally, the reduction of the $[TcO_4]^-$ anion by gluconic acid has been studied.[66]

Reduction of $K[ReO_4]$ by hydroquinone in concentrated hydrochloric acid followed by the addition of a saturated solution of magnesium chloride and extraction of the product in methyl ethyl ketone generates a solution of $H_2[ReOCl_5]$.[67]

In the classical synthesis of Zeise's salt, $K[PtCl_3(C_2H_4)]$, from $[PtCl_6]^{2-}$, ethanol serves as the reductant and also as the source of ethylene.[68] The reactions are described by Eqs. 9.25 and 9.26:

$$Na_2[PtCl_6] + 2EtOH \rightarrow H[PtCl_3(C_2H_4)] + 2NaCl + MeCHO + HCl + H_2O \quad (9.25)$$

$$H[PtCl_3(C_2H_4)] + KCl \rightarrow K[PtCl_3(C_2H_4)] + HCl \quad (9.26)$$

In other reactions where alcohols serve as reductants, the nickel(I) complex, $[Ni(catenand)]^+$, is formed by the reduction of $[Ni(catenand)]^{2+}$ by ethanol.[69] Albers and coworkers[70] have synthesized the ruthenium(II) polymers, $[RuCl_2(\eta^4\text{-cod})]_x$, in 30% to 40% yield from ruthenium chloride in boiling ethanol. The ethanol is oxidized to MeCHO, Eq. 9.27:

$$2RuCl_3 \cdot xH_2O + 2cod + EtOH \rightarrow 2[RuCl_2(\eta^4\text{-cod})]_x + 2HCl + MeCHO \quad (9.27)$$

The hydrido-complex, $[RuHCl(\eta^6\text{-}C_6Me_6)(PPh_3)]$, has been prepared by reduction of $[RuCl_2(\eta^6\text{-}C_6Me_6)]_2$ with 2-propanol under reflux, Eq. 9.28:[71]

$$[RuCl_2(\eta^6\text{-}C_6Me_6)]_2 + 2PPh_3 + 2Me_2CHOH + Na_2CO_3 \rightarrow 2[RuHCl(\eta^6\text{-}C_6Me_6)(PPh_3)] + 2Me_2CO + 2NaCl + H_2O + CO_2 \quad (9.28)$$

The rhodium(I) carbonyl complexes, *trans*-$[RhXCO(PR_3)_2]$, are formed by the reaction of the rhodium(III) complexes, $[RhX_3(PR_3)_3]$ and $[RhCl_2H(PR_3)_2]$, with methanol, ethanol, or 2-propanol.[72] The reaction takes place upon heating in the presence of a base. In a similar manner, the reaction of $Rh(NO_3)_3 \cdot 2H_2O$ with excess PPh_3 in ethanol or methanol occurs over 48 hours to yield *trans*-$[Rh(ONO_2)(CO)(PPh_3)_2]$.[73] In these reactions the alcohol plays the role of both reductant and carbonylating agent.

Olefins and even arenes are capable of reducing high oxidation state metal halides. For example, Persson and Anderson[74] have reported that allyltrimethylsilane, $H_2C{=}CHCH_2SiMe_3$, or cyclopentene interact with $MoCl_5$ in dimethoxyethane to give $[MoCl_4(dme)]$. A similar method for the preparation of $[MoCl_4(Et_2O)_2]$ using norbornene as the reductant has been reported.[75] Further, Hey-Hawkins and von Schnering[76] found that $[MoCl_4(dme)]$ was obtained from $MoCl_5$ in a toluene-dimethoxyethane mixture. It was suggested that the dimethoxyethane acted as the reductant but, based on the known reduction of Mo(V) derivatives with benzene,[77] it seems reasonable that the reductant is toluene.[74]

9.4 Sulfur-containing reductants

Gold(III) and platinum(IV) complexes act as oxidants towards a number of substrates and can be reduced by sulfur-containing reductants. At the beginning of this century Herrmann[78] established that alkyl sulfides reduce gold(III) chloride with formation of the gold(I) derivatives, $[AuCl(SR_2)]$. Today, these well-documented reactions[79] are carried out in aqueous or alcoholic solution, Eq. 9.29:

$$[AuCl_4]^- + 2R_2S + H_2O \rightarrow [AuCl(R_2S)] + R_2SO + 2H^+ + 3Cl^- \quad (9.29)$$

For example, Usón, Laguna and Laguna[80] have reported the reduction of hydrogen tetrachloroaurate(III), $H[AuCl_4]$, with the cyclic sulfide, SC_4H_8, in a mixture of water and ethanol. The dropwise addition of tetrahydrothiophene gave rise to a yellow precipitate of $[AuCl_3(SC_4H_8)]$ which was transformed into a white solid, $[AuCl(SC_4H_8)]$, as the addition of the alkyl sulfide was continued. $[AuCl(SC_4H_8)]$ was isolated in 95% yield.

Annibale and coworkers have reported a kinetic study of the reduction of Au(III) to Au(I) by various dialkyl sulfides.[81] In their evaluation of the reactions, the authors concluded: (*i*) The first step of the interaction establishes the substitution equilibria; (*ii*) The second step involves transfer of a chlorine atom from Au(III) to free sulfide. This leads to a digonal gold(I) complex and a halosulfonium salt, $[R_2SCl]Cl$, as the primary reaction products, Scheme 9.5:

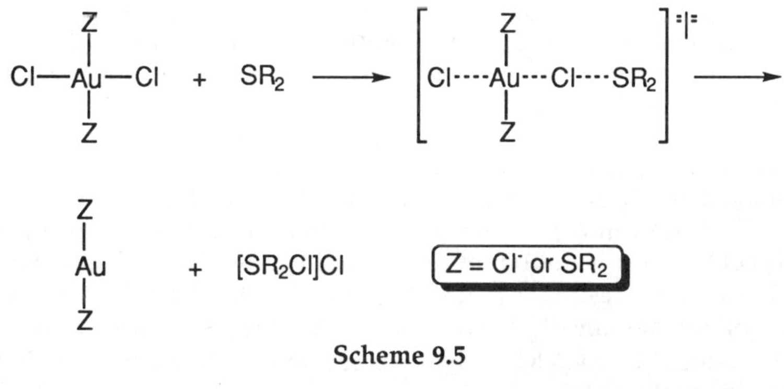

Scheme 9.5

The latter product was not detected but NMR experiments showed[81] that on dissolution of $[AuCl_3(SMe_2)]$ in methanol-*d*, $[SMe_2(OMe)]^+$ formed and, in the presence of water, $[SMe_2(OMe)]^+$ readily gave $Me_2S{=}O$; (*iii*) Reactivity of dialkyl sulfides as reductants for gold(III) increases as the positive charge on the complex increases:

$$[AuCl_4]^- << [AuCl_3(SR_2)] < [AuCl_2(SR_2)_2]^+;$$

(*iv*) Reactivity also increases with increasing basicity of the sulfides:

$$SMe_2 < SEt_2 < S(iso\text{-}Pr)_2.$$

Thioethers reduce platinum(IV) complexes in a manner similar to the reduction of gold(III) complexes. Chugaev and Benevolensky[82] have studied the interaction between Me_2S and $H_2[PtCl_6]$ in aqueous solution which results in the precipitation of a mixed platinum(II)/platinum(IV) complex, $[Pt(Me_2S)_4][PtCl_6]$. The reduction of platinum(IV) can be achieved also by dialkyl sulfoxides. For example, it has been reported[83] that on treatment of $K_2[PtCl_6]$ with dimethyl sulfoxide in boiling water, a mixture of *cis*-$[PtCl_2(DMSO)_2]$ and $K[PtCl_3(DMSO)]$ is formed. Further, Moiseev and coworkers[84] have established that dimethyl sulfoxide in aqueous solution also reduces Pt(III) compounds. Thus, the tetramer, $[Pt(CH_3CO_2)_3]_4$, upon reaction with dimethyl sulfoxide, is converted to the platinum(II) complex, $[Pt(CH_3CO_2)_2(H_2O)(DMSO)]$, and the dimeric platinum(I) sulfoxide complex, $[Pt(CH_3CO_2)(DMSO)_2]_2{\cdot}H_2O$.

Sulfur dioxide also reduces Au(III) and Pt(IV) complexes. Bubbling SO_2 through an aqueous solution of $H[AuCl_4]$ leads to reduction of Au(III) to Au(I) and formation of a colorless solution containing only gold(I)-chloride species. When a solution of *N*-ethyl-1,3-imidazoline-2-thione (L) is added to the solution, solid [AuClL] is obtained.[85] Additionally, Nyholm[86] has shown that the complex, $[PtCl_2I_2(AsMePh_2)_2]$, reacts with SO_2 in chloroform solution at room temperature to form the platinum(II) complex, $[PtI_2(AsMePh_2)_2]$.

Thiourea acts as a mild reductant ($E° = 0.427$ *V*) for transition metal complexes and is oxidized to formamidine disulfide. Thiourea is an excellent ligand for many metal ions, especially for those of the platinum metals.[88] Therefore, in most of its redox reactions, thiourea acts simultaneously as a reductant and as a ligand. For example, OsO_4 reacts with thiourea in sulfuric acid to form $[Os(thio)_6]^{3+}$ and formamidine disulfide.[89] Addition of $[NH_4][TcO_4]$ to a solution of thiourea in a mixture of ethanol and hydrochloric acid results[90] in the formation of the homoleptic thiourea complex of technetium(III), $[Tc(thio)_6]^{3+}$. An X-ray structure study of the isolated product, $[Tc(thio)_6]Cl_3{\cdot}4H_2O$, has shown that the six thiourea ligands are *S*-bonded to the metal.

Freeman and coworkers[91] have reported the reduction of $H[AuCl_4]$ in aqueous solution by thiourea which yields the gold(I) complex, $[Au(thio)_2]Cl$. Kukushkin, Vorobev-Desyatovskii and Patrabansh[92] have investigated the reactions of the *cis* and *trans* isomers of $[PtCl_4(amine)_2]$ (amine = NH_3,

$EtNH_2$, py) with an excess of thiourea in acidic aqueous solutions. In these reactions thiourea plays both the role of reductant and of a ligand. With the *cis* isomers of [$PtCl_4$(amine)$_2$], thiourea forms the greenish-yellow complex, [Pt(thio)$_4$]Cl_2, and with the *trans* isomers, the colorless complexes, *trans*-[Pt(amine)$_2$(thio)$_2$]Cl_2, are formed. Thus, the reaction of thiourea[92] with isomeric complexes of the type, [$PtCl_4$(amine)$_2$], may be used to determine the geometry.

Sodium dithionite, $Na_2S_2O_4$, commonly is used to reduce high-valent technetium and iron complexes. For example, the reaction of an alkaline aqueous solution of [NH_4][TcO_4] and the tetradentate amido-thiol ligand, H_4L (Figure 9.5), with sodium dithionite, followed by addition of tetraphenyl-arsonium chloride, yields [Ph_4As][TcO(L)].[93]

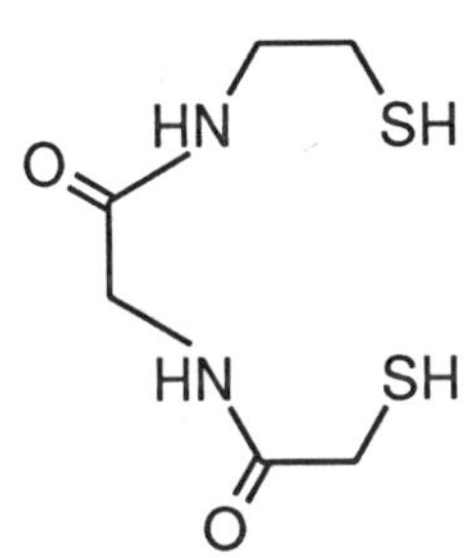

H_4L

Figure 9.5

The complex, [Bu_4N][TcO(L*)$_2$] (H_2L* = 2-aminobenzenethiol), is formed in a similar manner.[94]

The iron(III) complex, [FeL_2]$^-$, (L^{2-} = pyridine-2,6-dicarbothioate) is reduced[95] in aqueous solution by sodium dithionite with formation of the dianionic iron(II) complex, [FeL_2]$^{2-}$. Wade and Castro[96] have reported the synthesis of [Fe(tpp-*d*)] by reduction of the iron(III) compound, [FeCl(tpp-*d*)] with $Na_2S_2O_4$ in an alkaline aqueous-isopropanol medium. Reduction of sodium nitroprusside, Na[Fe(CN)$_5$(NO)]$\cdot$2H_2O, with sodium dithionite, followed by addition of 2,2'-bipyridyl and tetraphenylphosphonium bromide, yields a precipitate of [Ph_4P][Fe(CN)$_3$(NO)(bipy)].[97]

Mineey and Traylor[98] have obtained a complex of sodium dithionite and 18-crown-6. The compound, provisionally formulated as $Na_2S_2O_4$/18-crown-6, is freely soluble in dimethylformamide, dimethyl sulfoxide, and alcohols. A solution of this material was used to reduce [Ph_4P]$_2$[FeS(PhS)$_2$]$_2$ to [Ph_4P]$_2$[FeS(PhS)]$_4$.[99]

Grinberg and coworkers[100,101] carried out the reduction of potassium hexachloroplatinate by rongalite, $NaHSO_2CH_2OH$, in aqueous acetate buffer solutions, Eq. 9.30:

$$2K_2[PtCl_6] + NaHSO_2CH_2OH + 2H_2O \rightarrow 2K_2[PtCl_4] + NaHSO_4 + 4HCl + CH_2O \quad (9.30)$$

Rongalite and its analogs are effective reductants for complexes of uranium and the transuranium elements.[101] For example, the reduction of uranyl compounds takes place according to Eq. 9.31:

$$UO_2^{2+} + NaHSO_2CH_2OH + 2H^+ \rightarrow U^{4+} + NaHSO_3 + CH_2O + H_2O \quad (9.31)$$

It has been established[102] that the reduction of the tellurium(IV) compound, **I**, by excess hydrated sodium sulfide results in the formation of the tellurium(II) compound, **II**, Figure 9.6:

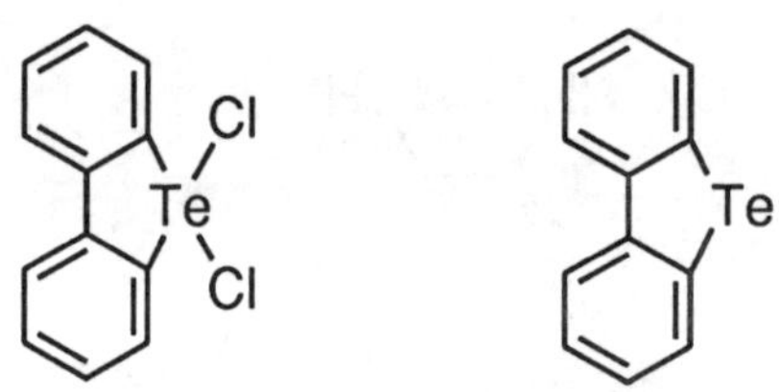

I **II**

Figure 9.6

Reduction of the gold(III) carbene complex, $[AuI_3\{C(Ph)(NMe_2)\}]$, by sodium thiosulfate has been investigated.[103] The gold(I) complex, $[AuI\{C(Ph)(NMe_2)\}]$, is formed. Elding, Elmroth, and Skibsted[104] have studied the kinetics and mechanisms of the reactions between the amine- and haloaminegold(III) complexes, $[Au(NH_3)_4]^{3+}$, *trans*-$[AuCl_2(NH_3)_2]^+$, and *trans*-$[AuBr_2(NH_3)_2]^+$, with sodium thiocyanate in water. On the basis of potentiometric p*H* measurements and data from sequential-mixed-stopped-flow spectrophotometry, it was concluded that the mechanisms involve competitive electron transfer and ligand substitution processes, as shown in Scheme 9.6. In the reduction step, the thiocyanate attacks a coordinated SCN^- ligand with subsequent reductive elimination of $(SCN)_2$.[104]

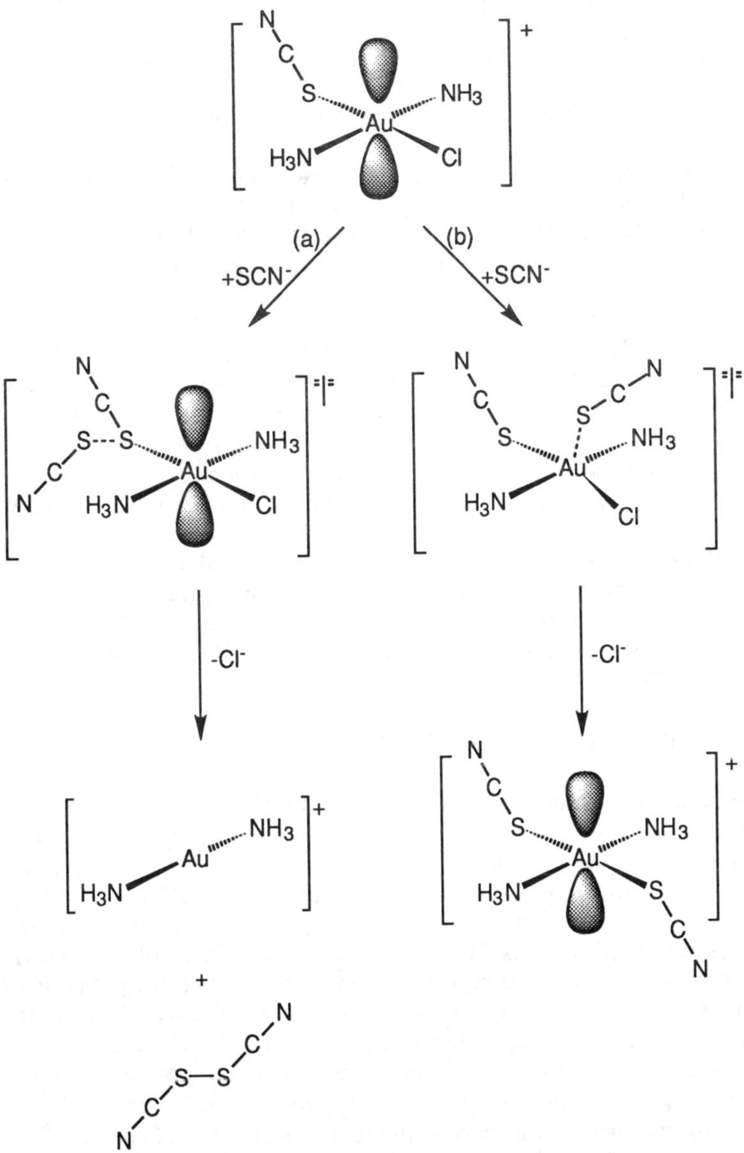

Scheme 9.6

Reduction of the copper(II) complex, $[Cu(2,9\text{-}Me\text{-}phen)_2]^{2+}$, by potassium thiocyanate in aqueous solution has been described.[105] The reaction yields $[Cu(2,9\text{-}Me\text{-}phen)_2]^+$ and dithiocyanate, $(SCN)_2$.

9.5 Low oxidation state metal ions

The use of low oxidation state metal ions or their complexes as reductants has limited applicability in synthetic coordination chemistry. These reductants typically must be synthesized and the products of their oxidation often form complexes themselves, leading to complicated mixtures of by-products.

Tin(II) chloride is used as a two-electron reductant. For example, the reduction of $[NH_4][TcO_4]$ by tin(II) chloride in 0.1 *M* HCl in the presence of penicillamine gives the complex, $[TcO(penicillaminato)_2]^-$, (penicillaminato = $Me_2C(S^-)CH_2(NH_2)(CO_2)^-$).[106] Further, halide complexes of platinum(IV) can be reduced with $SnCl_2{\cdot}2H_2O$. Chatt and coworkers[107] have reported the reduction of the terahalide complex, $[PtBr_4L_2]$ (L = *p*-styryldimethylarsine, *o*-allylphenyldimethylarsine), in acetone and ethanol, respectively, which results in the formation of $[PtBr_2L_2]$.

Reduction involving Sn(II) can be carried out in organic solvents. Lockhart and Haitko[108] have reported the formation of aliphatic and aromatic copper(I) carboxylates by treatment of copper(II) carboxylates, $Cu(O_2CR)_2$, (R = Me, Ph, $C_{17}H_{35}$), with tin(II) 2-ethylhexanoate in dichloromethane, Eq. 9.32:

$$2Cu(O_2CR)_2 + SnL_2 \rightarrow 2Cu(O_2CR) + SnL_2(O_2CR)_2 \qquad (9.32)$$

Reduction reactions that involve titanium(III) species are common in preparative organic chemistry[109] but relatively rare in coordination chemistry. However, Floriani and coworkers[110-112] have described the reduction of vanadium complexes, [VO(salophen)] and [VO(acacen)], by the titanium(III) solvento-complex, $[TiCl_3(THF)_3]$. Reaction of these vanadyl complexes in THF results in formation of [VCl(salophen)(THF)] and [VCl(acacen)(THF)]. Both compounds have been structurally characterized and an X-ray diffraction study of [VCl(salophen)(THF)] indicated pseudooctahedral coordination for V(III). Further, the Cl^- and THF ligands were found to be *trans* to each other in the axial positions, while the equatorial plane was defined by the salophen ligand, Figure 9.7.[111] The proposed mechanism[112] for these conversions is shown in Scheme 9.7.

Chromium(II) and vanadium(II) compounds are promising one-electron reductants.[113-118] For example, tris(picolinato)vanadate(II), $[V(pic)_3]^-$, (E° = -0.41 *V*)[119] will reduce Fe(III) in iron oxide by a heterogeneous outer-sphere reaction.[113-115]

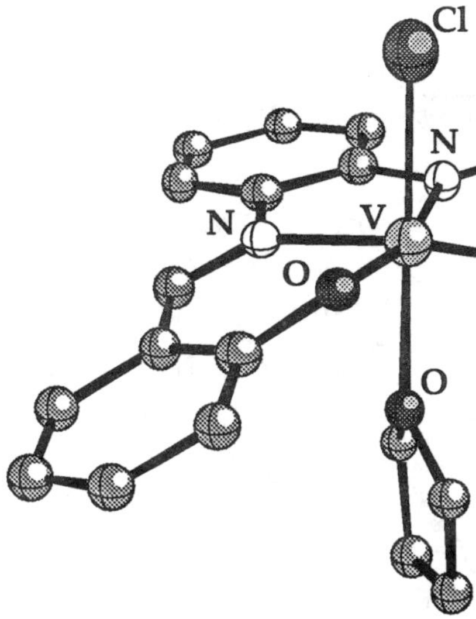

Figure 9.7

Reduction of the cobalt(III) ammine complexes, $[CoF(NH_3)_5]^{2+}$ and $[Co(CO_2H)(NH_3)_5]^{2+}$, by the copper(I) complexes, $[CuL_4][ClO_4]$, (L = py, MeCN), in dimethyl sulfoxide has been studied.[120] The active species in the reduction processes are the complexes, $[Cu(H_2O)_n(DMSO)_{4-n}]^+$.

Low oxidation state metal ions are used in selective reduction of coordinated ligands. In particular, Armor[121] has shown that chromium(II) chloride reduces the nitrosyl ligand in $[Ru(NH_3)_5NO]^{3+}$, Eq. 9.33:

$$[Ru(NH_3)_5NO]^{3+} + 6Cr^{2+} + 5H^+ \rightarrow [Ru(NH_3)_6]^{2+} + 6Cr^{3+} + H_2O \quad (9.33)$$

On treatment[122] of $[Ph_3PCH_2Ph]_2[PtCl_3(NO_2)]$ with titanium(III) sulfate in water/MeCN solution, the solvento-complex $[Ph_3PCH_2Ph][PtCl_3(MeCN)]$, is obtained, Eq. 9.34:

$$[PtCl_3(NO_2)]^{2-} + Ti^{3+} + MeCN \rightarrow [PtCl_3(MeCN)]^- + TiO^{2+} + NO \quad (9.34)$$

Further progress in the development of low oxidation state metal complexes as reductants in synthetic coordination chemistry will require compatibility with non-aqueous solvents.

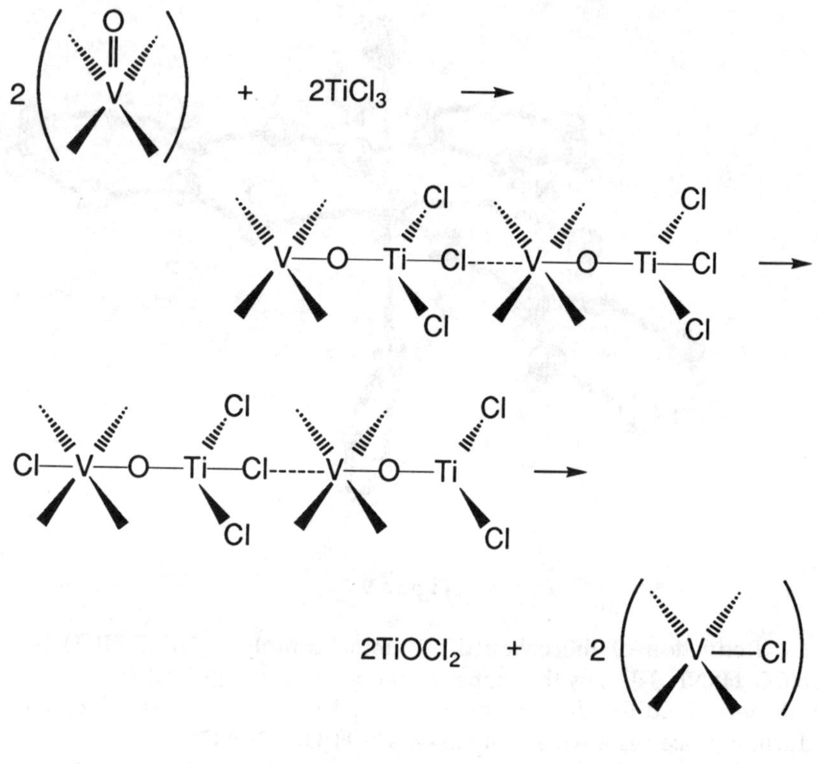

Scheme 9.7

The compounds now used as reductants include complexes of titanium(III): $[TiCl_3(THF)_3]$,[123a] $[TiCl_3(MeCN)_3]$,[124,125] and $Ti(SO_3CF_3)_3$;[126] of vanadium(III): $[VCl_3(THF)_3]$;[123] of vanadium(II): $VCl_2(THF)_x$[127] and $V(O_3SCF_3)_2(dme)_2$;[128] of chromium(II): $[Me_4N][CrX_3]$ (X = Cl, Br),[129] $[CrCl_2(THF)_2]$,[130] and $Cr(SO_3Cl)_2$;[131] of molybdenum(III): $[MoCl_3(THF)_3]$,[132] as well as other complexes. The complexes mentioned here are soluble in selected organic solvents and are useful for syntheses in non-aqueous media.

9.6 Tertiary phosphines and phosphites

Tertiary phosphines are used widely as deoxygenating reagents in preparative coordination chemistry. Their reactivity toward the oxygen atom is due to the high energy of the P=O bond in the corresponding phosphoranes (120-150 kcal/mol).[133] Chatt was one of the first to use tertiary phosphines as both reductants and ligands for metal ions. For example, in 1962, together with Rowe,[134] Chatt synthesized the rhenyl complex, $[ReOCl_3(PPh_3)_2]$. The reaction occurs in aqueous HCl, Eq. 9.35:[134,135]

$$Re_2O_7 + 6PPh_3 + 6HCl \rightarrow 2[ReOCl_3(PPh_3)_2] + 2OPPh_3 + 3H_2O \quad (9.35)$$

The deoxygenation of Re_2O_7 by triethylphosphine proceeds differently in pyridine, Eq. 9.36:[136]

$$Re_2O_7 + PEt_3 + 4py \rightarrow [ReO_2(py)_4][ReO_4] + OPEt_3 \quad (9.36)$$

Re_2O_7 was found, in pyridine, to form $[ReO_3(py)_3][ReO_4]$. Its reaction with triethylphosphine results in reduction of the cation whereas the $[ReO_4]^-$ anion is inert to reduction by triethylphosphine, as demonstrated by the reaction of Re_2O_7 with excess PEt_3. Additionally, no reaction takes place when a pyridine solution of $[NH_4][ReO_4]$ is mixed with PEt_3. Deoxygenation of $[ReO_3(py)_3][ReO_4]$ leads to formation of the mixed-valent Re^{5+}/Re^{7+} salt, $[ReO_2(py)_4][ReO_4]$.

Additional examples of deoxygenation of Re=O complexes have been reported by Herrmann and coworkers[137-140] who have shown that a homobinuclear complex of rhenium(V) is formed as a result of partial deoxygenation of $[(\eta^5\text{-}Cp^*)ReO_3]$ by triphenylphosphine in tetrahydrofuran, Scheme 9.8:

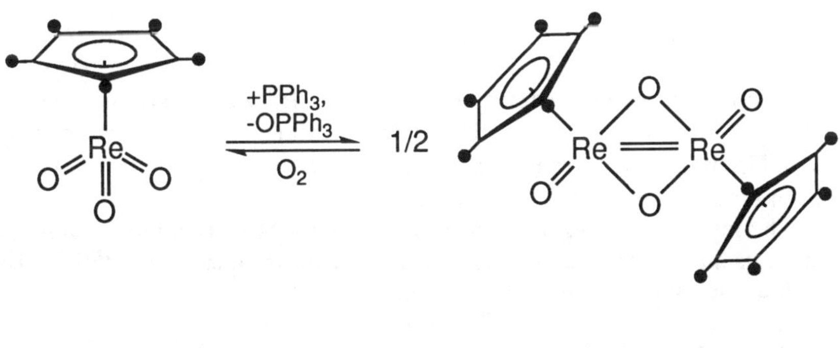

Scheme 9.8

Reduction of a rhenium(V) oxo complex with $P(OEt)_3$ has been described.[141] $[ReOCl_3(py)_2]$ was refluxed with a twenty-fold molar excess of triethylphosphite in benzene which resulted in a two-electron reduction of the metal center and formation of $[ReCl_3\{P(OEt)_3\}_3]$. Further, a similar reaction with *cis*-$[ReOCl_3(PEt_3)_2]$ yields *cis,mer*-$[ReCl_3(PEt_3)\{P(OEt)_3\}_2]$.[141]

It has been established[142,143] that interaction of $[ReOCl_3(PPh_3)_2]$ and 1,2-bis-(diphenylphosphino)methane, dppm, in a 1:1 molar ratio in refluxing $CHCl_3$ leads to the substitution product, *fac*-[$ReOCl_3$(dppm-*P,P'*)]. Under the same experimental conditions but with a 1:2 molar ratio, the reaction gives the rhenium(III) derivative, Scheme 9.9:

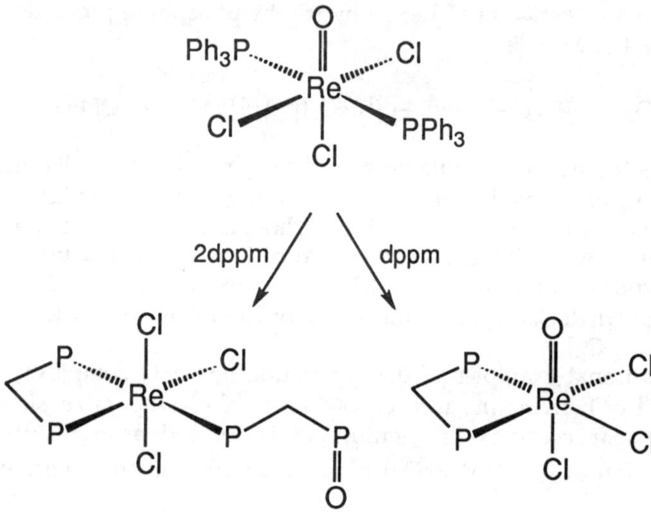

Scheme 9.9

The structure of *mer*-[$ReCl_3$(dppm-*P,P'*)(dppm-O-*P*)] was determined by X-ray diffraction[143] (for clarity only one carbon atom of each aromatic ring is shown), Figure 9.8. An example of the reduction of $[TcO_4]^-$ with phosphines has been reported by Refosco and coworkers.[144]

The kinetics of oxygen atom transfer from a Mo(VI) complex to tertiary phosphines has been studied by Holm, Reynolds, and Berg.[145-147] They found that the rate constant of the reaction, Eq. 9.37:

$$[MoO_2(S_2CNEt_2)_2] + PR_3 \rightarrow [MoO(S_2CNEt_2)_2] + OPR_3 \qquad (9.37)$$

increases monotonically with increasing nucleophilicity of the phosphines in the order: $PEt_3 > PPhEt_2 > PPh_2Et > PPh_3$. Reaction with PEt_3, the most basic phosphine, had a rate constant eight times greater than that of the least basic phosphine, PPh_3.

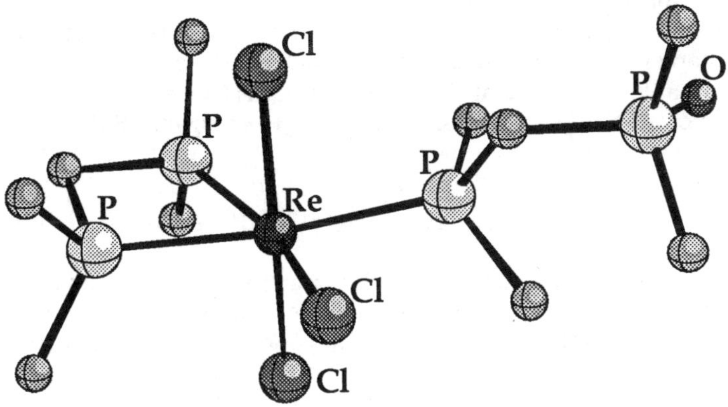

Figure 9.8

On the basis of these investigations the authors[145-147] proposed the following mechanism for the deoxygenation reaction, Scheme 9.10:

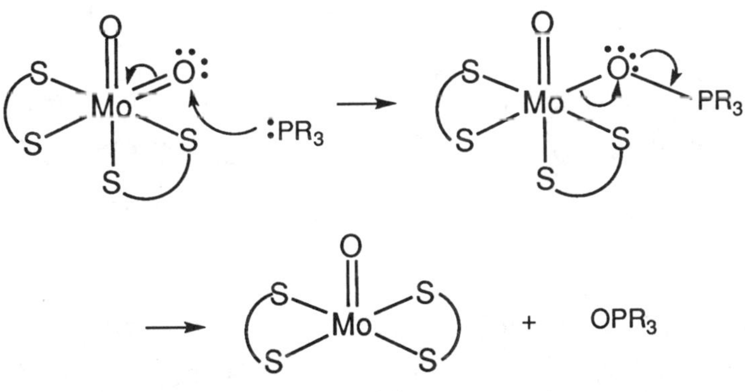

Scheme 9.10

Direct attack of a sterically hindered phosphine at the metal center is unlikely.[145] A likely mechanism involves attack on the oxygen atom, possibly at a vacant π*-orbital on the Mo=O fragment, and a concerted two-electron transfer to Mo, followed by Mo–O bond cleavage and P–O bond formation. Elimination of OPR_3 and rearrangement to the square-pyramidal complex completes the reaction.

Deutsch and coworkers[148] arrived at a similar conclusion concerning the mechanism for deoxygenation of Tc=O bonds by tertiary phosphines. They were able to show that the reduction of $[TcO(acacen)(H_2O)]^+$ results in formation of a Tc(III) complex containing PPh_3 as a ligand, Eq. 9.38:

$$[TcO(acacen)(H_2O)]^+ + 3PPh_3 \rightarrow [Tc(acacen)(PPh_3)_2]^+ + OPPh_3 + H_2O \quad (9.38)$$

One possible mechanism for this reaction involves substitution of a coordinated water molecule by a phosphine, followed by the transfer of an oxygen atom to a second phosphine with coordination of a third phosphine, Scheme 9.11:

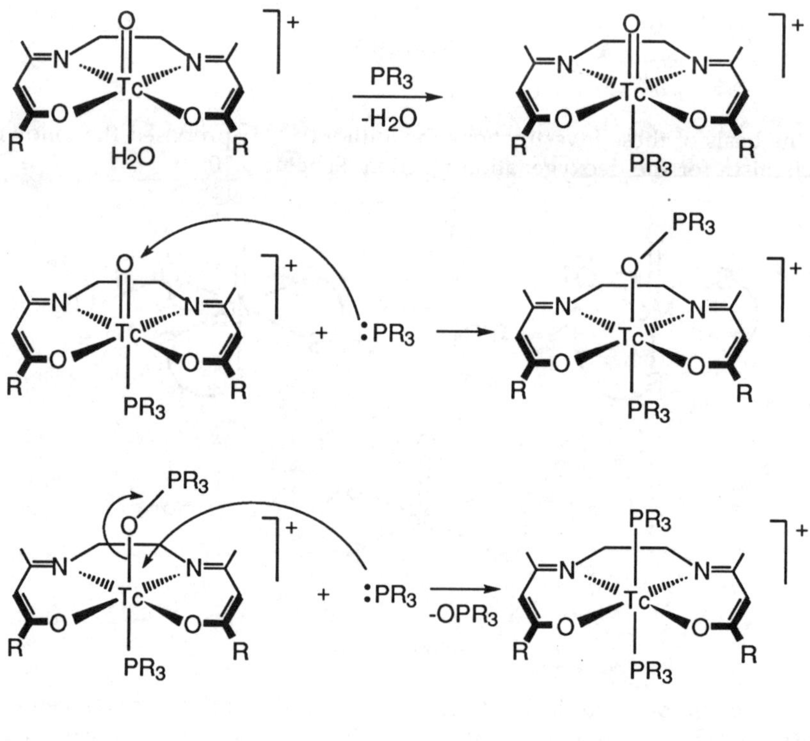

Scheme 9.11

Products of the two-electron reduction of oxo-complexes can react with the oxo-complexes themselves to form compounds of intermediate formal oxidation state. For example, deoxygenation of the chromium(IV) oxo complex, [CrO(tpp)], by triphenylphosphine results in formation of the chromium(II) complex, [Cr(tpp)].[149] The complex, [Cr(tpp)], then reacts with [CrO(tpp)] and forms the dimeric chromium(III) complex, $[Cr(tpp)]_2(\mu\text{-}O)$.[149]

Transfer of an oxygen atom from the *cis*-dioxomolybdenum complex, $[MoO_2(S_2CNR_2)_2]$, to PPh_3 results in the formation of $[MoO(S_2CNR_2)_2]$.[150] Depending on the nature of the substituents, $[MoO(S_2CNR_2)_2]$ can enter into an equilibrium reaction with the starting material, Eq. 9.39:

$$[MoO_2(S_2CNR_2)_2] + [MoO(S_2CNR_2)_2] \rightleftharpoons [Mo_2O_3(S_2CNR_2)_4] \qquad (9.39)$$

Holm and coworkers[151] have studied the reduction of $[MoO_2L]$ (L = 3-*t*-Bu-2-(salicylideneamino)benzenethiolato(2-)) with $PMePh_2$ in the presence of 2,2'-bipyridyl. The molybdenum(IV) product, [MoOL(bipy)], has been isolated and structurally characterized, Figure 9.9:

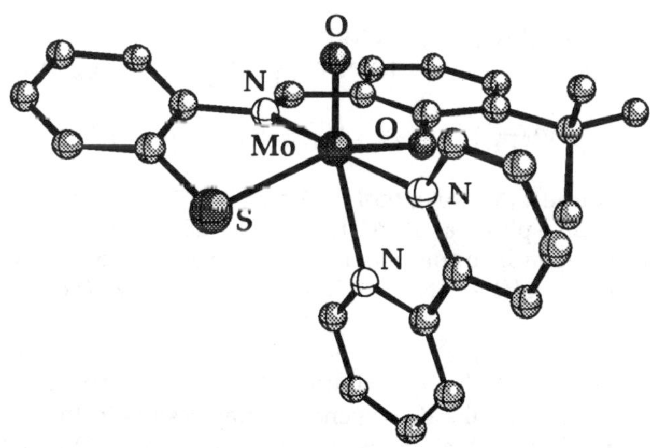

Figure 9.9

In the absence of a trapping reagent, in dimethylformamide, the μ-oxo complex, $[Mo_2O_3(L)_2(DMF)_2]$, is formed.

In addition to deoxygenation reactions that result in oxidation state changes of the metal center, deoxygenation of coordinated ligands may occur without change in the oxidation state of the metal. For example, the nitrite ligand in $[Co(NO_2)(salophen)py]$ is reduced to a nitrosyl group upon treatment with PPh_3 in dichloroethane, Eq. 9.40:[152]

$$[Co(NO_2)(salophen)py] + PPh_3 \rightarrow [Co(NO)(salophen)] + OPPh_3 + py \quad (9.40)$$

The planarity of the salophen ligand and the six-coordinate nature of $[Co(NO_2)(salophen)py]$ suggest that oxidation takes place without prior coordination of PPh_3 to the cobalt center.[152]

Reactions of the ruthenium and osmium nitro complexes, $[M(NO_2)(CO)_2(PPh_3)_2]$, with triphenylphosphine generate the nitrosyl complexes, $[M(NO)_2(PPh_3)_2]$, triphenylphosphine oxide, and CO.[153]

Pipes and Meyer[154] have investigated the electrochemical oxidation of $[OsNO_2(trpy)(bipy)][PF_6]$. A two-electron oxidation occurs in the presence of PPh_3 and results in the formation of $[OsNO(trpy)(bipy)]^{3+}$ and $OPPh_3$. The reaction sequence proposed to occur in the presence of PPh_3 is shown in Eqs. 9.41 - 9.43:

$$[OsNO_2(trpy)(bipy)]^{+} - e^{-} \rightarrow [OsNO_2(trpy)(bipy)]^{2+} \quad (9.41)$$

$$[OsNO_2(trpy)(bipy)]^{2+} + PPh_3 \rightarrow [OsN^{\bullet}O(trpy)(bipy)]^{2+} + OPPh_3 \quad (9.42)$$

$$[OsN^{\bullet}O(trpy)(bipy)]^{2+} - e^{-} \rightarrow [OsNO(trpy)(bipy)]^{3+} \quad (9.43)$$

The complexes, $[M(qo)_n]$, (qoH = 5-methoxy-2-nitrosophenol) react with excess triphenylphosphine at 20 °C to produce $OPPh_3$ and metal complexes whose nature depends upon the metal. For example, nitrosophenolato complexes of nickel(II), $[Ni(qo)_2]$, and zinc(II), $[Zn(qo)_2]$, react with triphenylphosphine in chloroform or pyridine to form iminophosphorane complexes (Figure 9.10).[155] The reaction of hydrochloric acid with these products yields the free ligand, Figure 9.10. Charalambous, Kensett, and Jenkins[155] observed that the deoxygenation step results in the formation of a nitrene intermediate, which subsequently reacts with PPh_3 to form the iminophosphorane complex. In contrast, reaction of $[Cu(qo)_2]$ with excess PPh_3 leads to reduction of the metal center, generation of $[Cu(qo)(PPh_3)_2]$, and formation of phenazine, Figure 9.11. The formation of phenazine also proceeds *via* the intermediate generation of nitrenes.

Reaction of the nitrosylcarbonyl complex, $[Mo(CO)_2NOCp]$, with PPh_3 results in the formation of $[Mo(NCO)(CO)(PPh_3)_2Cp]$, along with other reaction products.[156] The reaction pathway involves deoxygenation of the nitrosyl ligand and generation, from the nitrosyl, of an organometallic nitrene which then captures carbon monoxide to form isocyanate.

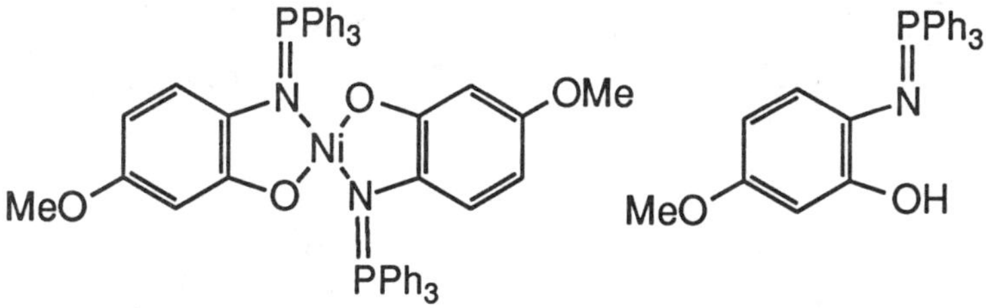

Figure 9.10

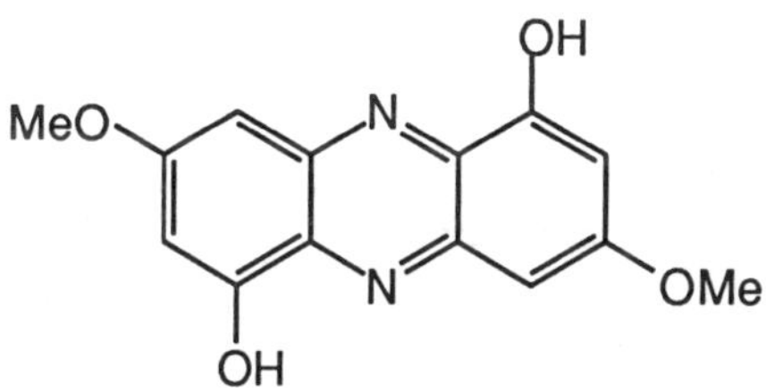

Figure 9.11

Tertiary phosphines can reduce peroxo (dioxygen) complexes. For example, Wilke, Schott and Heimbach[157] have shown that $[M(PPh_3)_4]$ (M = Ni, Pd, Pt) complexes catalyze the oxidation of triphenylphosphine by oxygen. The reaction, in benzene, involves the formation of peroxo complexes, $[MO_2(PPh_3)_2]$, Eq. 9.44:

$$[M(PPh_3)_4] + O_2 \rightarrow [MO_2(PPh_3)_2] + 2PPh_3 \qquad (9.44)$$

Further reduction of the peroxo complexes by PPh_3 occurs.[158] The sequence of reactions is shown in Scheme 9.12. Tarafder and Miah[159] have established that treatment of $[Zr(O_2)(\text{pyridine-2-carboxylato})_2]$ with EPh_3 (E = P, As) in THF results in the formation of the oxo complex, $[Zr(O)(\text{pyridine-2-carboxylato})_2]$, and $OEPh_3$. A possible mechanism for the reaction includes nucleophilic attack of EPh_3 on the metal center, followed by decomposition of the intermediate so formed, Scheme 9.13.

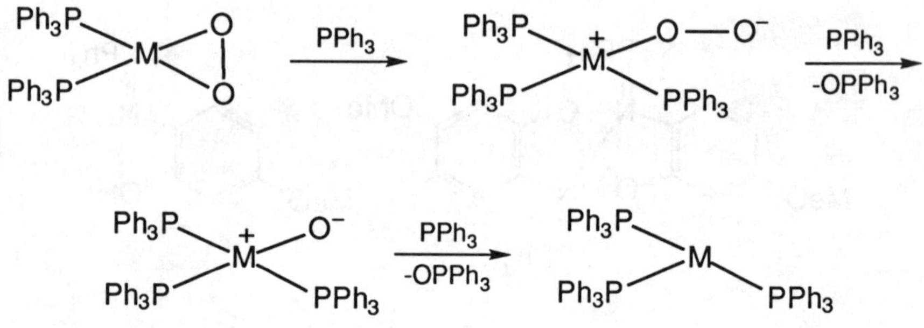

Scheme 9.12

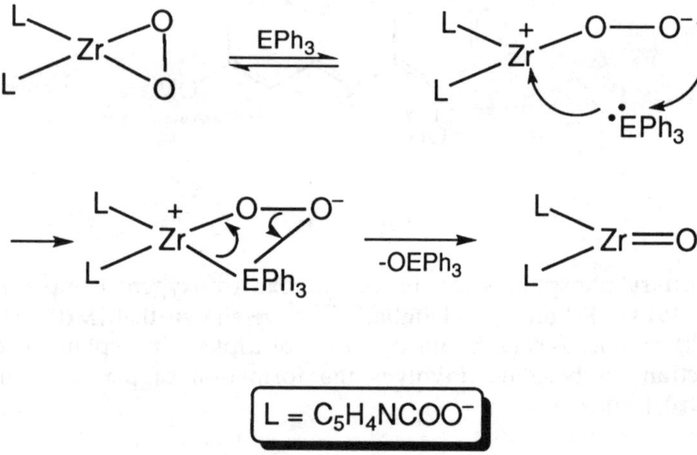

Scheme 9.13

Coordinated phosphines can undergo intramolecular redox reactions. For example, the hydroperoxorhodium complex, $[RhCl(OOH)(acac)(PPh_3)_2]$, was obtained[160] from a deoxygenation reaction carried out in chloroform in the presence of excess triphenylphosphine. The hydrooxorhodium intermediate, $[Rh(OH)Cl(acac)(PPh_3)_2]$, reacts with $CHCl_3$ to form the rhodium dichloride product, Scheme 9.14:

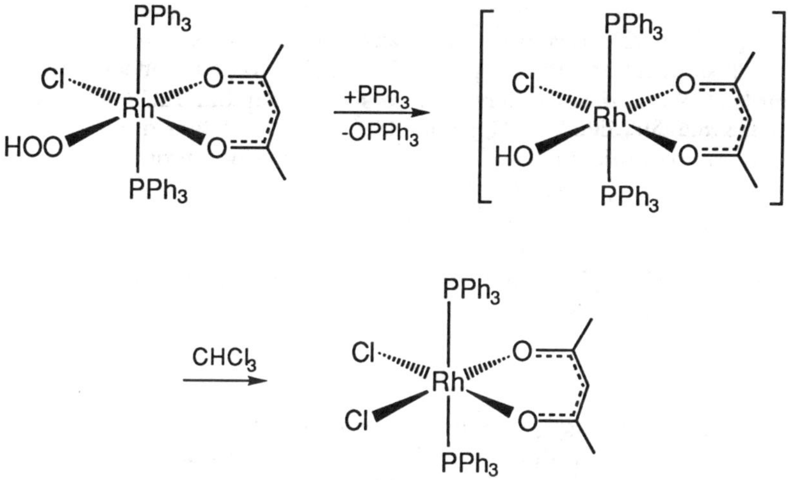

Scheme 9.14

Investigations[160] have shown that the reaction mechanism is intramolecular in nature and involves the transfer of an oxygen atom to a coordinated triphenylphosphine ligand, followed by liberation of triphenylphosphine oxide, Scheme 9.15:

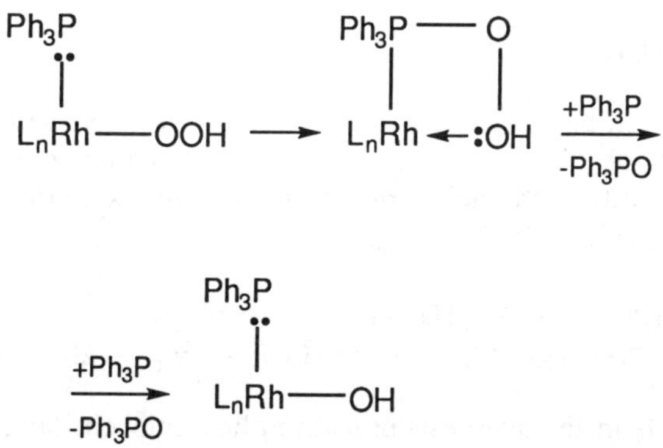

Scheme 9.15

A solid-state thermal conversion of the platinum(II) complex,

$[Pt(NO_3)_2(PPh_2CF_3)(DMSO)]$, has been reported.[161] Deoxygenation of the nitrate ligands by coordinated phosphine results in formation of the complex, $[Pt(NO_2)_2(OPPh_2CF_3)(DMSO)]$, which was characterized spectroscopically.

Phosphines are able to reduce high oxidation state metal halides. For example, Hoye and coworkers[162] have shown that $Na_2[PtCl_6]$ is reduced by three equivalents of $P(CH_2OH)_3$, Eq. 9.45:

$$Na_2[PtCl_6] + 3P(CH_2OH)_3 + H_2O \rightarrow [PtCl_2\{P(CH_2OH)_3\}_2] + O{=}P(CH_2OH)_3 + 2HCl + 2NaCl \quad (9.45)$$

$[PtCl_2\{P(CH_2OH)_3\}_2]$ also has been obtained in the reaction of $K_2[PtCl_4]$ and $P(CH_2OH)_3$ in water.[162]

9.7 Other reductants

Some reductants are rarely used in preparative coordination chemistry and others have only limited application. Such reductants are discussed here. Hydrazine and its salts are strong reductants capable of generating noble metals from their complexes. However, if the amount of reductant is carefully controlled, the metal center in a noble metal complex can be reduced to a lower oxidation state. For example, the synthesis of potassium tetrachloroplatinate(II) by reduction of $K_2[PtCl_6]$ with hydrazine sulfate, $N_2H_4{\cdot}H_2SO_4$, is well known. Grinberg has shown,[163] in his classic experiments, that the reaction mechanism involves the formation of platinum black which reacts with potassium hexachloroplatinate(IV), Eq. 9.46:

$$Pt + [PtCl_6]^{2-} + 2Cl^- \rightarrow 2[PtCl_4]^{2-} \quad (9.46)$$

This reduction can be carried out not only with hydrazine salts, but also with hydrazine hydrate in water.[63] When anhydrous hydrazine is used to reduce $[PtCl_6]^{2-}$ in absolute methanol, a platinum(II) complex containing N_2H_4 as a ligand[63] is obtained, Eq. 9.47:

$$2[NH_4]_2[PtCl_6] + 7N_2H_4 \rightarrow 2[PtCl_2(N_2H_4)_2] + 4NH_4Cl + N_2 + 2N_2H_4{\cdot}2HCl \quad (9.47)$$

The use of N_2H_4 in the synthesis of sodium hexachloroiridate(III), $Na_3[IrCl_6]$, from $Na_2[IrCl_6]$ has been investigated, Eq. 9.48:[164]

$$4Na_2[IrCl_6] + N_2H_4{\cdot}HCl + 4NaCl \rightarrow 4Na_3[IrCl_6] + N_2 + 6HCl \quad (9.48)$$

The reaction of $Na_2[MoO_4]$ with N_2H_4 in concentrated hydrochloric acid, followed by addition of NH_4Cl, results in the formation of $[NH_4][MoOCl_5]$.[165] The nitride complex of technetium, $[Tc(N)L_2]$ (L^- = diphenyldithiophosphinate), was obtained by reaction of a $[TcO_4]^-$ salt with hydrazine, followed by addition of the ligand, L.[166] There are several examples of reduction of metal salts by hydrazine in non-aqueous solvents. For example, MoO_3 reacts with $N_2H_4{\cdot}HX$ (X = F, Cl, Br, I) in DMF in the presence of the corresponding HX, and yields $[MoOXQ_2]$ after addition of the dithiocarbamate ligand, HQ.[167]

Hydroxylamine is used as a reductant in the synthesis of coordination compounds.[168,169] The complex, $[RuCl(NO)(py)_4][PF_6]$, undergoes a one-electron electrochemical oxidation in aqueous solution and this complex is obtained by reduction of $[RuCl(NO)(py)_4][ClO_4]_2$ with hydroxylamine, followed by addition of ammonium hexafluorophosphate.[168] Like hydrazine, hydroxylamine can act as both a reductant and as a ligand. Stetsenko has shown[169] that Alexander's base, $[Pt(NH_2OH)_4][OH]_2$, is obtained in high yield from potassium tetrachloroplatinate by reduction with excess hydroxylamine hydrochloride in the presence of sodium carbonate.

Sarker, Mohammad, and Subramanian[170] have reported the reductive nitrosylation of sodium tungstate, $Na_2[WO_4]$, by $NH_2OH{\cdot}HCl$ in pyridine, which results in the formation of $[W(NO)_2Cl_2(py)_2]$. Reduction is accompanied by coordination of the NH_2OH oxidation products to tungsten. Other useful nitrogen-containing reductants are sodium azide and lithium nitride. Both have been used in synthetic procedures.[171,172]

Sodium hypophosphite, $NaH_2PO_2{\cdot}H_2O$, is used in electroplating because it is a strong reductant.[173] $NaH_2PO_2{\cdot}H_2O$ typically reduces activated metal ions to compounds in lower oxidation states, rather than to metals. For example, the reduction of vanadium(V) ions to vanadium(IV) and the reduction of thallium(III) ions to thallium(I) have been investigated.[174,175] $RhCl_3{\cdot}3H_2O$ reacts with sodium hypophosphite[173] to form rhodium(I) phosphate, RhH_2PO_3OH, which has been isolated as a solid.

The superoxide ion can act either as a reductant or as an oxidant, depending on reaction conditions. As a reductant, the reaction of superoxide ion does not depend on the presence of protons, Eq. 9.49, but to act as an oxidant protons are necessary, Eq. 9.50:

$$O_2^- - e^- \rightleftharpoons O_2 \tag{9.49}$$

$$O_2^- + e^- + 2H^+ \rightleftharpoons H_2O_2 \tag{9.50}$$

Thus, reduction of substrates is anticipated upon treatment with superoxide ion in aprotic media. For example, on treatment of the nickel(III) complex,

$[Ni(Q)(H_2O)][ClO_4]$ (H_2Q = 1,4,7,10,13-pentaazacyclohexadecane-14,16-dione), with KO_2 in anhydrous dimethyl sulfoxide, the nickel(II) derivative, [Ni(Q)], is obtained.[176]

In order to increase the concentration of superoxide ions in solution, reactions can be carried out in the presence of crown ethers. Complexation of K^+ ions with these macrocyclic ligands leads to the formation of [K/crown ether]O_2 and such species are commonly more soluble in organic solvents than KO_2 itself (Chapter 2). This, in turn, allows reactions to be completed more rapidly. Thus, the complex, $[Cu(phen)_2][ClO_4]_2$, in a solution of dimethyl sulfoxide, is reduced by superoxide [K/dicyclohexyl-18-crown-6]O_2 (one equivalent), Eq. 9.51:[177]

$$[Cu(phen)_2][ClO_4]_2 + [K/dicyclohexyl\text{-}18\text{-}crown\text{-}6]O_2 \rightarrow [Cu(phen)_2][ClO_4] + [K/dicyclohexyl\text{-}18\text{-}crown\text{-}6][ClO_4] + O_2 \quad (9.51)$$

Similarly, Hasegawa, Imamura, and Fujimoto[178] have established that molybdenum(V) tetraphenylporphyrins, [MoO(tpp)X] (X = Cl, Br, SCN), are reduced by [K/18-crown-6]O_2 in a mixture of dichloromethane and dimethyl sulfoxide, forming the molybdenum(IV) derivative, [MoO(tpp)]. The mechanism for the reduction appears to proceed through an intermediate peroxo complex, $[MoO(O_2)(tpp)]^-$, that is stable in solution at -80 °C, but which is converted into [MoO(tpp)] at room temperature.

Red or white phosphorus, as well as arsenic and antimony, can be used to reduce zirconium(IV) complexes.[179] The reduction of the hydride complex of zirconium(IV), $[ZrH_2Cp_2]$, was carried out in toluene at room temperature and the tetrameric compound, $(ZrCp_2)_4$, was obtained.

Ferrocene can serve as a mild reductant. The potential for the $FeCp_2^+/FeCp_2$ couple in acetonitrile is +0.31 *V* (*vs.* SCE, 25 °C[3]) and the potential can be decreased or increased by introduction of substituents onto the Cp rings.[180] Gerard's salt,[181] $[PtCl_4(NH_3)_2]$, is reduced in methanol by $FeCp_2$, Eq. 9.52:

$$[PtCl_4(NH_3)_2] + 2FeCp_2 \rightarrow [PtCl_2(NH_3)_2] + 2[FeCp_2]Cl \quad (9.52)$$

Cobaltocenes, $CoCp_2$ and $CoCp^*_2$ are effective one-electron reductants for coordination compounds. The potential for the $CoCp_2^+/CoCp_2$ couple is -0.9 *V* (*vs.* SCE, 25 °C, DMF) and, for the $CoCp^*_2{}^+/CoCp^*_2$ couple, the potential is -1.48 *V* (*vs.* SCE, 25 °C, DMF).[3] An important advantage of these cobaltocenes is their high solubility in hydrocarbons and other organic solvents. For example, Koch and coworkers[182] have reduced the ruthenium(IV) complex, $[Ru(SC_6H_2Pr_3\text{-}iso\text{-}2,4,6)_3(MeCN)_2]^+$ (-0.23 *V*, *vs.* SCE, MeCN), with $CoCp_2$ in THF solution to give the Ru(III) analog, Scheme 9.16:

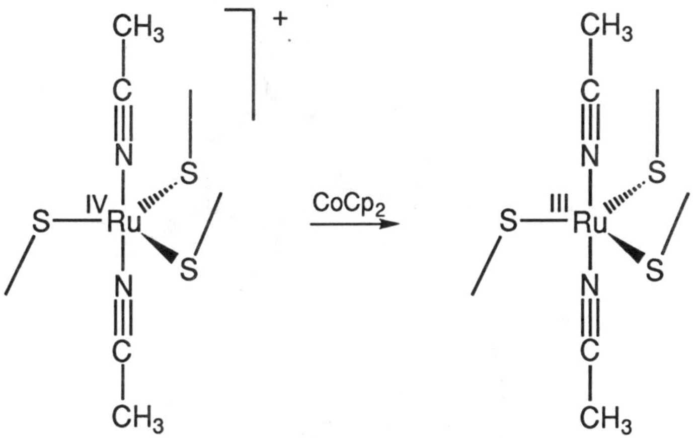

Scheme 9.16

The product has been structurally characterized, Figure 9.12.

Calderazzo and coworkers[183] have reduced the solvento-complexes, $[MCl_4(THF)_2]$ (M = Ti, Zr, Hf), with $CoCp_2$ in tetrahydrofuran solution to give $[CoCp_2][TiCl_4(THF)_2]$ or $[CoCp_2][MCl_5(THF)]$ (M = Zr or Hf).

Reduction of the cobalt complexes, $[Co(\eta^5\text{-}C_5R_5)(NO)L][PF_6]$ (R/L = Ph/P(OMe)$_3$ or Ph/P(OPh)$_3$), with $CoCp_2$ in tetrahydrofuran leads to *P*-donor ligand elimination to give $[Co(\eta^5\text{-}C_5R_5)(\mu\text{-}NO)]_2$.[184] Connelly, Raven, and Geiger[184] have identified $[Co(\eta^5\text{-}C_5R_5)(NO)L]$ as the primary reduction product by e.s.r. spectroscopy at -196 °C. In contrast, the rhodium analogs, $[Rh(\eta^5\text{-}C_5R_5)(NO)L]$ (L = P(OMe)$_3$ or P(OPh)$_3$), are stable at room temperature and have been isolated from the reaction between $[Rh(\eta^5\text{-}C_5R_5)(NO)L][PF_6]$ and $CoCp_2$ in benzene, Eq. 9.53:

$$[Rh(\eta^5\text{-}C_5R_5)(NO)L][PF_6] + CoCp_2 \rightarrow [Rh(\eta^5\text{-}C_5R_5)(NO)L] + [CoCp_2][PF_6] \quad (9.53)$$

The reaction occurs quickly (in about ten minutes) and partial evaporation of the solvent *in vacuo* gives the solid product, $[Rh(\eta^5\text{-}C_5R_5)(NO)L]$.

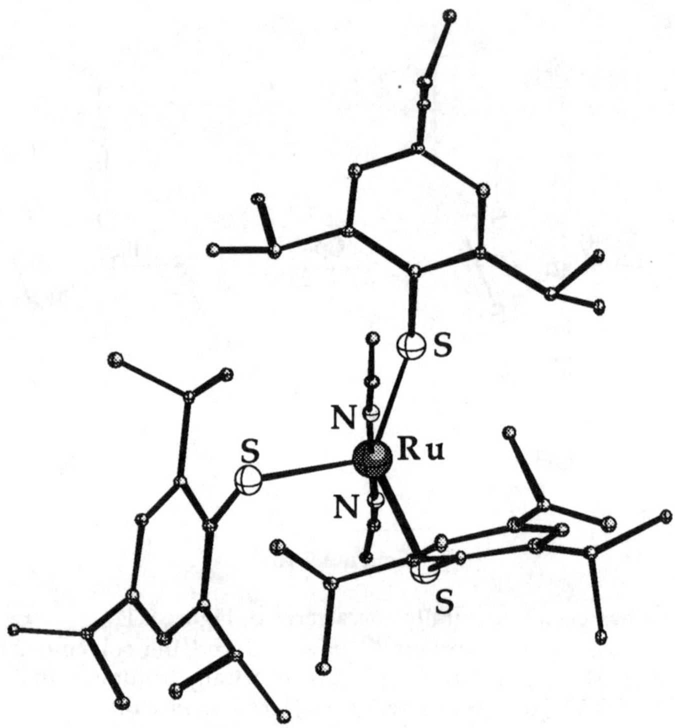

Figure 9.12

Further, $CoCp_2$ can be used to reduce metal clusters. For example, Walton, Moehring, and Fanwick[185] have shown that cobaltocene may be employed to reduce $[Re_2H_7(PPh_3)_4Au(PPh_3)][PF_6]$. The reaction is carried out in acetone and is complete within several minutes. Addition of methanol gives a precipitate of $[Re_2H_7(PPh_3)_4Au(PPh_3)]$.

Thus, although a limited set of reductants are employed in the synthesis of coordination compounds, the result is a rich chemistry. The choice of a specific reductant is determined by the nature of the complex and the conditions under which the reaction is to be carried out. This Chapter has focused on preparative reactions carried out in solution. The extremely important role of the solvent is obvious because the solvent and the reductant are inseparably related.

References

1. M. Hudlicky, *Reductions in Organic Chemistry* (Horwood, Chichester, 1984).
2. F. A. Cotton and G. Wilkinson, *Advanced Inorganic Chemistry* Fifth Edition (John Wiley & Sons, New York, 1988) p. 601.
3. D. Astruc, *Angew. Chem. Int. Ed. Engl.* **27** (1988) 643.
4. R. Bartoszewicz, V. Mecznikowska-Stoljarczik, and B. Opszondek, *Methods for the Reduction of Organic Compounds* [in Russian] (IL, Moscow, 1960) p. 43.
5. V. G. Syrkin, *Metal Carbonyls* [in Russian] (Khimia, Moscow, 1984) p. 196.
6. J. Li, A. D. Hunter, R. McDonald, B. D. Santarsiero, S. G. Bott, and J. L. Atwood, *Organometallics* **11** (1992) 3050.
7. K. M. Phahl and J. Ellis, *Organometallics* **3** (1984) 230.
8. C. Loubster and S. Lotz, *Inorg. Synth.* **29** (1992) 174.
9. M. Pfeffer, J. Fisher, and A. Mitschler, *Organometallics* **3** (1984) 1531.
10. T. S. Coolbaugh, D. Santasiero B., and R. H. Grubbs, *J. Am. Chem. Soc.* **106** (1984) 6310.
11. W. P. Weiner, F. J. Hollander, and R. G. Bergman, *J. Am. Chem. Soc.* **106** (1984) 7462.
12. T. Yoshida and S. Otsuka, *Inorg. Synth.* **19** (1979) 101.
13. Yu. N. Kukushkin, *The Chemistry of Coordination Compounds* [in Russian] (Vyshaya Shkola, Moscow, 1985) p. 455.
14. M. J. Fernández-Trujillo, M. G. Basallote, P. Valerga, and M. C. Puerta, *J. Chem. Soc., Dalton Trans.* (1993) 923.
15. D. J. Sikora and M. D. Rausch, *J. Organomet. Chem.* **276** (1984) 21.
16. C. M. Archer, J. R. Dilworth, R. M. Thompson, M. McPartlin, D. C. Povey, and J. D. Kelly, *J. Chem. Soc., Dalton Trans.* (1993) 461.
17. L. D. Field, A. F. Masters, M. Gibson, D. R. Latimer, T. W. Hambley, and I. E. Buys, *Inorg. Chem.* **32** (1993) 211.
18. T. Cuenca and P. Royo, *J. Organomet. Chem.* **293** (1985) 61.
19. M. E. Riehl, S. R. Wilson, and G. S. Girolami, *Inorg. Chem.* **32** (1993) 218.

20. P. M. Morse, S. R. Wilson, and G. S. Girolami, *Inorg. Chem.* **29** (1990) 3200.

21. B. K. Coltrain and S. C. Jackels, *Inorg. Chem.* **20** (1981) 2032.

22. K. B. Yatsimirskii, V. V. Pavlishchuk, and E. V. Rybak-Akimova, *Zh. Neorg. Khim.* **30** (1985) 1601.

23. T. V. Talalaeva and K. A. Kocheshkov, *Methods of Organometallic Chemistry. Lithium, Sodium, Potassium, Rubidium, and Cesium,* [in Russian] vol. 2, (Nauka, Moscow, 1971) p. 1009.

24. E. T. Kaiser and L. Kevan, eds, *Radical Ions* (Interscience, New York, 1968).

25. F. Billiau, G. Folcher, H. Marguet-Ellis, P. Rigny, and E. Saito, *J. Am. Chem. Soc.* **103** (1981) 5603.

26. J. M. Maher, R. P. Beatty, and N. J. Cooper, *Organometallics* **1** (1982) 215.

27. M. Y. Darensbourg and S. Slater, *J. Am. Chem. Soc.* **103** (1981) 5914.

28. J. A. Connor, E. J. James, C. Overton, and N. El Murr, *J. Organomet. Chem.* **218** (1981) C31.

29. E. Lindner, H. Behrens, and S. Birkle, *J. Organomet. Chem.* **15** (1968) 165.

30. A. Antiñolo, M. F. Lappert, and D. J. W. Winterborn, *J. Organomet. Chem.* **272** (1984) C37.

31. P. Jutzi and B. Heilscher, *J. Organomet. Chem.* **291** (1985) C25.

32. T. Mashiko, C. A. Reed, K. J. Haller, and W. R. Scheidt, *Inorg. Chem.* **23** (1984) 3192.

33. A. J. Kunin, E. J. Nanni, and R. Eisenberg, *Inorg. Chem.* **24** (1985) 1852.

34. (a) W. Rüdorff, *Adv. Inorg. Chem. Radiochem.* **1** (1959) 223; (b) R. Csuk, B. I. Glänzer, and A. Fürstner, *Adv. Organomet. Chem.* **28** (1988) 85.

35. Yu. N. Novikov and M. E. Volpin, *Uspekhi Khim.* **40** (1971) 1568.

36. (a) K. A. Jensen, B. Nygaard, G. Elisson, and P. H. Nielsen, *Acta Chem. Scand.* **19** (1965) 768; (b) C. Ungurenasu and M. Palie, *J. Chem. Soc., Chem. Commun.* (1975) 388.

37. (a) M. A. Bennet, J. S. Drage, K. D. Griffiths, N. K. Roberts, G. B. Robertson, and W. A. Wickramasinghe, *Angew. Chem.* **100** (1988) 1002; (b) M. A. Schwindt, T. Lejon, and L. S. Hegedus, *Organometallics* **9** (1990) 2814.

38. M. C. Manning and W. C. Trogler, *Inorg. Chim. Acta* **50** (1981) 247.

39. K. A. Kocheshkov, A. N. Nesmeyanov, and M. M. Nad', *Dokl. AN SSSR* **26** (1940) 53.

40. A. J. Lindsay, R. P. Tooze, M. Motevalli, M. B. Hursthouse, and G. Wilkinson, *J. Chem. Soc., Chem. Commun.* (1984) 799.

41. A. J. Lindsay, R. P. Tooze, M. Motevalli, M. B. Hursthouse, and G. Wilkinson, *J. Chem. Soc., Chem. Commun.* (1984) 1383.

42. K.-H. Thiele and A. Krüger, *Z. Anorg. Allg. Chem.* **527** (1985) 95.

43. A. Krüger and K.-H. Thiele, *Z. Anorg. Allg. Chem.* **532** (1986) 53.

44. H. Lehmkuhl, H. Nehl, R. Benn, and R. Mynott, *Angew. Chem.* **98** (1986) 628.

45. J. R. Bleeke and D. A. Moore, *Inorg. Chem.* **25** (1986) 3522.

46. R. D. Feltham, G. Elbaze, R. Ortega, C. Eck, and J. Dubrawski, *Inorg. Chem.* **24** (1985) 1503.

47. M. I. Bruce, C. M. Jensen, and N. L. Jones, *Inorg. Synth.* **26** (1989) 259.

48. H.-H. Wang, L. H. Pignolet, P. E. Reedy, M. M. Olmstead, and A. L. Balch, *Inorg. Chem.* **26** (1987) 377.

49. F. Calderazzo and R. Poli, *Gazz. Chim. Ital.* **115** (1985) 573.

50. D. J. Cole-Hamilton, *J. Chem. Soc., Chem. Commun.* (1980) 1213.

51. G. Booth and J. Chatt, *J. Chem. Soc.* (1962) 2099.

52. J. Kriege-Simondsen, G. Elbaze, M. Dartiguenave, R. D. Feltham, and Y. Dartiguenave, *Inorg. Chem.* **21** (1982) 230.

53. J. Kriege-Simondsen and R. D. Feltham, *Inorg. Chim. Acta* **71** (1983) 185.

54. R. D. Feltham and J. C. Kriege, *J. Am. Chem. Soc.* **101** (1979) 5064.

55. D. T. Doughty, G. Gordon, and R. P. Stewart, *J. Am. Chem. Soc.* **101** (1979) 2645.

56. S. Bhaduri, B. F. G. Johnson, C. J. Savory, J. A. Segal, and R. H. Walter, *J. Chem. Soc., Chem. Commun.* (1974) 809.

57. S. A. Bhaduri, I. Bratt, B. F. G. Johnson, A. Khair, J. A. Segal, R. Walters, and C. Zuccaro, *J. Chem. Soc., Dalton Trans.* (1981) 234.

58. W. Bos, J. J. Bour, J. J. Steggerda, and L. H. Pignolet, *Inorg. Chem.* **24** (1985) 4298.

59. W. P. Bosman, W. Bos, J. M. M. Smits, P. T. Beurskens, J. J. Bour, and J. J. Steggerda, *Inorg. Chem.* **25** (1986) 2093.

60. D. E. Fjare and W. L. Gladfelter, *J. Am. Chem. Soc.* **106** (1984) 4799.

61. (a) G. Vasapollo, P. Giannoccaro, C. F. Nobile, and F. Allegretta, *J. Organomet. Chem.* **270** (1984) 109; (b) G. Vasapollo, C. F. Nobile, P. Giannoccaro, and F. Allegretta, *J. Organomet. Chem.* **277** (1984) 417.

62. M. Cameron, B. G. Gowenlock, and G. Vasapollo, *Chem. Soc. Rev.* **19** (1990) 355.

63. N. G. Klyuchikov, *Guide to Inorganic Synthesis* [in Russian], (Goskhimizdat, Moscow, 1953).

64. C.-M. Che, M. Jamal, C.-K. Poon, and W.-C. Chung, *Inorg. Chem.* **24** (1985) 2868.

65. A. Bujewski, H. Warachim, K. Grzezicki, and J. Dobrowolski, *Thermochim. Acta* **92** (1985) 779.

66. V. Yatirajam and M. Lakshimi Kantam, *Indian J. Chem.* **23A** (1984) 771.

67. U. Abram, *Zentralinst. Kernforsch. Rossendorf Dresden* [Ber.], No. 538, Jahresbericht 1983: Bereich Radioaktive Isotope (1984) 24.

68. *Synthesis of Complex Compounds of the Platinum Metals*, ed. I. I. Chernyaev, Handbook [in Russian], (Nauka, Moscow, 1964) p. 24.

69. C. O. Dietrich-Buchecker, J. M. Kern, and J. P. Sauvage, *J. Chem. Soc., Chem. Commun.* (1985) 760.

70. M. O. Albers, T. V. Ashworth, H. E. Oosthuizen, and E. Singleton, *Inorg. Synth.* **26** (1989) 68.

71. M. A. Bennett and J. L. Latten, *Inorg. Synth.* **26** (1989) 180.

72. E. V. Leonova, V. Kh. Syundyukova, F. S. Denisov, A. A. Koridze, V. S. Khandkarova, and A. Z. Rubezhov, *Methods of Organometallic Chemistry. Cobalt, Nickel, and the Platinum Metals* [in Russian] (Nauka, Moscow, 1978) p. 376.

73. B. T. Heaton, J. A. Iggo, C. Jacob, H. Blanchard, M. B. Hursthouse, I. Ghatak, M. E. Harman, R. G. Somerville, W. Heggie, P. R. Page, and I. Villax, *J. Chem. Soc., Dalton Trans.* (1992) 2533.

74. C. Persson and C. Anderson, *Inorg. Chim. Acta* **203** (1993) 235.

75. L. Castellani and M. C. Gallazzi, *Transit. Metal Chem.* **10** (1985) 194.

76. E. Hey-Hawkins and H. G. von Schnering, *Z. Naturforsch.* **46b** (1991) 307.

77. M. L. Larson and F. W. Moore, *Inorg. Chem.* **3** (1964) 285.

78. (a) F. Herrmann, *Ber.* **38** (1905) 2813; (b) F. Herrmann, *Ber.* **39** (1906) 3812.

79. See for example, (a) F. Gasparrini, M. Giovannoli, D. Misiti, G. Natile, and G. Palmieri, *Tetrahedron* **40** (1984) 165; (b) F. Gasparrini, M. Giovannoli, D. Misiti, G. Natile, and G. Palmieri, *Tetrahedron* **39** (1983) 3181; (c) F. Gasparrini, M. Giovannoli, D. Misiti, G. Natile, and G. Palmieri, *J. Org. Chem.* **55** (1990) 1323; (d) F. Gasparrini, M. Giovannoli, D. Misiti, G. Natile, G. Palmieri, and L. Maresca, *J. Am. Chem. Soc.* **115** (1993) 4401; (e) D. B. Dyson, R. V. Parish, C. A. McAuliffe, R. G. Pritchard, R. Fields, and B. Beagley, *J. Chem. Soc., Dalton Trans.* (1989) 907.

80. R. Usón, A. Laguna, and M. Laguna, *Inorg. Synth.* **26** (1989) 85.

81. G. Annibale, L. Canovese, L. Cattalini, and G. Natile, *J. Chem. Soc., Dalton Trans.* (1980) 1017.

82. L. A. Chugaev and I. Benevolensky, *Izv. Sekt. Platiny* **4** (1926) 127.

83. V. N. Spevak, *Ph.D. Dissertation*, Lensovet Technological Institute, Leningrad, U.S.S.R. (Russian Federation), 1971.

84. R. I. Rudii, N. V. Cherkashina, Ya. V. Salin, and I. I. Moiseev, *Izv. AN SSSR, Ser. Khim.* (1983) 1866.

85. M. S. Hussain and A. Isab, *J. Coord. Chem.* **14** (1985) 17.

86. R. S. Nyholm, *J. Chem. Soc.* (1950) 843.

87. E. A. Klimkovich, L. V. Nashivanko, and Yu. I. Usatenko, *Izv. Vyssh. Uchebn. Zaved., Khim., Khim. Tekhnol.* **16** (1973) 43.

88. N. V. Vorobev-Desyatovskii, Yu. N. Kukushkin, and V. V. Sibirskaya, *Koord. Khim.* **11** (1985) 1299.

89. R. D. Sauerbrunn and E. B. Sandell, *J. Am. Chem. Soc.* **75** (1953) 3554.

90. M. J. Abrams, A. Davison, R. Faggiani, A. G. Jones, and C. J. L. Lock, *Inorg. Chem.* **23** (1984) 3284.

91. D. W. Freeman, F. G. Baglin, L. M. Wilkes, and J. J. MacDougall, *Spectroscopy* **1** (1986) 32.

92. Yu. N. Kukushkin, N. V. Vorobev-Desyatovskii, and K. M. Patrabansh, *Zh. Obsch. Khim.* **55** (1985) 1759.

93. D. Brenner, A. Davison, J. Lister-James, and A. G. Jones, *Inorg. Chem.* **23** (1984) 3793.

94. G. Bandoli and T. I. A. Gerber, *Inorg. Chim. Acta* **126** (1987) 205.

95. U. Hildebrand, J. Lex, K. Taraz, S. Winkler, W. Ockels, and H. Budzikiewicz, *Z. Naturforsch.* **39b** (1984) 1607.

96. R. S. Wade and C. E. Castro, *Inorg. Chem.* **24** (1985) 2862.

97. J. Fiedler and J. Masek, *Inorg. Chim. Acta* **105** (1985) 83.

98. T. Mineey and T. G. Traylor, *Bioinorg. Chem.* **9** (1978) 409.

99. D. Coucouvanis, C. N. Murphy, and E. Simhon, *Inorg. Synth.* **21** (1982) 23.

100. A. A. Grinberg, E. M. Marshak, and K. N. Lyubomirova, *Zh. Neorg. Khim.* **10** (1965) 717.

101. V. V. Budanov, *Chemistry and Technology of Reducing Agents Based on Sulfoxylic Acid. Rongalite and Its Analogs* [in Russian] (Khimiya, Moscow, 1984) p. 115.

102. T. N. Srivastava and S. Mehrotra, *Synth. React. Inorg. Met.-Org. Chem.* **15** (1985) 709.

103. M. Böck, *Dissertation for the Degree of Doctor of Natural Science*, Faculty of Chemistry, Biology, and Earth Scheince, Technical University of Munich (1984).

104. S. Elmroth, L. H. Skibsted, and L. I. Elding, *Inorg. Chem.* **28** (1989) 2703.

105. G. Davies and D. J. Loose, *Inorg. Chem.* **15** (1976) 694.

106. D. L. Johnson, A. R. Fritzberg, B. L. Hawkins, S. Kasina, and D. Eshima, *Inorg. Chem.* **23** (1984) 4204.

107. M. A. Bennett, J. Chatt, G. J. Erskine, J. Lewis, R. F. Long, and R. S. Nyholm, *J. Chem. Soc., A* (1967) 501.

108. T. P. Lockhart and D. A. Haitko, *Polyhedron* **4** (1985) 1745.

109. J. E. McMurray, *Acc. Chem. Res.* **7** (1974) 281.

110. S. Gambarotta, M. Mazzanti, C. Floriani, and M. Zehnder, *J. Chem. Soc., Chem. Commun.* (1984) 1116.

111. M. Mazzanti, S. Gambarotta, C. Floriani, A. Chiesi-Villa, and C. Guastini, *Inorg. Chem.* **25** (1986) 2308.

112. M. Mazzanti, C. Floriani, A. Chiesi-Villa, and C. Guastini, *Inorg. Chem.* **25** (1986) 4158.

113. A. M. Lannon, A. G. Lappin, and M. G. Segal, *Inorg. Chem.* **23** (1984) 4167.

114. M. G. Segal and R. M. Sellers, *J. Chem. Soc., Faraday Trans. I* **78** (1982) 1149.

115. M. G. Segal and R. M. Sellers, *J. Chem. Soc., Chem. Commun.* (1980) 991.

116. N. A. Lewis and A. M. Ray, *Inorg. Chem.* **23** (1984) 3649.

117. J. R. Hanson and M. S. Mehta, *J. Chem. Soc.*, C (1969) 2349.

118. L. E. Bennett and H. Taube, *Inorg. Chem.* **7** (1968) 254.

119. R. C. Mercier and M. R. Paris, *Bull. Soc. Chim. Fr.* (1965) 3577.

120. J. McB. Harrowfield, L. Spiccia, and D. W. Watts, *Aust. J. Chem.* **36** (1983) 1923.

121. J. Armor, *Inorg. Chem.* **12** (1973) 1959.

122. I. A. Krol, V. Yu. Kukushkin, Z. A. Starikova, and V. M. Tkachuk, *Zh. Obsch. Khim.* **58** (1988) 2625.

123. (a) L. E. Manzer, *Inorg. Synth.* **21** (1982) 135; (b) J. R. Rambo, S. L. Bartley, W. E. Streib, and G. Christou, *J. Chem. Soc., Dalton Trans.* (1994) 1813.

124. G. D. McDonald, M. Thompson, and E. M. Larsen, *Inorg. Chem.* **7** (1968) 648.

125. M. W. Duckworth, G. W. A. Fowles, and R. A. Hoodless, *J. Chem. Soc.* (1963) 5665.

126. A. D. Hugi, L. Helm, and A. E. Merbach, *Inorg. Chem.* **26** (1987) 1763.

127. J. Darkwa, D. M. Giolando, C. J. Murphy, and T. B. Rauchfuss, *Inorg. Synth.* **27** (1990) 51.

128. J. C. Dobson, M. Sano, and H. Taube, *Inorg. Chem.* **30** (1991) 456.

129. C. Belitto, D. Fiorani, and S. Viticoli, *Inorg. Chem.* **24** (1985) 1939.

130. F. H. Kohler and W. Prossdorf, *Z. Naturforsch.* **32b** (1977) 1026.

131. S. A. A. Zaidi, M. Shakir, and M. Aslam, *Indian J. Chem.* **A22** (1983) 1078.

132. J. L. Dilworth and R. L. Richards, *Inorg. Synth.* **20** (1980) 119.

133. (a) J. I. G. Cadogan, *Quart. Rev.* **22** (1968) 222; (b) D. E. C. Corbridge, *Phosphorus: An Outline of its Chemistry, Biochemistry and Technology*, Fourth Edition (Elsevier, Amsterdam, 1990) and references therein.

134. J. Chatt and G. A. Rowe, *J. Chem. Soc.* (1962) 4019.

135. G. W. Parshall, *Inorg. Synth.* **17** (1977) 110.

136. J. W. Johnson, J. F. Brody, G. B. Ansell, and S. Zentz, *Inorg. Chem.* **23** (1984) 2415.

137. W. A. Herrmann, R. Serrano, and H. Bock, *Angew. Chem. Int. Ed. Engl.* **23** (1984) 383; *Angew. Chem.* **96** (1984) 364.

138. W. A. Herrmann, R. Serrano, U. Küsthardt, M. L. Ziegler, E. Guggolz, and Th. Zahn, *Angew. Chem. Int. Ed. Engl.* **23** (1984) 515; *Angew. Chem.* **96** (1984) 498.

139. W. A. Herrmann, R. Serrano, U. Küsthardt, E. Guggolz, B. Nuber, and M. L. Ziegler, *J. Organomet. Chem.* **287** (1985) 329.

140. W. A. Herrmann, E. Voss, U. Küsthardt, and E. Herdtweck, *J. Organomet. Chem.* **294** (1985) C37.

141. L. E. Rhodes, K. G. Caulton, W. K. Rybak, and J. J. Ziolkowsky, *Polyhedron* **5** (1986) 1891.

142. R. Rossi, A. Marchi, L. Marvelli, L. Magon, M. Peruzzini, U. Casellato, and R. Graziani, *Inorg. Chim. Acta* **204** (1993) 63.

143. X. L. R. Fontaine, E. H. Fowles, T. P. Layzell, B. L. Shaw, and M. Thornton-Pett, *J. Chem. Soc., Dalton Trans.* (1991) 1519.

144. F. Refosco, F. Tisato, G. Bandoli, C. Bolzati, A. Dolmella, A. Moresco, and M. Nicolini, *J. Chem. Soc., Dalton Trans.* (1993) 605.

145. M. S. Reynolds, J. M. Berg, and R. H. Holm, *Inorg. Chem.* **23** (1984) 3057.

146. R. H. Holm and J. M. Berg, *Acc. Chem. Res.* **19** (1986) 363.

147. R. H. Holm, *Chem. Rev.* **87** (1987) 1401.

148. S. S. Jurisson, K. Dancey, M. McPartlin, P. A. Tasker, and E. Deutsch, *Inorg. Chem.* **23** (1984) 4743.

149. J. T. Groves, W. J. Kruper, Jr., R. C. Haushalter, and W. M. Butler, *Inorg. Chem.* **21** (1982) 1363.

150. D. B. McDonald and J. I. Shulman, *Anal. Chem.* **47** (1975) 2023.

151. J. A. Craig, E. W. Harlan, B. S. Snyder, M. A. Whitener, and R. H. Holm, *Inorg. Chem.* **28** (1989) 2082.

152. B. S. Tovrog, S. E. Diamond, and F. Mares, *J. Am. Chem. Soc.* **101** (1979) 270.

153. K. R. Grundy, K. R. Laing, and W. R. Roper, *J. Chem. Soc., Chem. Commun.* (1970) 1500.

154. D. W. Pipes and T. J. Meyer, *Inorg. Chem.* **23** (1984) 2466.

155. J. Charalambous, M. J. Kensett, and J. M. Jenkins, *J. Chem. Soc., Chem. Commun.* (1977) 400.

156. A. T. McPhail, G. R. Knox, C. G. Robertson, and G. A. Sim, *J. Chem. Soc., A* (1971) 205.

157. G. Wilke, H. Schott, and P. Heimbach, *Angew. Chem. Int. Ed. Engl.* **6** (1967) 92.

158. A. Sen and J. Halpern, *J. Am. Chem. Soc.* **99** (1977) 8337.

159. M. T. H. Tarafder and M. A. L. Miah, *Inorg. Chem.* **25** (1986) 2265.

160. H. Suzuki, S. Matsuura, Y. Moro-oka, and T. Ikawa, *J. Organomet. Chem.* **286** (1985) 247.

161. Yu. N. Kukushkin, E. I. Grigorov, K. A. Khokhryakov, I. G. Maslennikov, and E. S. Postnikova, *Koord. Khim.* **10** (1984) 543.

162. P. A. T. Hoye, P. G. Pringle, M. B. Smith, and K. Worboys, *J. Chem. Soc., Dalton Trans.* (1993) 269.

163. A. A. Grinberg, *Physical Chemistry of Complex Compounds. Selected Works* [in Russian] (Nauka, Leningrad, 1972) p. 130.

164. M. S. Ram and D. M. Stanbury, *Inorg. Chem.* **24** (1985) 2954.

165. A. Syamal and M. R. Maurya, *Inorg. Synth.* **26** (1989) 36.

166. U. Abram, H. Spies, W. Görner, R. Kirmse, and J. Stach, *Inorg. Chim. Acta* **109** (1985) L9.

167. S. Vasanthi, K. S. Nagaraja, and M. R. Udupa, *Transit. Met. Chem.* **9** (1984) 382.

168. K. Aoyagi, M. Mukaida, H. Kakihana, and K. Shimizu, *J. Chem. Soc., Dalton Trans.* (1985) 1733.

169. A. I. Stetsenko, *Zh. Neorg. Khim.* **6** (1961) 1769.

170. S. Sarkar, I. B. Mohammad, and P. Subramanian, *Chem. Lett.* (1985) 1633.

171. M. S. Ram and D. M. Stanbury, *Inorg. Chem.* **24** (1985) 4233.

172. M. Kilner and G. Parkin, *J. Organomet. Chem.* **302** (1986) 181.

173. Ya. I. Valsyunene and A. Yu. Yuryavichus, *Zh. Neorg. Khim.* **29** (1984) 2600.

174. K. K. Sengupta, B. B. Pal, and D. C. Mukherjee, *Z. Phys. Chem. (Frankfurt-am-Main)* **72** (1970) 230.

175. J. Poor and L. Treindl, *Chem. Zvesti* **23** (1969) 816.

176. R. Machida, E. Kimura, and Y. Kushi, *Inorg. Chem.* **25** (1986) 3461.

177. J. S. Valentine and A. B. Curtis, *J. Am. Chem. Soc.* **97** (1975) 224.

178. K. Hasegawa, T. Imamura, and M. Fujimoto, *Inorg. Chem.* **25** (1986) 2154.

179. H. Köpf and T. Klapötke, *Z. Naturforsch.* **40b** (1985) 447.

180. K. N. Brown, P. T. Gulyas, P. A. Lay, N. S. McAlpine, A. F. Masters, and L. Phillips, *J. Chem. Soc., Dalton Trans.* (1993) 835.

181. A. Peloso, *J. Chem. Soc., Dalton Trans.* (1984) 249.

182. S. P. Satsangee, J. H. Hain, Jr., P. T. Cooper, and S. A. Koch, *Inorg. Chem.* **31** (1992) 5160.

183. P. Biagini, F. Calderazzo, G. Pampaloni, and P. F. Zanazzi, *Gazz. Chim. Ital.* **117** (1987) 27.

184. N. G. Connelly, S. J. Raven, and W. E. Geiger, *J. Chem. Soc., Dalton Trans.* (1987) 467.

185. G. A. Moehring, P. E. Fanwick, and R. A. Walton, *Inorg. Chem.* **26** (1987) 1861.

Chapter 10. BORON AND ALUMINUM HYDRIDES IN PREPARATIVE COORDINATION CHEMISTRY

In the late 1940s and early 1950s work from Schlesinger's research group led to a number of patents[1] that describe sodium borohydride, $Na[BH_4]$, and lithium aluminum hydride, $Li[AlH_4]$. Over the intervening years it has been demonstrated that these compounds are complementary reagents in synthesis: $Na[BH_4]$ is a mild reducing agent, whereas, $Li[AlH_4]$ is a powerful reductant.[1-3] In the late 1950s and early 1960s these reagents were introduced into preparative coordination chemistry and today they are commonly used as reductants.

Brown has outlined[4] several ways to modify the reducing abilities of complex hydrides. These include: (*i*) the use of different solvents, (*ii*) structural modification of the complex hydride, and (*iii*) introduction of activating agents to the reaction system. This Chapter focuses on work that was completed after the publication of Brown's book[4] and deals specifically with the reactivity of boron and aluminum hydrides with metal complexes, where the emphasis is on synthetically useful methods.

The substitution of hydride ligands in $[BH_4]^-$ and $[AlH_4]^-$ anions by other groups is an effective approach to the modification of the reducing abilities of these reagents. Many substituted hydrides have been synthesized by replacement of H^- in the $[BH_4]^-$ anion. For example, alkali metal trialkylborohydrides[5-8] and trialkoxyborohydrides,[8,9] as well as sodium and lithium cyanoborohydrides[10-14] and dicyanoborohydrides,[12,15] have been prepared in this way.

The reducing abilities of substituted borohydrides cover a broad range. Based on rate data for the reaction of *n*-octyl chloride with $Na[BH_4]$ and $Li[BHEt_3]$, Eq. 10.1:

$$n\text{-}C_8H_{17}Cl + H^- \rightarrow n\text{-}C_8H_{18} + Cl^- \qquad (10.1)$$

nucleophilicity factors[16] were calculated.[17] The substitution of three hydride ligands in $[BH_4]^-$ by ethyl groups was found to increase the reducing ability 10^4 times. Based upon a large set of data[15,17-20] the nucleophilicities of the common, complex hydrides towards typical substrates may be ordered as follows:

$$NaH > K[BH(OPr\text{-}i)_3] > Li[BHEt_3] > Li[AlH_4] > Na[BH_4]$$
$$> Na[BH_3CN] > Na[BH_2(CN)_2]$$

Further, the reactivity of $[AlH_4]^-$ was shown[21] to be modified by substitution of hydride ligands by alkyl- or alkoxy-groups. The substituted aluminohydrides, $Na[AlH_2Et_2]$, $Li[AlH(OBu\text{-}t)_3]$, and $Na[AlH_2(OC_2H_4OMe)_2]$, are widely used in preparative chemistry. Sodium dihydridobis(2-methoxyethoxy)aluminate, $Na[AlH_2(OC_2H_4OMe)_2]$, is probably the most commonly employed and data reported in the book by Strouf, Casensky and Kubanek[22] demonstrate that this compound may be used in place of $Na[AlH_4]$ or $Li[AlH_4]$ in many reactions. Indeed, in some syntheses, $Na[AlH_2(OC_2H_4OMe)_2]$ is more suitable and more effective than the $M[AlH_4]$ reagents.[22] The use of $Na[AlH_2(OC_2H_4OMe)_2]$ has two major advantages in comparison with the use of $M[AlH_4]$ reductants; no strict safety measures are required in routine experimental work with $Na[AlH_2(OC_2H_4OMe)_2]$, and this compound is soluble in ethers and aromatic hydrocarbons due to solvation of the sodium cations by the oxygen atoms of the alkoxy-groups, Figure 10.1:

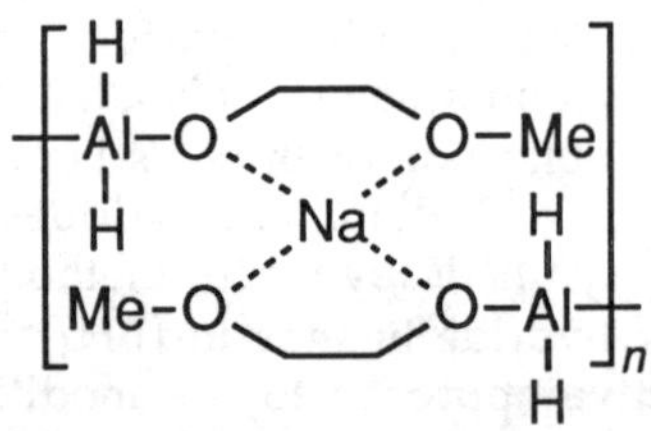

Figure 10.1

The high reactivity of sodium dihydridobis(2-methoxyethoxy)aluminate is attributed[22] largely to its enhanced solubility in hydrocarbon solvents in comparison to the solubilities of the traditional $M[AlH_4]$ reagents.

The replacement of hydride by substituents of different types in boron and aluminum hydrides makes possible the preparation of an entire spectrum of reagents for selective reductions.

10.1 Applications of boron and aluminum hydrides in the synthesis of metal hydrides

The synthesis of metal-hydride complexes is discussed in detail in a text[23] and in reviews.[24-28] The available data show that ion exchange between metal complexes and hydride donors is a simple and effective route to metal hydrides.

Lithium aluminum hydride and sodium borohydride are commonly used in the synthesis of metal-hydride complexes. Diethyl ether, tetrahydrofuran or diglyme, in which $Li[AlH_4]$ is readily soluble, are used as solvents while, with $Na[BH_4]$, these solvents, as well as ethanol or isopropanol, may be employed. Methanol is rarely used because $Na[BH_4]$

reacts relatively rapidly with this solvent.

It has been shown[29,30] that $FeCl_2$ in ethanol reacts with $Na[BH_4]$ in the presence of ligands, L = $P(OEt)_3$[29] and $PPh(OEt)_2$,[30] to form $[Fe(H)_2L_4]$ complexes. In a related reaction, [FeH(Cl)(dppe)$_2$] may be generated by treatment of $FeCl_2$ in the presence of 2 equivalents of dppe with $Na[BH_4]$ in ethanol, Eq. 10.2:[31]

$$FeCl_2 + Na[BH_4] + 2dppe \rightarrow [FeH(Cl)(dppe)_2] + NaCl + 1/2B_2H_6 \quad (10.2)$$

The reaction of $[RuCl_2(CO)_2(PPh_3)_2]$ with $Li[AlH_4]$ in tetrahydrofuran leads to substitution of the chloride ligands and formation of $[Ru(H)_2(CO)_2(PPh_3)_2]$.[32]

The formation of dihydride complexes from dihalide complexes generally requires a considerable excess of $Na[BH_4]$ or $Li[AlH_4]$; with ratios close to two, the formation of monohydride complexes is often observed. For example, the reaction of [OsCl$_2$(dppm)$_2$] with a two-fold excess of lithium aluminum hydride leads to the formation of [OsH(Cl)(dppm)$_2$]. Replacement of both chloride ligands and formation of [Os(H)$_2$(dppm)$_2$] requires a ten-fold excess of $Li[AlH_4]$.[33] In a related example, the dihydride complex, $[Ru(H)_2(PMe_2Ph)\{PPh(CH_2CH_2PPh_2)_2\}]$, was prepared by reaction of $[RuCl_2(PMe_2Ph)\{PPh(CH_2CH_2PPh_2)_2\}]$ with a 20-fold excess of $Na[BH_4]$ in refluxing ethanol.[34] Dihydride and polyhydride complexes are conveniently prepared with more-active hydride donors. Thus, the reaction of $[IrCl_2(PMe_3)Cp^*]$ with 1.5 equivalents of sodium borohydride in isopropanol has been shown[35] to yield the monohydride complex, $[IrCl(H)(PMe_3)Cp^*]$. The dihydride can be prepared under more forcing conditions with a large excess of $Na[BH_4]$. It has been shown,[35] however, that $[Ir(H)_2(PMe_3)Cp^*]$ is obtained through reaction of $[IrCl_2(PMe_3)Cp^*]$ with the strong hydride donor, $Na[AlH_2(OC_2H_4OMe)_2]$. The reaction was carried out in benzene in the presence of a 2.5-fold excess of $Na[AlH_2(OC_2H_4OMe)_2]$, Scheme 10.1.

The synthesis of the complex, $[Ru(H)_2(PPh_3)(\eta^6\text{-}C_6Me_6)]$, by reaction of $[Ru(O_2CCF_3)_2(PPh_3)(\eta^6\text{-}C_6Me_6)]$ with two equivalents of the hydride, $Na[AlH_2(OC_2H_4OMe)_2]$, in benzene has been described.[36] The same reagent reacts with $[Rh(O_2CCF_3)_2(PPr_3)Cp^*]$ in a benzene/diethyl ether mixture to form $[Rh(H)_2(PR_3)Cp^*]$.[37] Substitution of the bromide ligand in the rhodium(I) complex, $[RhBr(Ph)(PMe_3)Cp^*]$, occurs[38] upon reaction with the hydride donor, Li[BH(Bu-*sec*)$_3$], in THF at 35 °C, Eq. 10.3:

$$[RhBr(Ph)(PMe_3)Cp^*] + Li[BH(Bu\text{-}sec)_3] \rightarrow [RhH(Ph)(PMe_3)Cp^*] + LiBr + B(Bu\text{-}sec)_3 \quad (10.3)$$

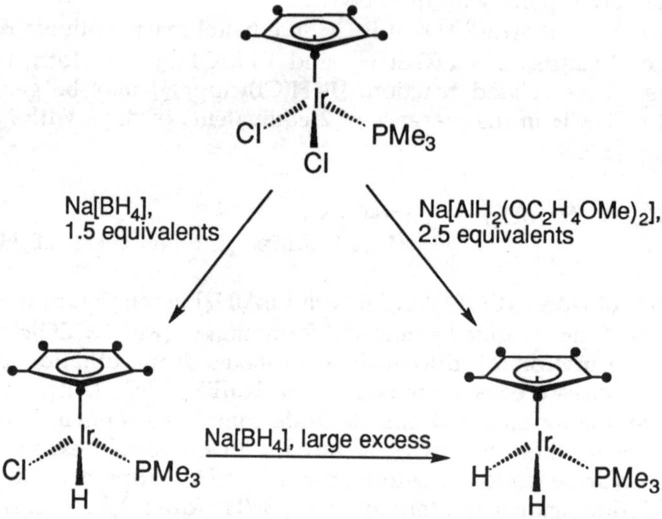

Scheme 10.1

Pathways for the synthesis of polyhydride complexes have been discussed in a review.[28] Polyhydrides are commonly prepared by reaction of halide complexes with an excess of sodium borohydride in ethanol.[39] The synthesis of $[W(H)_4L_4]$ (L = $PMePh_2$ and $PEtPh_2$) is a three step process (the yield for each step is shown), Scheme 10.2:

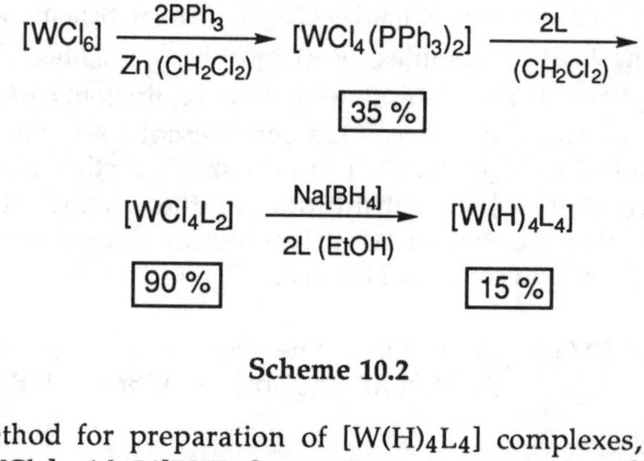

Scheme 10.2

A simple method for preparation of $[W(H)_4L_4]$ complexes, based on the reaction of $[WCl_6]$ with $Li[BHEt_3]$ in THF, has been proposed.[39] The one-step synthesis provides yields in the range of 50 - 70%. One advantage of this method is that tungsten tetrahydrides can be prepared directly from $[WCl_6]$, without the need to isolate the intermediate $[WCl_4L_2]$ complexes, Scheme 10.3:

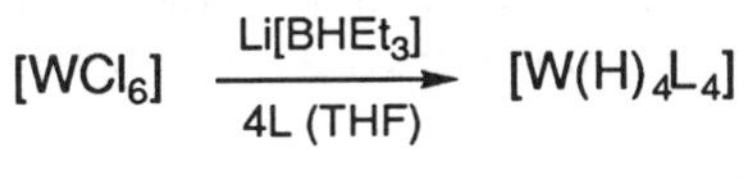

Scheme 10.3

Lithium triethylborohydride is a strong hydride donor and, because it contains only one available hydride, it exhibits little or no tendency to form chelated or borohydride-bridged complexes. The borane-derived by-products of hydrolysis are soluble in alcohol, making isolation and purification of the tungsten tetrahydrides convenient.[39,40] Gillbert and Bergman[41] have similarly reported the use of lithium triethylborohydride, in a mixture of hexane and benzene, in the synthesis of the polyhydride, $[Ir(H)_4Cp]$.

The reaction of complexes containing anionic ligands with alkali metal borohydrides can result in the formation of complexes with coordinated borohydride.[42,43] There are three main classes of metal-borohydride complexes, Figure 10.2:[43,44]

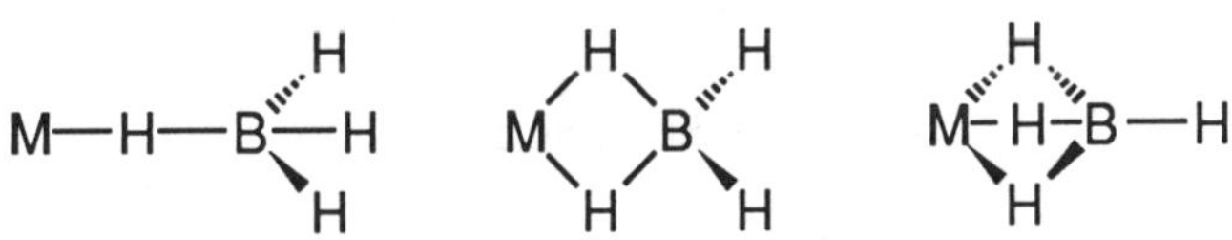

Figure 10.2

The majority of borohydride complexes have been prepared by ion exchange in non-aqueous solvents. For example, the nitrato copper(I) complex, $[Cu(NO_3)(PPh_3)_2]$, reacts[45] with sodium borohydride in a mixture of dichloromethane and ethanol, Eq. 10.4:

$$[Cu(NO_3)(PPh_3)_2] + Na[BH_4] \rightarrow [Cu(BH_4)(PPh_3)_2] + NaNO_3 \qquad (10.4)$$

Reaction of the iron(II) complex, $[FeCl_2(dmpe)]$, with $Na[BH_4]$ in a mixture of

THF and 2-propanol leads to the formation of $[Fe(H)(\eta^2\text{-}BH_4)(dmpe)_2]$.[46] It has been shown[44] that the oligomeric complex, $[RuCl_2L]_x$ (L = $PhP(C_3H_6PPh_2)_2$), reacts with a large excess of $Na[BH_4]$ in THF to yield $[Ru(H)(\eta^2\text{-}BH_4)L]$.

Factors that determine the stability of borohydride complexes have been discussed.[43] Thermodynamic considerations include the ease with which a coordinated $[BH_4]^-$ anion can reduce the metal center, Eq. 10.5:

$$M^{n+}(BH_4)^- \rightarrow M^{(n-1)+} + 1/2B_2H_6 + 1/2H_2 \qquad (10.5)$$

and the stability of the complex with respect to formation of a metal hydride and diborane, Eq. 10.6:

$$M^{n+}(BH_4)^- \rightarrow M^{n+}(H)^- + 1/2B_2H_6 \qquad (10.6)$$

The presence of bulky supporting ligands, leading to steric hindrance, may play an important role in kinetically stabilizing the coordinated borohydride. If the reaction coordinate for the decomposition requires expansion of coordination number, *e.g.*, for B–H bond scission, Scheme 10.4:

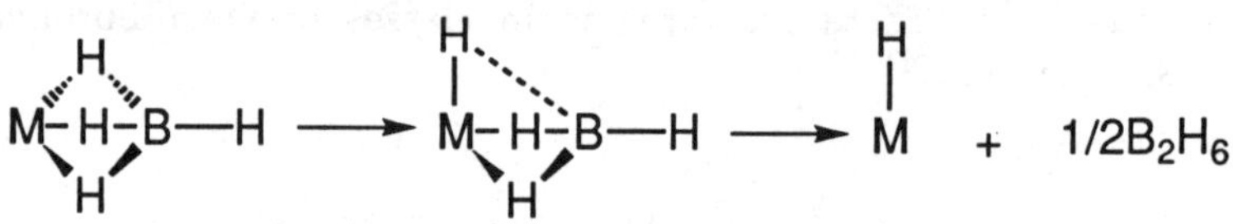

Scheme 10.4

then bulky supporting ligands may sterically retard this process.

A number of borohydride complexes are known to react with bases to form hydride complexes. For example, $[Zr(BH_4)_2Cp_2]$ reacts with one or two moles of trimethylamine in benzene to form $[ZrH(BH_4)Cp_2]$ or $[Zr(H)_2Cp_2]_n$, respectively.[47] The reaction of $[Nb(BH_4)Cp_2]$ with triphenylphosphine similarly produces $[NbH(PPh_3)Cp_2]$.[48] The reaction of $[RuH(\eta^2\text{-}BH_4)L]$[44] (*vide supra*) with bases, B, in the presence of ligands, A, in THF yields dihydride complexes, Eq. 10.7:

$$[RuH(\eta^2\text{-}BH_4)L] + B + A \rightarrow [Ru(H)_2AL] + B{\cdot}BH_3 \qquad (10.7)$$

where B = NEt_3, $P(OMe)_3$, NaOMe; A = CO, $P(OMe)_3$, PPh_3

Semenenko and coworkers[49] have proposed a method for the synthesis of tungsten and molybdenum polyhydride complexes, $[M(H)_4(PR_3)_4]$ (M = W, Mo; PR_3 = phosphine or phosphite). The method involves ion exchange between $[MCl_6]$ and $Ca[BH_4]_2$ in dimethoxyethane and results in the formation of the borohydride complexes, $[M(\eta^2\text{-}BH_4)_x]$, which were identified *in situ* by IR spectroscopy. Addition of PR_3 leads to the decomposition of the η^2-borohydride complexes and isolation of the polyhydrides, $[M(H)_4(PR_3)_4]$.

Examples of reactions between transition metal complexes and $Na[BH_3CN]$ or $Na[BH_2(CN)_2]$, which lead to cyanoborohydride[50] and dicyanoborohydride[51] complexes, have been described. The synthesis of complexes with coordinated aluminohydride ligands has also been reported.[52]

10.2 Reduction without formation of M–H bonds

Although the transformation of a metal-halide complex into the analogous hydride, in one sense, might be considered as a reduction,[24] examination of oxidation states suggests otherwise, at least within this formalism, Scheme 10.5:[53]

$$[RuCl_2(PPh_3)_3] \xrightarrow{Na[BH_4],\ PPh_3} [RuH_2(PPh_3)_4]$$

$$[IrCl(CO)(PPh_3)_2] \xrightarrow{Na[BH_4],\ PPh_3} [IrH(CO)(PPh_3)_3]$$

Scheme 10.5

On the other hand,[24] the replacement of chloride ligands by hydride ligands should result in higher electron density at the metal center, a common criterion for a reduction.[54] Such reactions have been termed "nucleophilic ionic hydrogenation".[55] Others[24] have used such terms as reduction and substitution interchangeably. In this section reactions are discussed which can be unambiguously identified as redox processes.

Chatt and Rowe showed[56] that the platinum(0) complex, $[Pt(dppe)_2]$, can be prepared by reduction of $[Pt(dppe)_2]Cl_2$ with sodium borohydride in aqueous solution, Eq. 10.8:

$$[Pt(dppe)_2]Cl_2 + 2Na[BH_4] \rightarrow [Pt(dppe)_2] + 2NaCl + B_2H_6 + H_2 \quad (10.8)$$

Similarly, another platinum(0) complex, $[Pt\{P(OC_6H_4OMe\text{-}2)_3\}_3]$, results from the reduction of $[PtCl_2\{P(OC_6H_4OMe\text{-}2)_3\}_2]$ in the presence of the phosphite with sodium borohydride in acetonitrile.[57] With ethylene present, this reaction gives rise to the olefin complex, $[Pt(C_2H_4)\{P(OC_6H_4OMe\text{-}2)_3\}_2]$, which has been structurally characterized (for clarity, only one carbon atom of each phenyl ring is shown), Figure 10.3:

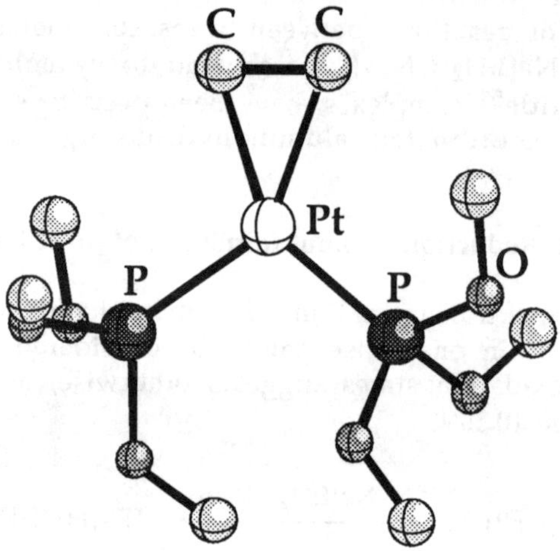

Figure 10.3

Vanadium(0), chromium(0), molybdenum(0), and tungsten(0) complexes of the type $[M(dmpe)_3]$ have been prepared by treatment of the appropriate metal chlorides with three equivalents of ligand and $Na[BH_4]$ or $Li[AlH_4]$ in THF.[58] Reductive cleavage[59] of the dimeric complexes, $[Mn(CO)_5]_2$, $[Co(CO)_4]_2$, $[Co(CO)_3(PPh_3)]_2$, $[Mo(CO)_3Cp]_2$, and $[Fe(CO)_2Cp]_2$, can be effected with potassium hydride in tetrahydrofuran or in a mixture of THF and HMP, Eq. 10.9:

$$[M(CO)_nL]_2 + 2KH \rightarrow 2K[M(CO)_nL] + H_2 \qquad (10.9)$$

The reduction products are isolated in quantitative yields. It has been shown[59] that alkali metal trialkylborohydrides can also be used in reductive cleavage reactions, but the resultant trialkylborane by-products are difficult to separate and lead to problems in product isolation.

At least some of the reduction reactions discussed above proceed by formation of intermediate metal-hydride complexes. For example, it has been shown[60] that the chromium complexes, $[CrX(NO)_2(\eta^5\text{-}Cp)]$ (X = NO_3, NO_2, Cl, I, η^1-Cp, BF_4), react with $Na[AlH_2(OC_2H_4OMe)_2]$ in benzene or dichloromethane to produce the dimeric product, $[Cr(NO)_2(\eta^5\text{-}Cp)]_2$. The reaction[60] proceeds *via* the thermally unstable hydridochromium complex, $[Cr(H)(NO)_2(\eta^5\text{-}Cp)]$, which dimerizes to generate the product with concomitant loss of molecular hydrogen.

10.3 Application of hydrides as reductants for coordinated carbonyl ligands

Studies dealing with the reduction of coordinated carbonyl groups have focused mainly upon the elucidation of a mechanism for the Fisher-Tropsch process, *i.e.* the catalytic reductive oligomerization of carbon monoxide to form hydrocarbon mixtures. The stoichiometric reduction of a carbonyl group to form a methyl ligand was first reported by Treichel and Shubkin in 1967.[61] It was found that the complexes, $[M(CO)_2(PPh_3)Me(\eta^5\text{-}Cp)]$ (M = Mo, W), result from the reaction of $Na[BH_4]$ with $[M(CO)_3(PPh_3)(\eta^5\text{-}Cp)]^+$ in THF. It was suggested[61] that the reaction proceeds *via* intermediate formation of formyl and hydroxymethyl complexes, Scheme 10.6:

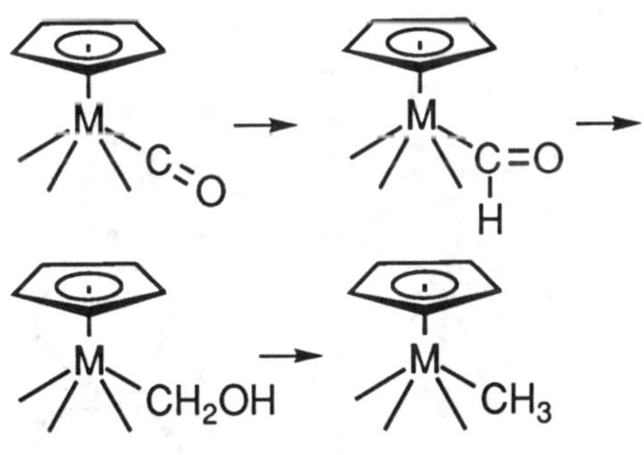

Scheme 10.6

In 1976, reactions of $Na[BH(OMe)_3]$ and $K[BH(OPr\text{-}i)_3]$ with various metal carbonyl complexes were reported.[62] The results are summarized below, Scheme 10.7. The majority of metal formyl complexes have been prepared in this manner.[63] For example, reduction of one carbonyl ligand in the ruthenium(II) complex, $[Ru(CO)_2(dppe)_2][SbF_6]_2$, was accomplished through treatment with $Na[BH(OEt)_3]$ in nitromethane or dichloromethane at -30 °C. The formyl complex, $[Ru(CHO)(CO)(dppe)_2][SbF_6]$, resulted.[64] The osmium(II) complex, $[Os(CHO)(CO)(dppe)_2][SbF_6]$, was synthesized similarly[65] and has been characterized by X-ray crystallography, Figure 10.4.

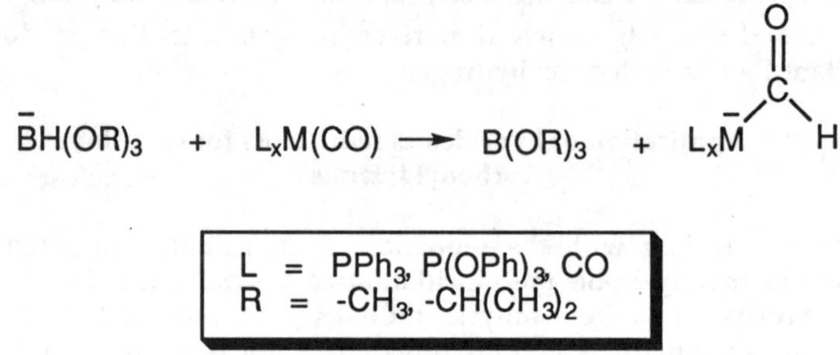

Scheme 10.7

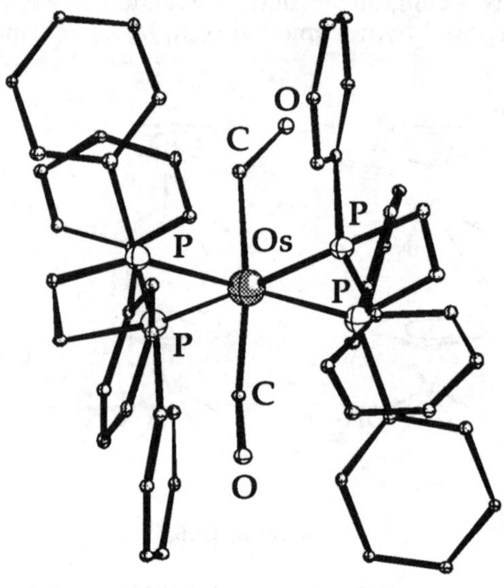

Figure 10.4

The structure clearly shows the presence of mutually *trans* carbonyl and formyl groups, the two being easily distinguished by an Os–C–O angle of

130.7° for the formyl ligand. The formyl hydrogen was not located in the X-ray structure, but the presence of a formyl group was verified by IR and NMR methods.[65]

It is often the case that metal formyl complexes are thermally unstable and, even with slight increases in temperature, are subject to fast transformations into other products. For example, the reaction[66] of $[Re(CO)_2NO(\eta^5\text{-}Cp)][PF_6]$ with $K[BH(OPr\text{-}i)_3]$ in THF at -78 °C yields the neutral formyl complex, $[Re(CHO)(CO)(NO)(\eta^5\text{-}Cp)]$, which decomposes at room temperature with formation of the hydride complex, $[ReH(CO)NO(\eta^5\text{-}Cp)]$, Scheme 10.8:

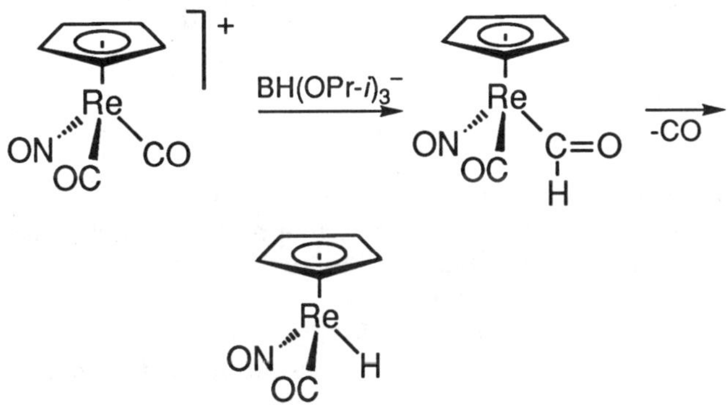

Scheme 10.8

The synthesis and further transformation of formyl complexes *via* hydrido-acyl intermediates into η^2-acetaldehyde complexes has been reported.[67] The molybdenum(II) carbonyl complex, $[Mo(CO)_3Me(\eta^5\text{-}Cp)]$, was found to react with $Li[BHEt_3]$ in THF at -66 °C to yield the formyl complex, $[Mo(CHO)(CO)_2Me(\eta^5\text{-}Cp)]^-$. Slow warming allowed the spectroscopic observation of successive rearrangements of the formyl complex into hydrido-acyl and, finally, η^2-acetaldehyde derivatives, Scheme 10.9.

In many cases, the synthesis of formyl complexes is performed in tetrahydrofuran using strong monohydride donors, typically alkali metal trialkylborohydrides. Although $Li[BHEt_3]$ is commonly employed in preparative coordination chemistry,[63,68] it is not always possible to convert carbonyl complexes into formyl complexes with this reagent. In such cases a stronger hydride donor, such as $K[BH(Bu\text{-}sec)_3]$, is required.[63,69]

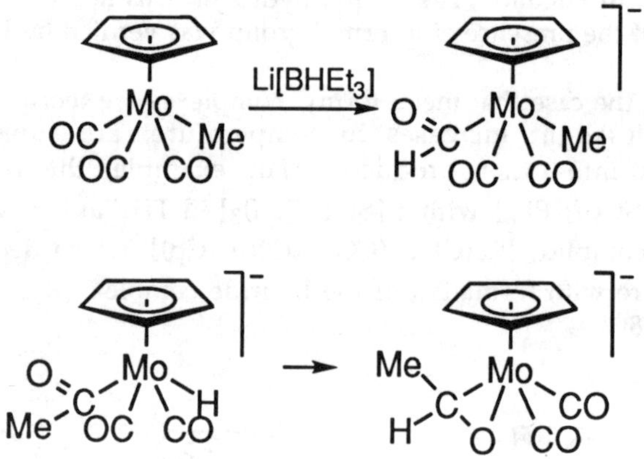

Scheme 10.9

Many formyl complexes are thermally unstable (*vide supra),* and, therefore, the separation of trialkylborane by-products of low volatility from a reaction mixture can be problematic. In such cases, formyl complexes are generally produced and studied *in situ.*

Gladysz has noted[63] that K[BH(OPr-*i*)$_3$] is a valuable reagent for the synthesis of metal formyl complexes. Potassium hydride[70] is not so useful in such syntheses: reaction rates are low, and any formyl intermediates are likely to decompose as fast as they form. Lithium aluminum hydride may function as a donor of more than one hydride and so reactions with this reagent may result in the formation of metal hydride products.[63]

The solvent employed may directly affect the extent of reduction achieved upon treatment of a carbonyl complex with a hydride source. Representative examples have been described by Sweet and Graham[71] who studied the reduction of a coordinated carbonyl group in the complex, $[Re(NO)(CO)_2(\eta^5\text{-Cp})]^+$, with sodium borohydride in tetrahydrofuran, water, or a mixture of the two. It was found that formyl, hydroxymethyl, or methyl complexes can be isolated from the reactions, depending upon the choice of solvent, Scheme 10.10. It was concluded that $[BH_4]^-$ serves as a donor of three hydride ions to $[Re(NO)(CO)_2(\eta^5\text{-Cp})]^+$ in THF solution whereas, in aqueous solution, it provides only one hydride ion. The first stage of the reaction mechanism was suggested to be similar both in aqueous media and in THF solution, Scheme 10.11.

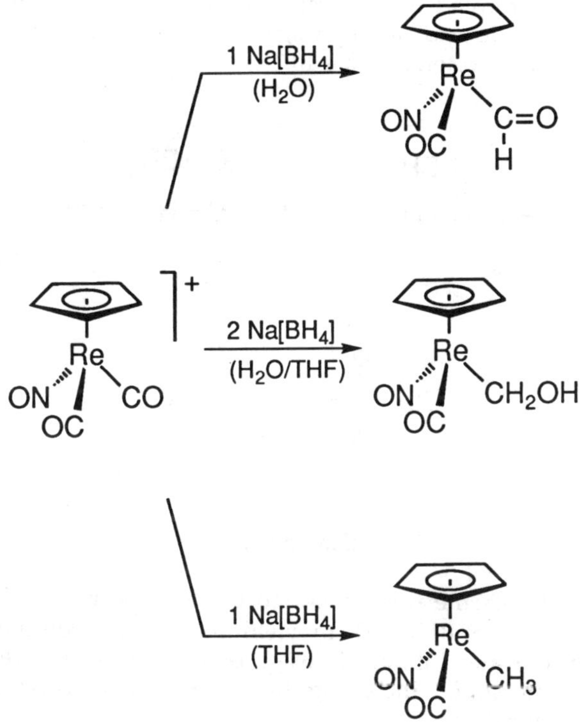

Scheme 10.10

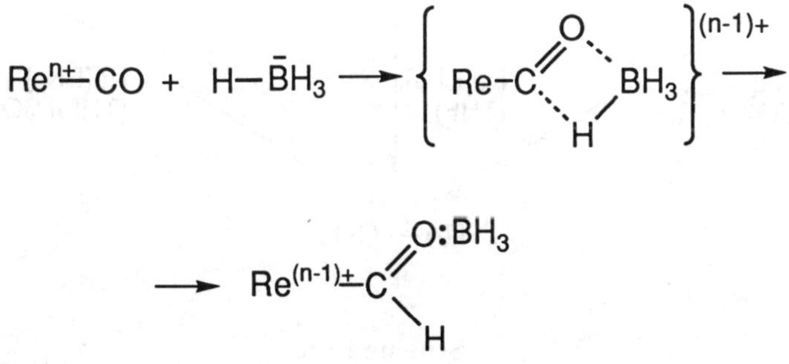

Scheme 10.11

The intermediate formed in this step is decomposed by water to form the formyl complex, boric acid, and hydrogen. In anhydrous tetrahydrofuran the reduction proceeds further, Scheme 10.12:

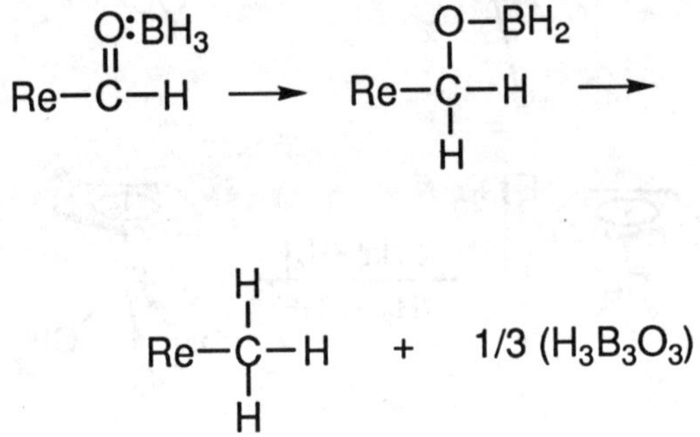

Scheme 10.12

This mechanistic scheme was supported by the observation that the formyl complex reacts with BH_3·THF to yield the methyl complex.[72] When a tetrahydrofuran/water mixture is used as the solvent, the reaction can be stopped at each stage, including the step of hydroxymethyl complex formation, Scheme 10.13:

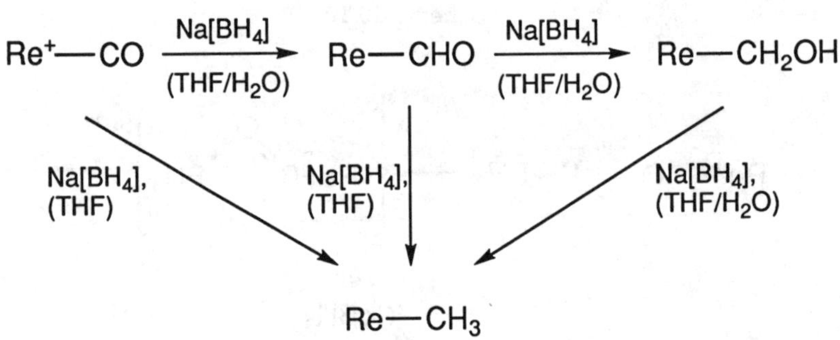

Scheme 10.13

Lapinte and Astruc[73] have studied the effects of solvent on the extent to which reduction of a carbonyl ligand occurs in $[Fe(CO)_3(\eta^5\text{-}Cp^*)][PF_6]$ upon treatment with sodium borohydride. The complex, $[FeH(CO)_2(\eta^5\text{-}Cp^*)]$, was shown to form in aqueous solution or in a THF/water mixture. It was suggested[73] that it is formed by decomposition of an unstable formyl complex or its borane adduct. In dichloromethane, however, the reduction leads to formation of the hydroxymethyl product, $[Fe(CH_2OH)(CO)_2(\eta^5\text{-}Cp^*)]$, while the methyl complex, $[Fe(CO)_2Me(\eta^5\text{-}Cp^*)]$, is formed in anhydrous THF.

Hydroxymethyl complexes are often the most difficult members of the series to prepare and isolate. In some cases, such complexes can be synthesized through reaction of carbonyl complexes with $Na[BH_3CN]$ in alcoholic solutions. For example, the reduction of $[M(CO)_3(\eta^5\text{-}Cp)]^+$ (M = Ru, Fe) with a four-fold excess of $Na[BH_3CN]$ in methanol at 25 °C produced the neutral hydroxymethyl complexes, $[M(CO)_2(CH_2OH)(\eta^5\text{-}Cp)]$, in yields of 55% (Ru) and 45% (Fe).[74] It has been shown[75] that $[Fe(CO)_2(CH_2OH)(\eta^5\text{-}Cp)]^+$ reacts further with $Na[BH_3CN]$ in methanol or ethanol to produce the alkoxymethyl derivative, Scheme 10.14:

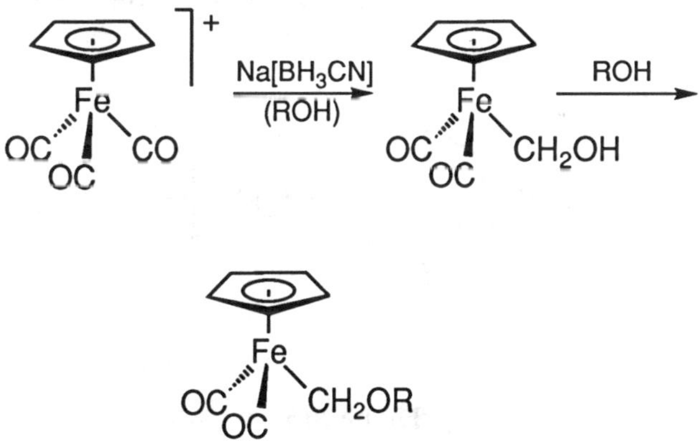

Scheme 10.14

Nelson has similarly shown[76] that the neutral hydroxymethyl complex, $[Ru(CO)_2(CH_2OH)(\eta^5\text{-}Cp^*)]$, is formed by reduction of $[Ru(CO)_3(\eta^5\text{-}Cp^*)][BF_4]$ with a three-fold excess of $Na[BH_3CN]$ in methanol. The reaction of the starting material with an equimolar quantity of $Na[BH_3CN]$ in methanol yields the alkoxymethyl complex, $[Ru(CO)_2(CH_2OMe)(\eta^5\text{-}Cp^*)]$, as the major product. It has been mentioned[75] that, despite many successful applications of $Na[BH_3CN]$ as a reagent for the synthesis of hydroxymethyl complexes, in certain cases other reactions occur. For example, $[Mo(CO)_4(\eta^5\text{-}Cp)][PF_6]$ and $[Mn(CO)_6][PF_6]$ react with $Na[BH_3CN]$ in methanol to yield the hydrides, $[Mo(H)(CO)_3(\eta^5\text{-}Cp)]$ and $[MnH(CO)_5]$. In addition, the reactions of $[Fe(CO)_2(PPh_3)(\eta^5\text{-}Cp)]^+$ and $[Mn(CO)_3(PPh_3)(\eta^5\text{-}Cp)]^+$ with $Na[BH_3CN]$ result in the formation of the cyanide complexes, $[Fe(CO)(CN)(PPh_3)(\eta^5\text{-}Cp)]$ and $[Mn(CO)_2(CN)(PPh_3)(\eta^5\text{-}Cp)]$.[75]

10.4 Reactions of boron and aluminum hydrides with unsaturated hydrocarbyls and related ligands

Unactivated, unsaturated hydrocarbons are not routinely subject to attack by nucleophiles. Complexation of alkenes, alkynes, or arenes to a positively charged metal center may lead to activation of the hydrocarbyl group towards nucleophilic attack and, in particular, to addition of hydride ion. The effect of the $[L_nM]$ fragment on the reactivity of the unsaturated ligand is largely due to the transfer of electron density from the unsaturated hydrocarbyl to the metal center upon coordination, a process that is most effective when the metal center is positively charged, Figure 10.5:

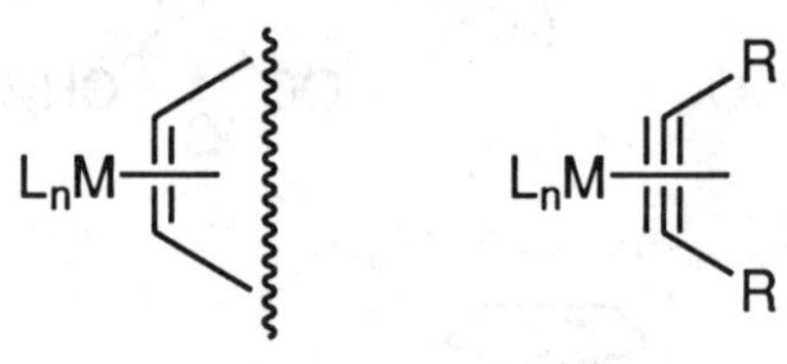

Figure 10.5

The effect of coordination to a metal ion has been found[77] to be similar to that caused by insertion of an electron-withdrawing substituent into a hydrocarbon chain.

Green and coworkers have reported[78,79] examples of nucleophilic hydride addition to coordinated η-alkyne and η-arene ligands in transition metal complexes. The transformation of a η^2-diphenylacetylene complex of molybdenum(II) upon treatment with $K[BHBu_3]$ is representative, Scheme 10.15:

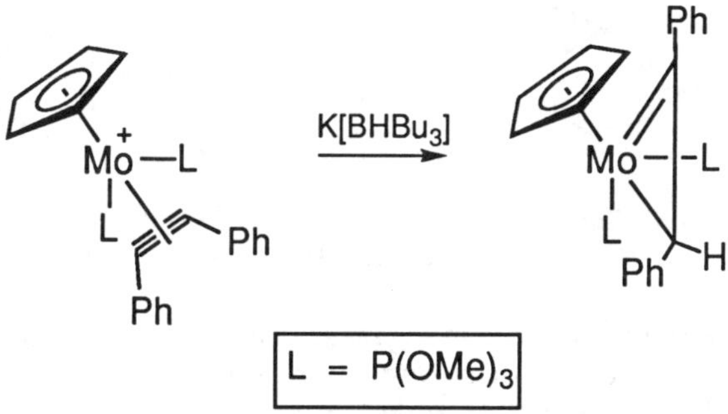

Scheme 10.15

The addition of one or even two hydride ions to a coordinated arene can be accomplished depending upon the degree of activation induced by the metal center and the reducing ability of the hydride donor. Commonly, addition of the first nucleophile to a coordinated arene proceeds under relatively mild conditions. For example, addition[80] of hydride ion to $[ML(\eta^6\text{-}C_6H_6)]^{2+}$, where ML = $Ru(\eta^6\text{-}C_6Me_6)$, $Rh(\eta^5\text{-}Cp^*)$, $Rh(\eta^5\text{-}C_5H_4Et)$, or $Ir(\eta^5\text{-}Cp^*)$, is facile, Scheme 10.16:

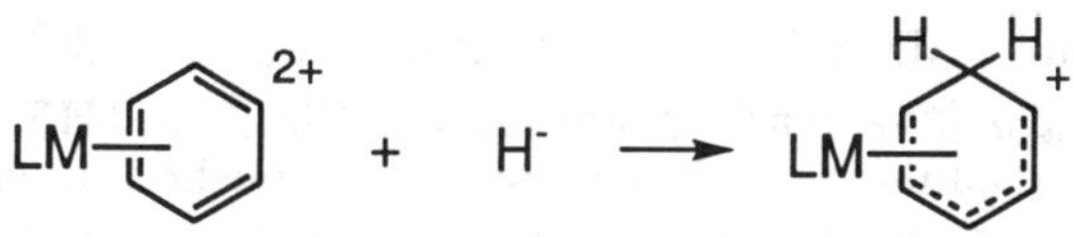

Scheme 10.16

Activation of coordinated benzene by a cationic metal fragment allows the use of a relatively weak hydride donor such as sodium borohydride in water or in dimethoxyethane.[80,81] Addition of a second hydride ion is more difficult due to the partial deactivation that results from the first step and there are few reported examples of such addition reactions. Strong hydride donors, such as $Na[AlH_2(OR)_2]$ (R = Me, Et, C_2H_4OMe) and $Na[BHEt_3]$, have been used[80] to generate a η^4-cyclohexadiene complex from a η-cyclohexadienyl complex, Scheme 10.17:

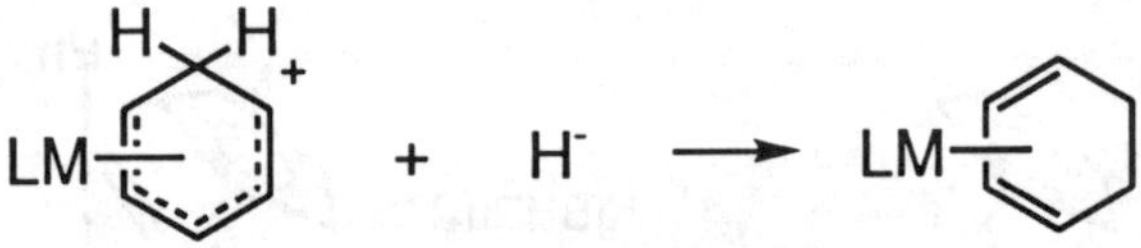

Scheme 10.17

It has been reported[82] that the manganese complexes, $[Mn(C_6H_5R)(CO)_3]^+$ (R = Me, OMe, Ph), react with strong hydride donors, such as $Li[BHEt_3]$ or $K[BH(OPr\text{-}i)_3]$, with reduction of the arene fragment through nucleophilic addition of two hydride ions, Scheme 10.18:

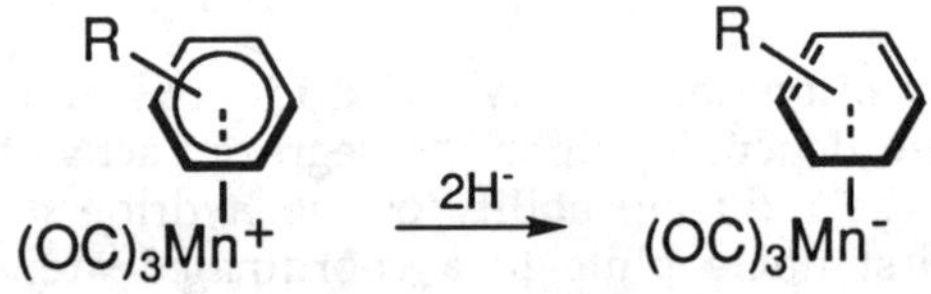

Scheme 10.18

Based on ^1H NMR spectroscopy it was shown that the addition of the first hydride was much faster than that of the second.

The complex, $[Mn(CO)_3(\eta^4\text{-}C_6H_6R)]$, was prepared[83] by treatment of $[Mn(CO)_3(\eta^6\text{-}C_6H_6)]^+$ with a Grignard reagent, RMgX, in THF. Introduction of the second nucleophile (hydride ion) was accomplished by activation of the metal carbonyl fragment through substitution of a carbonyl ligand by a nitrosyl group. Addition of hydride ion to the resultant cationic complex was achieved through treatment with sodium borohydride, Scheme 10.19.

Thus, the double nucleophilic addition of hydride ion is accomplished by either use of strong hydride donors or additional activation of the complex by structural modification.

If a complex contains both carbonyl and unsaturated hydrocarbyl ligands, hydride addition may follow both of the available pathways. Not all of the factors controlling the directions taken by such reactions have been elucidated.[84] Generally,[77,84,85] formyl complexes are the kinetic products, whereas, the attack of hydride on unsaturated hydrocarbyl ligands gives the thermodynamic products.

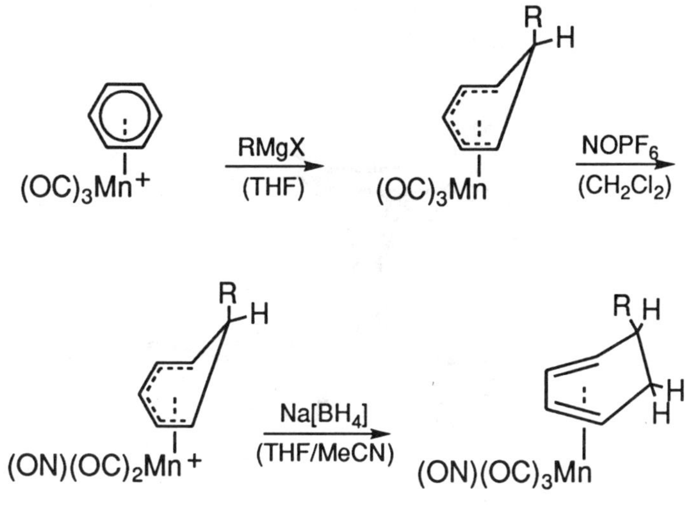

Scheme 10.19

10.5 Reactions of boron and aluminum hydrides with other coordinated ligands

There are only a few reported examples of the reactions of other coordinated ligands with boron and aluminum hydrides. For example, it has been demonstrated[86] that the reaction of $[W(N_2)_2(PMe_2Ph)_4]$ with $Na[AlH_2(OC_2H_4OMe)_2]$ leads to reduction of the dinitrogen ligand and formation of ammonia (0.59 moles per mole of complex) and small amounts of hydrazine. It has been shown[87] that the reaction of the silyltungsten complex, $[W(SiHClMe)(CO)_2(PMe_3)(\eta^5\text{-}Cp^*)]$, with $Li[AlH_4]$ in diethyl ether results in substitution of chloride by hydride at silicon, Scheme 10.20:

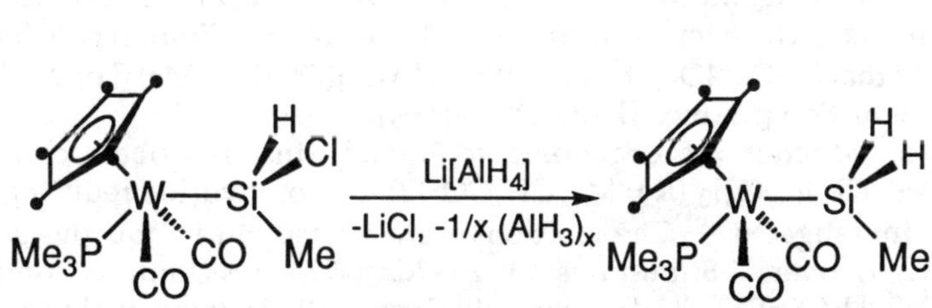

Scheme 10.20

The X-ray crystal structure of the resultant tungsten silyl-complex, [W(SiH_2Me)$(CO)_2$(PMe_3)(η^5-Cp*)], has been determined, Figure 10.6:

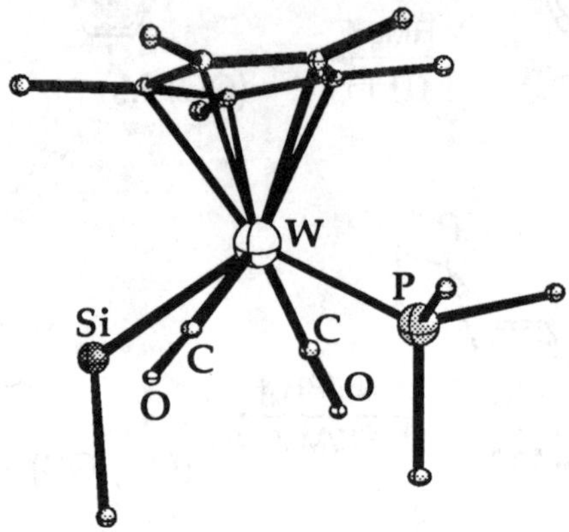

Figure 10.6

Herberich and Linn have reported[88] that [W(H)($CHMeCO_2Me$)$(\eta^5\text{-Cp})_2$] reacts with Li[AlH_4] to give the olefin complex, [W(CH_2CHMe)$(\eta^5\text{-Cp})_2$]. Similar functional olefin complexes were reduced to produce the alkyl derivatives.

The reaction of the nitrosyl complex, $[Cr(NO)_2(\eta^5\text{-Cp})]_2$, with Na[$AlH_2(OC_2H_4OMe_2)$] was found[60] to lead to the formation of amide complexes: [$Cr_2(NO)_3(NH_2)(\eta^5\text{-Cp})_2$] (15%), [$Cr_2(NO)_2(NH_2)_2(\eta^5\text{-Cp})_2$] (2%) and [$Cr_2(NO)_2(NH_2)(OH)(\eta^5\text{-Cp})_2$], (3%). The reduction of the nitrosyl ligand in $[Cr(NO)_2(\eta^5\text{-Cp})]_2$ presumably proceeds through nucleophilic attack at the bridging nitrosyl ligand by a hydride ion.[60] The monoamido complex could, thus, form *via* a two-step mechanism, Scheme 10.21. Similarly, it has been observed[60] that [$FeCl(NO)_3$] reacts with Na[$AlH_2(OC_2H_4OMe)_2$] or Na[BH_4] to yield the reduction product, [$Fe_2(NO)_4(NH_2)_2$].

The reduction of organonitrile ligands in the cobalt complexes, [$Co(NH_3)_5(NCR)$][ClO_4]$_3$ (R = Me, CH_2CO_2Et), by borohydride reducing agents has been investigated.[89] The reactions were carried out in aqueous solution, at p*H ca.* 9, over 7-8 minutes at 20 °C. The alkylamine complexes, [$Co(NH_3)_5(NH_2CH_2R)$][ClO_4]$_3$, were obtained with isolated yields of about 50%. The coordination of the organonitriles to the metal center led to an acceleration in the reduction rates of about four orders of magnitude.

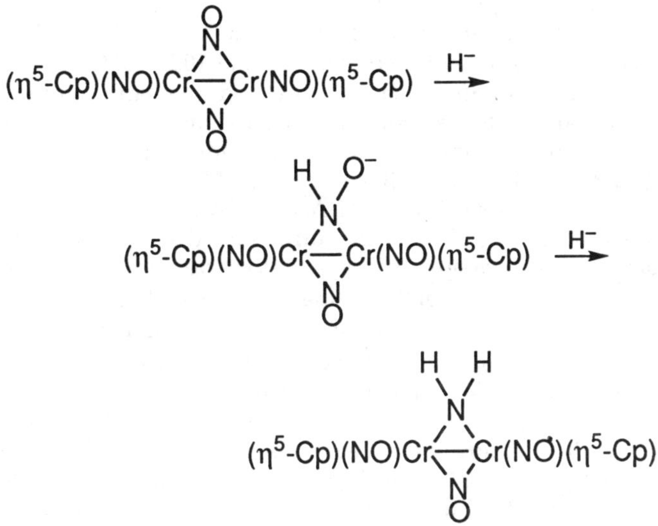

Scheme 10.21

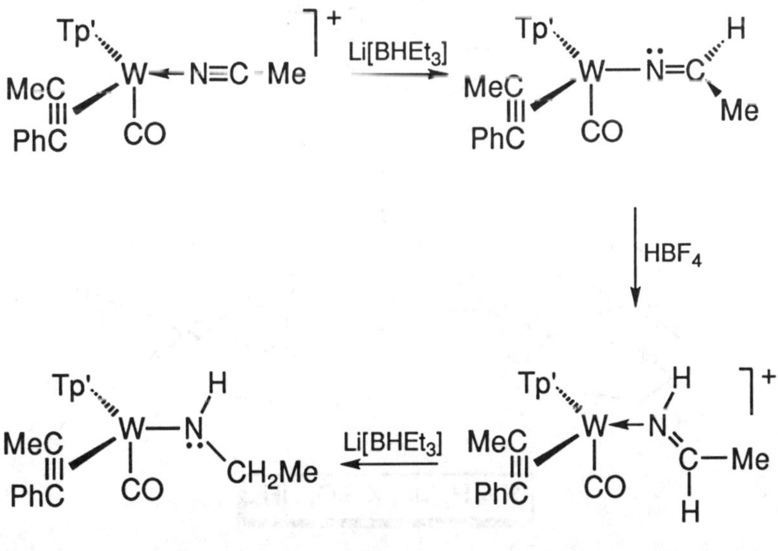

Scheme 10.22

Feng and Templeton[90] have studied the stepwise reduction of coordinated acetonitrile in [WTp'(CO)(PhCCMe)(NCMe)][BF_4] (Tp' = hydrotris(3,5-dimethylpyrazolyl)borate), Scheme 10.22 (*vide supra*).

Rhodes and Venanzi[91] have observed the ruthenium(II)-assisted conversion of coordinated acetonitrile to ethylamine by hydride addition. They reported that the sodium borohydride reduction of the ruthenium(II) complex, [Ru{MeC($CH_2AsPh_2)_3$}$(MeCN)_3$][CF_3SO_3]$_2$, in methanol yields the hydride, [RuH$(NH_2CH_2Me)_2${MeC$(CH_2AsPh_2)_3$}][CF_3SO_3]$_2$. The coordination of MeCN to ruthenium polarizes the nitrile ligand and activates the cyanide carbon atom toward nucleophilic attack. Succesive addition of hydride from [BH_4]$^-$ and protons from MeOH to the cyanide carbon and the nitrogen, respectively, yields the amine complex. Labeling studies[91] confirmed that [BH_4]$^-$ is the source of hydride delivered to the cyanide carbon.

Detailed studies[92,93] of Cu(II)-assisted template condensations and some subsequent reactions (*e.g.* reduction by sodium borohydride) leading to novel macrocyclic ligands have been reported. It has been shown that treatment of complexes containing ligands bearing pendant –OR (R = H, Me) substituents with Na[BH_4] in a methanol/water mixture leads to reduction of the coordinated ligand, Scheme 10.23:

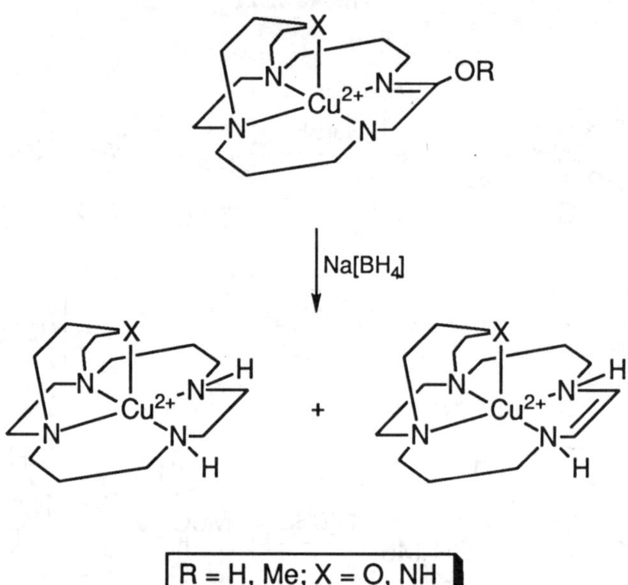

Scheme 10.23

The reaction of $[Ni_2(\mu\text{-}SR)_2(\eta^5\text{-}Cp)_2]$ (R = alkyl) with $Li[AlH_4]$ has been reported[94] to yield RH as the major organic product, together with traces of RSSR and a metal-containing fragment tentatively formulated as $[Ni_2(\mu\text{-}S)_2(\eta^5\text{-}Cp)_2]^{2-}$. Studies by ^1H NMR spectroscopy suggest the involvement of a nickel-hydride intermediate in the reaction.

The possible applications of boron and aluminum hydrides in preparative coordination chemistry are far from exhausted and there remains much to be done in investigations of reductive transformations of coordinated ligands.

References

1. (a) H. I. Schlesinger and H. C. Brown, *U. S. Patents* 2,461,661; 2,461,662; 2,461,663 (1949); (b) H. I. Schlesinger and A. E. Finholt, *U. S. Patents* 2,567,972; 2,576,311 (1951).
2. H. C. Brown, *Hydroboration* (W. A. Benjamin, New York, 1962) p. 41.
3. (a) W. Büchner and H. Niederprüm, *Pure Appl. Chem.* **49** (1977) 733; (b) N. N. Maltseva and V. S. Hain, *Sodium Borohydride* [in Russian] (Nauka, Moscow, 1985).
4. H. C. Brown, *Boranes in Organic Chemistry* (Cornell Univ., London, 1972).
5. J. A. Gladysz, *Aldrichimica Acta* **12** (1979) 13.
6. E. C. Ashby, *Inorg. Synth.* **17** (1977) 1.
7. H. C. Brown and S. Krishnamurthy, *J. Am. Chem Soc.* **94** (1972) 7159.
8. A. Hajós, *Complex Hydrides and Related Reducing Agents in Organic Synthesis* (Akad. Kiado, Budapest, 1979).
9. H. C. Brown, J. S. Cha, and B. Nazer, *Inorg. Chem.* **23** (1984) 2929 and references therein.
10. G. Wittig and P. Raff, *Lieb. Ann.* **573** (1951) 195.
11. R. C. Wade, E. A. Sullivan, J. R. Berschied, and K. F. Purcell, *Inorg. Chem.* **9** (1970) 2146.
12. J. Emri and B. Györi, *J. Chem. Soc., Chem. Commun.* (1983) 1303.
13. C. F. Lane, *Synthesis* (1975) 135.
14. R. O. Hutchins and N. R. Natale, *Org. Prep. Proc. Intern.* **11** (1979) 201.
15. B. F. Spielvogel, F. U. Ahmed, M. K. Das, and A. T. McPhail, *Inorg. Chem.* **23** (1984) 3263.
16. A. Streitwieser, Jr., *Solvolytic Displacement Reactions* (McGraw-Hill, New York, 1962) p. 9.
17. H. C. Brown and S. Krishnamurthy, *J. Am. Chem. Soc.* **95** (1973) 1669.
18. R. S. Bayoud, E. R. Biehl, and P. C. Reeves, *J. Organomet. Chem.* **174** (1979) 297.
19. M. M. Kreevoy and J. E. C. Hutchins, *J. Am. Chem. Soc.* **91** (1969) 4329.
20. D. A. Brown, W. K. Glass, and M. Turki Ubeid, *Inorg. Chim. Acta* **89** (1984) L47.

21. (a) C. F. Lane in *Aspects of Mechanism and Organometallic Chemistry*, ed. J. H. Brewster, (Plenum, New York, 1978) p. 181; (b) J. Málek and M. Cerny, *Synthesis* (1972) 217; (c) H. Haubenstock, *Topics in Stereochem.* **14** (1983) 231.

22. O. Strouf, B. Casensky, and V. Kubanek, *Sodium Dihydrido-bis(2-methoxyethoxo)aluminate (SDMA)* (Academia, Prague, 1985).

23. *Transition Metal Hydrides*, ed. E. L. Muetterties, (Marcel Dekker, New York, 1971).

24. H. D. Kaesz and R. B. Saillant, *Chem. Rev.* **72** (1972) 231.

25. (a) R. H. Crabtree, *The Organometallic Chemistry of the Transition Metals* (John Wiley & Sons, New York, 1988) p. 53; (b) G. G. Hlatky and R. H. Crabtree, *Coord. Chem. Rev.* **65** (1985) 41; (c) R. G. Pearson, *Chem. Rev.* **85** (1985) 41; (d) M. Y. Darensbourg and C. E. Ash, *Adv. Organomet. Chem.* **27** (1987) 1.

26. (a) L. M. Venanzi, *Coord. Chem. Rev.* **43** (1982) 251; (b) D. S. Moore and S. D. Robinson, *Chem. Soc. Rev.* **12** (1983) 415.

27. The chemistry of nonclassical η^2-dihydrogen complexes has been reviewed: (a) G. J. Kubas, *Acc. Chem. Res.* **21** (1988) 120; (b) R. H. Crabtree and D. G. Hamilton, *Adv. Organomet. Chem.* **28** (1988) 299; (c) R. H. Crabtree, *Acc. Chem. Rev.* **23** (1990) 95.

28. A. P. Borisov, V. D. Makhaev, and K. N. Semenenko, *Koord. Khim.* **6** (1980) 1139.

29. W. Kruse and R. H. Atalla, *J. Chem. Soc., Chem. Commun.* (1968) 921.

30. D. Titus, A. A. Orio, and H. B. Gray, *Inorg. Synth.* **13** (1972) 117.

31. P. Giannoccaro and A. Sacco, *Inorg. Synth.* **17** (1977) 69.

32. J. D. Cotton, M. I. Bruce, and F. G. A. Stone, *J. Chem. Soc. A* (1968) 2162.

33. S. D. Ibekwe and U. A. Raeburn, *J. Organomet. Chem.* **19** (1969) 447.

34. D. Michos, X.-L. Luo, and R. H. Crabtree, *Inorg. Chem.* **31** (1992) 4245.

35. K. Isobe, P. M. Bailey, and P. M. Maitlis, *J. Chem. Soc., Dalton Trans.* (1981) 2003.

36. H. Werner and H. Kletzin, *J. Organomet. Chem.* **228** (1982) 289.

37. H. Werner and J. Wolf, *Angew. Chem.* **94** (1982) 308.

38. W. D. Jones, and F. J. Feher, *J. Am. Chem. Soc.* **104** (1982) 4240.

39. R. H. Crabtree and G. G. Hlatky, *Inorg. Chem.* **21** (1982) 1273.

40. R. H. Crabtree and G. G. Hlatky, *Inorg. Chem.* **23** (1984) 2388.

41. T. M. Gillbert and R. G. Bergman, *Organometallics* **2** (1983) 1458.

42. (a) K. N. Semenenko, O. V. Kravchenko, and V. B. Polyakova, *Uspekhi Khimii* **42** (1973) 1; (b) V. N. Konoplev, N. N. Maltseva, and V. S. Khain, *Koord. Khim.* **18** (1992) 1143.

43. (a) T. J. Marks and J. R. Kolb, *Chem. Rev.* **77** (1977) 263; (b) U. Mirsaidov and T. N. Dimova, *Transition Metal Borohydrides* [in Russian] (Donish, Dushanbe, 1985).

48. (a) J. B. Letts, T. J. Mazanec, and D. W. Meek, *J. Am. Chem. Soc.* **104** (1982) 3898; (b) T. J. Mazanec, J. B. Letts, and D. W. Meek, *J. Chem. Soc., Chem. Commun.* (1982) 356.

45. H. J. Gysling, *Inorg. Synth.* **19** (1979) 92.

46. M. V. Baker and L. D. Field, *Appl. Organomet. Chem.* **4** (1990) 543.

47. B. D. James, R. K. Nanda, and M. G. H. Wallbridge, *Inorg. Chem.* **6** (1967) 1979.

48. C. R. Lucas and M. L. H. Green, *J. Chem. Soc., Chem. Commun.* (1972) 1005.

49. (a) K. N. Semenenko, V. D. Makhaev, and A. P. Borisov, *Koord. Khim.* **4** (1978) 146; (b) K. N. Semenenko, V. D. Makhaev, A. P. Borisov, and G. N. Boiko, *Koord. Khim.* **4** (1978) 1550; (c) A. P. Borisov, V. D. Makhaev, and K. N. Semenenko, *Koord. Khim.* **8** (1982) 778.

50. (a) D. G. Holah, A. N. Hughes, and N. I. Khan, *Can. J. Chem.* **62** (1984) 1016; (b) K. M. Melmed, T. Li, J. J. Mayerle, and S. J. Lippard, *J. Am. Chem. Soc.* **96** (1974) 69; (c) B. G. Segal and S. J. Lippard, *Inorg. Chem.* **13** (1974) 822; (d) B. G. Segal and S. J. Lippard, *Inorg. Chem.* **16** (1977) 1623.

51. P. G. Egan, D. G. Holah, and K. W. Morse, *Inorg. Chem.* **23** (1984) 2203.

52. (a) A. I. Sizov, G. L. Soloveichik, and B. M. Bulichev, *Koord. Khim.* **11** (1985) 339; (b) A. R. Barron, J. E. Salt, and G. Wilkinson, *J. Chem. Soc., Dalton Trans.* (1986) 1329; (c) A. R. Barron and G. Wilkinson, *Polyhedron* **5** (1986) 1897.

53. (a) R. Young and G. Wilkinson, *Inorg. Synth.* **17** (1977) 75; (b) G. Wilkinson, *Inorg. Synth.* **13** (1972) 127.

54. (a) C. K. Jørgensen, *Oxidation Numbers and Oxidation States* (Springer, New York, 1969) p. 11; (b) L. J. Chinn, *Selection of Oxidants in Synthesis* (Marcel Dekker, New York, 1971) p. 5; (c) R. Bartoszewicz, V. Mecznikowska-Stoljarczik, and B. Opszondek, *Methods for the Reduction of Organic Compounds* (IL, Moscow, 1960) p. 43.

55. D. N. Kursanov, Z. N. Parnes, M. I. Kalinkin, and N. M. Loim, *Ionic Hydrogenation* [in Russian] (Khimia, Moscow, 1979) p. 10.

56. J. Chatt and G. A. Rowe, *Nature* **191** (1961) 1191.

57. M. J. Baker, K. N. Harrison, A. G. Orpen, P. G. Pringle, and G. Show, *J. Chem. Soc., Dalton Trans.* (1992) 2607.

58. J. Chatt and H. R. Watson, *Nature* **189** (1961) 1003.

59. K. Inkrott, R. Goetze, and S. G. Shore, *J. Organomet. Chem.* **154** (1978) 337.

60. B. W. Hames, P. Legzdins, and J. C. Oxley, *Inorg. Chem.* **19** (1980) 1565.

61. P. M. Treichel and R. L. Shubkin, *Inorg. Chem.* **6** (1967) 1328.

62. C. P. Casey and S. M. Neumann, *J. Am. Chem. Soc.* **98** (1976) 5395.

63. J. A. Gladysz, *Adv. Organomet. Chem.* **20** (1982) 1.

64. G. Smith and D. J. Cole-Hamilton, *J. Chem. Soc., Chem. Commun.* (1982) 490.

65. G. Smith, D. J. Cole-Hamilton, M. Thornton-Pett, and M. B. Hursthouse, *J. Chem. Soc., Dalton Trans.* (1985) 387.

66. C. P. Casey, M. A. Andrews, D. R. McAlister, and J. E. Rinz, *J. Am. Chem. Soc.* **102** (1980) 1927.

67. J. T. Gauntlett, B. F. Taylor, and M. J. Winter, *J. Chem. Soc., Dalton Trans.* (1985) 1815.

68. (a) K. P. Darst and C. M. Lukehart, *J. Organomet. Chem.* **171** (1979) 65; (b) R. L. Pruett, R. C. Schoening, J. L. Vidal, and R. A. Fiato, *J. Organomet. Chem.* **182** (1979) C57; (c) J. A. Gladysz and J. C. Selover, *Tetrahedron Letters* (1978) 319; (d) W. Tam, M. Marsi, and J. A. Gladysz, *Inorg. Chem.* **22** (1983) 1413.

69. J. A. Gladysz and W. Tam, *J. Am. Chem. Soc.* **100** (1978) 2545.

70. S. R. Winter, G. W. Cornett, and E. A. Thompson, *J. Organomet. Chem.* **133** (1977) 339.

71. (a) J. R. Sweet and W. A. G. Graham, *J. Organomet. Chem.* **173** (1979) C9; (b) J. R. Sweet and W. A. G. Graham, *J. Am. Chem. Soc.* **104** (1982) 2811.

72. W. Tam, G.-Y. Lin, W.-K. Wong, W. A. Kiel, V. K. Wong, and J. A. Gladysz, *J. Am. Chem. Soc.* **104** (1982) 141.

73. C. Lapinte and D. Astruc, *J. Chem. Soc., Chem. Commun.* (1983) 430.

74. (a) Y. C. Lin, D. Milstein, and S. S. Wreford, *Organometallics* **2** (1983) 1461; (b) G. O. Nelson and C. E. Sumner, *Organometallics* **5** (1986) 1983.

75. T. Bodnar, E. Coman, K. Menard, and A. Cutler, *Inorg. Chem.* **21** (1982) 1275.

76. G. O. Nelson, *Organometallics* **2** (1983) 1474.

77. S. G. Davies, M. L. H. Green, and D. M. P. Mingos, *Tetrahedron* **34** (1978) 3047.

78. M. Green, *J. Organomet. Chem.* **300** (1986) 93.

79. M. L. H. Green, A. K. Hughes, P. Lincoln, J. J. Martin-Polo, P. Mountford, A. Sella, L.-L. Wong, J. A. Bandy, T. W. Banks, K. Prout, and D. J. Watkin, *J. Chem. Soc., Dalton Trans.* (1992) 2063.

80. (a) N. A. Bailey, E. H. Blunt, G. Fairhurst, and C. White, *J. Chem. Soc., Dalton Trans.* (1980) 829; (b) S. L. Grundy and P. M. Maitlis, *J. Chem Soc., Chem. Commun.* (1982) 379; (c) S. L. Grundy and P. M. Maitlis, *J. Organomet. Chem.* **272** (1984) 265.

81. (a) C. White and P. M. Maitlis, *J. Chem Soc. A* (1971) 3322; (b) D. Jones, L. Pratt, and G. Wilkinson, *J. Chem. Soc.* (1962) 4458.

82. (a) W. Lamanna and M. Brookhart, *J. Am. Chem. Soc.* **103** (1981) 989; (b) M. Brookhart and A. Lukacs, *J. Am. Chem. Soc.* **106** (1984) 4161.

83. Y. K. Chung, H. S. Choi, D. A. Sweigart, and N. G. Connelly, *J. Am. Chem. Soc.* **104** (1982) 4245.

84. S. G. Davies, J. Hibberd, S. J. Simpson, S. E. Thomas, and O. Watts, *J. Chem. Soc., Dalton Trans.* (1984) 701.

85. S. G. Davies, J. Hibberd, and S. J. Simpson, *J. Organomet. Chem.* **264** (1983) C16.

86. M. Hidai, T. Takahashi, I. Yokotake, and Y. Uchida, *Chem. Lett.* (1980) 645.

87. S. Schmitzer, U. Weis, H. Käb, W. Buchner, W. Malisch, T. Polzer, U. Posset, and W. Kiefer, *Inorg. Chem.* **32** (1993) 303.

88. G. E. Herberich and K. Linn, *J. Organomet. Chem.* **425** (1991) C4.

89. D. G. Butler, I. I. Creaser, S. F. Dyke, and A. M. Sargeson, *Acta Chem. Scand.* **A32** (1978) 789.

90. S. G. Feng and J. L. Templeton, *J. Am. Chem. Soc.* **111** (1989) 6477.

91. L. F. Rhodes and L. M. Venanzi, *Inorg. Chem.* **26** (1987) 2692.

92. K. A. Beveridge, A. McAuley, and C. Xu, *Inorg. Chem.* **30** (1991) 2074.

93. D. G. Fortier and A. McAuley, *J. Am. Chem. Soc.* **112** (1990) 2640.

94. N.-F. Ho, T. C. W. Mak, and T. Y. Luh, *J. Chem. Soc., Dalton Trans.* (1990) 3591.

Chapter 11. MOLECULAR REARRANGEMENTS OF COORDINATION COMPOUNDS

In coordination chemistry, as in organic chemistry, syntheses are sometimes complicated by molecular rearrangements. In this Chapter, these processes are discussed and classified. A molecular rearrangement is defined here as a transformation resulting in a change in connectivity without any change in atomic composition. Thus, geometric isomerization is an example of a molecular rearrangement. The isomerization of square-planar and octahedral complexes is discussed in Chapters 1 and 12 and is not considered further here. Changes in spin state, *i.e.* transformations of high-spin complexes into low-spin complexes, or *vice versa*, and related electronic transformations are not included in the definition of molecular rearrangement employed here. Enantiomerization, ligand rotations about metal-ligand bonds, and transformations due to conformational changes of coordinated ligands similarly are excluded. Haptotropic rearrangements and sigmatropic shifts are subjects of current interest but they are very specific types of molecular rearrangement applicable to a relatively narrow class of coordination compounds. These processes are not considered here but are discussed in detail in a number of comprehensive reviews.[1-7]

11.1 Rearrangements involving changes in metal-ligand binding sites

The most important member of this class of rearrangement is linkage isomerism which is a process that may take place in complexes containing ambidentate ligands, *i.e.*, ligands with more than one donor site but which bind in monodentate fashion. In 1894, Jørgensen[8] first reported linkage isomers in his studies of cobalt nitro-complexes and the isomerization reactions of these species have subsequently been investigated in detail. The rates of *O–O* and *O–N* isomerization in the cobalt(III) complexes $[Co(NO_2)(NH_3)_5]^{2+}$ and $[Co(ONO)(NH_3)_5]^{2+}$, for example, have been studied by NMR spectroscopy,[9] Scheme 11.1. The rate constants were found to be in the ratio, $k_e = 1.2k_i$, *i.e.*, *O–O* and *O–N* isomerization reactions occurred at comparable rates. It was established that, at equilibrium in aqueous solution, more than 99% of the nitro isomer of $[Co(NO_2)(NH_3)_5]^{2+}$ was present.[9]

Linkage isomerism has been the subject of several reviews.[5,7] Much recent work has dealt with nitro-nitrite isomerization[10] and thiocyanate-isothiocyanate isomerization.[11,12] Examples of linkage isomerism involving cyanate,[13] methanesulfinate,[14] terpyridyl,[15] alizarin,[16] dihydroxybenzoate,[17] and acetone[18] complexes have been reported also. Some additional examples are discussed in paragraphs that follow. Thus, the rate of M–CN/M–NC linkage isomerization at chromium has been studied,[19] Eq. 11.1.

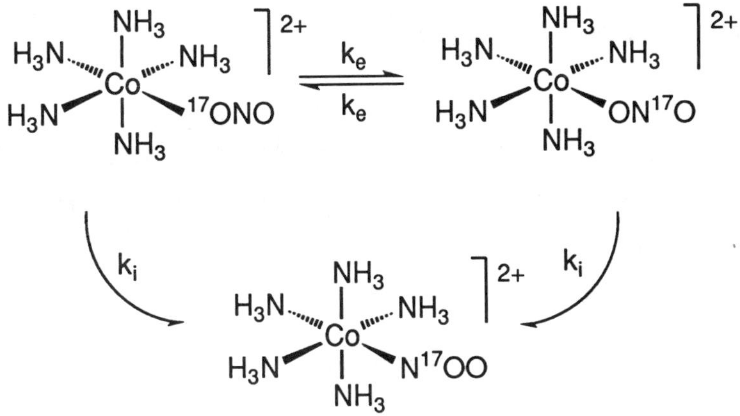

Scheme 11.1

$$[C{\equiv}N{-}Cr(NO)(H_2O)_4]^+ \rightarrow [N{\equiv}C{-}Cr(NO)(H_2O)_4]^+ \qquad (11.1)$$

and was observed to depend on the *pH* of the solution. Protonation of the cyanide ligand, *i.e.* formation of $[HCN{-}Cr(NO)(H_2O)_4]^{2+}$, was found to significantly lower the rate of isomerization in comparison with that associated with the unprotonated form.

The kinetics of *O–S* linkage isomerism in the cobalt(II) sulfite complexes, $[Co(OSO_2)(Tetren)]^+$ (Tetren = tetraethylenepentamine)[20] and $[Cr(OSO_2)(H_2O)_5]^+$,[21] have also been investigated. Conversion of these *O*-bonded complexes to the *S*-bonded isomers, $[Co(SO_3)(Tetren)]^+$ and $[Cr(SO_3)(H_2O)_5]^+$, was found to occur by first-order processes and at almost identical rates. The kinetics of the reaction between $[Pt(CN)_4Cl(OH)]^{2-}$ and HSO_3^- has been studied.[22] In the first step an oxygen-bonded sulfito-complex, $[Pt(CN)_4Cl(OSO_2H)]^{2-}$, is formed. In the second step, the intermediate complex undergoes a slower, rate-determining, intramolecular linkage isomerization to the thermodynamically more stable *S*-bonded isomer, $[Pt(CN)_4Cl(SO_3H)]^{2-}$, which is reduced rapidly to $[Pt(CN)_4]^{2-}$ and HSO_4^- in an inner-sphere two-electron process.

The fulminate complex, $[Et_4N][W(CNO)(CO)_5]$, isomerizes to give an isocyanate complex, $[Et_4N][W(NCO)(CO)_5]$, in a variety of solvents.[23] The solvent-dependent rate of isomerization was found to increase in the order: tetrahydrofuran < acetone < chloroform < dichloromethane. Isomerization did not take place in methanol due to stabilization of the fulminate by hydrogen bonding with the solvent.

Linkage isomerization of a cobalt(III) 5-methyltetrazolate complex has been investigated,[24-27] Scheme 11.2:

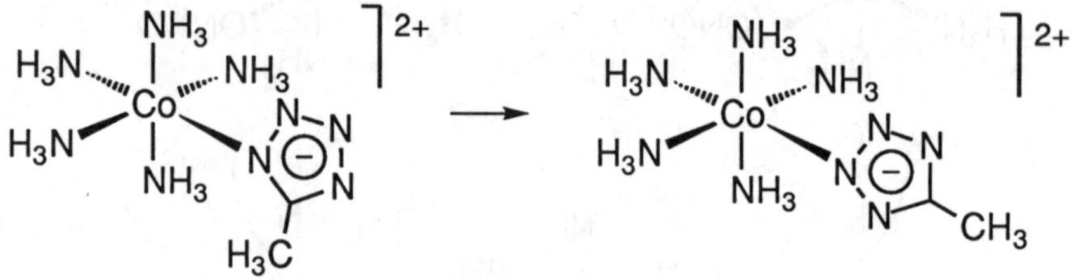

Scheme 11.2

The equilibrium was found to lie almost entirely to the right. Increasing the acidity of the medium led to an increase in the rate of isomerization which also was found to occur in the solid state, but at a significantly reduced rate.

Dialkyl sulfoxides typically form complexes with platinum(II) that involve binding through sulfur.[28-30] The formation of Pt–S(O)R_2, rather than Pt–OSR_2, can be explained in terms of the HSAB principle (Chapter 1); thus, platinum(II), a soft acid, preferentially interacts with sulfur, the soft base. The complex, *trans*-[PtCl_2{*O*-$(PhCH_2)_2SO$}(PCy_3)], was found[31] to be the kinetically controlled product of the reaction between the chloride-bridged complex, [$Pt_2(\mu\text{-}Cl)_2Cl_2(PCy_3)_2$], and dibenzyl sulfoxide. The sulfoxide is bound to platinum(II) *via* oxygen in *trans*-[PtCl_2{*O*-$(PhCH_2)_2SO$}(PCy_3)]. It was shown[31] that the *O*-bonded complex undergoes a change in geometry and linkage isomerization over time. As a result, the thermodynamic product, *cis*-[PtCl_2{*S*-$(PhCH_2)_2SO$}(PCy_3)], is formed, Scheme 11.3:

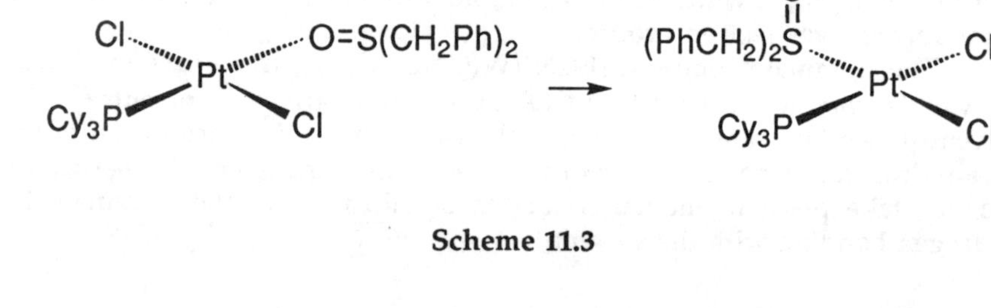

Scheme 11.3

Other examples of linkage isomerization of dialkyl sulfoxide complexes have been reported by Taube and coworkers.[32,33]

An unusual rearrangement reaction was observed upon treatment of complex **A** with lithium diisopropylamide, *i*-Pr_2NLi, which resulted in deprotonation of a C–H bond with formation of **B**. The latter was shown to rearrange to complex **C**,[34] Scheme 11.4:

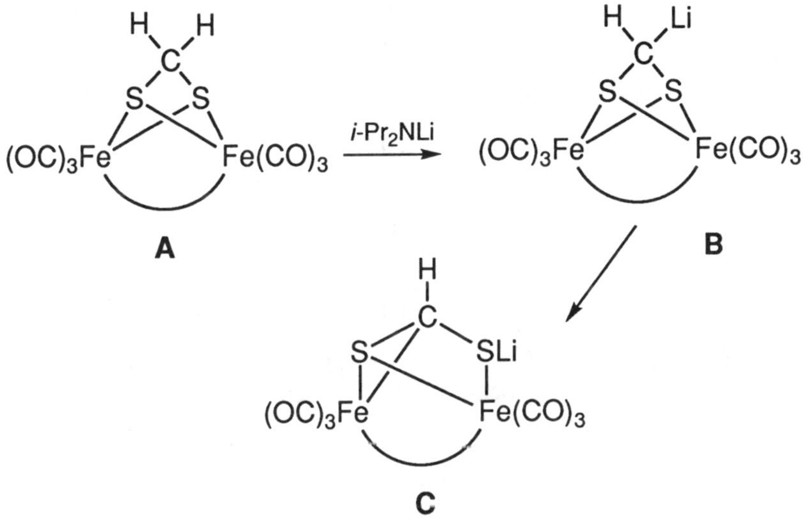

Scheme 11.4

The ligand, 1-diphenylphosphino-2-(2-pyridyl)ethane, in [RhCl(CO)(*P*-$Ph_2PC_2H_4Py)_2$], is coordinated to rhodium(I) *via* phosphorus. At 170 °C in air this ligand is oxidized to the phosphine oxide, followed by a molecular rearrangement,[35] Scheme 11.5. Prior to oxidation, the rhodium(I) center preferentially coordinates to the P(III) site, a soft base. However, after insertion of oxygen into the rhodium-phosphorus bond, coordination *via* the pyridine nitrogen is favored over coordination to the oxygen of the P(V) group.

Based on IR and NMR data, a rearrangement of the homobinuclear rhodium complexes, [Rh_2Cl_2{SCNC(O)R}(CO)(dppm)$_2$] (R = OEt, Ph), has been identified,[36] Scheme 11.6. This rearrangement provides an example of the equilibration of a carbonyl ligand between bridging and terminal sites.

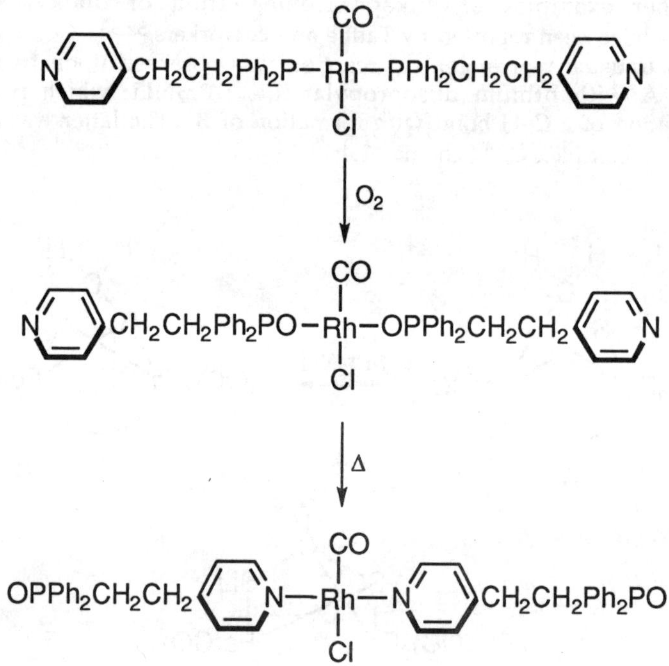

Scheme 11.5

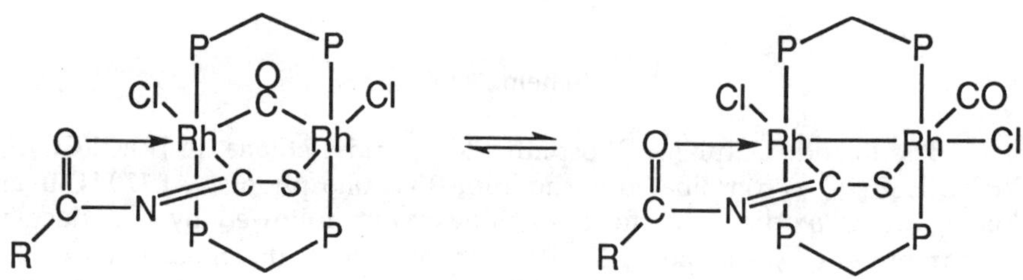

Scheme 11.6

11.2 Rearrangements involving changes in coordination polyhedra

Many four-coordinate complexes of nickel(II) undergo tetrahedral – square-planar rearrangement reactions. In the case of palladium(II) and platinum(II) complexes, tetrahedral intermediates may be invoked in mechanisms of geometric isomerization, but square-planar isomers are favored. Nickel(II) complexes of the type $[NiX_2(PR_{3-n}Ar_n)_2]$ (n = 0 - 3) have been widely used in studies of the equilibrium between tetrahedral (paramagnetic) and square-planar (diamagnetic) forms. For example, it was shown[37] that the complexes, $[NiX_2(P\text{-}PPh_2CH_2SR)_2]$ (X = Cl, Br, I; R = Me, Ph), exist in solution as both diamagnetic and paramagnetic forms. In general, only one form can be isolated from solution and only for the complex, $[NiBr_2(P\text{-}PPh_2CH_2SMe)_2]$, could both square-planar and tetrahedral isomers be obtained in the solid state.

In Tolman's review[38] electronic and steric ligand effects are identified as the key factors affecting the tetrahedral – square-planar equilibria of nickel(II) phosphine complexes. The role of electronic ligand effects has been discussed[39] and it was concluded that the difference in the stabilization energies of square-planar and tetrahedral isomers is decreased by ligands capable of forming strong π-bonds with a metal center. Thus, particularly for first row transition metals, the possibility of a tetrahedral – square-planar rearrangement is increased by the presence of π-acid ligands.

The influence of steric factors on the position of tetrahedral – square-planar equilibria has been discussed in detail.[38,39] In general, two trends are apparent: (*i*) the smaller the steric parameter, θ, of the phosphine ligand in a complex of the type $[NiX_2(PR_3)_2]$, the greater the stabilization of the square-planar form, and (*ii*) the stabilization of the square-planar form is sensitive to the nature of the anionic ligand, X^-, and decreases in the order: $Cl^- > Br^- > I^-$.

Solvent effects can influence the position of a tetrahedral – square-planar equilibrium. Thus, it has been found[40] that the nickel(II) alkyldiphenylphosphine complexes, $[NiX_2(PRPh_2)_2]$ (X = Cl, Br, I; R = Me, Et, Pr), may occur both in a diamagnetic, square-planar form and in a paramagnetic, tetrahedral form. The diamagnetic complexes are dark red or brown and the paramagnetic complexes are blue or green. It was observed that the use of more polar solvents favors the formation of the more polar (tetrahedral) isomer. The complexes, $[NiX_2(PBuPh_2)_2]$ (where X = Cl, Br, I), have been prepared in alcoholic media and were shown[41] to be paramagnetic. The magnetic susceptibility was found to decrease when the complexes were dissolved in benzene. A partial transformation of the paramagnetic form into the diamagnetic isomer explains the susceptibility data,[42] Scheme 11.7.

The interconversion of square-planar and tetrahedral nickel(II) complexes typically occurs at a fast rate and such equilibria have been studied by NMR spectroscopy.[43] A modified, extended-Hückel method has been used to calculate electronic energy surfaces for the glycinate complex, $[Pt(Gly)_2]$, in the transition from square-planar to tetrahedral configurations for both the ground and excited states. The calculations were carried out for various positions of the atomic nuclei during twisting of the molecule relative to the axis,[44] Figure 11.1.

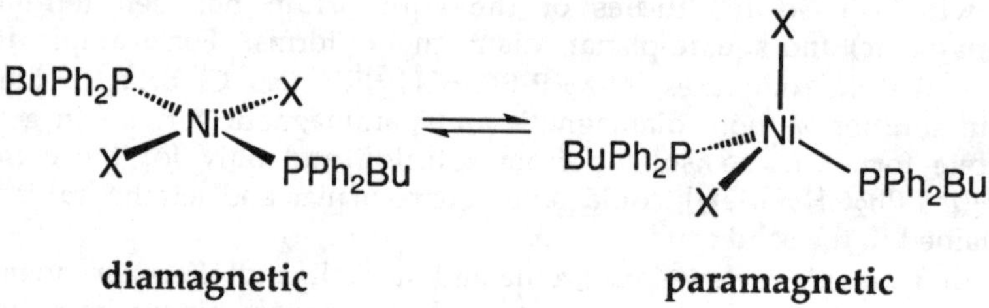

Scheme 11.7

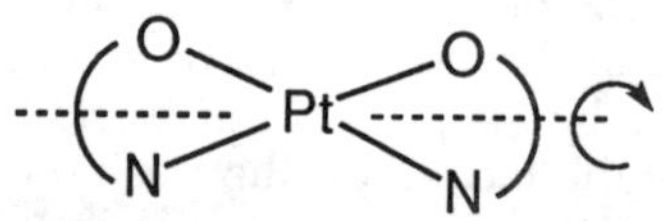

Figure 11.1

The calculations revealed that the maximum on the total electronic energy surface for ground state molecules occurs at the tetrahedral configuration. In contrast, for the excited state molecule, a tetrahedral geometry corresponds to a minimum on the total electronic energy surface. Because the energy of the tetrahedral complex in an excited state is similar to that of a tetrahedral complex in the ground state, a mechanism of isomerization involving a transition between these states is supported.

Ihara and Tsuchiya[45] have observed the rearrangement of square-planar, dicationic nickel(II) complexes to neutral, octahedral complexes by anion coordination. In the solid state, square-planar complexes of the type $[Ni(Abi)_4]X_2$ (Abi = 2-aminobenzimidazole; X = Cl, Br, or NO_3) are converted into octahedral complexes, with loss of water of crystallization upon heating, Eq. 11.2:

$$[Ni(Abi)_4]X_2 \cdot nH_2O \rightarrow [NiX_2(Abi)_4] + nH_2O \qquad (11.2)$$

The same type of rearrangement has been observed[46] for the square-planar benzimidazole complex, $[Ni(Bimd)_4][NO_3]_2 \cdot 2.5C_2H_5OH$. Upon heating, the ethanol of crystallization is lost and coordination of the nitrate counterions occurs to yield the octahedral complex, $[Ni(NO_3)_2(Bimd)_4]$.

Studies of solid state conversions of chloride and bromide complexes of the types $[Ni(dl\text{-}Bua)_2(H_2O)_2]X_2$ (Bua = 2,3-butanediamine)[47] and $[Ni(dl\text{-}Stien)_2(H_2O)_2]X_2$ (Stien = 1,3-diphenyl-1,2-diaminoethane; X = Cl, Br, NO_2)[48] have revealed similar reactions. Dehydration is the initial step in the process and results in conversion of the octahedral aqua-complexes into the square-planar complexes, $[Ni(dl\text{-}Bua)_2]X_2$ and $[Ni(Stien)_2]X_2$. Further heating results in rearrangement to form new octahedral complexes, Eq. 11.3:

$$[Ni(Stien)_2)]X_2 \rightarrow [NiX_2(Stien)_2)] \quad (11.3)$$

It has been found[49] that the Anderson rearrangement (Chapter 12) of $[bipyH_2][CoCl_4]$ when the solid state is heated results, initially, in the formation of a polymeric complex with chloride bridges and an octahedral geometry about the metal centers, Eq. 11.4:

$$[bipyH_2][CoCl_4] \rightarrow 1/n[CoCl_2(bipy)]_n + 2HCl \quad (11.4)$$

This polymeric complex then undergoes an intramolecular rearrangement resulting in conversion to the monomeric compound, $[CoCl_2(bipy)]$, with a tetrahedral structure.

The dynamics of the square-planar – octahedral isomerization of a nickel(II) complex with *N*-(2-piperidinoethyl)salicylaldiminato ligands has been studied by ultrasonic relaxation methods.[50] This rearrangement can be represented as follows, Scheme 11.8:

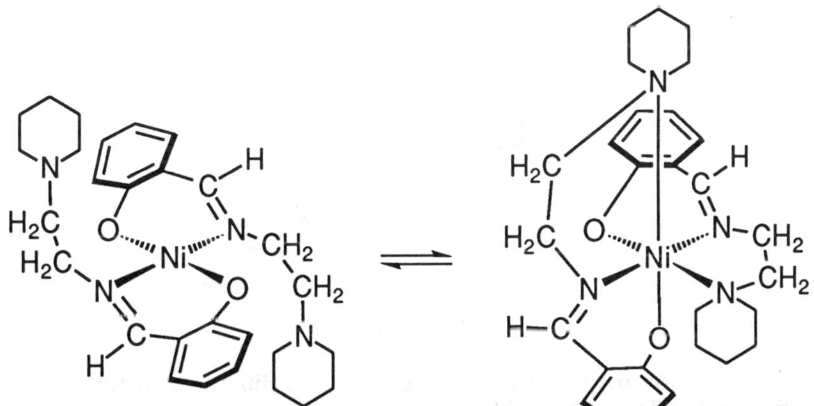

Scheme 11.8

A similar conversion has been observed for a five-coordinate palladium(II) complex with a tris(2-dimethylaminoethyl)amine ligand, which rearranges to give a four-coordinate complex with a pendant amino functionality,[51] Scheme 11.9:

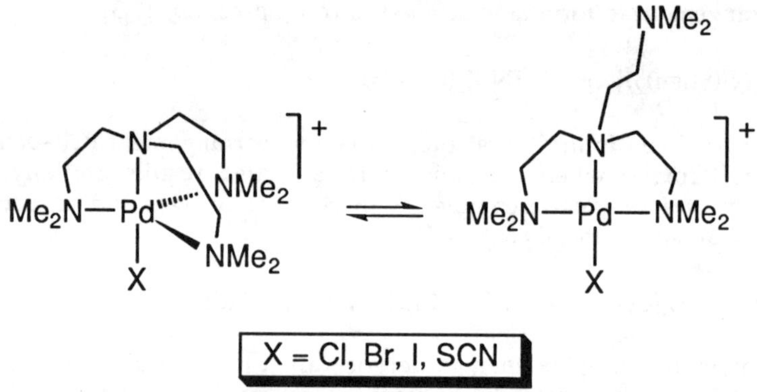

Scheme 11.9

Several nickel(II) complexes containing bidentate nitrogen-donor ligands are known to exhibit reversible thermochromism.[52] For example, [$Ni_2Cl_4Q_2$], where Q is shown below, Figure 11.2:

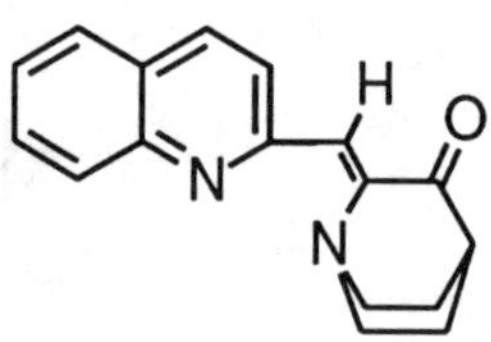

Figure 11.2

is yellow in color at room temperature, having a dimeric structure with two bridging chloride ligands. The coordination polyhedron about nickel is square-pyramidal. Upon heating to approximately 230 °C the chloride bridges are cleaved and a monomeric form, violet in color, is formed. The monomer is presumed to adopt a pseudotetrahedral geometry. The dimeric structure is restored upon cooling.

11.3 Rearrangement of coordinated ligands on a metal framework

Molecular rearrangements that involve a single ligand or several neighboring ligands may be induced by coordination to a metal ion. For example, the conversion of ethylenimine to *N*-(2-aminoethyl)ethylenimine *via* coordination to a transition metal was described almost simultaneously in the United States[53] and in Russia,[54-56] Scheme 11.10:

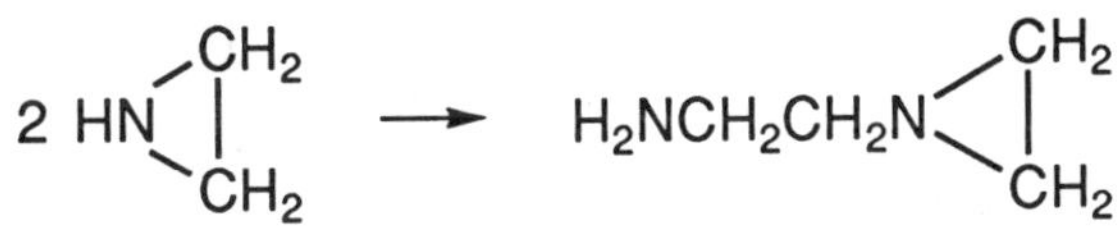

Scheme 11.10

A U.S. patent[53] disclosed that heating a closed vessel containing a complex of the type, $[MX_2(HNCH_2CR_2)_4]$ (M = Co(II), Ni(II), Cu(II), and Zn(II); X = Cl, Br, I, HSO_4, $1/2SO_4$, ClO_4, NO_3, HCOO; R = H, C_1-C_4 alkyl), at temperatures of 50-65 °C in a polar solvent (DMSO, DMF, CH_3OH, and mixtures of these solvents) leads to the formation of the complexes, $[MX_2(H_2NCH_2CH_2NCH_2CR_2)_2]$, Scheme 11.11:

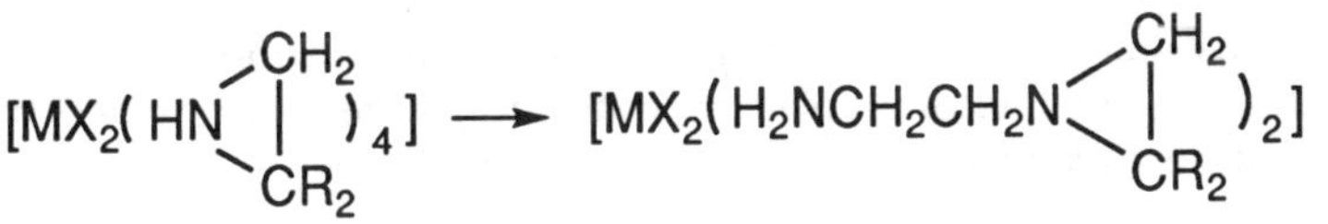

Scheme 11.11

Yields of products were reported to be greater than 95%. In a series of papers[55,56] it was shown that the conversion of ethylenimine is the initial step in a long series of reactions in which *N*-(2-aminoethyl)ethylenimine acts as an aminoalkylating agent. Treatment of the ethylenimine complex, $[Ni(HNCH_2CH_2)_4]^{2+}$, with ammonium thiocyanate results in cleavage of the ethylenimine ring[55] and formation of $[Ni(H_2NCH_2CH_2SCN)_4]^{2+}$. The reaction does not stop at this stage, however, and the 2-thiocyanato-ethylamine ligand is converted to coordinated aminothiazoline, Eq. 11.5:

$$[Ni(H_2NCH_2CH_2SCN)_4]^{2+} \rightarrow [Ni(NH_2\overline{C{=}NCH_2CH_2S})_4]^{2+} \quad (11.5)$$

Conversion of coordinated *N*-(2-aminoethyl)ethylenimine occurs in a similar manner.[55] The thiocyanate complex obtained in the first stage of the reaction undergoes a thiocyanogen-amine rearrangement, Scheme 11.12:

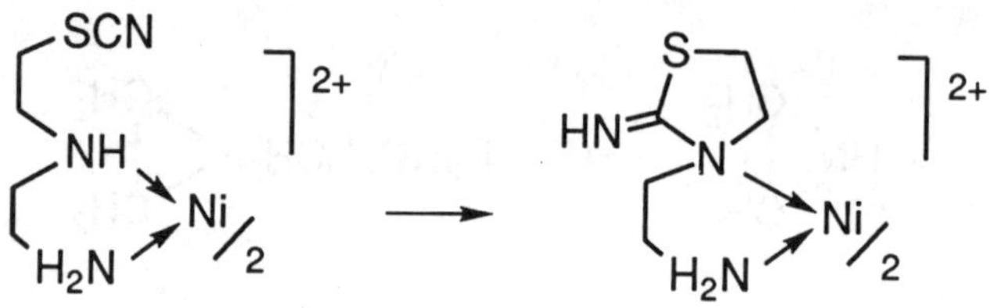

Scheme 11.12

The interaction between hydride and ylide ligands in the complex, $[W(H)(CH_2{=}PMe_2Ph)Cp_2][PF_6]$, has been studied.[57] On heating in acetone, the complex is transformed into $[W(Me)(PMe_2Ph)Cp_2][PF_6]$. Based on kinetic studies, two alternative mechanisms for the rearrangement were proposed, One involves the formation of an agostic methyl intermediate, $[W(CH_2)(\mu\text{-}H)Cp_2]^+$, and the other involves an equilibrium between a carbene hydride, $[W(CH_2)(H)Cp_2]^+$, and a methyl cation, $[W(CH_3)Cp_2]^+$.[57]

An isomerization involving migration of a phenyl group from zirconium to carbon, accompanied by transfer of a methanethiolate fragment from carbon to zirconium, has been observed,[58] Scheme 11.13:

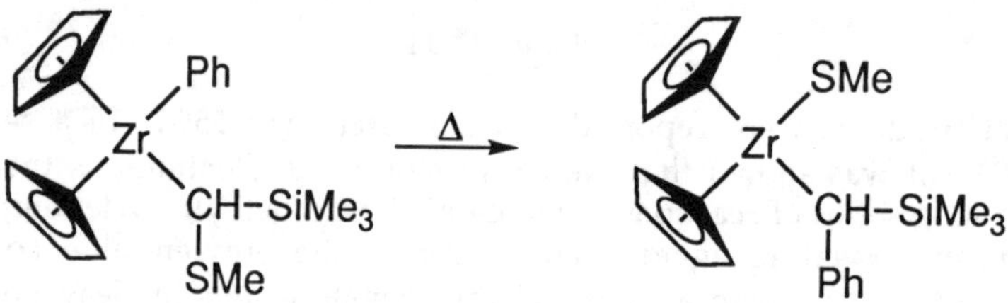

Scheme 11.13

In boiling toluene this transformation occurs over five days in a yield of 58%.

Transfer of coordinated triphenylphosphine to a carborane ligand, accompanied by hydride transfer from the carborane to the metal center, has been observed in nickel(II) complexes,[59] Eq. 11.6:

$$[Ni(C_2B_9H_{11})(PPh_3)_2] \rightarrow [Ni(H)(C_2B_9H_{10}PPh_3)(PPh_3)] \quad (11.6)$$

The resulting carborane ligand is present in the phosphonium form. The rearrangement takes place on prolonged heating in inert solvents.

The complex, $[Ni(PBu_3)_2(Me_5C_5P{=}NBu^t)]$, was formed by reaction of $Me_5C_5P{=}NBu^t$ with $[Ni(cod)(PBu_3)_2]$. Based on NMR data, an equilibrium between two forms in toluene at -30 °C was established, Scheme 11.14:

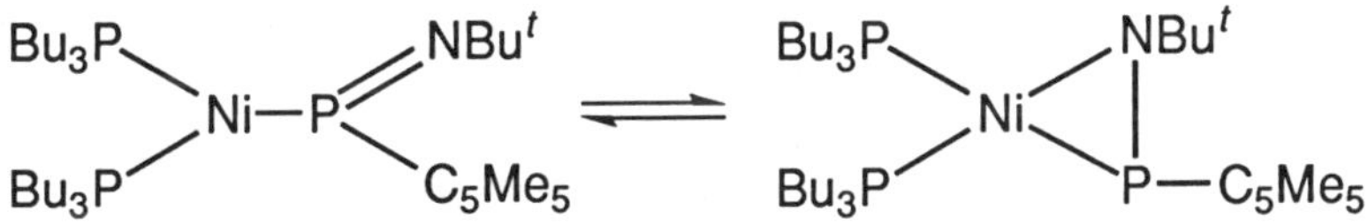

Scheme 11.14

Transfer of the pentamethylcyclopentadienyl group from phosphorus to nickel and loss of one of the tertiary phosphine ligands, yielding $[Ni(P{=}NBu^t)(PBu_3)Cp^*]$, occurs when the temperature is increased to -5 °C.[60]

The reactions of nickel(II) and copper(II) amino complexes with acetone have been studied extensively.[61] Condensation reactions of coordinated amines with acetone proceed relatively easily, but further reactions follow. The initially formed complexes, which contain Schiff bases, undergo intramolecular rearrangements to form complexes containing macrocyclic ligands, Scheme 11.15. This reaction is the basis for the synthesis of a large number of macrocyclic ligands.

11.4 Rearrangement of π-olefin complexes to σ-organometallics

The rearrangement of π-olefinic ligands to σ-alkyl ligands is central to many important catalytic processes, including olefin hydrogenation, hydroformylation and polymerization, as well as to palladium(II)-mediated olefin oxidation reactions.[62]

The conversion of a π-bonded olefinic ligand to σ-alkyl ligand requires the participation of an internal or external nucleophile, Scheme 11.16. The rearrangement of a platinum(II) ethylene hydride complex is an example of an intramolecular process,[63] Scheme 11.17 (L = $PMePh_2$).

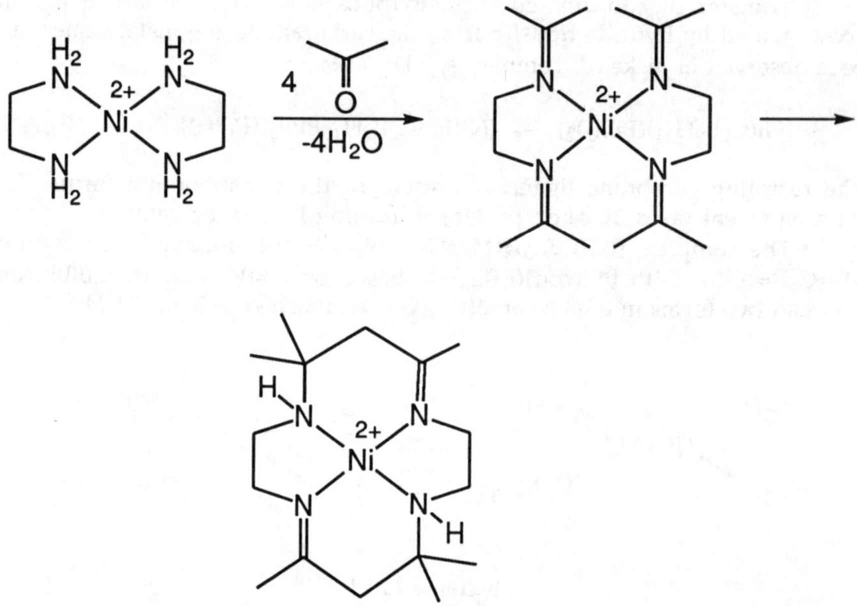

Scheme 11.15

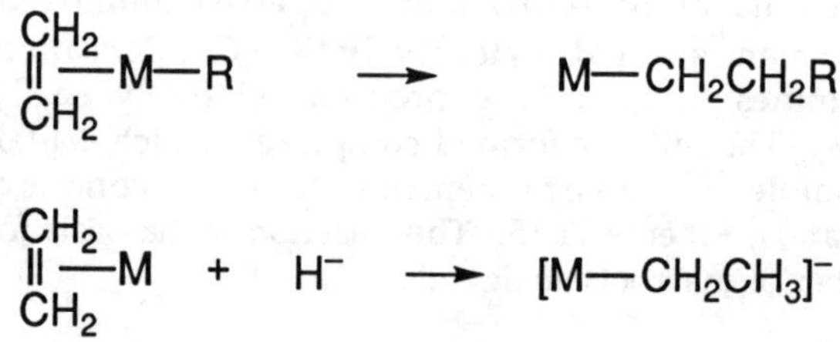

Scheme 11.16

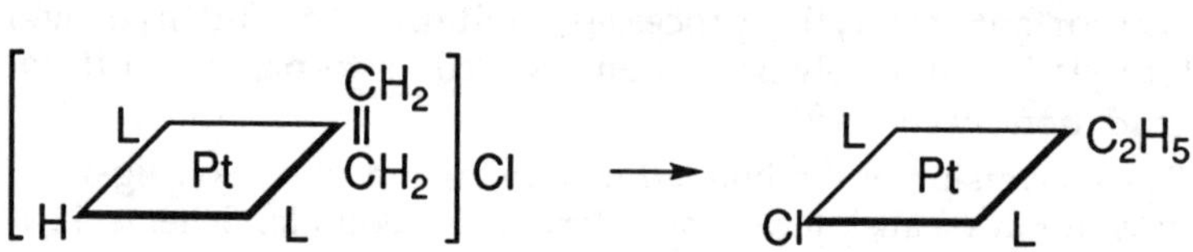

Scheme 11.17

Also, an ethylene hydride complex of ruthenium, $[RuH(\eta^2\text{-}C_2H_4)(PMe_3)(\eta^6\text{-}C_6H_6)][PF_6]$, is known.[64] An equilibrium between ethylene hydride and ethyl isomers has been established in solution, Scheme 11.18:

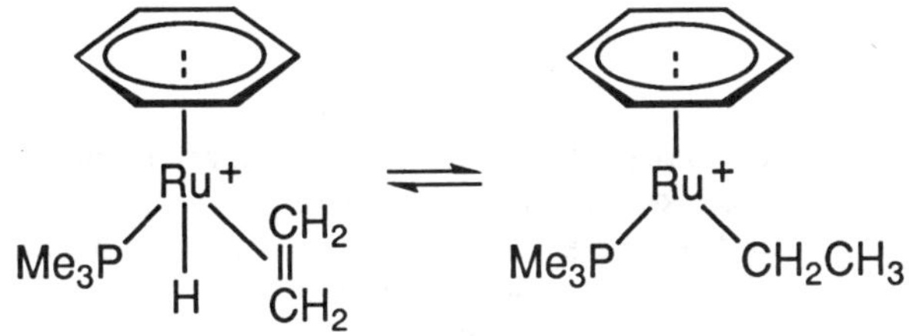

Scheme 11.18

Addition of phosphine shifts the equilibrium to the right because PMe_3 can occupy the free coordination site, generated by hydride migration, and, thus, inhibit the reverse σ–π rearrangement, in which an increase in coordination number occurs.

The kinetics and mechanism of the thermal transformation in cyclohexane of $[PtEt_2(PEt_3)_2]$ into $[Pt(C_2H_4)(PEt_3)_2]$ and ethane have been studied.[65] The reaction proceeds *via* formation of an ethylene hydride intermediate.

Through π–σ rearrangements, olefin hydride complexes may act as catalysts for olefin isomerization. A general mechanism is as follows, Scheme 11.19:

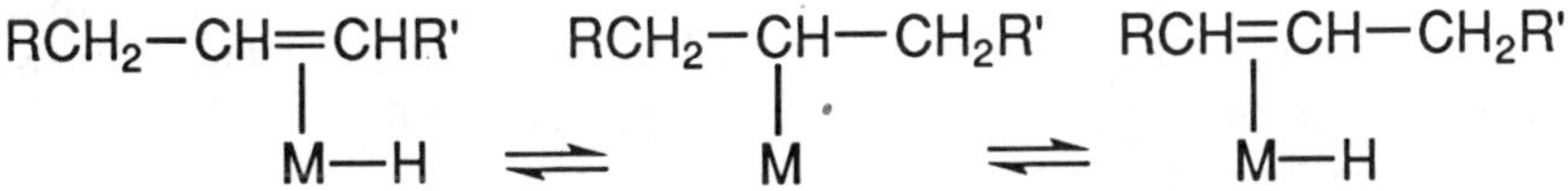

Scheme 11.19

Although transformations of ethyl complexes into hydride complexes and free ethylene are known,[66] such reactions are not the rearrangements that are the subject of this Chapter.

Olefin complexes undergo rearrangements in which hydridoallyl intermediates are involved, Scheme 11.20:

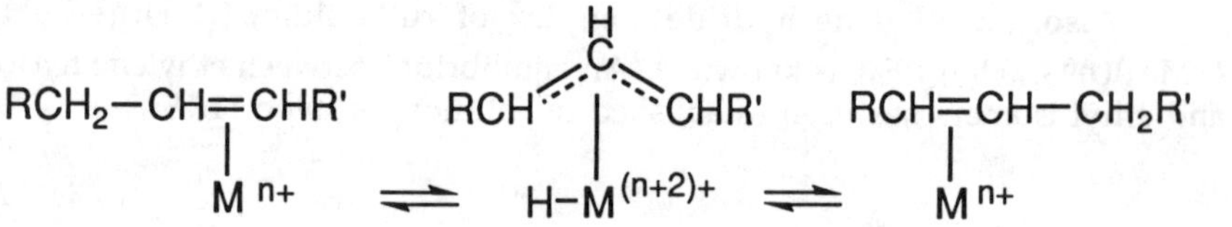

Scheme 11.20

Formation of a hydridoallyl complex in this manner increases both the oxidation state and the coordination number of the metal by two and, thus, reactivity is determined by the same factors as those that influence reactivity towards oxidative addition of simple substrates.

A number of dynamic processes characterize the solution chemistry of allyl complexes. In the case of palladium-allyl systems, the η^3-/η^1- interconversion is commonly recognized from variable temperature NMR experiments. The following example is illustrative of the solution chemistry observed,[67] Scheme 11.21:

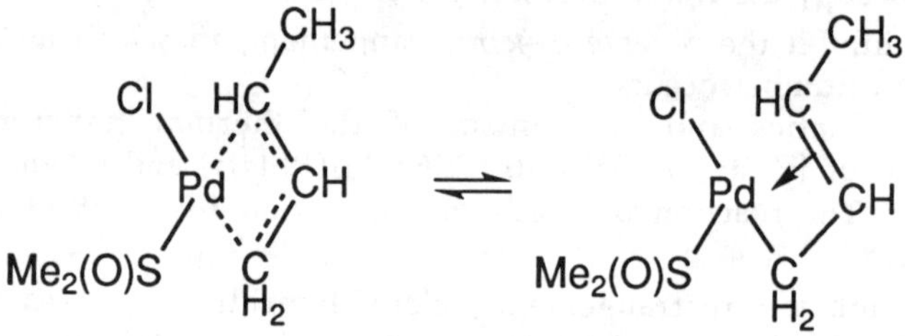

Scheme 11.21

A related π–σ-rearrangement is the isomerization of coordinated alkenes to metal-alkenyls *via* C–X oxidative addition. The following transformations are representative,[68,69] Eqs. 11.7 and 11.8:

$$[Pt(C_2F_3Br)(PMePh_2)_2] \rightarrow [PtBr(CF{=}CF_2)(PMePh_2)_2] \quad (11.7)$$

$$[Pt(C_2F_3Br)(AsPh_3)_2] \rightarrow [PtBr(CF{=}CF_2)(AsPh_3)_2] \quad (11.8)$$

An intramolecular mechanism for these reactions has been proposed[69] based on the insensitivity of the reaction rate to the nature of the solvent.

The rearrangement reactions of olefin complexes resemble the rearrangement of molecular hydrogen complexes into their dihydride isomers.[70] For example, it was shown by NMR and IR methods that, in the case of $[W(CO)_3(\eta^2\text{-}H_2)(PR_3)_2]$ (R = Pr, C_6H_{11}),[71] an equilibrium with the dihydride form was established in solution, Eq. 11.9:

$$[W(CO)_3(\eta^2\text{-}H_2)(PR_3)_2] \rightleftharpoons [W(CO)_3(H)_2(PR_3)_2] \qquad (11.9)$$

Such equilibration processes may be more common than currently realized in the chemistry of molecular hydrogen and dihydride complexes.

11.5 Rearrangement of acyl ligands to alkyl and carbonyl ligands

Transition metal acyl complexes are implicated in many catalytic reactions of carbon monoxide. Alkylation of a coordinated carbonyl group initially results in the formation of an acyl ligand, which then may rearrange,[72] Scheme 11.22:

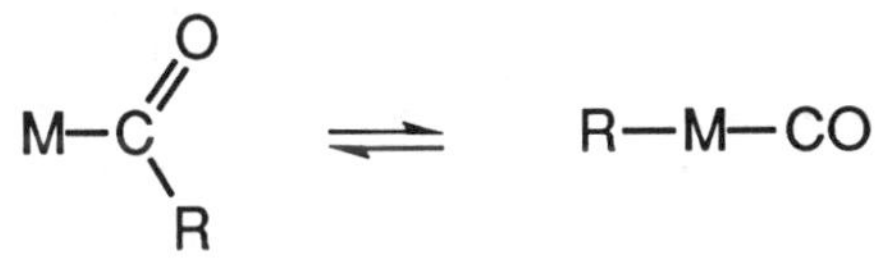

Scheme 11.22

Acyl complexes can be prepared directly by oxidative addition of acyl halides to transition metal complexes with suitable electronic and geometric properties.[73] The reactions of acetyl chloride with Wilkinson's catalyst, $[RhCl(PPh_3)_3]$, are illustrative,[74] Scheme 11.23:

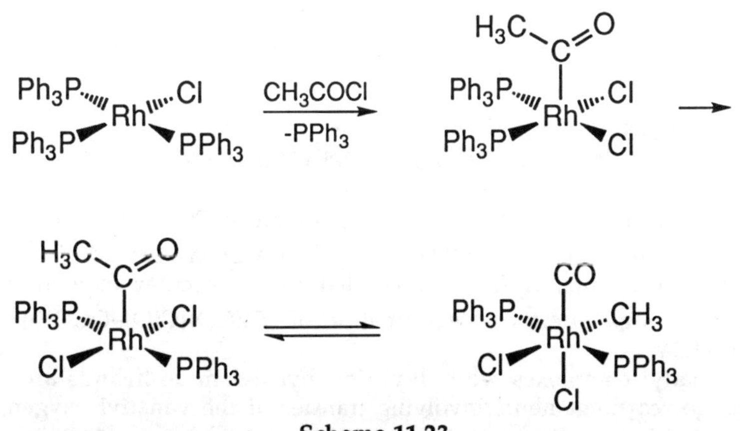

Scheme 11.23

This type of reaction, which has been the subject of many studies because of its potential utilization in the decarbonylation of acyl halides, has also been investigated with the analogous iridium(I) complex.[74] The reversibility of the carbonylation/decarbonylation process was first described by Heck and Breslow in 1960,[75] and work with a wide range of organic substrates has been reported.[74,76]

The mechanism of insertion of carbon monoxide into a metal-alkyl or metal-aryl bond involves migration of the alkyl (aryl) group to the coordinated carbon monoxide with concomitant formation of an acyl ligand. An extensive literature on this subject exists.[1-4]

11.6 Interligand oxygen transfer

Interligand oxygen transfer is a rearrangement that occurs between coordinated ligands on a metal framework. However, because of the importance of this class of reactions, it is considered here as a separate subject.

The synthesis of $K_2[Pt(NO_2)_6]$ involves oxidation of $K_2[Pt(NO_2)_4]$ with nitric acid in the presence of nitrogen oxides. The reaction proceeds *via* the intermediate, $K_2[Pt(NO_2)_4NO(NO_3)]$, where redistribution of oxygen between nitrosyl and nitrate ligands results in product formation.[77]

It was found[78] that heating solid $[Pt(NO_3)_2(PPh_2CF_3)_2]$ at 170 °C gave rise to a strong exothermic peak in the DTA and a slight mass loss in the TG. It was shown that the transformation was due to oxygen transfer from the nitrate ligands to the coordinated phosphines, Eq. 11.10:

$$[Pt(NO_3)_2(PPh_2CF_3)_2] \xrightarrow{\Delta} [Pt(NO_2)_2(OPPh_2CF_3)_2] \qquad (11.10)$$

The redistribution of oxygen between ligands in the dimethyl sulfoxide complex, $[Pt(NO_3)_2(PPh_2CF_3)(Me_2SO)]$, proceeds similarly,[79] Eq. 11.11:

$$[Pt(NO_3)_2(PPh_2CF_3)(Me_2SO)] \xrightarrow{\Delta} [Pt(NO_2)_2(OPPh_2CF_3)(Me_2SO)] + 1/2O_2 \qquad (11.11)$$

An unusual oxygen transfer from coordinated CO_2 to a carbonyl ligand in the complex, $Li[Fe(CO)_2(CO_2)(\eta^5\text{-}Cp)]$, has been studied.[80] Based on labeling experiments, it was shown that intermolecular oxygen transfer proceeds through intermediate formation of $[Fe(CO)\{C(O){-}OCO\}(\eta^5\text{-}Cp)]^-$, Scheme 11.24.

Vanadyl complexes with alkyltrimethylsilylamide ligands are known to undergo rearrangements involving transfer of the vanadyl oxygen, often accompanied by ligand reorganization, for example,[81] Scheme 11.25.

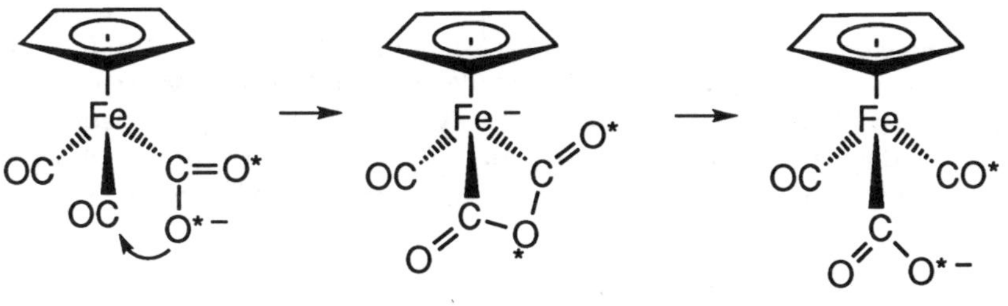

Scheme 11.24

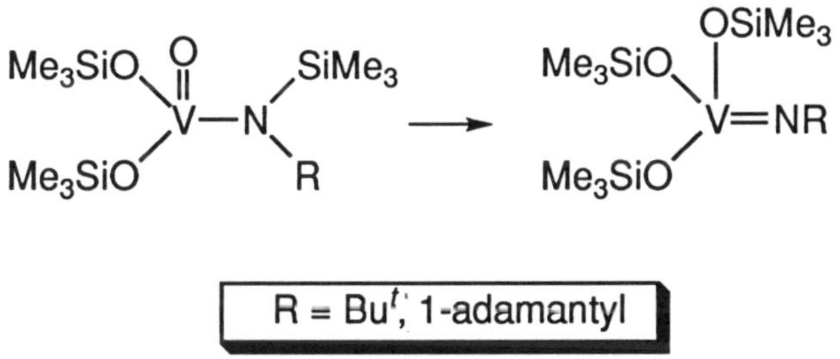

Scheme 11.25

Such rearrangement reactions are presumably driven by the high oxophilicity of silicon.

11.7 Conversion of phosphite complexes to phosphonate complexes

It has been shown[82] that coordinated trialkylphosphites in platinum(II), palladium(II), and rhodium(III) complexes are readily hydrolyzed to form $P(O)(OR)_2^-$ anions, without cleavage of the metal-phosphorus bond. The corresponding alcohols were obtained as the by-products from these reactions. The intramolecular transformation of alkylphosphites to the corresponding phosphonates also has been described.[83,84] For example, it has been shown[83] that thermolysis of $[Ru\{P(OMe)_3\}_5]$ at 120 °C in hexane in a sealed tube for 24 h results in quantitative transfer of a methyl group from the alkylphosphite to ruthenium, Eq. 11.12:

$$[Ru\{P(OMe)_3\}_5] \rightarrow [RuMe\{P(=O)(OMe)_2\}\{P(OMe)_3\}_4] \quad (11.12)$$

The complexes, $[M(AsMe_2)(CO)_3Cp]$ and $[M(SbMe_2)(CO)_3Cp]$ (M = Mo, W), react with $P(OMe)_3$ to form $[M(AsMe_2)(CO)_2\{P(OMe)_3\}Cp]$ and $[M(SbMe_2)(CO)_2\{P(OMe)_3\}Cp]$, respectively. Due to the high nucleophilicity of the $AsMe_2^-$ and $SbMe_2^-$ ligands, methyl migration from the phosphite occurs, with formation of the corresponding phosphonate,[85] Eq. 11.13:

$$[M(EMe_2)(CO)_2\{P(OMe)_3\}Cp] \rightarrow [M(EMe_3)(CO)_2\{P(=O)(OMe)_2\}Cp] \quad (11.13)$$

where E = As, Sb

Formation of the phosphorus-oxygen double bond is presumably a strong driving force in such rearrangement reactions.

References

1. S. Kawaguchi, *Variety in Coordination Modes of Ligands in Metal Complexes* (Springer, Heidelberg, 1987) p. 3.
2. R. B. Jordan, *Reaction Mechanisms of Inorganic and Organometallic Systems* (Oxford University, Oxford, 1991) p. 91.
3. K. G. Orrell, *Mechanisms of Inorg. and Organomet. Reactions* **8** (1994) 321.
4. K. G. Orrell, *Mechanisms of Inorg. and Organomet. Reactions* **7** (1993) 309.
5. J. L. Burmeister, *Coord. Chem. Rev.* **105** (1990) 77.
6. F. Basolo, *Coord. Chem. Rev.* **100** (1990) 47.
7. For early reviews on linkage isomerization: (a) J. L. Burmeister, *Coord. Chem. Rev.* **3** (1968) 225; (b) A. H. Norbury and A. I. P. Sinha, *Quart. Rev.* **24** (1970) 69; (c) A. D. Garnovsky, O. A. Osipov, and S. B. Bulgarevich, *Uspekhi Khimii (Russ. Chem. Rev.)* **41** (1972) 648; (d) A. H. Norbury, *Adv. Inorg. Chem. Radiochem.* **17** (1975) 231; (e) L. M. Shkolnikova and E. A. Shugam, *Kristallokhimia* **12** (1977) 169; (f) Yu. N. Kukushkin, *Koord. Khim.* **4** (1978) 1170; (g) A. M. Golub, H. Köhler, and V. V. Skopenko, eds., *Chemistry of Pseudohalides* (Elsevier, Amsterdam, 1986).
8. S. M. Jørgensen, *Z. Anorg. Chem.* **5** (1894) 147.
9. W. G. Jackson, G. A. Lawrence, P. A. Lay, and A. M. Sargeson, *J. Chem. Soc., Chem. Commun.* (1982) 70.
10. (a) N. Masciocchi, A. Kolyshev, V. Dulepov, E. Boldyreva, and A. Sironi, *Inorg. Chem.* **33** (1994) 2579; (b) E. Boldyreva, A. Virovets, L. Burleva, V. Dulepov, and N. Podberezskaya, *Russ. J. Struct. Chem., Russ. Ed.* **34** (1993) 128.
11. (a) M. J. Coyer, R. H. Herber, J. Chen, M. Croft, and S. P. Szu, *Inorg. Chem.* **33** (1994) 716; (b) M. J. Coyer, M. Croft, J. Chen, and R. H. Herber, *Inorg. Chem.* **31** (1992) 1752.
12. (a) D. A. Buckingham, C. R. Clark, and G. F. Linddell, *Inorg. Chem.* **31** (1992) 2909; (b) K. Eichele and R. Wasylishen, *Inorg. Chem.* **33** (1994) 2766; (c) A. D. Kirk and D. M. Kneeland, *Inorg. Chem.* **34** (1995) 1536.
13. (a) D. F.-T. Tuan and R. Huffmann, *Inorg. Chem.* **24** (1985) 871 and references therein; (b) T. Schönherr, *Inorg. Chem.* **25** (1986) 171.

14. W. G. Jackson, S. S. Jurisson, and M. A. O'Leary, *Inorg. Chem.* **32** (1992) 445.

15. E. R. Civitello, P. S. Dragovich, T. B. Karpishin, S. G. Novick, G. Bierach, J. F. O'Connell, and T. D. Westmoreland, *Inorg. Chem.* **32** (1992) 237.

16. (a) A. DelMedico, P. R. Auburn, E. S. Dodsworth, A. B. P. Lever, and W. J. Pietro, *Inorg. Chem.* **33** (1994) 1583; (b) A. DelMedico, S. S. Fiedler, A. B. P. Lever, and W. J. Pietro, *Inorg. Chem.* **34** (1995) 1507.

17. L. Cun and R. B. Jordan, *Inorg. Chem.* **29** (1990) 2937.

18. D. W. Powell and P. A. Lay, *Inorg. Chem.* **31** (1992) 3542.

19. D. I. Bustin, M. Rievaj, and J. Mocak, *Inorg. Chim. Acta* **27** (1978) 59.

20. A. C. Dash, A. A. El-Awady, and G. M. Harris, *Inorg. Chem.* **20** (1981) 3160.

21. P. A. Moritzen, A. A. El-Awady, and G. M. Harris, *Inorg. Chem.* **24** (1985) 313.

22. J. Berglund, R. Voigt, S. Fronaeus, and L. I. Elding, *Inorg. Chem.* **31** (1994) 3346.

23. F. Beck, W. P. Fehlhammer, and W. Beck, *J. Organomet. Chem.* **279** (1985) C22.

24. (a) W. R. Ellis and W. L. Purcell, *Inorg. Chem.* **21** (1982) 834; (b) W. L. Purcell, *Inorg. Chem.* **22** (1983) 1205; (c) R. J. Balahura, W. L. Purcell, M. E. Victoriano, M. L. Lieberman, V. M. Loyola, W. Fleming, and J. W. Fronabarger, *Inorg. Chem.* **22** (1983) 3602; (d) J. H. Hall, R. Lopez de la Vega, and W. L. Purcell, *Inorg. Chim. Acta* **102** (1985) 157.

25. S. Hubinger, J. H. Hall, and W. L. Purcell, *Inorg. Chem.* **32** (1993) 2394.

26. S. Hubinger and W. L. Purcell, *Inorg. Chem.* **30** (1991) 3707.

27. J. H. Hall and W. L. Purcell, *Inorg. Chem.* **29** (1990) 3806.

28. Yu. N. Kukushkin, *The Reactivity of Coordination Compounds* [in Russian] (Khimia, Leningrad, 1987) p. 288.

29. H. B. Kagan and B. Ronan, *Rev. Heteroatom Chem.* **7** (1992) 92.

30. G. Mestroni, E. Alessio, M. Calligaris, W. M. Attia, F. Quadrifoglio, S. Cauci, G. Sava, S. Zorzet, S. Pacor, C. Monti-Bragadin, M. Tamaro, and L. Dolzani, *Prog. Clinical Biochem. and Medicine* **10** (1989) 71.

31. J. A. Davies and A. Sood, *Inorg. Chem.* **24** (1985) 4213.

32. P. A. Lay, R. H. Magnuson, and H. Taube, *Inorg. Chem.* **28** (1989) 3001.

33. M. Sano and H. Taube, *Inorg. Chem.* **33** (1994) 705.

34. (a) D. Seyferth, G. B. Womack, L.-C. Song, M. Cowie, and B. W. Hames, *Organometallics* **2** (1983) 928; (b) D. Seyferth, G. B. Womack, M. Cowie, and B. W. Hames, *Organometallics* **2** (1983) 1696.

35. Yu. N. Kukushkin, A. I. Osokin, L. I. Danilina, and V. P. Kotelnikov, *Zh. Neorg. Khim.* **24** (1979) 2180.

36. J. A. E. Gibson and M. Cowie, *Organometallics* **3** (1984) 984.

37. T. Gerdau and R. Kramolowsky, *Z. Naturforsh.* **37b** (1982) 332.

38. C. A. Tolman, *Chem. Rev.* **77** (1977) 313.

39. Yu. N. Kukushkin, N. S. Panina, and V. N. Yakovlev, *Koord. Khim.* **11** (1985) 579.

40. R. G. Hayter and F. S. Humiec, *Inorg. Chem.* **4** (1965) 1701.

41. C. R. C. Coussmaker, M. H. Hutchinson, J. R. Mellor, L. E. Sutton, and L. M. Venanzi, *J. Chem. Soc.* (1961) 2705.

42. M. C. Browning, R. F. B. Davies, D. J. Morgan, L. E. Sutton, and L. M. Venanzi, *J. Chem. Soc.* (1961) 4816.

43. K. F. Purcell and J. C. Kotz, *Inorganic Chemistry* (W. B. Saunders, Philadelphia, 1977) p. 757.

44. F. S. Richardson, D. D. Shillady, and A. Waldrop, *Inorg. Chim. Acta* **5** (1971) 279.

45. Y. Ihara and R. Tsuchiya, *Bull. Chem. Soc. Japan* **53** (1980) 1614.

46. R. Tsuchiya, E. Kyuno, A. Uehara, S. Joba, and S. Ohba, *Chem. Lett.* (1976) 911.

47. Y. Ihara, T. Kamishima, and R. Tsuchiya, *Thermochim. Acta* **67** (1983) 23.

48. Y. Ihara and R. Tsuchiya, *Bull. Chem. Soc. Japan* **57** (1984) 2829.

49. A. Uehara, K. Ozutsumi, H. Furumoto, M. Suzuki, and R. Tsuchiya, *Thermochim. Acta* **87** (1985) 303.

50. A. F. Godfrey and J. K. Beattie, *Inorg. Chem.* **22** (1983) 3794.

51. S. N. Bhattacharya and C. V. Senoff, *Inorg. Chim. Acta* **41** (1980) 67.

52. D. R. Bloomquist and R. D. Wilett, *Coord. Chem. Rev.* **47** (1982) 125.

53. S. Motes, *U.S. Patent* 4170591 (1979).

54. Yu. N. Kukushkin, I. G. Kursky, V. B. Ukraintsev, and V. V. Strukov, *Zh. Neorg. Khim.* **25** (1980) 1270.

55. For reviews see: (a) V. B. Ukraintsev, S. V. Yakovlev, and Yu. N. Kukushkin, *Zh. Obsch. Khim.* **60** (1990) 11; (b) V. B. Ukraintsev, S. V. Yakovlev, and Yu. N. Kukushkin, *Zh. Obsch. Khim.* **59** (1989) 1798.

56. For review see: R. W. Hay, E. C. Constable, and L. A. P. Kane-Maguire, *Mechanisms of Inorg. and Organomet. Reactions* **8** (1994) 283.

57. J. C. Green, M. L. H. Green, and C. P. Morley, *Organometallics* **4** (1985) 1302.

58. E. A. Mintz, A. S. Ward, and D. S. Tice, *Organometallics* **4** (1985) 1308.

59. R. E. King, S. B. Miller, C. B. Knobler, and M. F. Hawthorne, *Inorg. Chem.* **22** (1983) 3548.

60. D. Gudat and E. Niecke, *J. Chem. Soc., Chem. Commun.* (1987) 10.

61. (a) D. A. House and N. F. Curtis, *J. Am. Chem. Soc.* **86** (1964) 223; (b) N. F. Curtis and D. A. House, *J. Am. Chem. Soc.* **86** (1964) 1331 and references therein.

62. (a) G. K. Anderson and J. A. Davies, *The Chemistry of the Metal-Carbon Bond*, eds. S. Patai and F. R. Hartley (John Wiley & Sons, New York, 1985) p. 334; (b) J. A. Davies, *The Chemistry of the Metal-Carbon Bond*, eds. S. Patai and F. R. Hartley (John Wiley & Sons, New York, 1985) p. 361; (c) J. A. Davies in *Comprehensive Organometallic Chemistry*, eds. E. W. Abel, F. G. A. Stone, and G. Wilkinson (Pergamon, Oxford, 2nd Ed., 1995) Vol. 9, p. 291.

63. (a) H. C. Clark and C. R. Jablonski, *Inorg. Chem.* **13** (1974) 2213; (b) H. C. Clark, C. R. Jablonski, J. Halpern, A. Mantovani, and T. A. Weil, *Inorg. Chem.* **13** (1974) 1541.

64. H. Werner and R. Werner, *J. Organomet. Chem.* **174** (1979) C63.

65. (a) T. J. McCarthy, R. G. Nurro, and G. M. Whitesides, *J. Am. Chem. Soc.* **103** (1981) 3396; (b) N. Koga, S. Obara, K. Kitaura, and K. Morokuma, *J. Am. Chem. Soc.* **107** (1985) 7109.

66. (a) G. Alibrandi, D. Minniti, R. Romeo, G. Cum, and R. Gallo, *J. Organomet. Chem.* **291** (1985) 133; (b) R. L. Brainard and G. M. Whitesides, *Organometallics* **4** (1985) 1550.

67. G. L. Statton and K. C. Ramey, *J. Am. Chem. Soc.* **88** (1966) 1327.

68. V. A. Mukhedkar, B. J. Kavathekar, and A. J. Mukhedkar, *J. Inorg. Nucl. Chem.* **37** (1975) 483.

69. J. Burgess, M. M. Hunt, and R. D. W. Kemmitt, *J. Organomet. Chem.* **134** (1977) 131.

70. The chemistry of nonclassical η^2-dihydrogen complexes has been reviewed: (a) G. J. Kubas, *Acc. Chem. Res.* **21** (1988) 120; (b) R. H. Crabtree and D. G. Hamilton, *Adv. Organomet. Chem.* **28** (1988) 299; (c) R. H. Crabtree, *Acc. Chem. Rev.* **23** (1990) 95.

71. G. J. Kubas, R. R. Rayan, and D. A. Wrobleski, *J. Am. Chem. Soc.* **108** (1986) 1339.

72. (a) J. A. Davies and C. T. Eagle, *Chemistry of The Platinum Group Metals: Recent Developments*, ed. F. R. Hartley (Elsevier, Amsterdam, 1991) p. 236; (b) J. A. Davies and R. J. Shaver, *The Chemistry of the Metal-Carbon Bond*, eds. S. Patai and F. R. Hartley (John Wiley & Sons, New York, 1985) p. 295; (c) Yu. N. Kukushkin, *Koord. Khim.* **12** (1986) 147.

73. R. H. Crabtree, *The Organometallic Chemistry of the Transition Metals* (Wiley-Interscience, New York, 1988) p. 121.

74. (a) M. A. Bennett and J. C. Jeffery, *Inorg. Chem.* **19** (1980) 3763; (b) M. A. Bennett, J. C. Jeffery, and G. B. Robertson, *Inorg. Chem.* **20** (1981) 323.

75. R. F. Heck and D. S. Breslow, *J. Am. Chem. Soc.* **82** (1960) 4438.

76. M. Kubota, D. A. Phillips, and J. E. Jacobsen, *J. Coord. Chem.* **10** (1980) 125.

77. I. I. Chernyaev, L. A. Nazarova, and A. S. Mironova, *Zh. Neorg. Khim.* **6** (1961) 2444.

78. K. A. Khokhriakov, E. I. Grigorov, N. S. Panina, and V. N. Yakovlev, *Koord. Khim.* **13** (1987) 237.

79. Yu. N. Kukushkin, E. I. Grigorov, K. A. Khokhriakov, I. G. Maslennikov, and E. S. Postnikova, *Koord. Khim.* **10** (1984) 543.

80. G. R. Lee and N. J. Cooper, *Organometallics* **4** (1985) 794.

81. W. A. Nugent and R. L. Harlow, *J. Chem. Soc., Chem. Commun.* (1979) 342.

82. E. N. Yourchenko, A. D. Troitskaya, L. S. Gracheva, and L. Ya. Alt, *Izv. AN SSSR, Ser. Khim.* (1975) 2153 and references therein.

83. R. F. Alex and R. K. Pomeroy, *Organometallics* **1** (1982) 453.

84. S. Aygen and R. van Eldik, *Organometallics* **6** (1987) 1080.

85. W. Malisch and R. Janta, *Angew. Chem. Int. Ed. Engl.* **17** (1978) 221.

Chapter 12. SOLID STATE THERMAL SYNTHESIS OF COORDINATION COMPOUNDS

Studies of solid state thermal transformations of metal complexes have led to the development of new and often simple methods for the synthesis of some classes of coordination compounds. Reactions that proceed in the absence of solvent frequently lead to formation of products that differ from those formed in solution. In solution there exist possibilities for solvent molecules, or their deprotonated forms, to function as ligands or reagents. This possibility is excluded when a reaction occurs in the solid state.

Thermal analysis is widely used in the study of coordination compounds. Routine instruments record sample mass, rate of mass change, and heat flow as the sample temperature is increased, usually in a linear fashion. There are thermal analysis methods that allow the study of gas evolution, changes in magnetic properties, changes in electric conductivity, and changes in other properties of the sample. Thermal analysis provides information on the temperature range over which an intermediate is thermally stable. By stopping the temperature increase in this range, it may be possible to isolate the products of a thermal transformation in high yield. Indeed, there are cases where compounds undergo clean thermal transformations in quantitative yields. The atmospheric environment of the sample is crucially important in thermal transformations. Two classes of atmosphere are generally recognized: oxidative (air or oxygen) and inert (nitrogen or argon). Thermal transformations often proceed differently under different classes of atmospheres and attention must be paid to this variable in duplicating experimental conditions.

Isomerization of coordination compounds is one important class of thermal transformation. In the case of inert complexes such isomerization may be irreversible and so allow the preparation of specific isomers. Some transformations of this type (*e.g.* isomerization of square-planar complexes) were introduced in Chapter 1 and additional examples will be discussed in Section 12.10. In this Chapter the use of solid state thermal transformations in synthetic coordination chemistry is discussed. Work on the use of thermal methods as analytical tools in coordination chemistry can be found elsewhere.[1,2]

12.1 The Anderson rearrangement

Dehydrohalogenation reactions of complex salts that contain metal halide anions and onium cations were first discovered by Anderson in 1855.[3,4] Almost immediately after the discovery of pyridine, Anderson prepared pyridinium hexachloroplatinate(IV), $[pyH]_2[PtCl_6]$, and observed its conversion in aqueous solution to $[PtCl_4(py)_2]$.[3] Werner explained this transformation in terms of a two step process, Scheme 12.1:

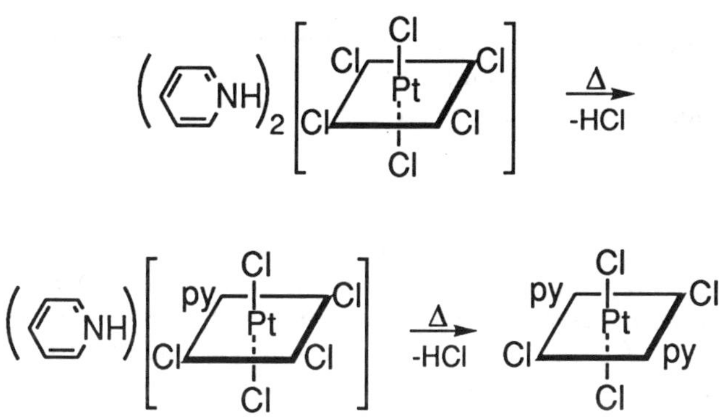

Scheme 12.1

Currently, a host of related solid state transformations are known.[4] The Anderson rearrangement is, in general, a substitution of inner sphere ligands by outer sphere ligands. It has been demonstrated that heating a wide range of salts, $[amineH]_2[MX_4]$ (M = Pd, Pt; X = Cl, Br), in the solid state results in the formation of *cis*-$[MX_2(amine)_2]$[5] (see Peyrone's rule, Chapter 1), Eq. 12.1:

$$[amineH]_2[MX_4] \rightarrow cis\text{-}[MX_2(amine)_2] + 2HX \quad (12.1)$$

where M = Pt, Pd; X = Cl or Br

Thus, such substitution reactions proceed both in the solid state and in solution in the manner predicted by the *trans*-effect. Peyrone's rule (Chapter 1) sometimes appears to be violated since isolated products may have undergone geometric isomerization subsequent to the initial thermal transformation.

At the onset temperature of the Anderson rearrangement, measured by thermogravimetry, the reaction rate is already significant. Consequently, the onset temperatures of rearrangements can be used to indirectly estimate the relative rates of similar processes.

The relationship between the onset temperatures of the Anderson rearrangement and the pK_a of the amines was first shown for $[amineH]_2[PtCl_4]$ complexes.[5] From reported experimental data it was shown that the compounds could be divided into two groups: compounds with an onset temperature for the rearrangement of 180-200 °C, which were found to contain amines with $pK_a > 9.0$, and compounds with an onset temperature for the rearrangement of 130-150 °C, which were found to contain amines with $pK_a < 6.0$.

The Anderson rearrangement reactions of $[amineH]_2[PtCl_4(OH)_2]$ (amine = ammonia, methylamine, pyridine, 3-methylpyridine, piperidine, quinoline) complexes,[6] have been studied, Eq. 12.2:

$$[amineH]_2[PtCl_4(OH)_2] \rightarrow [PtCl_4(amine)_2] + 2H_2O \quad (12.2)$$

In these cases the proton is transferred not to a chloride ligand, to form HCl, but to the hydroxyl ligand, with elimination of water. Rearrangement of these complexes proceeds at much lower temperatures than those found for $[amineH]_2[PtCl_6]$.[6,7] In general, the Anderson rearrangement occurs through proton transfer to the ligand of highest basicity. An illustrative example is the solid state thermal transformation of *cis*-$[amineH]_2[PtX_2(NO_2)_2]$ (amine = substituted benzimidazoles; X = Cl, Br), Eq. 12.3:

$$[amineH]_2[PtX_2(NO_2)_2] \rightarrow [PtX_2(amine)_2] + 2HNO_2 \quad (12.3)$$

In these reactions proton transfer occurs not to the halide ligand but to the more basic nitro group. A study of the thermal transformations of *trans*-$[amineH]_2[PtBr_2(NO_2)_2]$ (amine = benzimidazole) has shown that once one of the nitro groups has been substituted, to form $[amineH][PtBr_2(NO_2)(amine)]$, the second nitro group is more strongly bound, due to the relatively weak *trans*-effect of benzimidazole. Thus, although the basicity of the nitro group is higher than that of bromide, the bromide ligand undergoes substitution and *cis*-$[PtBr(NO_2)(amine)_2]$ is formed.[8]

The dehydrohalogenation and dimerization of *trans*-$[PtCl_2L(DMSO)]$ (L = 3-methyl-4-acetyl-5-aminopyrazole) has been studied,[9] Scheme 12.2. The product has been characterized by X-ray crystallography. The dehydrohalogenation is, in fact, an unusual variant of the Anderson rearrangement. Thus, it is a tertiary, inner-sphere nitrogen, and not a quaternary, outer-sphere nitrogen, that is subject to deprotonation and Pt–N bond formation.

The Anderson rearrangement is not restricted to platinum and palladium complexes, although they have provided most of the experimental data on this class of reactions. For example, $[AuCl_3(amine)]$ complexes (amine = quinoline, benzimidazole, triazole), were prepared from the salt, $[amineH]_2[AuCl_4]$, by heating in the solid state.[10] Similarly, $[phenH_2][RhCl_2(CO)_2]_2$ is transformed into the bridged complex, $[RhCl(CO)_2(\mu\text{-phen})RhCl(CO)_2]$, in almost quantitative yield through heating in the solid state.[11] The rhodium(III) salt, $[NH_4]_3[RhCl_6]$, has been shown[12] to produce $[NH_4]_2[RhCl_5(NH_3)]$, with evolution of HCl, within the temperature range 320-365 °C. The Anderson rearrangement of the rhenium complex, $[PPh_3H]_2[ReCl_6]$, is a simple and convenient synthesis of *trans*-$[ReCl_4(PPh_3)_2]$,[13] while heating $[bipyH_2][ReCl_6]$ in the solid state at 270-300 °C under nitrogen produces [ReCl4(bipy)] in quantitative yield.[14]

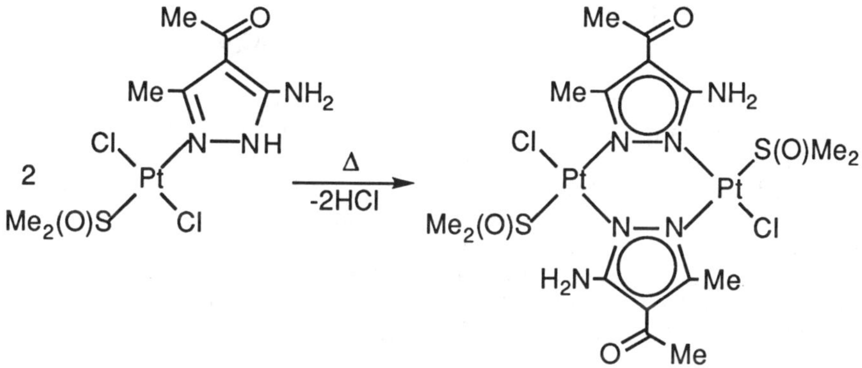

Scheme 12.2

The thermal transformation of the rhenium(IV) complex, $[PPh_3R']_2[ReCl_6]$ (R' = $MeC(O)CH_2CMe_2$), has been investigated.[15] The starting material was prepared by heating $[PPh_3H]_2[ReCl_6]$ in acetone in the presence of hydrochloric acid. The first stage of the thermal transformation was found to be an endothermic process that occurred at 190 °C and that involved liberation of two equivalents of alkyl halide, R'Cl, accompanied by coordination of triphenylphosphine to rhenium, Eq. 12.4:

$$[PPh_3R']_2[ReCl_6] \rightarrow [ReCl_4(PPh_3)_2] + 2R'Cl \qquad (12.4)$$

The product, $[ReCl_4(PPh_3)_2]$, was found to be stable on further heating to a temperature of 245 °C at which point oxidation of the PPh_3 ligands (see Section 12.9), with formation of $[ReCl_4(OPPh_3)_2]$, occurred, Eq. 12.5:

$$[ReCl_4(PPh_3)_2] + O_2 \rightarrow [ReCl_4(OPPh_3)_2] \qquad (12.5)$$

The complex, $[ReCl_4(OPPh_3)_2]$, was found to be thermally unstable above 245 °C and slowly decomposed.

Some unusual thermal reactions have been observed that can be considered formally as *anti*-Anderson rearrangements. For example, the complexes, $[Cr(NH_2Me)_6]Br[C_2B_9H_{12}]_2$, $[CrBr(NH_2Me)_5][C_2B_9H_{12}]_2$, and $[CrCl(NH_2Me)_5][C_2B_9H_{12}]_2$, were shown[16] to undergo proton transfer from the 3-position of the $C_2B_9H_{12}^{2-}$ anion to the coordinated amine with formation of the free onium cation. The vacant coordination site is occupied by the $C_2B_9H_{11}^{2-}$ anion, which is π-bonded to the metal center, Eq. 12.6:

$$[CrBr(NH_2Me)_5][C_2B_9H_{12}]_2 \rightarrow [MeNH_3]_2[Cr(NH_2Me)_3Br(C_2B_9H_{11})_2] \qquad (12.6)$$

Several general conclusions can be drawn from studies of the Anderson rearrangement: (*i*) the solid state rearrangement is general for coordination compounds of a variety of metals, (*ii*) for complexes of a given metal, with ligands of the same general type, it has been observed that the onset temperature of the Anderson rearrangement decreases as the basicity of the ligand entering from the onium cation decreases, (*iii*) if the complex contains several different ligands, the proton is transferred from the onium cation to the ligand of highest basicity. In general, the basicities of free and coordinated ligands follow the same order.

12.2 Dehydration and deaquation

Many coordination compounds are isolated as solids through crystallization from aqueous or aqueous-organic solutions and contain water either within the crystal lattice or as a coordinated ligand. Water in a crystal lattice is called water of crystallization and its removal is referred to as dehydration. The removal of coordinated water from a coordination compound is known as deaquation.

The removal of water of crystallization is typically an endothermic process. The energy required for dehydration is consumed by the van der Waals forces and by the evaporation of liberated water. The evaporation temperature is often within the range of 60-100 °C and amounts to about 40 kJ mol^{-1}. A typical DTA experiment reveals the dehydration process as a strong endothermic peak.

The number of molecules of water of crystallization in a lattice can vary greatly. For example, $[Co(en)_3]_2[Co(DH)_2SO_3(NCS)_3]\cdot 22H_2O$ (DH_2 = dimethyl glyoxime), crystallizes with 22 molecules of water per formula unit and $[Co(en)_3]_2[Co(DH)_2SO_3(NO_2)]_3\cdot 19H_2O$ with 19 molecules per formula unit. Dehydration of compounds containing more than one water molecule per formula unit may be a stepwise process and the dehydration of $CuSO_4\cdot 5H_2O$ is a classic example. Water is removed in three steps: the first and second steps each involve loss of two equivalents of water from the hydrate. The third step corresponds to loss of the last equivalent of H_2O.[17]

A stepwise loss of water of crystallization might be interpreted in terms of some inequivalence between water molecules in the lattice of a crystalline complex. However, such an interpretation of thermogravimetric data cannot be made unambiguously because a rearrangement of the crystal lattice during dehydration may affect the manner in which any remaining water molecules are lost.

The removal of coordinated water molecules — deaquation — is more complicated than simple dehydration and involves metal-ligand bond breaking and generation of a vacant coordination site. Coordination of an anion, for example, Eqs. 12.7-12.8:[18]

$$[Cu(DAF)_2(H_2O)_2][NO_3]_2 \rightarrow [Cu(NO_3)_2(DAF)_2] + 2H_2O \qquad (12.7)$$

where DAF = diaminofurazan

$$trans\text{-}[Ni(amine)_2(H_2O)_2]X_2 \rightarrow trans\text{-}[NiX_2(amine)_2] + 2H_2O \qquad (12.8)$$

where amine = propane-1,3-diamine, X = I;[19] amine = *l*-1,2-cyclohexanediamine, X = Cl, Br, NO_3 .[20]

results in ring closure through coordination of a remote donor group (*c.f.* Chapter 6), or formation of bridged complexes. Accordingly, deaquation involves both the breaking and forming of metal-ligand bonds. Energetically, these two processes tend to cancel each other because one is endothermic and the other is exothermic. However, the evaporation of water is endothermic and so, overall, deaquation is typically an endothermic process. This generality is, of course, only valid when the thermolysis experiment is performed in an open container, such that water can evaporate. It has been shown[21] that when deaquation is performed in a sealed capillary tube, from which evaporation of water cannot take place, it is possible to recognize effects due to $M–OH_2$ bond breaking and M–X bond formation. Thus, it was found[21] that deaquation of $[Co(NH_3)_5(H_2O)][X]_3$ to produce $[CoX(NH_3)_5][X]_2$ (X = Cl, Br, I), is endothermic in an open system but exothermic in a closed system.

Deaquation reactions of complexes such as $[Co(NH_3)_5(H_2O)][Co(CN)_6]$[22] are of interest because, in this case, the group that will occupy the vacant site created by loss of water from the cation is a coordinated cyanide ligand of the anion. Thus, formation of a bridged complex results, Scheme 12.3:

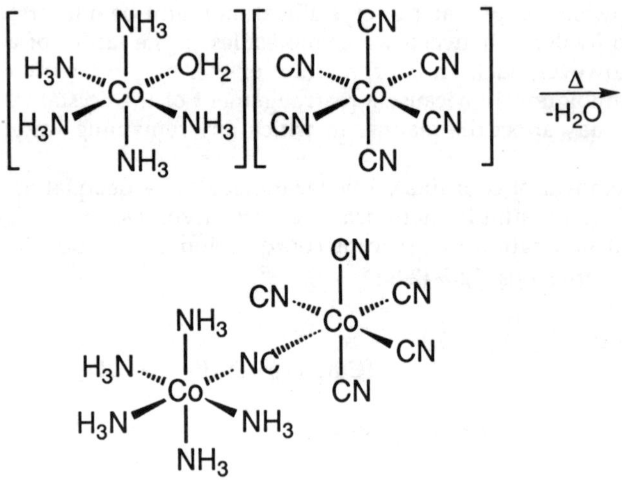

Scheme 12.3

There are many reports of solid state deaquation reactions that result in the formation of bridged complexes. Such complexes are formed also by the removal of coordinated water from neutral and/or anionic mononuclear complexes, for example, Eqs. 12.9-12.10:

$$2[CoCl_3(NH_3)_2(H_2O)] \rightarrow 2H_2O + [Cl_2(NH_3)_2Co(\mu\text{-}Cl)_2CoCl_2(NH_3)_2] \quad (12.9)$$

$$2K[CoCl_4(NH_3)(H_2O)] \rightarrow 2H_2O + K_2[Cl_3(NH_3)Co(\mu\text{-}Cl)_2CoCl_3(NH_3)] \quad (12.10)$$

Reactions of this general type are also characteristic of aqua-hydroxo-complexes. For example, $[Co(OH)(NH_3)_4(H_2O)]Br_2$ loses water in an endothermic process at 120 °C,[23] Eq. 12.11:

$$2[Co(OH)(NH_3)_4(H_2O)]Br_2 \rightarrow 2H_2O + [(NH_3)_4Co(\mu\text{-}OH)_2Co(NH_3)_4]Br_4 \quad (12.11)$$

The loss of a water molecule from a coordinated ligand has been described by Olczak-Kobza.[24] It was found that the solid state thermal decomposition of the bis(salicyloaldoximate) complex, $[Pb(HL)_2]$ (H_2L = *o*-(OH)(CH=NOH)C_6H_4), occurs in two steps, with loss of a ligand followed by dimerization and extrusion of one equivalent of water, Eqs. 12.12-12.13:

$$[Pb(HL)_2] \rightarrow [PbL] + H_2L \qquad (12.12)$$

$$2[PbL] \rightarrow [O(PbO\text{-}C_6H_4\text{-}C{\equiv}N)_2] + H_2O \qquad (12.13)$$

As a result, the nitrile dimer, $[O(PbO\text{-}C_6H_4\text{-}C{\equiv}N)_2]$, is formed.[24]

12.3 Syntheses involving abstraction of amine and other neutral ligands

Solid state deaquation reactions have been studied in detail because a large number of aqua-complexes are known. Metal amine complexes are also common and their thermal transformations have been discussed by Wendlandt and Smith.[25] It is generally found that there is little difference between aqua- and amine complexes with regard to routine thermal transformations. Indeed, certain complexes can be prepared more simply and more reliably by solid state thermal transformations of amino complexes than by routine solution procedures. For example, heating $[PtBr(NH_3)_5]Br_3$ within the temperature range 60-80 °C results in the formation of *trans*-$[PtBr_2(NH_3)_4]Br_2$, Scheme 12.4:

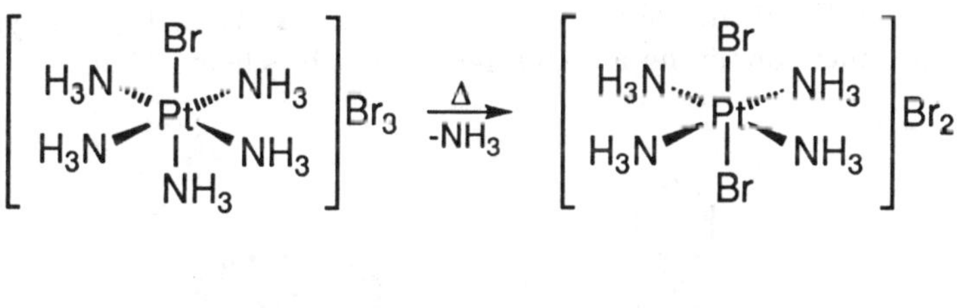

Scheme 12.4

The synthesis of $[PtBr_2(NH_3)_4]Br_2$ by oxidation with bromine in aqueous solution is seriously handicapped by the formation of a considerable amount of $[PtBr(OH)(NH_3)_4]$ as a by-product.

In a modification of the classical Reiset's synthesis,[26] solid $[Pt(NH_3)_4]Cl_2$ was heated at 190-195 °C *in vacuo* and found to produce *trans*-$[PtCl_2(NH_3)_2]$ quantitatively within a reasonable time (*ca.* 5 h) and without any detectable decomposition to undesirable by-products.

The reaction of $[Pt(py)_4]Cl_2$ with hydrochloric acid proceeds under reflux conditions. The resulting precipitate of *trans*-$[PtCl_2(py)_2]$ is periodically removed by filtration and this extends the time required for the synthesis up to an hour or more. However, on heating solid $[Pt(py)_4]Cl_2{\cdot}3H_2O$ at 140 °C, the complex *trans*-[$PtCl_2(py)_2$] is formed in a shorter time and in quantitative yield,[27] Scheme 12.5:

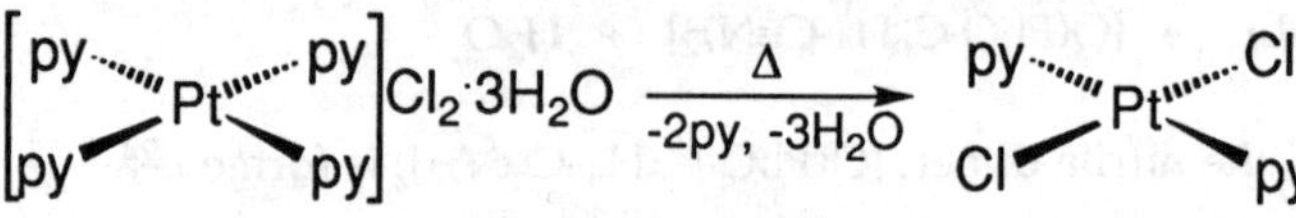

Scheme 12.5

Related *trans*-[$PtCl_2L_2$] complexes, *e.g.* where L = *N*-methylimidazole or thiazole,[28] have been prepared in a similar manner. The structure of the thiazole derivative has been confirmed by X-ray crystallography.

Syntheses involving amine displacement are not restricted to complexes of platinum. For example, heating the rhodium(I) complexes, [$Rh(CO)(amine)(PPh_3)_2$][NO_3] (amine = py, 4-Me-py), yields the nitrato derivatives, [$Rh(NO_3)(CO)(PPh_3)_2$].[29] It has been reported[30] that when the complexes, [$Co(DH)_2(amine)_2$][$InCl_4(amine)_2$] (DH_2 = dimethylglyoxime, amine = py, α-, β-, and γ-picoline), are heated in the solid state to 130-140 °C, they are converted into [$Co(DH)_2(amine)_2$][$InCl_4$] with liberation of free amine. Vacuum pyrolysis at 10^{-5} torr and 190-210 °C of the *bis*(pyridine)ruthenium porphyrin complex, [$Ru(OEP)(py)_2$], yields a binuclear product containing a metal-metal bond,[31] Scheme 12.6:

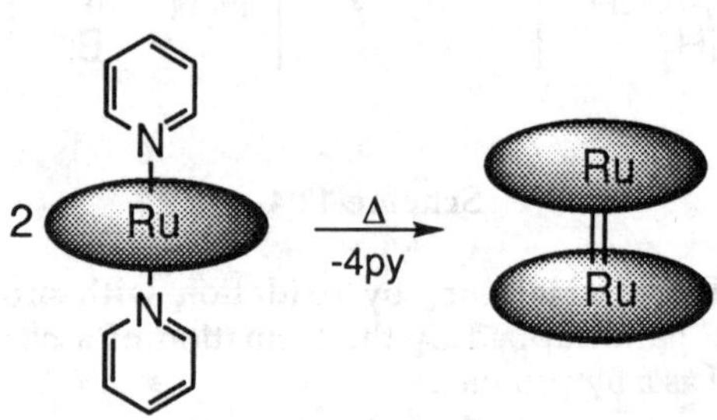

Scheme 12.6

Thermal transformations of complexes that involve loss of neutral ligands (*e.g.* H_2O, amines, *etc.*) are common. In the case of ligands that, as free molecules, are volatile, the individual steps in such transformations are usually quite distinct.[32] However, when less volatile molecules are involved, the thermal conversion becomes more complex and is often accompanied by thermolysis of coordinated ligands or ligands that were displaced but remained in a sample due to their low volatility.

The ease of amine displacement from cationic complexes depends upon the nature of the anionic counterion. It has been shown[33] that the solid state thermolysis of cationic platinum(II) complexes is sensitive to the nature of the counterion, Eq. 12.14:

$$[Pt(amine)(L_3)]X_2 \rightarrow [PtX(L_3)]X + amine \qquad (12.14)$$

where amine = aliphatic amine, pyridine or its derivatives;
L_3 = diethylenetriamine; X = Cl, Br, I, ClO_4

Thus, the solid state transformations begin at temperatures that reflect the nature of the counterions: $I^- < Br^- < Cl^-$; with perchlorate not being able to substitute for the amine ligand.

Acetonitrile, which usually functions as a weakly bound sp-*N*-donor ligand for soft metal ions, can be displaced easily upon heating in the solid-state. For example, the complexes, $Q[PtX_3(MeCN)]$ (Q = Et_4N, $(n\text{-}Bu)_4N$, Ph_3PCH_2Ph, Ph_4As, K; X = Cl, Br), undergo conversion in the temperature range 110-160 °C to the dimers, $Q_2[Pt_2(\mu\text{-}Cl)_2Cl_4]$, in almost quantitative yields.[34] The reaction where X = Cl is shown in Scheme 12.7:

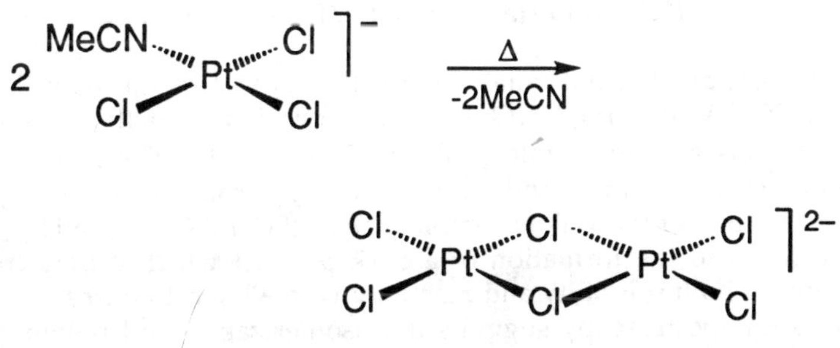

Scheme 12.7

Similarly, *cis*-$[PtX_2(MeCN)(R_2SO)]$ (X = Cl, Br; R = Me, Et), readily eliminates the acetonitrile ligand at 120-150 °C to form $[Pt_2X_2(\mu\text{-}X)_2(R_2SO)_2]$.[35] Thermolysis of *cis*-$[PtCl_2(MeCN)(Et_2SO)]$ leads to formation of the dimeric complex, $[Pt_2Cl_2(\mu\text{-}Cl)_2(Et_2SO)_2]$, which has been characterized by X-ray crystallography, Figure 12.1:

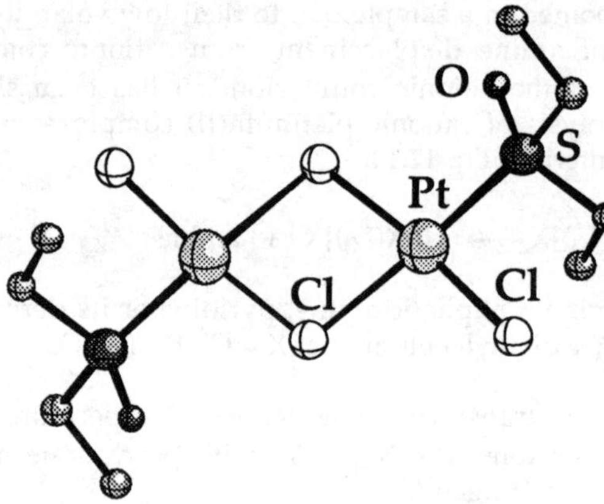

Figure 12.1

Among syntheses that involve solid-state thermal displacement of neutral ligands, the elimination of dialkyl sulfoxides,[36,37] ethylene,[38] dinitrogen[39] and trimethylphosphine[40] also have proved to be effective.

12.4 Solid state thermal polymerization

Differential thermal analysis of the mixed-metal complex, *cis*-$[Pt(SCN)(SCNAg)(NH_3)_2][NO_3]$, which contains formally unsaturated silver(I), reveals an exothermic peak at 134-144 °C.[41] The process occurs without any change in mass and with an enthalpy change of -38 ± 3 kJ mol^{-1}. Controlled heating of the yellow complex, *cis*-$[Pt(SCN)(SCNAg)(NH_3)_2][NO_3]$, at 134 °C leads to the formation of a dark product which differs from the yellow form in both solubility and reactivity with KI and thiourea.

Infrared spectroscopy suggests that isomerization and polymerization occur on heating *cis*-$[Pt(SCN)(SCNAg)(NH_3)_2][NO_3]$ in the solid state. A structure involving thiocyanate bridges between alternating platinum and silver ions has been proposed, Figure 12.2.

Cheng and Foxman[42] have studied the polymerization of the square-planar nickel(II) complex, $[NiBr_2\{P(C_2H_4CN)_3\}_2]$. This compound has pendant nitrile groups and the nickel ion has the potential to increase its coordination number by two. Accordingly, a thermal polymerization occurs in the solid state with formation of a blue polymeric product, Scheme 12.8.

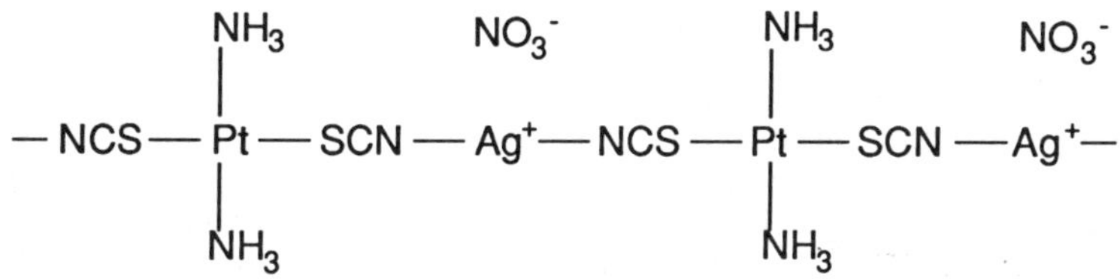

Figure 12.2

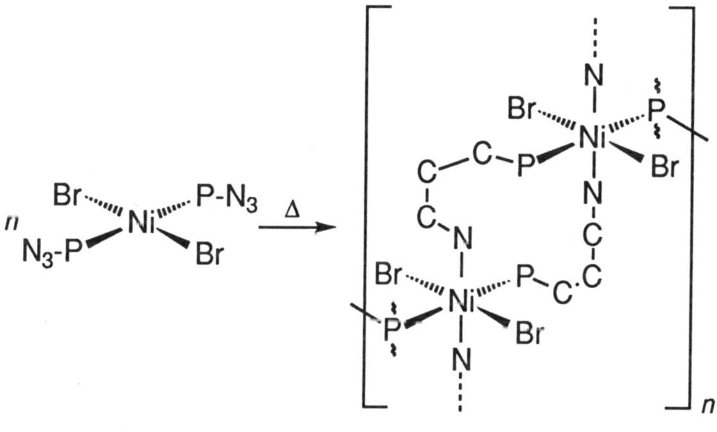

Scheme 12.8

When an acetone solution of the monomeric complex, $[NiBr_2\{P(C_2H_4CN)_3\}_2]$, was allowed to polymerize at various temperatures, the blue polymeric compound was always contaminated with an unidentified green substance. However, when the dry solid was allowed to polymerize at temperatures below 130 °C only the blue material was formed.[42]

These examples illustrate the conditions typically necessary for polymerization, *i. e.* coordinative unsaturation of the metal center along with the potential for bidentate or polydentate coordination of the ligands.

12.5 Syntheses that involve deprotonation of a coordinated ligand followed by ring-closure

Ring-closure reactions that involve deprotonation of coordinated ligands are well known. However, such reactions have been studied largely for solutions and data on solid state ring-closures of this type are fairly scarce. In metal halide complexes where a multidentate ligand coordinates to a metal center with only partial denticity, an increase in the denticity can often be achieved through solid state thermal dehydrohalogenation, resulting in ring-closure.

The thermal ring-closure reactions of the cobalt(II) complexes, Na[CoX(Hedta)] and $[CoX(Hgly)en_2]X_2$ (X = Cl, Br; Hedta = pentadentate ethylenediaminehydrogentetracetate; Hgly = monodentate glycine) have been reported.[43] Solid state thermolysis resulted in the loss of HX and ring-closure, with a corresponding increase in the denticity of the ligands, Scheme 12.9:

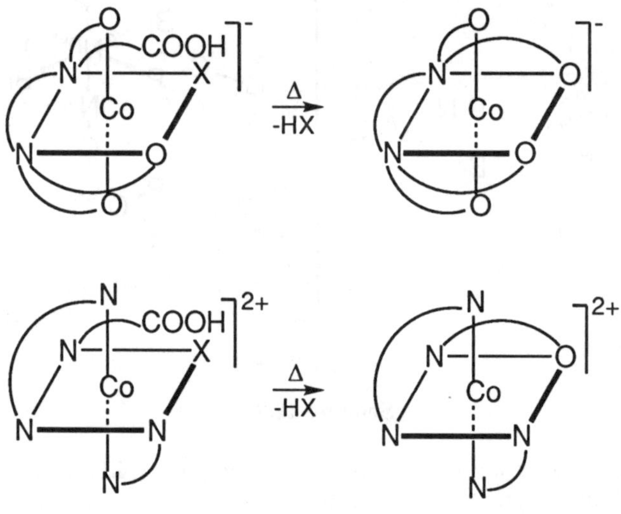

Scheme 12.9

The palladium(II) and platinum(II) complexes of oxyquinoline, $[MCl_2(OxH)_2]$, where the oxyquinoline ligands (OxH) are coordinated through nitrogen, are transformed into chelate complexes on heating in the solid state,[44] Eq. 12.15:

$$[MCl_2(OxH)_2] \rightarrow [M(Ox)_2] + 2HCl \tag{12.15}$$

In a similar way, halide complexes of Pd(II) and Pt(II) with appropriate ligands are subject to the Anderson rearrangement and ring-closure in a single step,[45] Scheme 12.10:

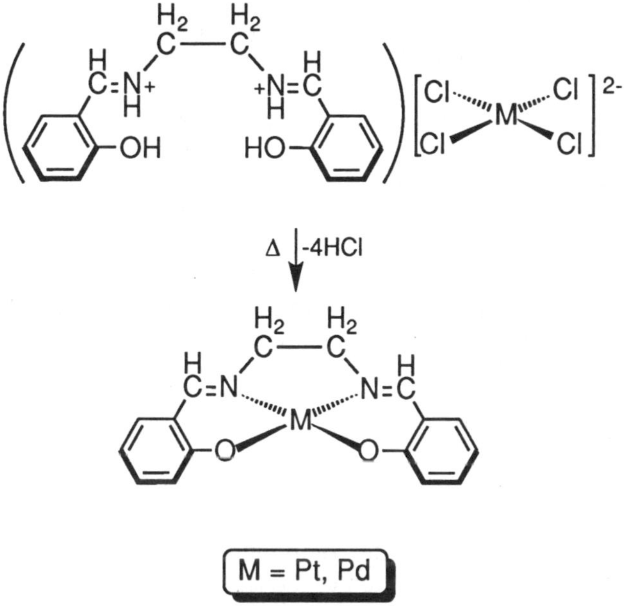

Scheme 12.10

Thermal analysis of *trans*-$[PtBr_2(NH_2CH_2COOH)_2]$ shows that, at a temperature of 250 °C, the complex eliminates two equivalents of HBr in an endothermic process, with the formation of the glycinato-complex, $[Pt(NH_2CH_2COO)_2]$.[46] A gradual evolution of HX was noted during thermolysis of the related complexes, *trans*-$[PtX_2(NH_2CH_2COOH)_2]$ (X = Cl, I) and *cis*-$[PtX(NH_2CH_2COO)(NH_2CH_2COOH)]$ (X = Cl, Br). The complete decomposition of these complexes appears to be preceded by dehydrohalogenation and ring-closure.[46]

Ring-closures may result not only from dehydrohalogenation but also through loss of water from two *cis*-ligands. On heating the rhodium(III) complex, $Na[Rh(OH)_2(HL)_2]\cdot H_2O$ (HL = deprotonated bidentate iminodiacetic acid), the water of crystallization is initially lost. Subsequently, there is proton transfer from a free carboxyl group to the coordinated hydroxyl. The resultant water molecule is lost and ring-closure results.[47]

12.6 Thermal deprotonation of coordinated ammine ligands

In an early study of the solid state thermal deprotonation of coordinated ammine ligands, Tsin-Shen and Tronev[48] showed that $ReCl_3{\cdot}4NH_3$ (a compound of uncertain structure) was transformed into a diamido-complex by loss of HCl, Eq. 12.16:

$$ReCl_3{\cdot}4NH_3 \rightarrow [ReCl(NH_2)_2(NH_3)_2] + 2HCl \qquad (12.16)$$

The same diamido-complex was prepared by treatment of $ReCl_3{\cdot}4NH_3$ with liquid ammonia. Further heating of $[ReCl(NH_2)_2(NH_3)_2]$ resulted in the formation of the imido-complex, $[Re_2(NH)Cl]_n$, and, ultimately, rhenium nitride, $[Re_2N]_n$.[48]

In a similar fashion,[49] heating $[ReCl_4(NH_3)]$ in the solid state results in a conversion of a coordinated ammine ligand into an amide and, then, into a nitride, Eqs. 12.17 - 12.18:

$$[ReCl_4(NH_3)] \rightarrow [ReCl_3(NH_2)] + HCl \ (250\ °C) \qquad (12.17)$$

$$[ReCl_3(NH_2)] \rightarrow [ReClN] + 2HCl \ (400\ °C) \qquad (12.18)$$

A related rhenium nitride compound, [ReFN], has been produced through the thermal decomposition of $[NH_4]_2[ReF_6]$ at 300 °C under argon,[49] Eq. 12.19:

$$[NH_4]_2[ReF_6] \rightarrow ReFN + NH_4F + 4HF \ (300\ °C) \qquad (12.19)$$

The first stage in this process is likely to be an Anderson rearrangement, followed by deprotonation of the coordinated ammine. The complex $[NH_4]_2[TcF_6]$ has been shown[50] to be subject to a similar transformation.

A technique for the preparation of aluminum nitride and gallium nitride films has been developed based upon the decomposition of ammine complexes.[51] The nitrides were prepared through thermal decomposition of $AlCl_3{\cdot}NH_3$ and $GaCl_3{\cdot}NH_3$ or $GaI_3{\cdot}NH_3$, for example, Eq. 12.20:

$$AlCl_3{\cdot}NH_3 \rightarrow AlN + 3HCl \qquad (12.20)$$

To prepare aluminum nitride, AlN, from $AlCl_3{\cdot}NH_3$, the thermal decomposition is performed at 930-1180 °C. The complex, $[Al(NH_3)_6]Br_3$, is transformed into AlN at a lower temperature (slightly above 250 °C).[52]

Deprotonation of a coordinated ammine ligand with formation of a bridged amido, (μ-NH_2), complex was described by Jepsen and House.[22] Thus, $[Co(NH_3)_5(H_2O)][Co(CN)_6]$ was found to lose water at 130-160 °C with formation of $[(NH_3)_5Co(\mu\text{-}CN)Co(CN)_5]$. Loss of HCN at 250 °C produced $[(NH_3)_4Co(\mu\text{-}NH_2)(\mu\text{-}CN)Co(CN)_4]$.

12.7 Solid state cyclometallation

Reactions of metal complexes whose ligands (typically *O*-, *N*-, *P*-, *S*-, and *As*- donors) contain hydrocarbon groups that are subject to metallation with the formation of metal-carbon σ-bonds are called cyclometallation reactions,[53,54] Scheme 12.11:

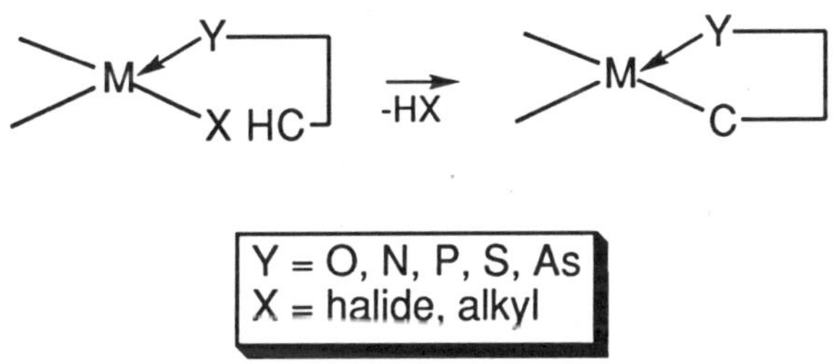

Scheme 12.11

Orthometallation refers specifically to the substitution of *ortho*-hydrogens of an aromatic ring of a ligand by a metal atom and is a specific type of cyclometallation, *e.g.*, Scheme 12.12:

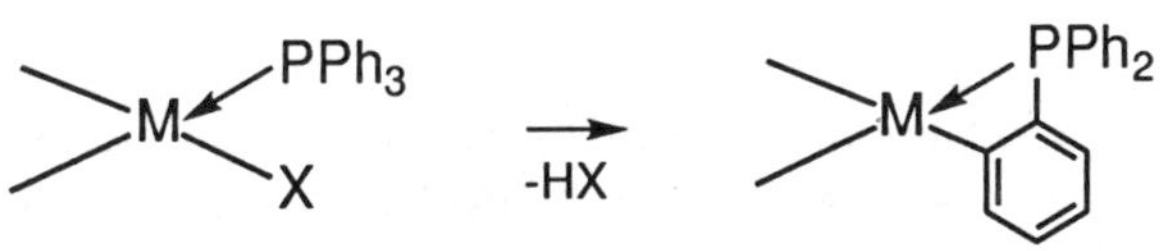

Scheme 12.12

Cyclometallation reactions that occur in solution are very well documented and such reactions often involve dehydrohalogenation. A solid state orthometallation reaction that occurs *via* dehydrohalogenation[55] has been reported, Scheme 12.13:

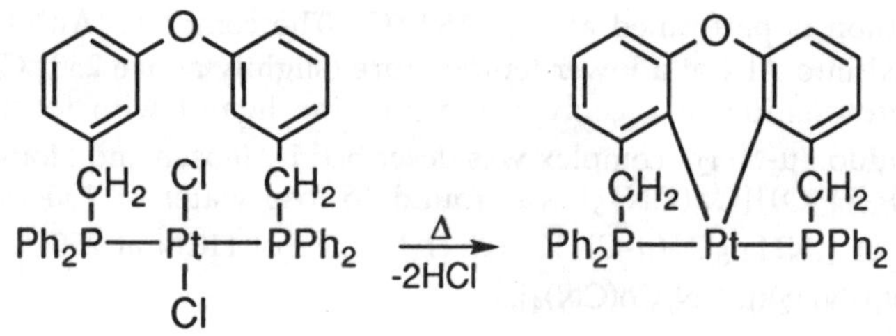

Scheme 12.13

The reaction involves orthometallation of two aromatic rings with formation of two five-membered metallocycles. DTA/TG and mass-spectrometric analyses were performed to study the thermal process under a nitrogen atmosphere. In general, dehydrohalogenation reactions are not well suited to study by thermal analysis methods because evolution of corrosive hydrogen halides is deleterious to the equipment.

Rauchfuss[56] has reported the rapid and quantitative cyclometallation of the platinum(II) complex of *o*-diphenylphosphinobenzaldehyde, (PCHO), *cis*-$[PtCl_2(PCHO)_2]$, upon vacuum thermolysis at 250 °C , Scheme 12.14:

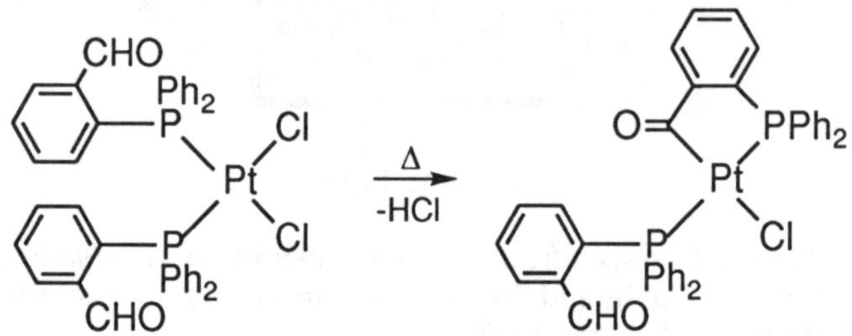

Scheme 12.14

Two distinct pathways for this reaction have been suggested: oxidative addition of the C–H bond followed by reductive elimination of HCl, and direct electrophilic attack on the formyl group by platinum(II) with displacement of a proton.

For complexes with *P*-donor ligands, metallation often occurs in solution with elimination of an alkane and such reactions also occur in the solid state. For example, the solid state cyclometallation of $[PtMe_2\{PMe_2(1\text{-naphthyl})\}_2]$ has been investigated.[57] When heated to 200 °C, the complex evolved methane (monitored by IR spectroscopy) and gave a product cyclometallated in the 8 (*peri*) position, containing a stable five-membered ring (metallation in the position *ortho-* with respect to phosphorus would lead to formation of a strained three-membered ring). It is noteworthy that ring size affects the stability of cyclometallated products and, hence, provides a possible route to control the direction of cyclometallation, if several pathways are possible.

Ring-closure requires a coordination site at the metal center and, therefore, coordinatively unsaturated complexes are particularly prone to metallation. The complexes, $[IrClXMe(PPh_3)_2]$ (X = Cl, Br, NCO), are subject to orthometallation[58] of coordinated PPh_3 on heating *in vacuo* at 180 °C with evolution of methane and formation of the dimeric complexes, $[Ir_2\{C_6H_4PPh_2\}_2Cl_2X_2(PPh_3)_2]$.

The 2-alkoxyphosphine complexes, *trans*-$[PtCl_2L_2]$ (L = $PPh_2\{C_6H_4OMe\text{-}2\}$, $PPh_2\{C_6H_4OEt\text{-}2\}$, $PBu_2\{C_6H_4OMe\text{-}2\}$), have been prepared.[59] These compounds undergo *O*-metallation upon heating in the solid state. A methyl or ethyl group is lost from the phosphine ligand and chloride is lost from the metal, resulting in formation of MeCl or EtCl and generation of the *bis*-chelate complexes containing five-membered rings.

Cyclometallation of the very bulky ligands, PBu_2R and PPh_2R (R = 2,3- or 2,6-dimethoxyphenyl), in *trans*-$[PtCl_2L_2]$ complexes proceeds rapidly[60] with formation of *cis*-$[Pt\{OC_6H_3(OMe)PPh_2\}_2]$ and *trans*-$[Pt\{OC_6H_3(OMe)PBu_2\}_2]$, each of which contains five-membered rings.

The majority of *N*- and *O*-donor ligands form five-membered rings in cyclometallation reactions. In contrast, *S*- and *P*-containing ligands show a tendency to form three- and four-membered rings, especially when formation of a five-membered ring is precluded.

Solid state cyclometallation that results in the formation of two different chelate rings has been demonstrated by Canty and Minchin,[61] Scheme 12.15:

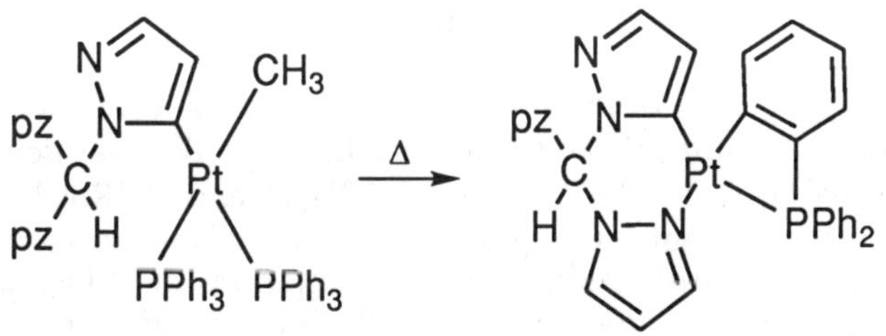

Scheme 12.15

The *bis*-chelate complex was prepared by slow heating *in vacuo* at 180 °C and contains metallated nitrogen and phosphorus donor ligands in six- and four-membered rings, respectively.

12.8 Thermal decarboxylation and disproportionation

Pyridinecarboxylic acids, *e.g.* 2-$HOOCC_5H_4N$, are known to eliminate CO_2 on heating in the solid state and, accordingly, the possibility of thermal decarboxylation of coordinated pyridinecarboxylic acids has been investigated. Thermal analysis of platinum(II) complexes containing 3-pyridinecarboxylic acid, $K[PtCl_3(NC_5H_4COOH)]$ and *cis*-$[PtCl_2(NC_5H_4COOH)_2]$,[62] did not provide any evidence of decarboxylation of the coordinated 3-pyridinecarboxylic acid. However, a disproportionation was revealed, Eq. 12.21:

$$2K[PtCl_3(NC_5H_4COOH)] \rightarrow cis\text{-}[PtCl_2(NC_5H_4COOH)_2] + K_2[PtCl_4] \quad (12.21)$$

Both dehydration and disproportionation occur upon heating $[Ni(acac)(tmen)(H_2O)_2]Br$ (tmen = *N,N,N',N'*-tetramethylethylenediamine) with formation of $[Ni(acac)_2(tmen)]$ and $[NiBr_2(tmen)]$.[63] It was suggested that the thermal transformation proceeds in three steps, Scheme 12.16:

$$2[Ni(acac)(tmen)(H_2O)_2]Br \xrightarrow[-4H_2O]{} 2[Ni(acac)(tmen)]Br \longrightarrow$$

$$[Ni(acac)(tmen)Br] \text{ or } [Ni(acac)(tmen)Br_2Ni(acac)(tmen)] \longrightarrow$$

$$[Ni(acac)_2(tmen)] + [NiBr_2(tmen)]$$

Scheme 12.16

A study of the decarboxylation of iminodiacetic acid (H_2L) in complexes of ruthenium(III) and rhodium(III) showed[64] that dehydration of $Na[RuCl_2(HL)_2]\cdot H_2O$, $Na[RuCl_3(HL)(H_2L)]\cdot 8H_2O$, and $Na[RuCl_3(HL)(H_2L)]\cdot 2H_2O$ takes place at about 100 °C. Decarboxylation of the carboxyl groups not participating in coordination, with elimination of CO_2, occurs at a temperature of about 200 °C and elimination of the coordinated (deprotonated) carboxyl groups occurs at temperatures of 340 °C and above. In these cases, both carbon monoxide and carbon dioxide are liberated.

There are many reports of the thermal degradation of carboxyl groups in organometallic complexes and these reports have been reviewed by Deacon, Faulks, and Pain.[65] For example, loss of carbon dioxide from the organotin and -lead cyanoacetates, $[M(O_2CCH_2CN)R_3]$ (M = Sn, R = Ph, Bu; M = Pb, R = Ph) upon heating *in vacuo* produced the corresponding $[M(CH_2CN)R_3]$ compounds. Disproportionation to form tetraethyltin was observed when triethyltin cyanoacetate was heated.[65]

12.9 Solid state thermal oxidation of tertiary phosphines

The oxidation of tertiary phosphines is a well known reaction (see Chapter 9) driven largely by the high P=O bond energy of the phosphoryl groups in the corresponding phosphine oxides.[66] The reaction of aliphatic phosphines with molecular oxygen follows a free radical mechanism and leads to the formation of a mixture of products of general formula, $R_nP(O)(OR)_{3-n}$. Aromatic phosphines, including PPh_3, are more difficult to oxidize and require stronger oxidants for quantitative oxidation.[66]

In 1963, the first report of solid state oxidation of tertiary phosphine complexes appeared.[67] It was found that solid $[CoCl_2(PEt_3)_2]$ was easily oxidized in air and transformed into a phosphine oxide complex. No experimental data were provided and the results have not been reproducible.[68] Indeed, it has been reported[68] that crystalline $[CoCl_2(PEt_3)_2]$ is stable not only in dry air but also in an atmosphere of pure oxygen at room temperature.

Thermal analysis of the cobalt(II)[69] and nickel(II)[70] complexes, $[MX_2(PPh_3)_2]$, has demonstrated solid state oxidation of triphenylphosphine to triphenylphosphine oxide, Eq. 12.22:

$$[MX_2(PPh_3)_2] + O_2 \rightarrow [MX_2(OPPh_3)_2] \qquad (12.22)$$

In air the oxidation appears as a strong exothermic peak in the DTA and as a mass increase in the TG. However, the same DTA/TG results were not obtained for samples heated in a helium atmosphere. Accordingly, it is oxygen from the air which serves as the oxidant for the conversion of P(III) to P(V). Based on X-ray powder diffraction, magnetochemical and thermal analyses, along with results from electronic and IR spectroscopies, it was shown[69] that the coordinated triphenylphosphine ligands in $[CoX_2(PPh_3)_2]$ (X = Cl, Br, I) oxidize in a stepwise fashion with formation of $[MX_2(OPPh_3)(PPh_3)]$ as an intermediate. Isothermal heating of the nickel(II) and cobalt(II) complexes, $[MX_2(PPh_3)_2]$, allows quantitative conversion to $[MX_2(OPPh_3)_2]$.[70]

The presence of a chelate ring, in general, should increase the thermal stability of metal complexes towards decomposition through a decrease in kinetic lability. Such trends have been observed in, for example, the solid state isomerization of platinum(II) complexes containing monodentate triphenylphosphine and bidentate dppm ligands.[71] A study[72] of thermally-initiated phosphine oxidation in nickel(II) and cobalt(II) complexes, $[MX_2(P\text{–}P)]$ and $[MX_2(P{=}P)]$ (X = Cl, Br, I; P–P = 1,2-bis(diphenylphosphino)ethane; P=P = *cis*-1,2-bis(diphenylphosphino)ethylene), has been reported. The application of DTA and TG methods in a static atmosphere of air showed that all of the complexes exhibited the same type of behavior.[72] Thus, TG revealed a mass increase that corresponded to a strong exothermic peak in the DTA. Isothermal treatment of the starting materials at temperatures close to those of the exothermic peaks identified by DTA led to quantitative formation of $[MX_2(OP\text{–}OP)]$ and $[MX_2(OP{=}OP)]$.

Analysis of the results obtained from solid state oxidation of nickel(II) and cobalt(II) complexes, $[MX_2(PPh_3)_2]$[70] and $[MX_2L_2]$ (where L_2 = P–P, P=P),[72] shows that the temperatures required for oxidation follow the order: chloride > bromide > iodide. Triphenylphosphine ligands are subject to oxidation at slightly lower temperatures than the bidentate ligands, P–P and P=P, while nickel(II) complexes are oxidized at higher temperatures than analogous cobalt(II) complexes, which may suggest a higher kinetic lability for the latter.

Solid state oxidation of coordinated phosphine ligands has also been reported for complexes of iron(III),[73] palladium(II),[74] and rhenium(V)[75] and such thermally initiated solid state oxidation reactions appear to be fairly widespread. The oxidation mechanism seems to include the formation of intermediates by reaction with molecular oxygen, followed by M–P and O–O bond breaking and P–O bond formation steps.[76]

12.10 Solid state geometric isomerization

The *cis-trans* isomerization of square-planar transition metal complexes, in particular those of platinum(II) and palladium(II), can be understood by comparison of the *trans*-effects of the individual ligands (Chapter 1). This approach predicts the favored direction of isomerization for a large group of complexes with ligands of different types. In this Section examples of geometric isomerization which are not adequately explained by the *trans*-effect are considered and it is shown that solid state effects can influence the outcome of such reactions.

Accumulated data on the geometric isomerization of thioether complexes of platinum(II) indicate that the direction of the rearrangement may be influenced significantly by the conditions under which it is carried out. Thus, when the tetrahydrothiophene complex, *trans*-$[PtBr_2(SC_4H_8)_2]$, is heated in the solid state, it undergoes *trans*→*cis* isomerization. In contrast, *cis*-$[PtBr_2(SC_4H_8)_2]$ in aqueous acetone, in the presence of KBr as catalyst, undergoes the reverse, *cis*→*trans*, rearrangement.[77] This early example demonstrates the dependence of the favored direction of a geometric rearrangement on reaction conditions.

Later work established[78] that other thioether complexes of platinum(II), *cis*- and *trans*-[$PtCl_2(py)(Me_2S)$], exhibit similar behavior and geometric rearrangements take place in different directions in the solid state and in nitromethane solution. The solution reaction is catalyzed by Me_2S. The favored directions are indicated in Scheme 12.17:

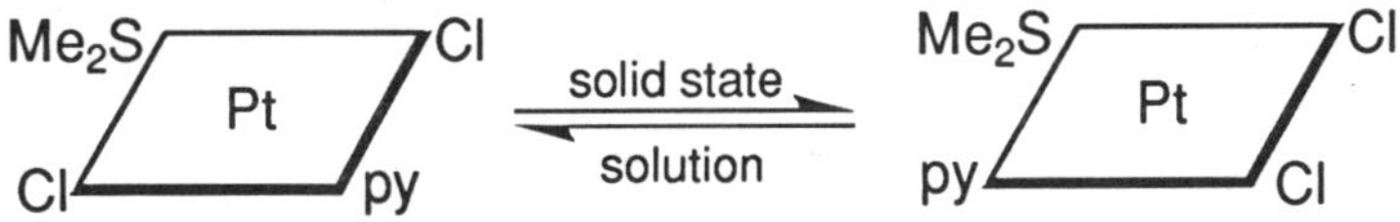

Scheme 12.17

The dependence of the direction of isomerization on reaction conditions is not restricted to thioether complexes. The mixed-ligand complexes, [$PtBr_2$(amine)(DMSO)] (amine = NH_3, NH_2Me, NH_2Et, piperidine, py), isomerize in the solid state and in solution in opposite directions.[79]

These observations indicate that isomerization is governed not only by the *trans*-effects of the ligands but also by external factors, in particular by differences in the crystal lattice energies of the isomers. Unfortunately, there are practically no published data on the crystal lattice energies of isomeric complexes. Such data can be estimated indirectly, however, from the melting points of the isomers since it is known that melting points correlate with crystal lattice energies.[80] It has been shown that the difference in the melting points of a pair of geometric isomers is a characteristic which can be used to predict the direction of an isomerization reaction. When solid [PtX_4LL'] (X = Cl and Br, L and L' = amines, dialkyl sulfoxides, thioethers) complexes are heated, geometric isomerization takes place to give the higher-melting isomer.[81] It was assumed that this isomer has the higher crystal lattice energy and that the formation of an isomer with a higher crystal lattice energy is the driving force for the rearrangement. Work on thioether complexes of platinum(II) is particularly informative in this regard.[82] Data on the thermal isomerization of [PtX_2L(thioether)] (X = Cl, Br, L = amine, thioether) complexes have been reported[83] and are compiled in Table 12.1.

When the compounds listed in Table 12.1 are heated in the solid state, both *cis*→*trans* and *trans*→*cis* isomerizations are observed. The common feature, however, is that these rearrangements favor the formation of the higher-melting isomer. A special case is the isomerization of the complexes, [$PtCl_2(R_2S)_2$] (R = Me, Et), where the melting points of the geometric isomers are the same. Heating either isomer of [$PtCl_2(R_2S)_2$] produces a molten mixture of both the *cis*- and *trans*-forms.

Table 12.1

Directions of isomerization and melting points (mp) of the isomers of [PtX_2L(thioether)][83]

Complex	Direction of isomerization	mp of isomer, °C	
		cis	*trans*
[$PtCl_2(Me_2S)_2$]	*cis* ⇌ *trans*	159	159
[$PtBr_2(Me_2S)_2$]	*cis* → *trans*	110-120	162
[$PtCl_2(Et_2S)_2$]	*cis* ⇌ *trans*	106-107	106-107
[$PtBr_2(Et_2S)_2$]	*cis* → *trans*	95	125
[$PtCl_2(SC_4H_8)_2$]	*cis* → *trans*	125	140
[$PtBr_2(SC_4H_8)_2$]	*trans* → *cis*	120	102
[$PtCl_2(EtSCH_2CO_2H)_2$]	*trans* → *cis*	130	90
[$PtCl_2(py)(SC_2H_4O)$]	*cis* → *trans*	110	170
[$PtCl_2(py)(Me_2S)$]	*trans* → *cis*	177	137-140

In addition to the thermal transformations described in the preceding sections, a variety of other reaction types may be induced by heat treatment. These include linkage isomerization,[84] thermal conversions of coordinated ligands,[85] reductive elimination reactions,[86] thermal conversions of counterions,[87] and the transformations of square-planar and octahedral complexes through counter-ion anation reactions[88] or *via* dehydration.[89]

References

1. (a) For general reviews on thermal analysis see: D. Dollimore, *Anal. Chem.* **66** (1994) 17R; (b) D. Dollimore, *Anal. Chem.* **64** (1992) 147R.

2. For reviews relating to thermal transformations of coordination compounds see: (a) D. Dollimore, in *Differential Thermal Analysis*, ed. R. C. McKenzie (Academic, New York, 1970) p. 395; (b) D. Dollimore, in *Differential Thermal Analysis*, ed. R. C. McKenzie (Academic, New York, 1970) p. 427.

3. Th. Anderson, *Ann. Chem. Pharm.* **96** (1855) 199.

4. For a review see: Yu. N. Kukushkin, V. F. Budanova, and G. N. Sedova, *Thermal Conversions of Coordination Compounds in the Solid Phase* [in Russian] (Leningrad State University, Leningrad, 1981).

5. (a) Yu. N. Kukushkin, G. N. Sedova, and A. N. Belyaev, *Zh. Neorg. Khim.* **23** (1978) 2121; (b) Yu. N. Kukushkin, G. N. Sedova, L. Ya. Pyzhova, R. A. Vlasova, and A. D. Garnovskii, *Zh. Neorg. Khim.* **24** (1979) 2265; (c) Yu. N. Kukushkin, R. A. Vlasova, L. Ya. Pyzhova, G. N. Sedova, and A. D. Garnovskii, *Zh. Neorg. Khim.* **25** (1980) 490; (d) Yu. N. Kukushkin, G. N. Sedova, and A. N. Belyaev, *Zh. Neorg. Khim.* **26** (1981) 429; (e) Yu. N. Kukushkin, G. N. Sedova, G. Kh. Khamnuev, and A. D. Garnovskii, *Zh. Neorg. Khim.* **26** (1981) 696.

6. Yu. N. Kukushkin, A. N. Belyaev, and V. I. Lobadyuk, *Zh. Neorg. Khim.* **26** (1981) 1840.

7. Yu. N. Kukushkin, G. N. Sedova, A. N. Belyaev, and L. V. Konovalov, *Zh. Neorg. Khim.* **25** (1980) 1036.

8. (a) Yu. N. Kukushkin, G. N. Sedova, R. A. Vlasova, and L. Ya. Pyzhova, *Zh. Neorg. Khim.* **23** (1978) 733; (b) Yu. N. Kukushkin, G. N. Sedova, and R. A. Vlasova, *Zh. Neorg. Khim.* **23** (1978) 737.

9. V. Yu. Kukushkin, E. A. Aleksandrova, V. M. Leovac, E. R. Iveges, V. K. Belsky, and V. E. Konovalov, *Polyhedron* **11** (1992) 2691.

10. Yu. N. Kukushkin, L. I. Danilina, N. P. Fedianin, and A. V. Iretskii, *Zh. Neorg. Khim.* **26** (1981) 3293.

11. Yu. N. Kukushkin, A. V. Iretskii, and L. I. Danilina, *Koord. Khim.* **10** (1984) 1393.

12. A. I. Ryumin, I. I. Smirnov, and A. G. Yankina, *Zh. Neorg. Khim.* **30** (1985) 2023.

13. K. V. Kotegov, G. N. Sedova, Z. A. Lovchikova, and Yu. N. Kukushkin *Zh. Neorg. Khim.* **24** (1979) 3301.

14. J. V. Caspar, B. P. Sullivan, and T. J. Meyer, *Inorg. Chem.* **23** (1984) 2104.

15. K. V. Kotegov, Z. A. Lovchikova, and Yu. N. Kukushkin, *Zh. Neorg. Khim.* **28** (1983) 2277.

16. Yu. N. Shevchenko, N. K. Davidenko, R. A. Svitsin, N. I. Yashina, and N. V. Egorova, *Zh. Neorg. Khim.* **25** (1980) 2180.

17. H. Oki, E. Kyuno, and R. Tsuchiya, *Bull. Chem. Soc. Japan* **43** (1970) 3263.

18. C. E. Stoner, A. L. Rheingold, and T. B. Brill, *Inorg. Chem.* **30** (1991) 360.

19. D. Das, A. Ghosh, C. Pariya, and N. R. Chaudhuri, *J. Chem. Res. (S)* (1994) 136.

20. Y. Ihara, S. Naride, and A. Uehara, *Thermochim. Acta* **200** (1992) 327.

21. W. W. Wendlandt, G. D'Ascenzo, and R. H. Gore, *J. Inorg. Nucl. Chem.* **32** (1970) 3404.

22. C. A. Jepsen and J. E. House, *J. Inorg. Nucl. Chem.* **43** (1981) 953.

23. W. W. Wendlandt and J. K. Fisher, *J. Inorg. Nucl. Chem.* **24** (1962) 1685.

24. M. Olczak-Kobza, *Thermochim. Acta* **133** (1988) 347.

25. W. W. Wendlandt and J. P. Smith, *The Thermal Properties of Transition Metal Ammine Complexes* (Elsevier, Amsterdam, 1967).

26. T. G. Appleton, A. J. Bailey, K. J. Barnham, and J. R. Hall, *Inorg. Chem.* **31** (1992) 3077.

27. Yu. N. Kukushkin and E. S. Postnikova, *Zh. Prikl. Khim.* **42** (1969) 2845.

28. M. Van Beusichem and N. Farrell, *Inorg. Chem.* **31** (1992) 634.

29. A. V. Iretskii and L. I. Danilina, *Zh. Neorg. Khim.* **33** (1988) 1209.

30. V. I. Tsapkov, M. V. Gandzii, Yu. M. Chumakov, V. I. Biyushkin, T. I. Malinovskii, G. Bkhusam, and N. M. Samus, *Koord. Khim.* **18** (1992) 850.

31. J. P. Collman, C. E. Barnes, T. J. Collins, P. J. Brothers, J. Gallucci, and J. A. Ibers, *J. Am. Chem. Soc.* **103** (1981) 7030.

32. E. Jóna, I. Horváth, M. Kubranová, and M. Koman, *Thermochim. Acta* **180** (1991) 307.

33. G. Favero, A. Peloso, and U. Vettori, *Inorg. Chim. Acta* **168** (1990) 189.

34. V. Yu. Kukushkin, V. M. Tkachuk, I. A. Krol, Z. A. Starikova, B. V. Zhadanov, and N. P. Kiseleva, *Zh. Obsch. Khim.* **61** (1991) 51.

35. V. Yu. Kukushkin, V. K. Belsky, V. E. Konovalov, R. R. Shifrina, A. I. Moiseev, and R. A. Vlasova, *Inorg. Chim. Acta* **183** (1991) 57.

36. (a) Yu. N. Kukushkin, V. P. Kotelnikov, and V. N. Spevak, *Zh. Neorg. Khim.* **22** (1977) 2208; (b) Yu. N. Kukushkin, I. V. Pakhomova, and E. N. Kalukova, *Zh. Obsch. Khim.* **43** (1973) 2594; (c) Yu. N. Kukushkin, V. F. Budanova, and G. N. Sedova, *Zh. Neorg. Khim.* **22** (1977) 1868; (d) E. A. Andronov, Yu. N. Kukushkin, T. M. Lukicheva, E. S. Postnikova, and Yu. S. Stcherbakov, *Zh. Prikl. Khim.* **50** (1977) 657.

37. (a) G. Alibrandi, R. Romeo, L. M. Scolaro, and M. L. Tobe, *Inorg. Chem.* **31** (1992) 5061; (b) G. Faraglia and S. Sitran, *Inorg. Chim. Acta* **176** (1990) 67.

38. I. V. Pakhomova, Yu. N. Kukushkin, L. V. Konovalov, and V. V. Strukov, *Zh. Neorg. Khim.* **29** (1984) 1000.

39. M. Kaneda, M. Suzuki, and A. Uehara, *Thermochim. Acta* **138** (1989) 161.

40. J. A. Labinger and J. A. Osborn, *Inorg. Synth.* **18** (1978) 62.

41. Yu. N. Kukushkin, S. N. Bakhireva, and R. B. Dushin, *Zh. Neorg. Khim.* **22** (1977) 1419.

42. K. Cheng and B. M. Foxman, *J. Am. Chem. Soc.* **99** (1977) 8102.

43. R. Tsuchiya, K. Manazaki, S. Sekiya, A. Uehara, and E. Kyuno, *Bull. Chem. Soc. Japan* **48** (1975) 2805.

44. Yu. N. Kukushkin, G. N. Sedova, R. A. Vlasova, and A. N. Belyaev, *Zh. Prikl. Khim.* **51** (1978) 2451.

45. Yu. N. Kukushkin, A. N. Belyaev, R. A. Vlasova, and V. I. Lobadyuk, *Zh. Neorg. Khim.* **27** (1982) 1500.

46. (a) N. A. Astakhova, V. S. Bondarenko, U. F. Kuklina, G. D. Malchikov, and N. A. Shestakova, *Zh. Neorg. Khim.* **23** (1978) 2739; (b) V. S. Bondarenko, U. F. Kuklina, G. D. Malchikov, and N. A. Shestakova, *Zh. Neorg. Khim.* **25** (1980) 1561.

47. F. González-Vilchez and M. Castillo-Martos, *Thermochim. Acta* **21** (1977) 127.

48. M. Tsin-Shen and V. G. Tronev, *Zh. Neorg. Khim.* **5** (1960) 861.

49. Yu. A. Buslaev, M. A. Glushkova, and A. M. Bolshakov, *Izv. AN SSSR, Neorg. Mater.* **9** (1973) 500.

50. D. E. LaValle, R. M. Steele, and W. T. Smith, *J. Inorg. Nucl. Chem.* **28** (1966) 260.

51. (a) Th. Renner, *Z. Anorg. Allg. Chem.* **298** (1959) 22; (b) N. I. Chetverikov, I. F. Chetverikova, V. N. Chernyaev, V. B. Novikov, and S. A. Popov, *Izv. AN SSSR, Neorg. Mater.* **11** (1975) 2087; (c) L. S. Sukhanova, E. B. Sokolov, V. K. Yanitskii, and Yu. P. Solovev, *Zh. Neorg. Khim.* **22** (1977) 1486.

52. C. J. F. J. Mariategui and L. R. Ochoa, *Bol. Soc. quim. Peru* **48** (1982) 190.

53. A. D. Ryabov, *Chem. Rev.* **90** (1990) 403 and references therein.

54. (a) D. W. Evans, G. R. Baker, and G. R. Newkome, *Coord. Chem. Rev.* **93** (1989) 155; (b) I. P. Rothwell, *Acc. Chem. Res.* **21** (1988) 153; (c) I. Omae, *Coord. Chem. Rev.* **83** (1988) 137; (d) V. V. Dunina, O. A. Zalevskaya, and V. M. Potapov, *Usp. Khim. (Russ. Chem. Rev.)* **57** (1988) 434.

55. C. Kahz and W. Marthy, *Proc. 17 International Conference on Coordination Chemistry*, Hamburg (1976) p. 353.

56. T. B. Rauchfuss, *J. Am. Chem. Soc.* **101** (1979) 1045.

57. J. M. Duff, B. E. Mann, B. L. Shaw, and B. Turtle, *J. Chem. Soc., Dalton Trans.* (1974) 139.

58. L. R. Smith and D. M. Blake, *J. Am. Chem. Soc.* **99** (1977) 3302.

59. C. E. Jones, B. L. Shaw, and B. L. Turtle, *J. Chem. Soc., Dalton Trans.* (1974) 992.

60. H. D. Empsall, P. N. Heys, and B. L. Shaw, *J. Chem. Soc., Dalton Trans.* (1978) 257.

61. A. J. Canty and N. J. Minchin, *J. Organomet. Chem.* **226** (1982) C14.

62. Yu. N. Kukushkin, T. O. Blumental, and A. N. Belyaev, *Zh. Neorg. Khim.* **26** (1981) 1297.

63. N. Hoshino, Y. Fukuda, and K. Sone, *Chem. Lett.* (1979) 437.

64. A. E. Shilov and A. A. Shteinman, *Coord. Chem. Rev.* **24** (1977) 97.

65. G. B. Deacon, S. J. Faulks, and G. N. Pain, *Adv. Organomet. Chem.* **25** (1986) 237.

66. D. E. C. Corbridge, *Phosphorus: An Outline of its Chemistry, Biochemistry and Technology*, 4th Ed. (Elsevier, Amsterdam, 1990) p. 43 and 320 and references therein.

67. K. A. Jensen, P. H. Nielsen, and C. Th. Pedersen, *Acta Chem. Scand.* **17** (1963) 1115.

68. D. D. Schmidt and J. T. Yoke, *J. Am. Chem. Soc.* **93** (1971) 637.

69. J. Rimbault, J. C. Pierrard, and R. Hugel, *Bull. Soc. Chim. France* (1976) 1705.

70. (a) Yu. N. Kukushkin, E. D. Ageeva, G. N. Sedova, G. B. Avetikyan, and T. M. Lukicheva, *Koord. Khim.* **3** (1977) 1872; (b) Yu. N. Kukushkin, E. D. Ageeva, G. N. Sedova, L. Peres Martin, and G. B. Avetikyan, *Koord. Khim.* **4** (1978) 1710.

71. Yu. N. Kukushkin, V. F. Budanova, G. N. Sedova, E. S. Postnikova, and V. P. Yustratov, *Koord. Khim.* **1** (1975) 1630.

72. Yu. N. Kukushkin, G. N. Sedova, E. D. Ageeva, G. B. Avetikyan, and L. Peres Martin, *Koord. Khim.* **5** (1979) 244.

73. Yu. N. Kukushkin, E. D. Ageeva, V. E. Aleshin, and G. N. Sedova, *Zh. Neorg. Khim.* **25** (1980) 1135.

74. (a) Yu. N. Kukushkin, G. N. Sedova, and R. A. Vlasova, *Zh. Neorg. Khim.* **24** (1979) 431; (b) Yu. N. Kukushkin, G. N. Sedova, and R. A. Vlasova, *Zh. Neorg. Khim.* **25** (1980) 495; (c) Yu. N. Kukushkin, P. G. Antonov, L. N. Mitronina, G. N. Sedova, and B. I. Ionin, *Zh. Neorg. Khim.* **23** (1978) 1028.

75. (a) Yu. N. Kukushkin, K. V. Kotegov, G. N. Sedova, and N. K. Kobilov, *Koord. Khim.* **4** (1978) 1384; (b) K. V. Kotegov, G. N. Sedova, N. K. Kobilov, and Yu. N. Kukushkin, *Koord. Khim.* **6** (1980) 897; (c) K. V. Kotegov, G. N. Sedova, Z. A. Lovchikova, and Yu. N. Kukushkin, *Zh. Neorg. Khim.* **24** (1979) 3301.

76. Yu. N. Kukushkin and G. N. Sedova, *Koord. Khim.* **6** (1980) 27.

77. E. A. Andronov, Yu. N. Kukushkin, and T. M. Lukicheva, *Zh. Obsch. Khim.* **50** (1980) 701.

78. V. Yu. Kukushkin, E. Yu. Pankova, and E. A. Afanasev, *Zh. Neorg. Khim.* **34** (1989) 1342.

79. V. Yu. Kukushkin and A. I. Moiseev, *Zh. Neorg. Khim.* **35** (1990) 1994.

80. (a) V. S. Urusov, *Theoretical Crystal Chemistry* [in Russian] (Nauka, Moscow, 1987) p. 276; (b) V. S. Urusov, *Crystallochemical Energetics* [in Russian] (Nauka, Moscow, 1975) p. 335.

81. V. Yu. Kukushkin, A. I. Moiseev, and E. O. Sidorov, *Zh. Obsch. Khim.* **59** (1989) 1958.

82. Yu. N. Kukushkin, O. F. Khodzhaev, V. V. Budanova, and N. A. Parpiev, *The Thermolysis of Coordination Compounds* [in Russian] (Fan, Tashkent, 1986) p. 198.

83. V. Yu. Kukushkin, *Zh. Neorg. Khim.* **34** (1989) 2807 and references therein.

84. (a) M. J. Coyer, R. H. Herber, J. Chen, M. Croft, and S. P. Szu, *Inorg. Chem.* **33** (1994) 716; (b) W. R. Ellis and W. L. Purcell, *Inorg. Chem.* **21** (1982) 834; (c) W. L. Purcell, *Inorg. Chem.* **22** (1983) 1205; (d) W. W. Wendland and J. H. Woodlock, *J. Inorg. Nucl. Chem.* **27** (1965) 259.

85. (a) J. A. Thomas and A. Davison, *Inorg. Chim. Acta* **190** (1991) 231; (b) Yu. S. Matveev and K. M. Dunaeva, *Koord. Khim.* **10** (1984) 511.

86. T. E. Busigina, V. K. Polovniak, and N. S. Akhmetov, *Ref. Zh. Khim.* (1983) 2B145.

87. (a) N. A. Santalova, G. A. Seisenbaeva, K. M. Dunaeva, and V. I. Spitsin, *Dokl. AN SSSR* **300** (1988) 1143; (b) T. Fujiwara and J. C. Bailar, Jr., *Inorg. Chem.* **25** (1986) 1806.

88. Y. Masuda, T. Masuda, H. Kume, and Y. Ihara, *Thermochim. Acta* **156** (1989) 137.

89. S. Koner and N. R. Chaudhuri, *Thermochim. Acta* **180** (1991) 177.

CONCLUSIONS

The solution to many important problems in coordination chemistry begins with synthesis but the development of a synthetic method is not necessarily an end in itself. Coordination compounds are employed in many areas of chemistry and other fields. Often problem solving begins with synthesis and the synthesis may be performed by those relatively unfamiliar with the discipline. The authors hope that this text, that deals with synthetic coordination chemistry, will be useful to those engaged in the preparation of coordination compounds for which a variety of end uses are likely.

Over the past 200 years many fundamental principles have been discovered that form the basis for the synthesis of coordination compounds. These principles are described fully in many texts dealing with inorganic chemistry and are discussed briefly in Chapter 1. The reader is referred to the cited references for a more complete discussion of the theoretical principles underpinning coordination chemistry.

In the first half of this century, the synthesis of coordination compounds was viewed as a topic that dealt largely with the isolation of complexes in the solid state. More recently, the concept of synthesis has expanded to include the generation of complexes in solution. This new direction in synthesis was largely brought about by the widespread application of non-aqueous solvents in coordination chemistry even though water remains an important solvent, especially in the generation of ionic compounds and neutral complexes with hydrophilic ligands. But due to the tendency of many metal ions to form aqua- and hydroxo-complexes, some syntheses cannot be performed in aqueous media and so non-aqueous solvents have become critically important. There are now many non-aqueous solvents available that exhibit a wide range of chemical and physical properties used in the synthesis of coordination compounds.

In coordination chemistry solubility is an important property that determines whether a complex will remain in solution (which frustratingly occurs during many attempts to isolate a solid) or will precipitate as a solid (which likewise occurs during many attempts to generate a solution). The eternal problem of the solubility of chemical compounds has not yet been solved completely. Our analysis of the current state-of-the-art in solubility, as it relates to synthetic coordination chemistry, is given in Chapter 2.

Solubility appears to depend less on the non-aqueous or aqueous solvent in use and more on two pertinent factors that are important in considerations of the solubility of coordination compounds in all solvents: the polarity of the solvent and its ability to function as a ligand or otherwise interact with a metal complex. Essentially, the behavior of water as a solvent for the synthesis of coordination compounds is consistent with the general behavior exhibited by other solvents. Therefore, there appears to be little difference between water and non-aqueous solvents for the generation of coordination compounds.

There is a growing tendency to apply the approaches and techniques used in synthetic organic chemistry to coordination chemistry brought about, in part, by the increasing use of metal complexes as reagents or catalysts in organic synthesis. Studies at the boundaries of organic and inorganic chemistry have resulted in cross-fertilization of ideas and adoption of methods and techniques by one discipline from the other. In this connection, it is noteworthy that reagents commonly used in preparative organic chemistry, such as specific dehydrating reagents, oxidants, and reductants, have been adopted in coordination chemistry. This has allowed the synthesis of new coordination compounds and the development of alternative routes to compounds previously prepared by more classical methods.

Apart from simple ligand substitution reactions, newer approaches to ligand exchange are now widely used including those that involve ligand abstraction by conversion to a poorly soluble product or through chemical transformation to a weak donor ligand. Such techniques have allowed the creative synthesis of mixed-ligand complexes as well as coordinatively unsaturated compounds.

This book has not dealt with the synthesis of complexes in the gas phase or in melts. These are important but specialized aspects of coordination chemistry which require separate treatment. The main objective of this book has been to present the concepts and methodology used in planning the synthesis of coordination compounds. We have described only briefly specific examples of synthetic procedures. A creative chemist studying synthesis will, when equipped with the necessary concepts and methodologies, work toward the optimal conditions for a particular synthesis.

The authors hope that this book will stimulate further interest in synthetic coordination chemistry both for those already working in this area and for those new to the discipline.

INDEX